Game-Theoretic Foundations for Probability and Finance

Game-Theoretic Foundations for Probability and Finance

GLENN SHAFER

Rutgers Business School

VLADIMIR VOVK

Royal Holloway, University of London

This edition first published 2019
© 2019 John Wiley & Sons, Inc.

The right of Glenn Shafer and Vladimir Vovk to be identified as the authors of this work has been asserted in accordance with law.

Registered Office
John Wiley & Sons, Inc., 111 River Street, Hoboken, NJ 07030, USA

Editorial Office
111 River Street, Hoboken, NJ 07030, USA

For details of our global editorial offices, customer services, and more information about Wiley products visit us at www.wiley.com.

Wiley also publishes its books in a variety of electronic formats and by print-on-demand. Some content that appears in standard print versions of this book may not be available in other formats.

Library of Congress Cataloging-in-Publication Data

Names: Shafer, Glenn, 1946- author. | Vovk, Vladimir, 1960- author.
Title: Game-theoretic foundations for probability and finance / Glenn Ray
    Shafer, Rutgers University, New Jersey, USA, Vladimir Vovk, University of
    London, Surrey, UK.
Other titles: Probability and finance
Description: First edition. | Hoboken, NJ : John Wiley & Sons, Inc., 2019. |
    Series: Wiley series in probability and statistics | Earlier edition
    published in 2001 as: Probability and finance : it's only a game! |
    Includes bibliographical references and index. |
Identifiers: LCCN 2019003689 (print) | LCCN 2019005392 (ebook) | ISBN
    9781118547939 (Adobe PDF) | ISBN 9781118548028 (ePub) | ISBN 9780470903056
    (hardcover)
Subjects: LCSH: Finance–Statistical methods. | Finance–Mathematical models.
    | Game theory.
Classification: LCC HG176.5 (ebook) | LCC HG176.5 .S53 2019 (print) | DDC
    332.01/5193–dc23
LC record available at https://lccn.loc.gov/2019003689

Cover design by Wiley
Cover image: © Web Gallery of Art/Wikimedia Commons

Contents

Preface

Probability theory has always been closely associated with gambling. In the 1650s, Blaise Pascal and Christian Huygens based probability's concept of expectation on reasoning about gambles. Countless mathematicians since have looked to gambling for their intuition about probability. But the formal mathematics of probability has long leaned in a different direction. In his correspondence with Pascal, often cited as the origin of probability theory, Pierre Fermat favored combinatorial reasoning over Pascal's reasoning about gambles, and such combinatorial reasoning became dominant in Jacob Bernoulli's monumental *Ars Conjectandi* and its aftermath. In the twentieth century, the combinatorial foundation for probability evolved into a rigorous and sophisticated measure-theoretic foundation, put in durable form by Andrei Kolmogorov and Joseph Doob.

The twentieth century also saw the emergence of a mathematical theory of games, just as rigorous as measure theory, albeit less austere. In the 1930s, Jean Ville gave a game-theoretic interpretation of the key concept of probability 0. In the 1970s, Claus Peter Schnorr and Leonid Levin developed Ville's fundamental insight, introducing universal game-theoretic strategies for testing randomness. But no attempt was made in the twentieth century to use game theory as a foundation for the modern mathematics of probability.

Probability and Finance: It's Only a Game, published in 2001, started to fill this gap. It gave game-theoretic proofs of probability's most classical limit theorems (the laws of large numbers, the law of the iterated logarithm, and the central limit theorem), and it extended this game-theoretic analysis to continuous-time diffusion processes using nonstandard analysis. It applied the methods thus developed to finance, discussing how the availability of a variance swap in a securities market might allow other options to be priced without probabilistic assumptions and studying a purely game-theoretic hypothesis of market efficiency.

The present book was originally conceived of as a second edition of *Probability and Finance*, but as the new title suggests, it is a very different book, reflecting the healthy growth of game-theoretic probability since 2001. As in the earlier book, we show that game-theoretic and measure-theoretic probability provide equivalent descriptions of coin tossing, the archetype of probability theory, while generalizing this archetype in different directions. Now we show that the two descriptions are equivalent on a larger central core, including all discrete-time stochastic processes that have only finitely many outcomes on each round, and we present an even broader array of new ideas.

We can identify seven important new ideas that have come out of game-theoretic probability. Some of these already appeared, at least in part, in *Probability and Finance*, but most are developed further here or are entirely new.

1. *Strategies for testing.* Theorems showing that certain events have small or zero probability are made constructive; they are proven by constructing gambling strategies that multiply the capital they risk by a large or infinite factor if the events happen. In *Probability and Finance*, we constructed such strategies for the law of large numbers and several other limit theorems. Now we add to the list the most fundamental limit theorem of probability – Lévy's zero-one law. The topic of strategies for testing remains our most prominent theme, dominating Part I and Chapters 7 and 8 in Part II.

2. *Limited betting opportunities.* The betting rates suggested by a scientific theory or the investment opportunities in a financial market may fall short of defining a probability distribution for future developments or even for what will happen next. Sometimes a scientist or statistician tests a theory that asserts expected values for some variables but not for every function of those variables. Sometimes an investor in a market can buy a particular payoff but cannot sell it at the same price and cannot buy arbitrary options on it. Limited betting opportunities were emphasized by a number of twentieth-century authors, including Peter Williams and Peter Walley. As we explained in *Probability and Finance*, we can combine Williams and Walley's picture of limited betting opportunities in individual situations with Pascal and Ville's insights into strategies for combining bets across situations to obtain interesting and powerful generalizations of classical results. These include theorems that are one-sided in some sense (see Sections 2.4 and 5.1).

3. *Strategies for reality.* Most of our theorems concern what can be accomplished by a bettor playing against an opponent who determines outcomes. Our games are determined; one of the players has a winning strategy. In *Probability and Finance*, we exploited this determinacy and an argument of Kolmogorov's to show that in the game for Kolmogorov's law of large numbers, the opponent has a strategy that wins when Kolmogorov's hypotheses are not satisfied.

In this book we construct such a strategy explicitly and discuss other interesting strategies for the opponent (see Sections 4.4, 4.5, and 4.7).

4. *Open protocols for science.* Scientific models are usually open to influences that are not themselves predicted by the models in any way. These influences are variously represented; they may be treated as human decisions, as signals, or even as constants. Because our theorems concern what one player can accomplish regardless of how the other players move, the fact that these signals or "independent variables" can be used by the players as they appear in the course of play does not impair the theorems' validity and actually enhances their applicability to scientific problems (see Chapter 10).

5. *Insuring against loss of evidence.* The bettor can modify his own strategy or adapt bets made by another bettor so as to avoid a total loss of apparently strong evidence as play proceeds further. The same methods provide a way of calibrating the p-values from classical hypothesis testing so as to correct for the failure to set an initial fixed significance level. These ideas have been developed since the publication of *Probability and Finance* (see Chapter 11).

6. *Defensive forecasting.* In addition to the player who bets and the player who determines outcomes, our games can involve a third player who forecasts the outcomes. The problem of forecasting is the problem of devising strategies for this player, and we can tackle it in interesting ways once we learn what strategies for the bettor win when the match between forecasts and outcomes is too poor. This idea, which came to our attention only after the publication of *Probability and Finance*, is developed in Chapter 12.

7. *Continuous-time game-theoretic finance.* Measure-theoretic finance assumes that prices of securities in a financial market follow some probabilistic model such as geometric Brownian motion. We obtain many insights, some already provided by measure-theoretic finance and some not, without any probabilistic model, using only the actual prices in the market. This is now much clearer than in *Probability and Finance*, as we use tools from standard analysis that are more familiar than the nonstandard methods we used there. We have abandoned our hypothesis concerning the effectiveness of variance swaps in stabilizing markets, now fearing that the trading of such instruments could soon make them nearly as liquid and consequently treacherous as the underlying securities. But we provide game-theoretic accounts of a wider class of financial phenomena and models, including the capital asset pricing model (CAPM), the equity premium puzzle, and portfolio theory (see Part IV).

The book has four parts.

- Part I, Examples in Discrete Time, uses concrete protocols to explain how game-theoretic probability generalizes classical discrete-time limit theorems. Most of these results were already reported in *Probability and Finance* in 2001, but our exposition has changed substantially. We seldom repeat word for word what we wrote in the earlier book, and we occasionally refer the reader to the earlier book for detailed arguments that are not central to our theme.

- Part II, Abstract Theory in Discrete Time, treats game-theoretic probability in an abstract way, mostly developed since 2001. It is relatively self-contained, and readers familiar with measure-theoretic probability will find it accessible without the introduction provided by Part I.

- Part III, Applications in Discrete Time, uses Part II's theory to treat important applications of game-theoretic probability, including two promising applications that have developed since 2001: calibration of lookbacks and p-values, and defensive forecasting.

- Part IV, Game-Theoretic Finance, studies continuous-time game-theoretic probability and its application to finance. It requires different definitions from the discrete-time theory and hence is also relatively self-contained. Its first chapter uses a simple concrete protocol to derive game-theoretic versions of the Dubins–Schwarz theorem and related results, while the remaining chapters use an abstract and more powerful protocol to develop a game-theoretic version of the Itô calculus and to study classical topics in finance theory.

Each chapter includes exercises, which vary greatly in difficulty; some are simple exercises to enhance the reader's understanding of definitions, others complete details in proofs, and others point to related literature, open problems, or substantial research projects. Following each chapter's exercises, we provide notes on the historical and contemporary context of the chapter's topic. But as a result of the substantial increase in mathematical content, we have left aside much of the historical and philosophical discussion that we included in *Probability and Finance*.

We are pleased by the flowering of game-theoretic probability since 2001 and by the number of authors who have made contributions. The field nevertheless remains in its infancy, and this book cannot be regarded as a definitive treatment. We anticipate and welcome the theory's further growth and its incorporation into probability's broad tapestry of mathematics, application, and philosophy.

Newark, New Jersey, USA   GLENN SHAFER AND VLADIMIR VOVK
and Egham, Surrey, UK
10 November 2018

Acknowledgments

For more than 20 years, game-theoretic probability has been central to both our scholarly lives. During this period, we have been generously supported, personally and financially, by more individuals and institutions than we can possibly name. The list is headed by two of the most generous and thoughtful people we know, our wives Nell Painter and Lyuda Vovk. We dedicate this book to them.

Among the many other individuals to whom we are intellectually indebted, we must put at the top of the list our students, our coauthors, and our colleagues at Rutgers University and Royal Holloway, University of London. We have benefited especially from interactions with those who have joined us in working on game-theoretic probability and closely related topics. Foremost on this list are the Japanese researchers on game-theoretic probability, led by Kei Takeuchi and Akimichi Takemura, and Gert de Cooman, a leader in the field of imprecise probabilities. In the case of continuous time, we have learned a great deal from Nicolas Perkowski, David Prömel, and Rafał Łochowski. The book's title was suggested to us by Ioannis Karatzas, who also provided valuable encouragement in the final stages of the writing.

At the head of the list of other scholars who have contributed to our understanding of game-theoretic probability, we place a number who are no longer living: Joe Doob, Jørgen Hoffmann-Jørgensen, Jean-Yves Jaffray, Hans-Joachim Lenz, Laurie Snell, and Kurt Weichselberger.

We also extend our warmest thanks to Victor Perez Abreu, Beatrice Acciaio, John Aldrich, Thomas Augustin, Dániel Bálint, Traymon Beavers, James Berger, Mark Bernhardt, Laurent Bienvenu, Nic Bingham, Jasper de Bock, Bernadette Bouchon-Meunier, Olivier Bousquet, Ivan Brick, Bernard Bru, Peter Carr, Nicolò Cesa-Bianchi, Ren-Raw Chen, Patrick Cheridito, Alexey Chernov, Roman Chychyla,

Fernando Cobos, Rama Cont, Frank Coolen, Alexander Cox, Harry Crane, Pierre Crépel, Mark Davis, Philip Dawid, Freddy Delbaen, Art Dempster, Thierry Denoeux, Valentin Dimitrov, David Dowe, Didier Dubois, Hans Fischer, Hans Föllmer, Yoav Freund, Akio Fujiwara, Alex Gammerman, Jianxiang Gao, Peter Gillett, Michael Goldstein, Shelly Goldstein, Prakash Gorroochurn, Suresh Govindaraj, Peter Grünwald, Yuri Gurevich, Jan Hannig, Martin Huesmann, Yuri Kalnishkan, Alexander Kechris, Matti Kiiski, Jack King, Elinda Fishman Kiss, Alex Kogan, Wouter Koolen, Masayuki Kumon, Thomas Kühn, Steffen Lauritzen, Gabor Laszlo, Tatsiana Levina, Chuanhai Liu, Barry Loewer, George Lowther, Gábor Lugosi, Ryan Martin, Thierry Martin, Laurent Mazliak, Peter McCullagh, Frank McIntyre, Perry Mehrling, Xiao-Li Meng, Robert Merton, David Mest, Kenshi Miyabe, Rimas Norvaiša, Ilia Nouretdinov, Marcel Nutz, Jan Obłój, André Orléan, Barbara Osimani, Alexander Outkin, Darius Palia, Dan Palmon, Dusko Pavlovic, Ivan Petej, Marietta Peytcheva, Jan von Plato, Henri Prade, Philip Protter, Steven de Rooij, Johannes Ruf, Andrzej Ruszczyński, Bharat Sarath, Richard Scherl, Martin Schweizer, Teddy Seidenfeld, Thomas Sellke, Eugene Seneta, John Shawe-Taylor, Alexander Shen, Yiwei Shen, Prakash Shenoy, Oscar Sheynin, Albert N. Shiryaev, Pietro Siorpaes, Alex Smola, Mete Soner, Steve Stigler, Tamas Szabados, Natan T'Joens, Paolo Toccaceli, Matthias Troffaes, Jean-Philippe Touffut, Dimitris Tsementzis, Valery N. Tutubalin, Miklos Vasarhelyi, Nikolai Vereshchagin, John Vickers, Mikhail Vyugin, Vladimir V'yugin, Bernard Walliser, Chris Watkins, Wei Wu, Yangru Wu, Sandy Zabell, and Fedor Zhdanov.

We thank Rutgers Business School and Royal Holloway, University of London, as institutions, for their financial support and for the research environments they have created. We have also benefited from the hospitality of numerous other institutions where we have had the opportunity to share ideas with other researchers over these past 20 years. We are particularly grateful to the three institutions that have hosted workshops on game-theoretic probability: the University of Tokyo (on several occasions), then Royal Holloway, University of London, and the latest one at CIMAT (Centro de Investigación en Matemáticas) in Guanajuato. We are grateful to the Web Gallery of Art and its editor, Dr. Emil Krén, for permission to use "Card Players" by Lucas van Leyden (Museo Nacional Thyssen-Bornemisza, Madrid) on the cover.

GLENN SHAFER AND VLADIMIR VOVK

Part I
_____

Examples in Discrete Time

Many classical probability theorems conclude that some event has small or zero probability. These theorems can be used as predictions; they tell us what to expect. Like any predictions, they can also be used as tests. If we specify an event of putative small probability as a test, and the event happens, then the putative probability is called into question, and perhaps the authority behind it as well.

The key idea of game-theoretic probability is to formulate probabilistic predictions and tests as strategies for a player in a betting game. The player – we call him Skeptic – may be betting not so much to make money as to refute a theory or forecaster – whatever or whoever is providing the probabilities. In this picture, the claim that an event has small probability becomes the claim that Skeptic can multiply the capital he risks by a large factor if the event happens.

There is nothing profound or original in the observation that you make a lot more money than you risk when you bet on an event of small probability, at the corresponding odds, and the event happens. But as Jean Ville explained in the 1930s [386, 387], the game-theoretic picture has a deeper message. In a sequential setting, where probabilities are given on each round for the next outcome, an event involving the whole sequence of outcomes has a small probability if and only if Skeptic has a strategy for successive bets that multiplies the capital it risks by a large factor when the event happens. In this part of the book, we develop the implications of Ville's insight. As we show, it leads to new generalizations of many classical results in probability theory,

Game-Theoretic Foundations for Probability and Finance, First Edition. Glenn Shafer and Vladimir Vovk.
© 2019 John Wiley & Sons, Inc. Published 2019 by John Wiley & Sons, Inc.

1

thus complementing the measure-theoretic foundation for probability that became standard in the second half of the twentieth century.

The charm of the measure-theoretic foundation lies in its power and simplicity. Starting with the short list of axioms and definitions that Andrei Kolmogorov laid out in 1933 [224] and adding when needed the definition of a stochastic process developed by Joseph Doob [116], we can spin out the whole broad landscape of mathematical probability and its applications. The charm of the game-theoretic foundation lies in its constructivity and overt flexibility. The strategies that prove classical theorems are computable and relatively simple. The mathematics is rigorous, because we define a precise game, with precise assumptions about the players, their information, their permitted moves, and rules for winning, but these elements of the game can be varied in many ways. For example, the bets offered to Skeptic on a given round may be too few to define a probability measure for the next outcome. This flexibility allows us to avoid some complications involving measurability, and it accommodates very naturally applications where the activity between bets includes not only events that settle Skeptic's last bet but also actions by other players that set up the options for his next bet.

Kolmogorov's 1933 formulation of the measure-theoretic foundation is abstract. It begins with the notion of a probability measure P on a σ-algebra $\mathcal{F}$ of subsets of an abstract space Ω, and it then proceeds to prove theorems about all such triplets $(\Omega, \mathcal{F}, P)$. Outcomes of experiments are treated as random variables – i.e. as functions on Ω that are measurable with respect to $\mathcal{F}$. But many of the most important theorems of modern probability, including Émile Borel's and Kolmogorov's laws of large numbers, Jarl Waldemar Lindeberg's central limit theorem, and Aleksandr Khinchin's law of the iterated logarithm, were proven before 1933 in specific concrete settings. These theorems, the theorems that we call *classical*, dealt with a sequence $y_1, y_2, \ldots$ of outcomes by positing or defining in one way or another a system of probability distributions: (i) a probability distribution for y_1 and (ii) for each n and each possible sequence $y_1, \ldots, y_{n-1}$ of values for the first $n - 1$ outcomes, a probability distribution for y_n. We can fit this classical picture into the abstract measure-theoretic picture by constructing a *canonical space* Ω from the spaces of possible outcomes for the y_n.

In this part of the book, we develop game-theoretic generalizations of classical theorems. As in the classical picture, we construct global probabilities and expected values from ingredients given sequentially, but we generalize the classical picture in two ways. First, the betting offers on each round may be less extensive. Instead of a probability distribution, which defines odds for every possible bet about the outcome y_n, we may offer Skeptic only a limited number of bets about y_n. Second, these offers are not necessarily laid out at the beginning of the game. Instead, they may be given by a player in the game – we call this player Forecaster – as the game proceeds.

Our game-theoretic results fall into two classes, finite-horizon results, which concern a finite sequence of outcomes $y_1, \ldots, y_N$, and asymptotic results, which concern an infinite sequence of outcomes $y_1, y_2, \ldots$. The finite-horizon results can be more directly relevant to applications, but the asymptotic results are often simpler.

Because of its simplicity, we begin with the most classical asymptotic result, Borel's law of large numbers. Borel's publication of this result in 1909 is often seen as the decisive step toward modern measure-theoretic probability, because it exhibited for the first time the isomorphism between coin-tossing and Lebesgue measure on the interval [0, 1] [54, 350]. But Borel's theorem can also be understood and generalized game-theoretically. This is the topic of Chapter 1, where we also introduce the most fundamental mathematical tool of game-theoretic probability, the concept of a supermartingale.

In Chapter 2, we shift to finite-horizon results, proving and generalizing game-theoretic versions of Jacob Bernoulli's law of large numbers and Abraham De Moivre's central limit theorem. Here we introduce the concepts of game-theoretic probability and game-theoretic expected value, which we did not need in Chapter 1. There zero was the only probability needed, and instead of saying that an event has probability 0, we can say simply that Skeptic becomes infinitely rich if it happens.

In Chapter 3, we study some supermartingales that are relevant to the theory of large deviations. Three of these, Kolmogorov's martingale, Doléans's supermartingale, and Hoeffding's supermartingale, will recur in various forms later in the book, even in Part IV's continuous-time theory.

In Chapter 4, we return to the infinite-horizon picture, generalizing Chapter 1's game-theoretic version of Borel's 1909 law of large numbers to a game-theoretic version of Kolmogorov's 1930 law of large numbers, which applies even when outcomes may be unbounded. Kolmogorov's classical theorem, later generalized to a martingale theorem within measure-theoretic probability, gives conditions under which an average of outcomes asymptotically equals the average of the outcomes' expected values. Kolmogorov's necessary and sufficient conditions for the convergence are elaborated in the game-theoretic framework by a strategy for Skeptic that succeeds if the conditions are satisfied and a strategy for Reality (the opponent who determines the outcomes) that succeeds if the conditions are not satisfied.

In Chapter 5, we study game-theoretic forms of the law of the iterated logarithm, including those already obtained in *Probability and Finance* and others obtained more recently by other authors.

1

Borel's Law of Large Numbers

This chapter introduces game-theoretic probability in a relatively simple and concrete setting, where outcomes are bounded real numbers. We use this setting to prove game-theoretic generalizations of a theorem that was first published by Émile Borel in 1909 [44] and is often called Borel's law of large numbers.

In its simplest form, Borel's theorem says that the frequency of heads in an infinite sequence of tosses of a coin, where the probability of heads is always p, converges with probability one to p. Later authors generalized the theorem in many directions. In an infinite sequence of independent trials with bounded outcomes and constant expected value, for example, the average outcome converges with probability one to the expected value.

Our game-theoretic generalization of Borel's theorem begins not with probabilities and expected values but with a sequential game in which one player, whom we call Forecaster, forecasts each outcome and another, whom we call Skeptic, uses each forecast as a price at which he can buy any multiple (positive, negative, or zero) of the difference between the outcome and the forecast. Here Borel's theorem becomes a statement about how Skeptic can profit if the average difference does not converge to zero. Instead of saying that convergence happens with probability one, it says that Skeptic has a strategy that multiplies the capital it risks by infinity if the convergence does not happen.

In Section 1.1, we formalize the game for bounded outcomes. In Section 1.2, we state Borel's theorem for the game and prove it by constructing the required

Game-Theoretic Foundations for Probability and Finance, First Edition. Glenn Shafer and Vladimir Vovk.
© 2019 John Wiley & Sons, Inc. Published 2019 by John Wiley & Sons, Inc.

strategy. Many of the concepts we introduce as we do so (situations, events, variables, processes, forcing, almost sure events, etc.) will reappear throughout the book.

The outcomes in our game are determined by a third player, whom we call Reality. In Section 1.3, we consider the special case where Reality is allowed only a binary choice. Because our results tell us what Skeptic can do regardless of how Forecaster and Reality move, they remain valid under this restriction on Reality. They also remain valid when we then specify Forecaster's moves in advance, and this reduces them to familiar results in probability theory, including Borel's original theorem.

In Section 1.4, we develop terminology for the case where Skeptic is allowed to give up capital on each round. In this case, a capital process that results from fixing a strategy for Skeptic is called a *supermartingale*. Supermartingales are a fundamental tool in game-theoretic probability.

In Section 1.5, we discuss how Borel's theorem can be adapted to test the calibration of forecasts, a topic we will study from Forecaster's point of view in Chapter 12. In Section 1.6, we comment on the computability of the strategies we construct.

1.1   A PROTOCOL FOR TESTING FORECASTS

Consider a game with three players: Forecaster, Skeptic, and Reality. On each round of the game,

- Forecaster decides and announces the price m for a payoff y,
- Skeptic decides and announces how many units, say M, of y he will buy,
- Reality decides and announces the value of y, and
- Skeptic receives the net gain $M(y - m)$, which may be positive, negative, or zero.

The players move in the order listed, and they see each other's moves.

We think of m as a forecast of y. Skeptic tests the forecast by betting on y differing from it. By choosing M positive, Skeptic bets y will be greater than m; by choosing M negative, he bets it will be less. Reality can keep Skeptic from making money. By setting $y := m$, for example, she can assure that Skeptic's net gain is zero. But if she does this, she will be validating the forecast.

We write $\mathcal{K}_n$ for Skeptic's capital after the nth round of play. We allow Skeptic to specify his initial capital $\mathcal{K}_0$, we assume that Forecaster's and Reality's moves are all between -1 and 1, and we assume that play continues indefinitely. These rules of play are summarized in the following protocol.

Protocol 1.1
>   Skeptic announces $\mathcal{K}_0 \in \mathbb{R}$.
>   FOR $n = 1, 2, \ldots$:

Forecaster announces $m_n \in [-1, 1]$.
Skeptic announces $M_n \in \mathbb{R}$.
Reality announces $y_n \in [-1, 1]$.
$\mathcal{K}_n := \mathcal{K}_{n-1} + M_n(y_n - m_n)$.

We call protocols of this type, in which Skeptic can test the consistency of forecasts with outcomes by gambling at prices given by the forecasts, *testing protocols*. We define the notion of a testing protocol precisely in Chapter 7 (see the discussion following Protocol 7.12).

To make a testing protocol into a game, we must specify goals for the players. We will do this for Protocol 1.1 in various ways. But we never assume that Skeptic merely wants to maximize his capital, and usually we do not assume that his gains $M_n(y_n - m_n)$ are losses to the other players.

We can vary Protocol 1.1 in many ways, some of which will be important in this or later chapters. Here are some examples.

- Instead of $[-1, 1]$, we can use $[-C, C]$, where C is positive but different from 1, as the move space for Forecaster and Reality. Aside from occasional rescaling, this will not change the results of this chapter.

- We can stop playing after a finite number of rounds. We do this in some of the testing protocols we use in Chapter 2.

- We can require Forecaster to set m_n equal to zero on every round. We will impose this requirement in most of this chapter, as it entails no loss of generality for the results we are proving.

- We can use a two-element set, say $\{-1, 1\}$ or $\{0, 1\}$, as Reality's move set instead of $[-1, 1]$. When we use $\{0, 1\}$ and require Forecaster to announce the same number $p \in [0, 1]$ on each round, the picture reduces to coin tossing (see Section 1.3).

As we have explained, our emphasis in this chapter and in most of the book is on strategies for Skeptic. We show that Skeptic can achieve certain goals regardless of how Forecaster and Reality move. Since these are worst-case results for Skeptic, they remain valid when we weaken Forecaster or Reality in any way: hiding information from them, requiring them to follow some strategy specified in advance, allowing Skeptic to influence their moves, or otherwise restricting their moves. They also remain valid when we enlarge Skeptic's discretion. They remain valid even when Skeptic's opponents know the strategy Skeptic will play; if a strategy for Skeptic reaches a certain goal no matter how his opponents move, it will reach this goal even if the opponents know it will be played.

We will present protocols in the style of Protocol 1.1 throughout the book. Unless otherwise stated, the players always have perfect information. They move in the order listed, and they see each other's moves. In general, we will use the term *strategy* as it is usually used in the study of perfect-information games: unless otherwise stated, a strategy is a pure strategy, not a mixed or randomized strategy.

1.2 A GAME-THEORETIC GENERALIZATION OF BOREL'S THEOREM

Skeptic might become infinitely rich in the limit as play continues:

$$\lim_{n\to\infty} \mathcal{K}_n = \infty. \tag{1.1}$$

Since Reality can always keep Skeptic from making money, Skeptic cannot expect to win a game in which (1.1) is his goal. But as we will see, Skeptic can play so that Reality and Forecaster will be forced, if they are to avoid (1.1), to make their moves satisfy various other conditions – conditions that can be said to validate the forecasts. Moreover, he can achieve this without risking more than the capital with which he begins.

The following game-theoretic bounded law of large numbers is one example of what Skeptic can achieve.

Proposition 1.2 *In Protocol 1.1, Skeptic has a strategy that starts with nonnegative capital ($\mathcal{K}_0 \geq 0$), does not risk more than this initial capital ($\mathcal{K}_n \geq 0$ for all n is guaranteed), and guarantees that either*

$$\lim_{n\to\infty} \frac{1}{n} \sum_{i=1}^{n}(y_i - m_i) = 0 \tag{1.2}$$

or (1.1) will happen.

Later in this chapter (Corollary 1.9), we will derive Borel's law of large numbers for coin tossing as a corollary of this proposition.

After simplifying our terminology and our protocol, we will prove Proposition 1.2 by constructing the required strategy for Skeptic. We will do this step by step, using a series of lemmas as we proceed. First we formalize the notion of forcing by Skeptic and show that a weaker concept suffices for the proof (Lemma 1.4). Then we construct a strategy that forces Reality to eventually keep the average $\sum_{i=1}^{n}(y_i - m_i)/n$ less than a given small positive number κ in order to keep Skeptic's capital from tending to infinity, and another strategy that similarly forces her to keep it greater than $-\kappa$ (Lemma 1.7). Then we average the two strategies and average further over smaller and smaller values of κ. The final average shares the accomplishments of the individual strategies (Lemma 1.6) and hence forces Reality to move $\sum_{i=1}^{n}(y_i - m_i)/n$ closer and closer to zero.

The strategy resulting from this construction can be called a *momentum* strategy. Whichever side of zero the average of the first $n - 1$ of the $y_i - m_i$ falls, Skeptic bets that the nth will also fall on that side: he makes M_n positive if the average so far is positive, negative if the average so far is negative. Reality must make the average converge to zero in order to keep Skeptic's capital from tending to infinity.

Forcing and Almost Sure Events

We need a more concise way of saying that Skeptic can force Forecaster and Reality to do something.

Let us call a condition on the moves $m_1, y_1, m_2, y_2, \ldots$ made by Skeptic's opponents an *event*. We say that a strategy for Skeptic *forces* an event E if it guarantees both of the two following outcomes no matter how Skeptic's opponents move:

$$\mathcal{K}_n \geq 0 \text{ for every } n, \tag{1.3}$$

$$\text{either } \mathcal{K}_n \to \infty \text{ or } E \text{ happens.} \tag{1.4}$$

When Skeptic has a strategy that forces E, we say that Skeptic *can force E*. Proposition 1.2 can be restated by saying that Skeptic can force (1.2). When Skeptic can force E, we also say that E is *almost sure*, or happens *almost surely*. The concepts of forcing and being almost sure apply to all testing protocols (see Sections 6.2, 7.2, and 8.2).

It may deepen our understanding to list some other ways of expressing conditions (1.3) and (1.4):

1. If $\mathcal{K}_n < 0$, we say that Skeptic is *bankrupt* at the end of the nth round. So the condition that (1.3) holds no matter how the opponents move can be expressed by saying that the strategy does not risk bankruptcy. It can also be expressed by saying that $\mathcal{K}_0 \geq 0$ and that the strategy risks only this initial capital.

2. If Skeptic has a strategy that forces E, and α is a positive number, then Skeptic has a strategy for forcing E that begins by setting $\mathcal{K}_0 := \alpha$. To see this, consider these two cases:
   - If the strategy forcing E begins by setting $\mathcal{K}_0 := \beta$, where $\beta < \alpha$, then change the strategy by setting $\mathcal{K}_0 := \alpha$, leaving it otherwise unchanged.
   - If the strategy forcing E begins by setting $\mathcal{K}_0 := \beta$, where $\beta > \alpha$, then change the strategy by multiplying all moves M_n it prescribes for Skeptic by α/β.

   In both cases, (1.3) and (1.4) will still hold for the modified strategy.

3. We can weaken (1.3) to the condition that there exists a real number c such that Skeptic's capital never falls below c. Indeed, if Skeptic has a strategy that guarantees that (1.4) holds and his capital never falls below a negative number c, then we obtain a strategy that guarantees (1.3) and (1.4) simply by adding $-c$ to the initial capital.

Let us also reiterate that a strategy for Skeptic that forces E has a double significance:

1. On the one hand, the strategy can be regarded as assurance that E will happen under the hypothesis that the forecasts are good enough that Skeptic cannot multiply infinitely the capital he risks by betting against them.
2. On the other hand, the strategy and E can be seen as a test of the forecasts. If E does not happen, then doubt is cast on the forecasts by Skeptic's success in betting against them.

Specializing the Protocol

To make the strategy we construct as simple as possible, we simplify Protocol 1.1 by assuming that Forecaster is constrained to set all the m_n equal to zero. Under this assumption, Skeptic's goal (1.2) simplifies to the goal

$$\lim_{n \to \infty} \overline{y}_n = 0, \tag{1.5}$$

where $\overline{y}_n$ is the average of $y_1 \ldots y_n$, and the protocol reduces to the following testing protocol, where Reality is Skeptic's only opponent.

Protocol 1.3
> Skeptic announces $\mathcal{K}_0 \in \mathbb{R}$.
> FOR $n = 1, 2, \ldots$:
> > Skeptic announces $M_n \in \mathbb{R}$.
> > Reality announces $y_n \in [-1, 1]$.
> > $\mathcal{K}_n := \mathcal{K}_{n-1} + M_n y_n$.

Whenever we modify a testing protocol by constraining Skeptic's opponents but without constraining Skeptic or changing the rule by which his capital changes, we say that we are *specializing* the protocol. We call the modified protocol a *specialization*.

If Skeptic can force an event in one protocol, then he can force it in any specialization, because his opponents are weaker there. The following lemma confirms that this implication also goes the other way in the particular case of the specialization from Protocol 1.1 to Protocol 1.3.

Lemma 1.4   *Suppose Skeptic can force (1.5) in Protocol 1.3. Then he can force (1.2) in Protocol 1.1.*

Proof By assumption, Skeptic has a Protocol 1.3 strategy that multiplies its positive initial capital infinitely if (1.5) fails. Consider the Protocol 1.1 strategy that begins with the same initial capital and, when Forecaster and Reality move $m_1, m_2, \ldots$

and $y_1, y_2, \ldots$, moves $\frac{1}{2}M_n$ on the nth round, where M_n is the Protocol 1.3 strategy's nth-round move when Reality moves

$$\frac{y_1 - m_1}{2}, \frac{y_2 - m_2}{2}, \ldots. \tag{1.6}$$

(The y_i and m_i being in $[-1, 1]$, the $(y_i - m_i)/2$ are also in $[-1, 1]$.) When Reality moves (1.6) in Protocol 1.3, the strategy there multiplies its capital infinitely unless

$$\lim_{n \to \infty} \frac{1}{n} \sum_{i=1}^{n} \frac{y_i - m_i}{2} = 0. \tag{1.7}$$

The nth-round gain by the new strategy in Protocol 1.1, $\frac{1}{2}M_n(y_n - m_n)$, being the same as the gain $M_n((y_n - m_n)/2)$ in Protocol 1.3 when Reality moves (1.6), this new strategy will also multiply its capital infinitely unless (1.7) happens. But (1.7) is equivalent to (1.2). □

Situations, Events, and Variables

We now introduce some terminology and notation that is tailored to Protocol 1.3 but applies with some adjustment and elaboration to other testing protocols. Some basic concepts are summarized in Table 1.1.

We begin with the concept of a *situation*. In general, this is a finite sequence of moves by Skeptic's opponents. In Protocol 1.3, it is a sequence of moves by Reality – i.e. a sequence of numbers from $[-1, 1]$. We use the notation for sequences described in the section on terminology and notation at the end of the book: omitting commas, writing □ for the empty sequence, and writing ω_n for the nth element of an infinite sequence ω and ω^n for its prefix of length n. When Skeptic and Reality are playing the nth round, after Reality has made the moves $y_1, \ldots, y_{n-1}$, they are *in the situation* $y_1 \ldots y_{n-1}$. They are in the *initial situation* □ during the first round of play. We write $\mathbb{S}$ for the set of all situations, including □, and we call $\mathbb{S}$ the *situation space*.

Table 1.1 Basic concepts in Protocol 1.3

Concept	Definition	Notation
Situation	Sequence of moves by Reality	$s = y_1 \ldots y_n$
Situation space	Set of all situations	$\mathbb{S}$
Initial situation	Empty sequence	□
Path	Complete sequence of moves by Reality	$\omega = y_1 y_2 \ldots$
Sample space	Set of all paths	Ω
Event	Subset of the sample space	$E \subseteq \Omega$
Variable	Function on the sample space	$X : \Omega \to \mathbb{R}$

An infinite sequence $y_1 y_2 \ldots$ of elements of $[-1, 1]$ is a *path*. This is a complete sequence of moves by Reality. We write Ω for the set of all paths, and we call Ω the *sample space*. We call a subset of Ω an *event*. We often use an uppercase letter such as E to denote an event, but we also sometimes use a statement about the path $y_1 y_2 \ldots$. As stated earlier, an event is a condition on the moves by Skeptic's opponents. Thus (1.5) is an event, but (1.3) and (1.4) are not events.

We call a real-valued function on Ω a *variable*. We often use an uppercase letter such as X to denote a variable, but we also use expressions involving $y_1 y_2 \ldots$. For example, we use $\bar{y}_n$ to denote the variable that maps $y_1 y_2 \ldots$ to $\bar{y}_n$. We can even think of y_1 as a variable; it is the variable that maps $y_1 y_2 \ldots$ to y_1.

Processes and Strategies for Skeptic

We call a real-valued function on $\mathbb{S}$ a *process*. Given a process S and a nonnegative integer n, we write S_n for the variable $\omega \in \Omega \mapsto S(\omega^n) \in \mathbb{R}$. The variables $S_0, S_1, \ldots$ fully determine the process S. We sometimes define a process by specifying a sequence $S_0, S_1, \ldots$ of variables such that $S_n(\omega)$ depends only on ω^n. In measure-theoretic probability it is conventional to call a sequence of variables $S_0, S_1, \ldots$ such that $S_n(\omega)$ depends only on ω^n an *adapted process*. We drop the adjective *adapted* because all the processes we consider in this book are adapted.

We call a real-valued function $\mathcal{A}$ on $\mathbb{S} \setminus \{\square\}$ a *predictable process* if for all $\omega \in \Omega$ and $n \in \mathbb{N}$, $\mathcal{A}(\omega^n)$ depends only on ω^{n-1}. Strictly speaking, a predictable process is not a process, because it is not defined on $\square$. But like a process, a predictable process can be specified as a sequence of variables, in this case the sequence $\mathcal{A}_1, \mathcal{A}_2, \ldots$ given by $\mathcal{A}_n : \omega \in \Omega \mapsto \mathcal{A}(\omega^n) \in \mathbb{R}$.

A strategy ψ for Skeptic in Protocol 1.3 can be represented as a pair $(\psi^{\text{stake}}, \psi^{\text{M}})$, where $\psi^{\text{stake}} \in \mathbb{R}$ and ψ^{M} is a predictable process. Here ψ^{stake} is the value ψ specifies for the initial capital $\mathcal{K}_0$, and $\psi^{\text{M}}(\omega^n)$ is the move M_n it specifies in the situation ω^{n-1}. The predictability property is required because in this situation Skeptic does not yet know ω_n, Reality's move on the nth round.

The strategies for Skeptic form a vector space: (i) if $\psi = (\psi^{\text{stake}}, \psi^{\text{M}})$ is a strategy for Skeptic and β is a real number, then $\beta\psi = (\beta\psi^{\text{stake}}, \beta\psi^{\text{M}})$ is a strategy for Skeptic and (ii) if $\psi^1 = (\psi^{1,\text{stake}}, \psi^{1,\text{M}})$ and $\psi^2 = (\psi^{2,\text{stake}}, \psi^{2,\text{M}})$ are strategies for Skeptic, then $\psi^1 + \psi^2 = (\psi^{1,\text{stake}} + \psi^{2,\text{stake}}, \psi^{1,\text{M}} + \psi^{2,\text{M}})$ is as well.

A strategy $\psi = (\psi^{\text{stake}}, \psi^{\text{M}})$ for Skeptic determines a process whose value in s is Skeptic's capital in s when he follows ψ. This process, which we designate by $\mathcal{K}^\psi$, is given by

$$\mathcal{K}_0^\psi = \psi^{\text{stake}}$$

and

$$\mathcal{K}_n^\psi = \mathcal{K}_{n-1}^\psi + \psi_n^{\text{M}} y_n \tag{1.8}$$

for $n \geq 1$. We refer to $\mathcal{K}^\psi$ as ψ's *capital process*.

If β_1 and β_2 are nonnegative numbers adding to 1, and ψ^1 and ψ^2 are strategies that both begin with capital α, then playing the convex combination $\beta_1 \psi^1 + \beta_2 \psi^2$ can be thought of as dividing the capital α between two accounts, putting $\beta_1 \alpha$ in the first and $\beta_2 \alpha$ in the second, and playing $\beta_1 \psi^1$ with the first account and $\beta_2 \psi^2$ with the second. The resulting capital process is $\mathcal{K}^{\beta_1 \psi^1 + \beta_2 \psi^2} = \beta_1 \mathcal{K}^{\psi^1} + \beta_2 \mathcal{K}^{\psi^2}$.

We can sometimes also form infinite convex combinations of strategies and their capital processes. If $\beta_1, \beta_2, \ldots$ are nonnegative real numbers adding to 1, $\psi^1, \psi^2, \ldots$ all begin with capital α, and the sum $\sum_{k=1}^{\infty} \beta_k \psi^k$ converges in $\mathbb{R}$, then the sum $\sum_{k=1}^{\infty} \beta_k \mathcal{K}^{\psi^k}$ also converges in $\mathbb{R}$ (by induction on (1.8)) and is the capital process for $\sum_{k=1}^{\infty} \beta_k \psi^k$. In this case, Skeptic is dividing the initial capital among a countably infinite number of accounts and playing $\beta_k \psi^k$ with the kth account.

The averaging of strategies is perhaps the most frequently used technique in game-theoretic probability. In the case of an infinite convex combination, however, we must keep in mind the condition that $\sum_{k=1}^{\infty} \beta_k \psi^k$ must converge in $\mathbb{R}$. This means that $\sum_{k=1}^{\infty} \beta_k \psi^{k,\text{stake}}$ must converge in $\mathbb{R}$ and that $\sum_{k=1}^{\infty} \beta_k \psi^{k,\text{M}}(s)$ must converge in $\mathbb{R}$ for every $s \in \mathbb{S} \setminus \{\square\}$. This is hardly guaranteed. If for every rate of growth there is a situation s such that the sequence $\psi^{1,\text{M}}(s), \psi^{2,\text{M}}(s), \ldots$ grows at least that fast, then there will be no sequence of coefficients $\beta_1, \beta_2, \ldots$ for which the convex combination exists.

Weak Forcing

If Skeptic has a nonnegative capital process that is unbounded on paths on which a given event E fails, then, as we shall see, he also has a nonnegative capital process that tends to infinity on these paths. This motivates the concept of weak forcing, which applies not only to Protocols 1.1 and 1.3 but also to all discrete-time testing protocols.

According to our definition of forcing, a strategy ψ for Skeptic *forces E* if $\mathcal{K}^{\psi} \geq 0$ and

$$\lim_{n \to \infty} \mathcal{K}_n^{\psi}(\omega) = \infty$$

for every path $\omega \notin E$. We now say that ψ *weakly forces E* if $\mathcal{K}^{\psi} \geq 0$ and

$$\sup_n \mathcal{K}_n^{\psi}(\omega) = \infty \qquad (1.9)$$

for every path $\omega \notin E$.

Lemma 1.5 *If Skeptic can weakly force E, then he can force E.*

Proof Suppose ψ is a strategy for Skeptic that weakly forces E. Define a strategy ψ' for Skeptic as follows.

- Play ψ starting from $\mathcal{K}_0 = \mathcal{K}_0^\psi$ until your (Skeptic's) capital equals or exceeds $\mathcal{K}_0 + 1$, continuing to play ψ indefinitely if your capital always remains below $\mathcal{K}_0 + 1$. Let m be the first round, if there is one, where your capital $\mathcal{K}_m$ equals or exceeds $\mathcal{K}_0 + 1$. Starting from round $m + 1$, play the strategy

$$M_n := \frac{\mathcal{K}_m - 1}{\mathcal{K}_m} \psi_n^{\mathrm{M}}, \tag{1.10}$$

which is a scaled-down version of ψ^{M}. This means that at the end of round m you have set aside one unit of capital and are now using as working capital only the remaining $\mathcal{K}_m - 1$. From round m onward, ψ risks a net loss of no more than $\mathcal{K}_m$. So the scaled-down strategy (1.10), which makes proportionally smaller bets and incurs proportionally smaller gains and losses, risks a net loss of no more than $\mathcal{K}_m - 1$.

- Continue with the strategy (1.10) until your working capital equals or exceeds $\mathcal{K}_m + 1$. Then again reduce this working capital by setting aside one unit of capital and further scale down the strategy so that only this reduced working capital is at risk.

- Continue in this way, setting aside another unit of capital and scaling down your bets accordingly every time your current working capital becomes a unit or more greater than it was when you last set aside one unit of capital.

The strategy ψ' defined in this way never risks losing more than its current working capital. So its capital process is nonnegative. On any path ω not in E, (1.9) happens, so ψ' sets aside a unit of capital infinitely many times, and so its capital tends to infinity. Thus it forces E.   □

The preceding proof relies on one simple feature of a testing protocol: Skeptic can always scale down a strategy by multiplying all its moves by a positive constant less than 1. When the strategy's capital is unbounded on a given path, the scaled-down version will also have unbounded capital on that path. According to the general definition of a testing protocol given in Chapter 7, scaling down is available to Skeptic in any testing protocol. So the argument works whenever the protocol allows play to continue indefinitely. In general, forcing and weak forcing are equivalent concepts.

In Protocols 1.1 and 1.3, there is another way of scaling down that converts a weakly forcing strategy into a forcing one. Suppose we stop a weakly forcing strategy if and when its capital reaches a certain fixed level. By forming a convex combination of such stopped strategies, we obtain another kind of scaled-down version of the weakly forcing strategy. If the stopped strategies stop at higher and higher levels of capital, then perhaps the convex combination's capital will tend to infinity. This can be made to work in Protocol 1.3 (Exercise 1.1) but does not always work in other testing protocols. As we noted earlier, we cannot always form a given convex combination of Skeptic's strategies.

The next lemma also relies on forming a convex combination of strategies and thus may not be valid for all testing protocols we consider in later chapters.

Lemma 1.6 *If Skeptic can weakly force each of a sequence $E_1, E_2, \ldots$ of events in Protocol 1.3, then he can weakly force $\bigcap_{k=1}^{\infty} E_k$ in Protocol 1.3.*

Proof Let ψ^k be a strategy that weakly forces E_k. Assume, without loss of generality, that ψ^k begins with unit capital. Because it avoids risk of bankruptcy, the absolute value of its move M_n is never greater than its current capital $\mathcal{K}_{n-1}$. This implies that the capital process $\mathcal{K}^{\psi^k}$ can at most double on each step:

$$\mathcal{K}_n^{\psi^k} \leq 2^n.$$

Because (1.8) and the nonnegativity of $\mathcal{K}^{\psi^k}$ imply that $\mathcal{K}_n^{\psi^k} - |\psi_n^{k,\mathrm{M}}| \geq 0$, or $|\psi_n^{k,\mathrm{M}}| \leq \mathcal{K}_n^{\psi^k}$, we can also say that

$$|\psi_n^{k,\mathrm{M}}| \leq 2^n$$

for all k and n, which implies the convergence needed to define a strategy ψ by taking a convex combination of the ψ^k. For example, we can define ψ by

$$\psi := \sum_{k=1}^{\infty} 2^{-k} \psi^k.$$

Since ψ^k weakly forces E_k, ψ also weakly forces E_k. So ψ weakly forces $\bigcap_{k=1}^{\infty} E_k$ □

The following lemma, which is very specific to Protocol 1.3, is central to our proof that Skeptic can force $\overline{y}_n \to 0$ in this protocol.

Lemma 1.7 *Suppose $\kappa > 0$. In Protocol 1.3, Skeptic can weakly force*

$$\limsup_{n \to \infty} \overline{y}_n \leq \kappa \tag{1.11}$$

and he can also weakly force

$$\liminf_{n \to \infty} \overline{y}_n \geq -\kappa. \tag{1.12}$$

Proof Assume without loss of generality that $\kappa \leq 1/2$. Let ψ be the strategy that sets $\mathcal{K}_0 := 1$ and $M_n := \kappa \mathcal{K}_{n-1}$ on each round. Its capital process $\mathcal{K}^{\psi}$ is given by $\mathcal{K}_0^{\psi} = 1$ and

$$\mathcal{K}_n^{\psi} = \mathcal{K}_{n-1}^{\psi}(1 + \kappa y_n) = \prod_{i=1}^{n}(1 + \kappa y_i). \tag{1.13}$$

Let $\omega = y_1 y_2 \ldots$ be a path such that $\sup_n \mathcal{K}_n^\psi(\omega) < \infty$. Then there exists a constant $C_\omega > 0$ such that $\mathcal{K}_n^\psi(\omega) \leq C_\omega$ and hence

$$\sum_{i=1}^{n} \ln(1 + \kappa y_i) \leq \ln C_\omega$$

for all n. Because $t - t^2 \leq \ln(1 + t)$ when $t \geq -1/2$, it follows that for all n,

$$\kappa \sum_{i=1}^{n} y_i - \kappa^2 \sum_{i=1}^{n} y_i^2 \leq \ln C_\omega$$

and hence

$$\kappa \sum_{i=1}^{n} y_i - \kappa^2 n \leq \ln C_\omega$$

or

$$\bar{y}_n \leq \frac{\ln C_\omega}{\kappa n} + \kappa.$$

So (1.11) holds on the path ω. Thus ψ weakly forces (1.11). The same argument, with $-\kappa$ in place of κ, establishes that Skeptic can weakly force (1.12). □

Completing the Proof

To complete the proof of Proposition 1.2, we combine the preceding lemmas.

Proof of Proposition 1.2 By Lemma 1.7, Skeptic can weakly force (1.11) and (1.12) in Protocol 1.3 for $\kappa = 2^{-k}$ for $k = 1, 2, \ldots$. Lemma 1.6 says that he can weakly force the intersection of these events, and this intersection is the event $\bar{y}_n \to 0$. So by Lemma 1.5, he can force $\bar{y}_n \to 0$ in Protocol 1.3, and by Lemma 1.4, he can force (1.2) in Protocol 1.1. □

1.3 BINARY OUTCOMES

Now consider the case where Reality has only a binary choice. In this special case, our framework takes on more familiar colors. If we keep Forecaster in the picture, his moves amount to probabilities about what will happen next. If we take him out of the picture by fixing a constant value for these probabilities, we obtain an even simpler and more classical picture: successive trials of an event with fixed probability.

Probability Forecasting

Because Proposition 1.2 asserts that Skeptic can achieve certain goals regardless of how his opponents move, it remains valid when we specialize Protocol 1.1 by restricting Reality to a binary choice and requiring Forecaster to choose a number between Reality's two options. Requiring Reality to choose from $\{0, 1\}$ and writing p_n instead of m_n for Forecaster's move, we obtain the following.

Protocol 1.8

> Skeptic announces $\mathcal{K}_0 \in \mathbb{R}$.
> FOR $n = 1, 2, \ldots$:
> > Forecaster announces $p_n \in [0, 1]$.
> > Skeptic announces $M_n \in \mathbb{R}$.
> > Reality announces $y_n \in \{0, 1\}$.
> > $\mathcal{K}_n := \mathcal{K}_{n-1} + M_n(y_n - p_n)$.

We call Forecaster's move p_n in this protocol a *probability* for $y_n = 1$, because it invites Skeptic to bet on this outcome at odds $p_n : 1 - p_n$. A positive sign for M_n means that Skeptic is betting on $y_n = 1$; a negative sign means he is betting on $y_n = 0$. The absolute value $|M_n|$ is the total stakes for the bet; the party betting on $y_n = 1$ puts up $|M_n| p_n$, the party betting on $y_n = 0$ puts up $|M_n|(1 - p_n)$, and the winner takes all.

Proposition 1.2 applies directly to Protocol 1.8, giving this corollary.

Corollary 1.9 *In Protocol 1.8, Skeptic has a strategy that starts with nonnegative capital ($\mathcal{K}_0 \geq 0$), does not risk more than this initial capital ($\mathcal{K}_n \geq 0$ for all n is guaranteed), and guarantees that either*

$$\lim_{n \to \infty} (\bar{y}_n - \bar{p}_n) = 0 \tag{1.14}$$

or $\lim_{n \to \infty} \mathcal{K}_n = \infty$, where $\bar{y}_n$ is the fraction of Reality's y_i equal to 1 on the first n rounds, and $\bar{p}_n$ is the average of the Forecaster's p_i on the first n rounds.

In other words, (1.14) happens almost surely.

Coin Tossing: The Archetype of Probability

The coin that comes up heads with probability p every time it is tossed is the archetype of probability theory. The most basic classical theorems, including the laws of large numbers and the central limit theorem, appear already in this picture and even in its simplest instantiation, where $p = 1/2$. All versions of probability theory, classical, measure-theoretic, and game-theoretic, have this archetype at its core.

We obtain a game-theoretic representation of coin tossing from Protocol 1.8 by interpreting $y_n = 1$ as heads and $y_n = 0$ as tails and requiring Forecaster to set $p_n = p$ for all n:

Protocol 1.10
PARAMETER: $p \in [0, 1]$
  Skeptic announces $\mathcal{K}_0 \in \mathbb{R}$.
  FOR $n = 1, 2, \ldots$:
    Skeptic announces $M_n \in \mathbb{R}$.
    Reality announces $y_n \in \{0, 1\}$.
    $\mathcal{K}_n := \mathcal{K}_{n-1} + M_n(y_n - p)$.

Then (1.14) simplifies to

$$\lim_{n \to \infty} \bar{y}_n = p. \tag{1.15}$$

The statement that this happens almost surely is Borel's theorem. But whereas Borel's formulation made it a theorem in measure theory, here it is a theorem in game theory. It is a statement about what Skeptic can achieve in a game. In Chapter 9, we will see how the measure-theoretic version can be derived from it.

There is no need, in the game-theoretic formulation, to say that the successive tosses of the coin are independent. To the extent that independence has any meaning here, it is already explicit in the protocol, because Forecaster's move on the nth round (his probability for $y_n = 1$ in the situation $y_1 y_2 \ldots y_{n-1}$) is p regardless of how $y_1, y_2, \ldots, y_{n-1}$ come out.

Another novel feature of the game-theoretic formulation is its openness to uses different from the usual uses of probability. The convergence (1.15) holds almost surely if Reality acts in some random or neutral way. But it also holds almost surely if Reality plays strategically. We are not required to think of the y_n as random outcomes, determined without regard to how someone might be betting.

1.4   SLACKENINGS AND SUPERMARTINGALES

We sometimes modify a protocol to give Skeptic the option of giving up money on each round. This does not affect whether or not he can force a given event, but it can simplify our mathematics.

Consider Protocol 1.3 for example. When Skeptic chooses $M_n \in \mathbb{R}$, he is choosing the linear payoff function $g(x) = M_n x$. One way of allowing him to give up money is to expand the set of his choices to include all payoff functions that are bounded above by some such linear function. This gives the following protocol.

Protocol 1.11
  Skeptic announces $\mathcal{K}_0 \in \mathbb{R}$.
  FOR $n = 1, 2, \ldots$:
    Skeptic announces $f_n \in \mathbb{R}^{[-1,1]}$ such that
      $\exists M \in \mathbb{R} \; \forall y \in [-1, 1] : f_n(y) \leq My$.
    Reality announces $y_n \in [-1, 1]$.
    $\mathcal{K}_n := \mathcal{K}_{n-1} + f_n(y_n)$.

We could alternatively allow Skeptic to decide whether and how much to give up after Reality makes her move on each round, but this would not change the set of capital processes available to Skeptic (see Exercise 1.7).

We call a testing protocol like Protocol 1.11, in which Skeptic is allowed to take an arbitrary loss on each round, *slack*, and we call a protocol deriving from another in the way Protocol 1.11 derives from Protocol 1.3 its *slackening*. A slack protocol is, of course, its own slackening.

Slackenings arise naturally when Skeptic does not need all his capital to force an event of interest. In our proof of Lemma 1.7, for example, we began with the capital process $\mathcal{K}^{\psi}$ in Protocol 1.3 given by (1.13), but then we effectively (though not explicitly, because we worked on a logarithmic scale) shifted attention to the process $\mathcal{T}$ given by

$$\mathcal{T}_n := \exp\left(\kappa \sum_{i=1}^{n} y_i - \kappa^2 \sum_{i=1}^{n} y_i^2 \right), \tag{1.16}$$

where $\kappa \in (0, 1/2]$. Because an empty sum is zero, (1.16) says in particular that $\mathcal{T}_0 = 1$. It can be verified (see Exercise 1.10) that $\mathcal{T}$ is a capital process for Skeptic in Protocol 1.11.

We call a capital process in the slackening of a testing protocol a *supermartingale* in the original protocol. Thus the process $\mathcal{T}$ given by (1.16), which is a capital process in Protocol 1.11, is a supermartingale but not a capital process in Protocol 1.3. When both $\mathcal{T}$ and $-\mathcal{T}$ are supermartingales, we call $\mathcal{T}$ a *martingale*. The martingales in a given testing protocol form a vector space, but the supermartingales generally do not. If a process is a supermartingale in one testing protocol, then it is also a supermartingale in any specialization. The same is true for a martingale.

The most general and useful of the three overlapping game-theoretic concepts we are discussing (capital process, martingale, and supermartingale) is the concept of a supermartingale. The concepts of forcing by Skeptic and being almost sure, defined in Section 1.2, can be defined using the concept of a supermartingale: Skeptic can force E (E is almost sure) if and only if there is a nonnegative supermartingale that tends to infinity on the paths where E fails. If there is a nonnegative supermartingale that tends to infinity on all paths on which both (i) E fails and (ii) another condition is satisfied, we say that E is *almost sure on paths where the condition is satisfied*. If E is almost sure on paths of which a given situation s is a prefix, we say that E is *almost sure in s*. If E is almost sure, then it is almost sure on all paths.

1.5 CALIBRATION

The convergence $\sum_{i=1}^{n}(y_i - m_i)/n \to 0$ is only one of many events that happen almost surely in Protocol 1.1. Suppose, for example, that we are interested in the average accuracy of Forecaster's m_n on some particular rounds, perhaps rounds where the forecasts fall in a particular range, or perhaps rounds where some other signal is received. The following protocol introduces a player named Monitor, who can help Skeptic focus his attention on particular rounds.

Protocol 1.12

>Skeptic announces $\mathcal{K}_0 \in \mathbb{R}$.
>FOR $n = 1, 2, \ldots$:
>>Forecaster announces $m_n \in [-1, 1]$.
>>Monitor announces $g_n \in \{\text{On}, \text{Off}\}$.
>>Skeptic announces $M_n \in \mathbb{R}$.
>>Reality announces $y_n \in [-1, 1]$.
>>$\mathcal{K}_n := \mathcal{K}_{n-1} + M_n(y_n - m_n)$.

Let $\text{On}(n)$ designate the set of i between 1 and n, inclusive, for which Monitor's signal g_i is On, so that $|\text{On}(n)| \to \infty$ is the event that Monitor announces On infinitely many times.

Proposition 1.13   *The event*

$$|\text{On}(n)| \to \infty \implies \frac{1}{|\text{On}(n)|} \sum_{i \in \text{On}(n)} (y_i - m_i) \to 0 \tag{1.17}$$

happens almost surely in Protocol 1.12.

Suppose, for example, that Monitor sets $g_n = \text{On}$ when and only when m_n is in some small subinterval $[a - \varepsilon, a + \varepsilon]$ of $[-1, 1]$. Applied to this case, Proposition 1.13 tells us that Skeptic can force Reality to make her average move asymptotically approximate a on those rounds where the forecasts approximate a. Lemma 1.6 then tells us that Skeptic can force Reality to do this simultaneously for a whole array of values of a, which we can choose finely spaced across $[-1, 1]$. In this sense, the forecasts $m_1, m_2, \ldots$ will almost surely be *calibrated* with respect to the outcomes $y_1, y_2, \ldots$. By including strategies for Monitors who respond to other signals, we can obtain calibration with respect to those signals as well. We will return to this topic in Chapter 12, where we will look at what Forecaster can do to assure calibration.

Proof of Proposition 1.13   Our proof of Proposition 1.2 also proves Proposition 1.13. To see this, note first that the arguments in the proof still apply if Skeptic's opponents are allowed to break off play at any time they please, provided that, aside from avoiding bankruptcy, Skeptic is required to achieve one of the two goals ($\mathcal{K}_n \to \infty$ and $\frac{1}{n} \sum_{i=1}^{n} (y_i - m_i) \to 0$) only if his opponents let play continue for infinitely many rounds. Applying this insight to the protocol obtained when we extract from Protocol 1.12 a protocol consisting of just those rounds where Monitor announces On, we see that Skeptic can force (1.17) in the extracted protocol. The strategy extends trivially to a strategy that forces (1.17) in the larger protocol: set $M_n = 0$ when $g_n = \text{Off}$.   □

1.6 THE COMPUTATION OF STRATEGIES

The theoretical results in this book are based on the explicit construction of strategies. All the strategies we construct are computable. We do not study their computational properties, but doing so could be interesting and useful.

Masayuki Kumon and Akimichi Takemura [229] construct a particularly simple and computationally efficient strategy that weakly forces $\overline{y}_n \to 0$ in Protocol 1.3: set $M_1 := 0$ and

$$M_n := \frac{\overline{y}_{n-1}}{2} \mathcal{K}_{n-1} \qquad (1.18)$$

for $n > 1$. If our computational model allows basic operations with real numbers, then the computation on each round for this strategy takes constant time, and this remains true when we combine it with the strategy in the proof of Lemma 1.5 to obtain a strategy for forcing.

In a different direction, we can ask about the rate at which Skeptic can increase his capital if the sequence of outcomes produced by Reality in Protocol 1.1 does not satisfy (1.2). For example, the construction in Section 1.2 shows that Skeptic has a computable strategy in Protocol 1.1 that keeps his capital nonnegative and guarantees that if (1.2) fails, then his capital increases exponentially fast – i.e.

$$\limsup_{n\to\infty} \frac{\log \mathcal{K}_n}{n} > 0. \qquad (1.19)$$

Similar questions have been studied in algorithmic probability theory: see [332, 333] in connection with the law of large numbers and [388] in connection with the law of the iterated logarithm and the recurrence property.

1.7 EXERCISES

Exercise 1.1 Complete the alternative proof of Lemma 1.5 sketched after its proof. ☐

Exercise 1.2 Suppose E is an event in Protocol 1.1 or Protocol 1.3, and suppose Skeptic can force E. Is it possible for E to fail after a finite number of rounds? ☐

Exercise 1.3 Consider the following protocol, obtained from Protocol 1.3 by constraining Skeptic to make his moves nonnegative.

> Skeptic announces $\mathcal{K}_0 \in \mathbb{R}$.
> FOR $n = 1, 2, \ldots$:
>   Skeptic announces $M_n \geq 0$.
>   Reality announces $y_n \in [-1, 1]$.
> $\mathcal{K}_n := \mathcal{K}_{n-1} + M_n y_n$.

Show that Skeptic can force $\limsup_{n\to\infty} \overline{y}_n \leq 0$. ☐

Exercise 1.4   (more difficult). Consider the following protocol, obtained from Protocol 1.3 by changing Reality's move space from $[-1, 1]$ to $[-1, \infty)$.

> Skeptic announces $\mathcal{K}_0 \in \mathbb{R}$.
> FOR $n = 1, 2, \ldots$:
> Skeptic announces $M_n \in \mathbb{R}$.
> Reality announces $y_n \in [-1, \infty)$.
> $\mathcal{K}_n := \mathcal{K}_{n-1} + M_n y_n$.

Show that if $0 < a_1 \leq a_2 \leq \cdots \in \mathbb{R}$ and $\lim_{n \to \infty} a_n = \infty$, then Skeptic can force $\sum_{i=1}^{n} y_i / a_n \to 0$ if and only if $\sum_{n=1}^{\infty} 1/a_n < \infty$. *Hint.* See [328].   □

Exercise 1.5   In Protocol 1.8, write $L_n(p_1, \ldots, p_n)$ for the product of Forecaster's probabilities for the actual outcomes on the first n rounds:

$$L_n(p_1, \ldots, p_n) := \prod_{i=1}^{n} (p_i y_i + (1 - p_i)(1 - y_i)).$$

The quantity $-\ln L_n(p_1, \ldots, p_n)$ is sometimes used to score Forecaster's performance; the larger this quantity, the worse he has performed [169]. Suppose Forecaster always chooses $p_n \in (0, 1)$, and suppose Skeptic plays so that $\mathcal{K}_n$ will always be positive no matter how the other players move. Suppose also that when Skeptic announces M_n, he also announces his own forecast, given by

$$q_n := p_n + \frac{M_n p_n (1 - p_n)}{\mathcal{K}_{n-1}}.$$

Verify that for all n, $q_n \in (0, 1)$ and

$$\mathcal{K}_n = \frac{L_n(q_1, \ldots, q_n)}{L_n(p_1, \ldots, p_n)}.$$ □

Exercise 1.6   Consider the following protocol, obtained from Protocol 1.10 by setting p equal to $1/2$:

> Skeptic announces $\mathcal{K}_0 \in \mathbb{R}$.
> FOR $n = 1, 2, \ldots$:
> Skeptic announces $M_n \in \mathbb{R}$.
> Reality announces $y_n \in \{0, 1\}$.
> $\mathcal{K}_n := \mathcal{K}_{n-1} + M_n(y_n - \frac{1}{2})$.

Show that the strategy for Skeptic that sets $\mathcal{K}_0 := 1$, $M_1 := 0$, and

$$M_n := 4 \frac{n-1}{n+1} \left(\bar{y}_{n-1} - \frac{1}{2} \right) \mathcal{K}_{n-1}$$

for $n = 2, 3, \ldots$ has the nonnegative capital process given by

$$\mathcal{K}_n = 2^n \frac{\left(\sum_{i=1}^{n} y_i\right)! \left(n - \sum_{i=1}^{n} y_i\right)!}{(n+1)!}.$$

Show that on a path where $\mathcal{K}_n$ is bounded, $\bar{y}_n$ converges to $1/2$. (This strategy for Skeptic, which proves the law of large numbers for the protocol, was devised by Jean Ville [see [91, section 3.3]].)   □

Exercise 1.7 As we learned in Section 1.4, Protocol 1.11 loosens Protocol 1.3 to allow Skeptic to give up money. The following two protocols also do this.

> Skeptic announces $\mathcal{K}_0 \in \mathbb{R}$.
> FOR $n = 1, 2, \ldots$:
>   Skeptic announces $M_n \in \mathbb{R}$.
>   Reality announces $y_n \in [-1, 1]$.
>   Skeptic announces $\mathrm{Gain}_n \in (-\infty, M_n y_n]$.
> $\mathcal{K}_n := \mathcal{K}_{n-1} + \mathrm{Gain}_n$.

> Skeptic announces $\mathcal{K}_0 \in \mathbb{R}$.
> FOR $n = 1, 2, \ldots$:
>   Skeptic announces $f_n \in \mathbb{R}^{[-1,1]}$ such that
> $\exists M \in \mathbb{R} \; \forall y \in [-1, 1] : f_n(y) \le \mathcal{K}_{n-1} + My$.
>   Reality announces $y_n \in [-1, 1]$.
> $\mathcal{K}_n := f_n(y_n)$.

Show that the set of capital processes in each of these protocols is the same as the set of capital processes in Protocol 1.11.   □

Exercise 1.8 Show that the process $\mathcal{T}$ given by

$$\mathcal{T}_n := \left(\sum_{i=1}^{n} y_i\right)^2 - n \tag{1.20}$$

is a supermartingale in Protocol 1.3 and generalize this result to Protocol 1.1.   □

Exercise 1.9 Prove the following two statements and generalize them to Protocol 1.1.

1. Suppose $\mathcal{T}$ is a supermartingale and $\mathcal{S}$ is a positive process in Protocol 1.3. Suppose $\mathcal{U}$ is a process such that

$$\mathcal{U}_n - \mathcal{U}_{n-1} \le \mathcal{S}_{n-1}(\mathcal{T}_n - \mathcal{T}_{n-1}) \tag{1.21}$$

for $n = 1, 2, \ldots$. Then $\mathcal{U}$ is a supermartingale in Protocol 1.3.

2. Suppose $\mathcal{T}$ is a positive supermartingale in Protocol 1.3, and $\mathcal{U}$ is a positive process such that

$$\frac{\mathcal{U}_n}{\mathcal{U}_{n-1}} \leq \frac{\mathcal{T}_n}{\mathcal{T}_{n-1}} \tag{1.22}$$

for $n = 1, 2, \ldots$. Then $\mathcal{U}$ is a supermartingale in Protocol 1.3. □

Exercise 1.10 Let $\kappa \in (0, 1/2]$. Show that the process $\mathcal{T}$ given by (1.16) is a supermartingale in Protocol 1.3 and generalize this result to Protocol 1.1. □

Exercise 1.11 Show that the strategy for Skeptic given by (1.18) weakly forces $\bar{x}_n \rightarrow 0$ in Protocol 1.3. □

Exercise 1.12 Does the strategy (1.18) in Protocol 1.3 guarantee (1.19)? □

1.8 CONTEXT

Testing Probabilities

As we reminded the reader in the first paragraph of the introduction to Part I, a probabilistic hypothesis can be tested by singling out an event to which it assigns low probability. If the event happens, then perhaps we can reject the hypothesis. This idea is often traced back to a note published by John Arbuthnot in 1710 [11]. Noting that male births had exceeded female births in London for 82 successive years, Arbuthnot argued that male and female births could not have equal chances. For details on this and other eighteenth- and nineteenth-century examples of hypothesis testing, see [170, 174, 367].

If a probabilistic hypothesis concerns successively observed outcomes $y_1 y_2 \ldots$, and if it gives a probability p_n for each y_n after $y_1 \ldots y_{n-1}$ are observed, then we might demand that a test take the actual outcomes $y_1 y_2 \ldots$ into account only via the actual probabilities $p_1 p_2 \ldots$ that the hypothesis and these $y_1 y_2 \ldots$ determined. This demand was first formulated in the 1980s by A. Philip Dawid, who called it the *prequential principle* [97]. The prequential principle is not necessarily satisfied when a test is defined by a low-probability event E defined globally by the whole sequence $y_1 y_2 \ldots$. It is, however, satisfied by game-theoretic tests, and it played an important role in the development of the theory in this part of the book (see [102]).

Classical and Modern Probability

The term *classical probability* is often used to refer to the mathematics and philosophy of probability as understood by Laplace; it relies on the notion of equally likely cases, and it is contrasted with *modern probability*, which stands as pure mathematics based on measure theory. Advances in mathematical probability during the early twentieth century, before Kolmogorov's formulation of the measure-theoretic foundation in 1933 [224], are often viewed as anticipations of the modern theory rather than as advances in the classical theory. See Lorraine Daston's *Classical Probability in the Enlightenment* [93], Jan von Plato's *Creating Modern Probability* [311], and the discussion in Hans Fischer's *A History of the Central Limit Theorem* [145, pp. 7–9].

In this book, we use *classical* in a somewhat different way. We see equally likely cases as only one possible underpinning for classical mathematical probability. Beginning at least in the correspondence between Pascal and Fermat in 1654, we see game-theoretic arguments (Pascal's backward recursion) as well as combinatorial arguments (Fermat's equally likely cases) in classical probability. We also include in classical probability the limit theorems proven before 1933, including Borel's law of large numbers, Cantelli's generalization of it (1916 [62]), Lindeberg's central limit theorem (1922 [251]), Khinchin's law of the iterated logarithm (1924 [217]), and even Kolmogorov's laws of large numbers (1929 [221] and 1930 [223]). All this work is reasonably rigorous as mathematics. While many of the theorems used countable additivity, none of them used the innovative features of Kolmogorov's measure-theoretic foundation, such as the definition of conditional expectation. All these classical theorems can be understood game-theoretically. So rather than using *modern probability* as a synonym for *measure-theoretic probability*, we suggest that measure theory and game theory both provide ways of making classical probability rigorous in the sense of modern mathematics.

Measure-Theoretic Probability

The standard terminology and notation used in measure-theoretic probability is sketched in the section on notation and terminology at the end of the book. Kolmogorov introduced the triple $(\Omega, \mathcal{F}, P)$, where $\mathcal{F}$ is a σ-algebra on Ω and P is a probability measure on $(\Omega, \mathcal{F})$. Before Kolmogorov's work it was customary to use σ-algebras to avoid the ill-behaved non-measurable sets whose existence can be derived from the axiom of choice, but Kolmogorov also gave σ-algebras a substantive role, using them to define the notion of conditional expectation. Doob's subsequent work, especially his 1953 book [116], enlarged σ-algebras' substantive role by using them to model the process of acquiring new information. The triple $(\Omega, \mathcal{F}, P)$ thus becomes a quadruple $(\Omega, \mathcal{F}, (\mathcal{F}_t), P)$, where $(\mathcal{F}_t)$ is a filtration.

Measure-theoretic probability comes closest to classical and game-theoretic probability when Ω is constructed from outcome spaces for a sequence of experiments or observations; as we noted in the introduction to Part I, Ω is often called a *canonical space* in this case [105, section IV.9]. When Ω is a canonical space, the filtration $\mathcal{F}$ is also constructed from the outcome spaces and hence does not play so foundational a role (see Chapter 9).

Laws of Large Numbers

The name *law of large numbers* was introduced in the nineteenth century by Siméon Denis Poisson as a name for the empirical regularities predicted by Bernoulli's theorem and its generalizations (see Chapter 2) [356]. The name was first applied to results like Borel's by Cantelli in 1916 [62], who developed Borel's result in the context of an infinite sequence of independent random variables that are not necessarily binary or even discrete.

In 1928 [218], Khinchin called results like Borel's and Cantelli's, which imply that the probability of

$$\exists n \geq N : \left| \frac{1}{n} \sum_{i=1}^{n} (y_i - m_i) \right| > \varepsilon,$$

where m_i is y_i's expected value, tends to zero with N for every $\varepsilon > 0$, *strong laws of large numbers* to distinguish them from Bernoulli's and Poisson's laws of large numbers, which say only that the probability of

$$\left| \frac{1}{N} \sum_{i=1}^{N} (y_i - m_i) \right| > \varepsilon$$

tends to zero with N for every $\varepsilon > 0$. Following Khinchin, many authors distinguish in general between *strong limit theorems*, which assert convergence almost surely, and *weak limit theorems*, which assert that the probability of a given divergence in a given number of trials tends to zero with the number of trials. This distinction is often useful, but it can also be misleading, because a weak limit theorem can often be proven under weak hypotheses and then it is no longer an obvious consequence of a corresponding strong limit theorem. For this reason, and also because we will be more interested in specific bounds on probabilities as a function of N than in the mere fact that these bounds tend to zero with N, we do not use the adjectives *strong* and *weak* for game-theoretic laws of large numbers. Note, however, that Kolmogorov's 1930 law of large numbers [223], which we make game-theoretic in Theorem 4.3, is often called his strong law, while his 1929 law of large numbers [222], which we make game-theoretic in Lemma 12.10, is often called his weak law.

Sample Space

Probability textbooks often call the set Ω in the triple $(\Omega, \mathcal{F}, P)$ the *sample space* [322, 358, 427]. This name is not completely natural for the game-theoretic picture, but we retain it as a way of connecting the picture to the classical and measure-theoretic pictures. We also write ω for a typical element of Ω. The objects Ω and ω are less fundamental in game-theoretic probability, however, than in measure-theoretic probability. In the discrete-time case, at least, they are constructed from the move spaces of Skeptic's opponents. Moreover, a situation s, being finite, is more basic in applications than a path ω. Situations arise in play, whereas paths are ideal concepts that serve to simplify our notation and our thinking about long sequences of play.

Event

This word has been used informally in probability theory for centuries; Abraham De Moivre used it in his *Doctrine of Chances*, which first appeared in 1718 [109]. Once the notion of a sample space Ω became standard, events were identified with subsets of Ω. In measure-theoretic probability, only measurable subsets of Ω (sets in the specified σ-algebra) are called events; these are the subsets to which probabilities are assigned. Usually all subsets that can be defined constructively can be included, but under the axiom of choice, the usual probability measures on the real numbers cannot be extended to assign probabilities to all subsets (see, for example, [40, pp. 45–46]). There are alternative foundations for mathematics that do not use the axiom of choice and under which all subsets of the real numbers are Lebesgue measurable and are therefore assigned probabilities by the usual probability measures (see [289]).

In discrete time, game-theoretic probability does not need the concept of measurability. The events and variables that we use to define supermartingales and prove theorems in discrete time are nevertheless measurable; the winning sets for the games we consider are even quasi-Borel (see Section 4.5). We do use measurability in Part IV's continuous-time theory.

Almost Sure

Almost everywhere and its equivalents in French (*presque partout*) and German (*fast überall*) were standard technical terms in measure theory in the early twentieth century. Maurice Fréchet is generally credited with transposing the measure-theoretic terminology into probability theory. As he explained in French in 1930 [150], he had distinguished between convergence in measure and convergence almost everywhere in his earlier work in functional analysis, and he noticed the distinction's relevance to probability when he undertook to teach probability at an advanced level in 1929. For probability, he also introduced the terms *convergence en probabilité* (convergence in probability) and *presque toujours* (almost always).

Kolmogorov and Paul Lévy followed Fréchet's lead but changed *always* to *sure*. Kolmogorov used *fast sicher* in his 1933 monograph [224], and Lévy used *presque sûrement* in his 1937 book [245].

Not all twentieth-century authors adopted *almost sure*. We find the older mathematical term *almost everywhere* in Harald Cramér's 1946 book [90] and Doob's 1953 book [116]. William Feller in his two-volume Fellertextbook [135, 136] contented himself with saying that certain events have probability one, just as Borel had done in 1909.

Variable and Random Variable

The use of *variable* to name uncertain quantities with probability distributions goes back to Laplace. Beginning in the late nineteenth century, *variable* was also widely used in mathematical statistics to name quantities that vary from one individual, experiment, or time period to another but for which probability distributions may or may not be defined. After World War II, the term *random variable* became the standard name in English for a variable with a probability distribution. In the measure-theoretic framework, a *random variable* is a measurable real-valued function on the sample space. This ensures that any random variable has a probability distribution, any sequence of random variables has a joint probability distribution, and any measurable function of random variables is a random variable [167, pp. 13–14].

We avoid the term *random variable*, because for many readers it would suggest that a probability measure or at least a σ-algebra has been specified. Instead, we call real-valued functions on our sample space simply *variables*, without imposing any measurability conditions.

Martingale

As Roger Mansuy has persuasively argued [261], this word derives from the name of the city Martigues on the French Mediterranean, referring both to the type of sails its inhabitants used and to their supposedly extravagant dress and behavior. It has been used since at least the eighteenth century, in common speech and in literature, as a name for the gambling strategy that doubles its bet every time it loses. It appears with this meaning in books on mathematical probability by the Marquis de Condorçet (published posthumously in 1805 [79]), Sylvestre François Lacroix (first published in 1816 [233]), and Louis Bachelier (published in 1912 [19]).

In his 1939 dissertation [387], Jean Ville used *martingale* somewhat differently. Working in a framework equivalent to our Protocol 1.10, Ville showed that for every event E that has measure-theoretic probability zero in that protocol, the player we call Skeptic has a strategy that multiplies the capital he risks infinitely if E happens. Ville began by calling any strategy for this player a martingale, even if it is not particularly reckless. Then, noting that the strategies are in a one-to-one correspondence with the player's capital processes, he shifted to calling the capital processes martingales. Called to military service in World War II just as his dissertation was being published, Ville did not return to martingales after the war. For more on his life and work before World War II, see [344].

Ville's work on martingales was inspired by Richard von Mises's effort to define randomness in terms of the stability of relative frequencies. In the case where there is a finite set of possible outcomes, von Mises had proposed that an infinite sequence of such outcomes be considered random if the relative frequency of each outcome converges in the entire sequence and in any infinite subsequence selected by an appropriate rule. As Ville realized, this is not enough; there are other aspects of randomness. It would not be acceptable, for example, for the relative frequency to converge from above or otherwise violate the law of the iterated logarithm. To model randomness faithfully, we must rule out these violations and any other event that can be described constructively and has probability zero under the classical theory. Ville showed that any such event can be ruled out by a gambling strategy; the event will not happen unless the strategy multiplies the capital it risks infinitely. Starting in the 1960s, Claus Peter Schnorr and others picked up this idea and developed it algorithmically (see [38]).

In the mid-1930s, working independently of Ville and motivated by classical problems in mathematical probability rather than by the issues raised by von Mises, Lévy extensively studied sequences $X_1, X_2, \ldots$ of random variables with the property that the expected value of X_n when $X_1, \ldots, X_{n-1}$ are known is equal to zero (see his 1937 book [245] and the historical article [272]). Like all classical probabilists, Lévy used gambling as his primary source of intuition, and so it is fully justified to say that he too was working with martingales. The random variable X_n is the gambler's gain from a fair bet on the nth trial, and the resulting capital process, say $S_n := \sum_{i=1}^{n} X_i$, is the martingale.

In the 1940s and 1950s, Doob made the concept of a martingale (he and Lévy adopted Ville's name for it) into a powerful tool within measure-theoretic probability (see especially Doob's 1953 book [116] and the historical article [255]). The game-theoretic concept of a martingale is close to the measure-theoretic concept of a local martingale (for details, see Section 14.1).

Doob's success in providing a measure-theoretic framework for continuous-time stochastic processes sealed the designation of measure theory as the mathematical foundation for probability. This emphasis on measure theory sometimes obscures, however, the game-theoretic meaning of martingales and supermartingales.

Supermartingale

The concepts of supermartingale and submartingale were introduced in Doob's 1953 book and in a 1952 article by J. Laurie Snell [360]. Snell's article was based on his doctoral dissertation, written under Doob's direction, and it cited Doob's book as forthcoming (see [118, p. 808]). Doob's original names for the two concepts were *semimartingale* and *lower semimartingale*. Gilbert Hunt suggested the names *supermartingale* and *submartingale* in 1960 [192], and Doob soon adopted them [117] (see Doob's 1997 interview with Snell [361, p. 307]).

Doob defined martingales, supermartingales, and submartingales in more than one way. According to one definition, a martingale is a sequence of random variables $\ldots, S_{n-1}, S_n, \ldots$ such that S_{n-1} is equal with probability one to the conditional expectation of S_n relative to the σ-algebra generated by S_{n-1} and earlier random variables in the sequence. Supermartingales and submartingales are defined by replacing the equality with an inequality [116, sections I.7 and II.7]. Another slightly different definition, now more widely used, begins with a filtration, a sequence of successively larger σ-algebras $\ldots, F_{n-1}, F_n, \ldots$, and requires that S_{n-1} be equal to (greater than or equal to in the supermartingale case) the conditional expectation of S_n relative to F_{n-1} [116, section VII.1]. This allows F_{n-1} to be larger than the σ-algebra generated by S_{n-1} and earlier random variables in the sequence. Doob formulated all these definitions in a general context in which time can be discrete or continuous. Their level of abstraction pushes the underlying intuition about betting far into the background.

In Lévy's and Doob's accounts, it is assumed that a martingale is integrable: the expected values of S_n exist and are finite. There is some variation in the case of supermartingales; some authorities (e.g. [199]) require that a supermartingale S_n be integrable whereas others (e.g. [319]) only require that S_n^- be integrable. Our game-theoretic picture avoids these complications. In discrete time, at least, we have no need to assume measurability or to assume that upper expected values are finite.

In the measure-theoretic picture, a martingale is the result of a strategy for betting at odds that are fair inasmuch as the gambler can take either side of any bet, while a supermartingale is the result of a strategy that may give money away by betting at odds that are less than fair. Our game-theoretic picture is more complicated in this respect, because it separates the possibility that the gambler may be offered only one side of a bet, as in Exercise 1.3, from the idea that he is giving money away by betting inefficiently. This poses a terminological choice. If Skeptic can take only one side of a bet but bets efficiently, should we call his capital process a martingale? We call it only a supermartingale, in part because this will be more natural for readers familiar with measure-theoretic terminology. When we want to distinguish a supermartingale that does not waste money from others that do, we call it *efficient* (see Section 7.2).

2

Bernoulli's and De Moivre's Theorems

In the preceding chapter, we studied the convergence of a frequency or an average, an event that depends on infinitely many successive outcomes. We turn now to related approximations involving only finitely many successive outcomes. Such approximations were first established by Jacob Bernoulli and Abraham De Moivre for independent trials of an event, such as a tossed coin falling heads, that happens on each trial with probability p. Bernoulli, in work published in 1713 [33], showed that the frequency that the coin will come up heads will be close to p with high probability in a large number of trials; this was the first law of large numbers. De Moivre, in 1733 [108], showed that the probabilities for deviations of the frequency from p can be described by the Gaussian probability distribution; this was the first central limit theorem.

Bernoulli's and De Moivre's theorems can be stated as follows, where $\bar{y}_N$ is the fraction of the first N tosses in which the coin falls heads, $\mathcal{N}_{0,1}$ is the standard Gaussian distribution, and the operators $\mathbb{P}$ and $\mathbb{E}$ give *global* probabilities and expected values, respectively, as opposed to the *local* probability p.

- *Bernoulli's theorem.* For any $\varepsilon > 0$,

$$\lim_{N \to \infty} \mathbb{P}(|\bar{y}_N - p| \geq \varepsilon) = 0. \tag{2.1}$$

Game-Theoretic Foundations for Probability and Finance, First Edition. Glenn Shafer and Vladimir Vovk.
© 2019 John Wiley & Sons, Inc. Published 2019 by John Wiley & Sons, Inc.

- *De Moivre's theorem.* When $a < b$,

$$\lim_{N \to \infty} \mathbb{P}\left(a < \frac{\bar{y}_N - p}{\sqrt{p(1-p)/N}} < b \right) = \int_a^b \mathcal{N}_{0,1}(dz). \qquad (2.2)$$

More generally, for a wide class of functions U that includes all piece-wise continuous functions that are zero outside a finite interval,

$$\lim_{N \to \infty} \mathbb{E}\left(U\left(\frac{\bar{y}_N - p}{\sqrt{p(1-p)/N}} \right) \right) = \int_{-\infty}^{\infty} U(z)\mathcal{N}_{0,1}(dz). \qquad (2.3)$$

These statements are valid classically, measure-theoretically, and game-theoretically. In the classical theory, global probabilities and global expected values were often defined combinatorially. In measure-theoretic probability, all probabilities are values of a probability measure, and expected values are derived from them as integrals. In game-theoretic probability, global probabilities and global expected values are defined in terms of the strategies available to Skeptic.

The game-theoretic interpretation of Bernoulli's and De Moivre's theorems uses Chapter 1's Protocol 1.10. Recall that when Protocol 1.10 is used to describe coin tossing, $y_n = 1$ means that the nth toss falls heads, and $y_n = 0$ means that the nth toss falls tails, so that $\bar{y}_N$ is the average of $y_1, \ldots, y_N$. Within the protocol, we interpret $\mathbb{P}$ and $\mathbb{E}$ as follows:

- $\mathbb{P}(E)$, for any statement E concerning the outcomes $y_1, \ldots, y_N$, is the infimum of $\mathcal{K}_0$ over strategies for Skeptic that guarantee $\mathcal{K}_N \geq 1$ when E happens and $\mathcal{K}_N \geq 0$ otherwise.
- $\mathbb{E}(X)$, for any variable X that depends only on $y_1, \ldots, y_N$, is the infimum of $\mathcal{K}_0$ over strategies for Skeptic that guarantee $\mathcal{K}_N \geq X$.

As we will see in this chapter, the two theorems hold with this interpretation of $\mathbb{P}$ and $\mathbb{E}$.

In Protocol 1.10, game-theoretic global probabilities and global expected values have classical properties. For example, $\mathbb{E}(X + Y) = \mathbb{E}(X) + \mathbb{E}(Y)$. But when Reality has more than two choices on each round, as in Protocols 1.1 and 1.3, the game-theoretic definitions produce functionals that do not always have these classical properties. So in general we call the game-theoretic quantities *global upper probabilities* and *global upper expected values*, and we replace the symbols $\mathbb{P}(E)$ and $\mathbb{E}(X)$ with $\overline{\mathbb{P}}(E)$ and $\overline{\mathbb{E}}(X)$.

Global upper probabilities are important in applications, because we can use them to test a forecaster or a theory with a finite number of observations. The upper probability $\overline{\mathbb{P}}(E)$ is the amount Skeptic must risk in order to obtain one monetary unit if E happens. If $\overline{\mathbb{P}}(E) \leq \alpha$, where α is a number close to zero, Skeptic bets on E, and

E happens, then Skeptic multiplies his capital by the large factor $1/\alpha$, and we can regard this as strong evidence against the forecaster or the theory.

In Section 2.1, we take a first look at the properties of global upper expected values and global upper probabilities, and we show that they agree with classical expected values and probabilities in the case of coin tossing. It follows from this and the classical proofs of Bernoulli's and De Moivre's theorems that these theorems are valid game-theoretically.

In Sections 2.2, 2.3, and 2.4, we generalize these theorems to protocols where Reality's choices are not binary. In Section 2.2, we generalize Bernoulli's theorem to Protocol 1.3. In Section 2.3, we generalize De Moivre's theorem to a protocol related to Louis Bachelier's model for option pricing. In Section 2.4, we state a version of De Moivre's theorem for Protocol 1.3 and refer the reader to [349] for a proof.

2.1 GAME-THEORETIC EXPECTED VALUE AND PROBABILITY

In Chapter 7, we will define global upper expected value and global upper probability very generally. In this section, we give simplified definitions that apply to events and variables that are settled in a fixed finite number of rounds.

To fix ideas, we first state our definitions in the context of this finite-horizon version of Protocol 1.3, in which play stops after N rounds.

Protocol 2.1
PARAMETER: $N \in \mathbb{N}$
    Skeptic announces $\mathcal{K}_0 \in \mathbb{R}$.
    FOR $n = 1, 2, \ldots, N$:
        Skeptic announces $M_n \in \mathbb{R}$.
        Reality announces $y_n \in [-1, 1]$.
        $\mathcal{K}_n := \mathcal{K}_{n-1} + M_n y_n$.

The situation space $\mathbb{S}$ for Protocol 2.1 is the set of all sequences of elements of $[-1, 1]$ of length N or less. The sample space Ω is the subset of $\mathbb{S}$ consisting of its elements that have length exactly N. Otherwise, we use the definitions and notational conventions given in Chapter 1. In particular (cf. Table 1.1):

- An *event* is a subset of the sample space. We write E^c for the complement of an event E; $E^c := \Omega \setminus E$.
- A *variable* is a real-valued function on the sample space.
- A *supermartingale* is a capital process for Skeptic in the slackening of the protocol, the modification in which he is allowed to give up money on each round.
- A *martingale* is a supermartingale $\mathcal{M}$ such that $-\mathcal{M}$ is also a supermartingale.

We write **T** for the set of all supermartingales in Protocol 2.1.

Global Upper Expected Value

By definition, the *global upper expected value* of a bounded below variable X in Protocol 2.1 is

$$\overline{\mathbb{E}}(X) := \inf\{\mathcal{T}_0 | \mathcal{T} \in \mathbf{T} \text{ and } \mathcal{T}_N \geq X\}. \tag{2.4}$$

We would obtain an equivalent definition if we took the infimum instead over all capital processes in the protocol, because for every supermartingale $\mathcal{T}$ there is a capital process $\mathcal{K}^\psi$ that begins with the same initial capital and satisfies $\mathcal{K}_N^\psi \geq \mathcal{T}_N$.

The definition (2.4) is the finite-horizon simplification of (7.19), the definition of $\overline{\mathbb{E}}(X)$ that we will study in Chapter 7. Because the two definitions agree when X is determined by the first N rounds of the protocol, it is also legitimate to say that $\overline{\mathbb{E}}(X)$ as defined here is X's global upper expected value in Protocol 1.3.

We call the functional $\overline{\mathbb{E}}$ the protocol's *global upper expectation*. The following lemma, the proof of which we leave to the reader (Exercise 2.1), lists some of its properties.

Lemma 2.2 *Global upper expectation has the following properties in Protocol 2.1.*

1. $\overline{\mathbb{E}}(X_1 + X_2) \leq \overline{\mathbb{E}}(X_1) + \overline{\mathbb{E}}(X_2)$.
2. *If $c \in \mathbb{R}$, then $\overline{\mathbb{E}}(X + c) = \overline{\mathbb{E}}(X) + c$.*
3. *If $c \geq 0$, then $\overline{\mathbb{E}}(cX) = c\overline{\mathbb{E}}(X)$.*
4. *If $X_1 \leq X_2$, then $\overline{\mathbb{E}}(X_1) \leq \overline{\mathbb{E}}(X_2)$.*
5. *If $X(\omega) = c$ for all $\omega \in \Omega$, then $\overline{\mathbb{E}}(X) = c$.*

We will study these properties in detail in Section 6.1.

Along with the global upper expected value of a bounded variable X, we are sometimes also interested in its *global lower expected value*, defined by

$$\underline{\mathbb{E}}(X) := -\overline{\mathbb{E}}(-X). \tag{2.5}$$

By Property 5 (or Property 3) in Lemma 2.2, $\overline{\mathbb{E}}(0) = 0$, and from this and Property 1, we see that

$$\underline{\mathbb{E}}(X) \leq \overline{\mathbb{E}}(X). \tag{2.6}$$

Roughly speaking, $\overline{\mathbb{E}}(X)$ is the least initial capital with which Skeptic can obtain final capital $\mathcal{K}_N$ equal to or exceeding X. In this sense, it is the lowest price at which the game enables him to buy X. Buying $-X$ for $\overline{\mathbb{E}}(-X)$ being the same as selling X for $-\overline{\mathbb{E}}(-X)$, (2.5) says that $\underline{\mathbb{E}}(X)$ is the highest price at which the game enables him to sell X.

When $\underline{\mathbb{E}}(X)$ and $\overline{\mathbb{E}}(X)$ are equal, we write $\mathbb{E}(X)$ for their common value and call it X's *global expected value*.

Lemma 2.3 *If M is a bounded martingale, then $\mathbb{E}(M_N) = M_0$.*

Proof By the definition of global upper expected value, (2.4), $\overline{\mathbb{E}}(M_N) \leq M_0$. Since $-M$ is also a martingale, we also have $\overline{\mathbb{E}}(-M_N) \leq -M_0$, or $\underline{\mathbb{E}}(M_N) \geq M_0$. It follows from (2.6) that $\underline{\mathbb{E}}(M_N) = \overline{\mathbb{E}}(M_N) = M_0$. $\qquad\qquad\square$

Global Upper Probability

In the game-theoretic account, just as in the classical and measure-theoretic accounts, probability is a special case of expected value.

Given an event E in Protocol 2.1, we call the global upper expected value $\overline{\mathbb{E}}(\mathbf{1}_E)$ the *global upper probability* of E, and we call the global lower expected value $\underline{\mathbb{E}}(\mathbf{1}_E)$ the *global lower probability* of E. When convenient we write $\overline{\mathbb{P}}(E)$ for $\overline{\mathbb{E}}(\mathbf{1}_E)$, $\underline{\mathbb{P}}(E)$ for $\underline{\mathbb{E}}(\mathbf{1}_E)$, and $\mathbb{P}(E)$ for $\mathbb{E}(\mathbf{1}_E)$ if it exists.

Lemma 2.4 *The following relations hold for any event E in Protocol 2.1:*

$$\overline{\mathbb{P}}(E) + \overline{\mathbb{P}}(E^c) \geq 1, \tag{2.7}$$

$$\underline{\mathbb{P}}(E) = 1 - \overline{\mathbb{P}}(E^c), \tag{2.8}$$

$$0 \leq \underline{\mathbb{P}}(E) \leq \overline{\mathbb{P}}(E) \leq 1. \tag{2.9}$$

Proof These three relations follow readily from the five properties in Lemma 2.2. We obtain (2.7) from Properties 1 and 5:

$$1 = \overline{\mathbb{E}}(1) = \overline{\mathbb{E}}(\mathbf{1}_E + \mathbf{1}_{E^c}) \leq \overline{\mathbb{E}}(\mathbf{1}_E) + \overline{\mathbb{E}}(\mathbf{1}_{E^c}) = \overline{\mathbb{P}}(E) + \overline{\mathbb{P}}(E^c).$$

We obtain (2.8) using Property 2:

$$\underline{\mathbb{P}}(E) = \underline{\mathbb{E}}(\mathbf{1}_E) = -\overline{\mathbb{E}}(-\mathbf{1}_E) = -\overline{\mathbb{E}}(\mathbf{1}_{E^c} - 1) = -\overline{\mathbb{E}}(\mathbf{1}_{E^c}) + 1 = 1 - \overline{\mathbb{P}}(E^c).$$

The second inequality in (2.9) follows from (2.7) and (2.8), while the first and third inequalities from Properties 4 and 5. $\qquad\qquad\square$

As explained in the introduction to this chapter, $\overline{\mathbb{P}}(E)$ being small can be taken to mean that E is unlikely. By (2.9), $\overline{\mathbb{P}}(E)$ being small implies that $\underline{\mathbb{P}}(E)$ is just as small or smaller. We can express the assertion that $\overline{\mathbb{P}}(E)$ is small by saying that the global lower probability $\underline{\mathbb{P}}(E^c)$ is close to 1; this is our way of saying that E^c is likely.

According to our definitions,

$$\overline{\mathbb{P}}(E) := \inf\{\mathcal{T}_0 | \mathcal{T} \in \mathbf{T} \text{ and } \mathcal{T}_N \geq \mathbf{1}_E\}. \tag{2.10}$$

A supermartingale $\mathcal{T}$ that satisfies the condition $\mathcal{T}_N \geq \mathbf{1}_E$ is nonnegative: it satisfies $\mathcal{T}_n \geq 0$ for $n = 0, \ldots, N$ (see Exercise 2.4). Our usual way of proving that an event E satisfies $\overline{\mathbb{P}}(E) \leq \alpha$ is to construct a nonnegative supermartingale $\mathcal{T}$ such that $\mathcal{T}_0 = \alpha$ and $\mathcal{T}_N \geq \mathbf{1}_E$. We say that such a supermartingale *witnesses* $\overline{\mathbb{P}}(E) \leq \alpha$.

Global Expected Values for Coin Tossing

The game-theoretic definitions of global upper and lower expected value can be used in any of the protocols we studied in Chapter 1. In particular, they can be used for coin tossing. In this setting, according to the following proposition, $\overline{\mathbb{E}}(X) = \underline{\mathbb{E}}(X)$ for every variable X. Every variable, in other words, has a global expected value.

Proposition 2.5   *Suppose X is a variable in Protocol 1.10 that is determined in the first N rounds. Then*

$$\mathbb{E}(X) = \sum_{t \in \{0,1\}^N} p^{\#t}(1-p)^{N-\#t} X(t), \tag{2.11}$$

where $\#t$ is the number of 1s in t.

Proof Consider the process $\mathcal{M}$ defined by $\mathcal{M}_N := X$ and then by backward recursion for $n = N - 1, \ldots, 0$:

$$\mathcal{M}(y_1 \ldots y_n) := p\mathcal{M}(y_1 \ldots y_n 1) + (1-p)\mathcal{M}(y_1 \ldots y_n 0). \tag{2.12}$$

It follows from (2.12) that

$$\mathcal{M}(y_1 \ldots y_n) = \mathcal{M}(y_1 \ldots y_{n-1}) + L_n(y_n - p)$$

for $n = 1, \ldots, N$, where

$$L_n := \mathcal{M}(y_1 \ldots y_{n-1} 1) - \mathcal{M}(y_1 \ldots y_{n-1} 0). \tag{2.13}$$

Thus $\mathcal{M}$ is the capital process for the strategy that sets $\mathcal{K}_0 := \mathcal{M}_0$ and makes the move L_n defined by (2.13) on round n. It is clear that $-\mathcal{M}$ is the capital process for the strategy that sets $\mathcal{K}_0 := -\mathcal{M}_0$ and makes the move $-L_n$ on round n. So $\mathcal{M}$ is a martingale.

If also follows from (2.12), by backward induction, that

$$\mathcal{M}(y_1 \ldots y_n) = \sum_{t \in \{0,1\}^{N-n}} p^{\#t}(1-p)^{N-n-\#t} X(y_1 \ldots y_n t) \tag{2.14}$$

for $n = N - 1, \ldots, 0$. When $n = 0$, (2.14) reduces to the right-hand side of (2.11). This is the value of $\mathcal{M}_0$, and by Lemma 2.3, it is equal to $\mathbb{E}(X)$. $\square$

The global expected values given by Proposition 2.5 are the same as the expected values given by classical probability for variables that are functions of N independent tosses of a coin that falls heads with probability p each time (see, e.g. [135, section IX.3]). As we noted in the introduction to this chapter, this establishes that Bernoulli's and De Moivre's theorems are valid game-theoretically. In the next three sections, we look at game-theoretic generalizations.

Exercise 2.6 generalizes Proposition 2.5 to the case where a local probability p_n is chosen according to a strategy for Forecaster that specifies it as a function of the situation $y_1 \ldots y_{n-1}$. This generalization accords with classical probability's rules for calculating expected values for variables depending on a finite sequence of dependent trials. For a further generalization, see Theorem 9.3.

2.2 BERNOULLI'S THEOREM FOR BOUNDED FORECASTING

The following proposition, which can be proven using any number of different nonnegative supermartingales, is the game-theoretic generalization of Bernoulli's theorem to Protocol 2.1.

Proposition 2.6   *For any $\varepsilon > 0$ and $\delta > 0$, there exists $N_{\varepsilon,\delta} \in \mathbb{N}$ such that*

$$\overline{\mathbb{P}}(|\bar{y}_N| \geq \varepsilon) \leq \delta$$

in any instantiation of Protocol 2.1 with $N \geq N_{\varepsilon,\delta}$.

(When a protocol has one or more parameters, we call any protocol obtained by fixing their values an *instantiation*.)

The two following lemmas constitute alternative proofs of Proposition 2.6.

Lemma 2.7   *For any $\varepsilon > 0$ and any value of the parameter N in Protocol 2.1,*

$$\overline{\mathbb{P}}(|\bar{y}_N| \geq \varepsilon) \leq \frac{1}{\varepsilon^2 N}. \tag{2.15}$$

Proof Consider the quadratic supermartingale $\mathcal{T}$ defined by (1.20); $\mathcal{T}_n := (\sum_{i=1}^{n} y_i)^2 - n$. In Protocol 2.1, $\mathcal{T}$ begins at 0 and is never less than $-N$. So the process $\mathcal{U}$ defined by

$$\mathcal{U}_n := \frac{\mathcal{T}_n + N}{\varepsilon^2 N^2} = \frac{n^2(\bar{y}_n)^2 - n + N}{\varepsilon^2 N^2}$$

is a nonnegative supermartingale that begins at $1/\varepsilon^2 N$. Moreover, $\mathcal{U}_N \geq 1$ when $|\bar{y}_N| \geq \varepsilon$. Since $\mathcal{U}$ is nonnegative, this means that $\mathcal{U}_N \geq \mathbf{1}_{|\bar{y}_N| \geq \varepsilon}$. Thus $\mathcal{U}$ witnesses (2.15). $\qquad\square$

Lemma 2.8 *For any $\varepsilon > 0$ and any value of the parameter N in Protocol 2.1,*

$$\overline{\mathbb{P}}(|\bar{y}_N| \geq \varepsilon) \leq 2\exp\left(-\frac{\varepsilon^2 N}{4}\right). \tag{2.16}$$

Proof Consider the nonnegative exponential supermartingale $\mathcal{T}$ defined by (1.16) for $\kappa \in (0, 1/2]$. We have

$$\mathcal{T}_N \geq \exp(\kappa N\varepsilon - \kappa^2 N)$$

when $\bar{y}_N \geq \varepsilon$, and hence the nonnegative supermartingale

$$\exp(-\kappa N\varepsilon + \kappa^2 N)\mathcal{T}, \tag{2.17}$$

which begins at $\exp(-\kappa N\varepsilon + \kappa^2 N)$, witnesses

$$\overline{\mathbb{P}}(\bar{y}_N \geq \varepsilon) \leq \exp(-\kappa N\varepsilon + \kappa^2 N). \tag{2.18}$$

The right-hand side of (2.18) is minimized as a function of κ when $\kappa = \varepsilon/2$. Assuming that $0 < \varepsilon \leq 1$ (without loss of generality, because the upper probability of $\bar{y}_N \geq \varepsilon$ is zero when $\varepsilon > 1$), we see that this value of κ falls in the allowed range $(0, 1/2]$. Substituting $\varepsilon/2$ for κ in (2.17) and (2.18), we see that the nonnegative supermartingale $\mathcal{U}$ defined by

$$\mathcal{U}_n := \exp\left(-\frac{\varepsilon^2 N}{4} + \frac{\varepsilon}{2}\sum_{i=1}^{n} y_i - \frac{\varepsilon^2}{4}\sum_{i=1}^{n} y_i^2\right)$$

witnesses

$$\overline{\mathbb{P}}(\bar{y}_N \geq \varepsilon) \leq \exp\left(-\frac{\varepsilon^2 N}{4}\right).$$

By symmetry, the nonnegative supermartingale $\mathcal{V}$ given by

$$\mathcal{V}(y_1 \ldots y_n) = \mathcal{U}(-y_1 \ldots - y_n)$$

witnesses

$$\overline{\mathbb{P}}(\bar{y}_N \leq -\varepsilon) \leq \exp\left(-\frac{\varepsilon^2 N}{4}\right).$$

It follows that the nonnegative supermartingale $\mathcal{U} + \mathcal{V}$ witnesses (2.16). $\quad\square$

The two upper bounds on $\overline{\mathbb{P}}(|\bar{y}_N| \geq \varepsilon)$, (2.15) and (2.16), are compared in Exercise 2.7. As we see there, $2\exp(-\varepsilon^2 N/4)$ is generally but not always a tighter bound than $1/\varepsilon^2 N$. In Chapter 3, we show that it can be strengthened to $2\exp(-\varepsilon^2 N/2)$; this is a special case of Hoeffding's inequality (see Corollary 3.8 and Eq. (3.16)).

Coin Tossing Again

Just as Borel's theorem generalized in Chapter 1 from Protocol 1.3, where the price on each round is zero, to Protocol 1.1, where Forecaster sets the price m_n, the inequalities (2.15) and (2.16) can be translated into bounds on $\overline{\mathbb{P}}(|\overline{y}_N - \overline{m}_N| \geq \varepsilon)$ in the finite-horizon version of Protocol 1.1 (see Exercise 2.8).

We can obtain corresponding inequalities for Protocol 1.10. Here, by Proposition 2.5, the upper probabilities are probabilities:

$$\mathbb{P}(|\overline{y}_N - p| \geq \varepsilon) \leq \frac{1}{\varepsilon^2 N} \tag{2.19}$$

and

$$\mathbb{P}(|\overline{y}_N - p| \geq \varepsilon) \leq 2 \exp\left(-\frac{\varepsilon^2 N}{4}\right); \tag{2.20}$$

again see Exercise 2.8. From either of these inequalities, we immediately obtain (2.1), which we have been calling Bernoulli's theorem in this chapter.

2.3 A CENTRAL LIMIT THEOREM

In this section, we prove a central limit theorem that casts light on Louis Bachelier's model for pricing financial options. This theorem is a generalization of the special case of De Moivre's theorem where $p = 1/2$, and the proof we give brings out the game-theoretic core shared by De Moivre's theorem and models for option pricing.

In the model Bachelier proposed in 1900 [18, 20], the stock price underlying a financial option follows Brownian motion, and the option's price is its payoff's expected value with respect to a Gaussian distribution. In this model the volatility is measured on the absolute scale (using the standard deviation of price changes) rather than the relative scale (using the standard deviation of returns) usually used nowadays. The feature of Brownian motion essential to Bachelier's model is its constant volatility, which enables a trader to hedge a change in the option's value by trading only in the stock. In practice the volatility of stock prices is not constant, but the variance swaps traded in many of today's financial markets allow traders to remedy this deficiency [191, section 26.16]. A *variance swap* (on the absolute scale) allows a trader to buy the sum of squared daily price changes over some future interval of time. By buying the swap at the beginning of trading and selling it at the end of trading on a particular day, the trader pays the difference between the opening and closing prices of the swap and receives in return the squared change in the stock's price for that day. The theorem we prove in this section casts light on how this allows the trader to hedge an option at the Gaussian price.

We use the following protocol, where D_{n-1} represents the price of the variance swap at the beginning of the nth trading period, and y_n represents the change in the stock's price during the period. The trader (represented by Skeptic) holds M_n shares of the underlying stock during the period, for a net gain (perhaps negative) of $M_n y_n$. He holds V_n units of the variance swap, paying $V_n D_{n-1}$ for them at the beginning of the period and receiving $V_n(y_n^2 + D_n)$ when he sells them at the end of the period. For simplicity, we scale the stock price so that D_0, the price of the variance swap at the outset, is equal to 1. Because we want the trading to last for many periods, this scaling requires that the price changes y_n and $D_n - D_{n-1}$ be very small.

Protocol 2.9
PARAMETER: $\delta \in (0, 1)$
    Skeptic announces $\mathcal{K}_0 \in \mathbb{R}$.
    $D_0 := 1$.
    FOR $n = 1, 2, \ldots$:
        Skeptic announces $M_n, V_n \in \mathbb{R}$.
        Reality announces $y_n \in [-\delta, \delta]$.
        Forecaster announces $D_n \geq 0$ such that $|D_n - D_{n-1}| \leq \delta^2$.
        $\mathcal{K}_n := \mathcal{K}_{n-1} + M_n y_n + V_n(y_n^2 + D_n - D_{n-1})$.
        If $D_n = 0$, $N := n$ and the game stops.
ADDITIONAL CONSTRAINT:
    Forecaster first announces $D_n := 0$ for some $n \leq \delta^{-5/2}$.

The constraints on Forecaster in this protocol force N to be large but not too large:

$$\delta^{-2} \leq N \leq \delta^{-5/2}. \tag{2.21}$$

Proposition 2.10   *Suppose U is a bounded continuous function. Then for any $\varepsilon > 0$, there exists $\delta \in (0, 1)$ such that*

$$\left| \overline{\mathbb{E}}(U(S_N)) - \int_{-\infty}^{\infty} U(z) \mathcal{N}_{0,1}(dz) \right| < \varepsilon \tag{2.22}$$

in the instantiation of Protocol 2.9 with that value of δ, where $S_N := \sum_{n=1}^{N} y_n$.

In the option-pricing interpretation of Proposition 2.10, S_N is the change in the stock's price from the beginning to the end of the protocol and $U(S_N)$ is the option's payoff. The proposition tells us that Skeptic can superhedge the option (i.e. guarantee himself a payoff at least as great) starting with initial capital within ε of $\int_{-\infty}^{\infty} U(z) \mathcal{N}_{0,1}(dz)$. The proposition makes the assumption, made by the Black–Scholes model and Bachelier's model in a different way, that the stock's overall volatility from the purchase of the option to its maturity is itself priced at the outset.

Although Proposition 2.10 assumes U is bounded and continuous, the following proof also works in the case where U is unbounded but has a constant slope to the left and to the right of some finite interval. This covers the case of call and put options.

Proof of Proposition 2.10 Set $S_n := \sum_{i=1}^{n} y_i$ for $n = 0, \ldots, N$ and $v_n := D_n - D_{n-1}$ for $n = 1, \ldots, N$. For any sequence $A_0, \ldots, A_N$, set $\Delta A_n := A_n - A_{n-1}$ for $n = 1, 2, \ldots, N$ (so that $v_n = \Delta D_n$). Assume, without loss of generality, that U is an infinitely differentiable function, constant outside a finite interval (see Exercise 2.9).

Consider the smoothed function $\overline{U}(\cdot, D)$ obtained from U by adding a Gaussian perturbation with mean zero and variance D to its argument and calculating the expected value:

$$\overline{U}(s, D) := \int_{-\infty}^{\infty} U(s + z)\mathcal{N}_{0,D}(dz) = \int_{-\infty}^{\infty} U(z)\mathcal{N}_{s,D}(dz). \tag{2.23}$$

It is well known that U satisfies the heat equation,

$$\frac{\partial \overline{U}}{\partial D} = \frac{1}{2}\frac{\partial^2 \overline{U}}{\partial s^2}, \tag{2.24}$$

for all $s \in \mathbb{R}$ and all $D > 0$ (see Section 2.6).

We will construct a martingale $\mathcal{M}_0, \mathcal{M}_1, \ldots$ such that $\mathcal{M}_0 = \overline{U}(0, 1)$ and

$$\mathcal{M}_n \approx \overline{U}(S_n, D_n) \tag{2.25}$$

for $n = 1, \ldots, N$. Because

$$\overline{U}(S_0, D_0) = \overline{U}(0, 1) = \int_{-\infty}^{\infty} U(z)\mathcal{N}_{0,1}(dz)$$

and

$$\overline{U}(S_N, D_N) = \overline{U}(S_N, 0) = U(S_N),$$

the proposition will then follow from Lemma 2.3, which tells us that $\mathbb{E}(\mathcal{M}_N) = \mathcal{M}_0$.

To find a strategy that begins with $\mathcal{K}_0 = \overline{U}(0, 1)$ and produces a martingale $\mathcal{M}$ satisfying (2.25), set

$$\Delta \overline{U}_n := \overline{U}(S_n, D_n) - \overline{U}(S_{n-1}, D_{n-1})$$

for $n = 1, \ldots, N$, and consider the second-order Taylor expansion

$$\Delta \overline{U}_n \approx \frac{\partial \overline{U}}{\partial s}\Delta s + \frac{\partial \overline{U}}{\partial D}\Delta D + \frac{1}{2}\frac{\partial^2 \overline{U}}{\partial s^2}(\Delta s)^2 + \frac{\partial^2 \overline{U}}{\partial s\partial D}\Delta s\Delta D + \frac{1}{2}\frac{\partial^2 \overline{U}}{\partial D^2}(\Delta D)^2, \tag{2.26}$$

where $\Delta s := y_n$ and $\Delta D := v_n$. Substituting these values for Δs and ΔD and using the heat equation, (2.24), we obtain

$$\Delta \overline{U}_n \approx \frac{\partial \overline{U}}{\partial s}y_n + \frac{\partial \overline{U}}{\partial D}(y_n^2 + v_n) + \frac{\partial^2 \overline{U}}{\partial s\partial D}y_n v_n + \frac{1}{2}\frac{\partial^2 \overline{U}}{\partial D^2}v_n^2. \tag{2.27}$$

The last two terms on the right-hand side are of order δ^3 and δ^4, respectively, and hence can be neglected along with the yet higher order terms in the full Taylor expansion. Therefore, (2.27) reduces to

$$\Delta \overline{U}_n \approx \frac{\partial \overline{U}}{\partial s} y_n + \frac{\partial \overline{U}}{\partial D}(y_n^2 + v_n).$$

So Skeptic's strategy is simple; he sets $M_n := \partial \overline{U}/\partial s$ and $V_n := \partial \overline{U}/\partial D$ for $n = 1, \ldots, N$, the partial derivatives being taken at $(s, D) = (S_{n-1}, D_{n-1})$.

This argument will be rigorous once we have dealt carefully with the remainder terms in the Taylor expansion. To this end, we write

$$\Delta \overline{U}(S_n, D_n) = \frac{\partial \overline{U}}{\partial s}(S_{n-1}, D_{n-1})\Delta S_n + \frac{\partial \overline{U}}{\partial D}(S_{n-1}, D_{n-1})\Delta D_n$$

$$+ \frac{1}{2}\frac{\partial^2 \overline{U}}{\partial s^2}(S_n', D_n')(\Delta S_n)^2 + \frac{\partial^2 \overline{U}}{\partial s \partial D}(S_n', D_n')\Delta S_n \Delta D_n$$

$$+ \frac{1}{2}\frac{\partial^2 \overline{U}}{\partial D^2}(S_n', D_n')(\Delta D_n)^2 \qquad (2.28)$$

for $n = 1, \ldots, N$, where (S_n', D_n') is a point strictly between (S_{n-1}, D_{n-1}) and (S_n, D_n). Applying Taylor's formula to $\partial^2 \overline{U}/\partial s^2$, we find

$$\frac{\partial^2 \overline{U}}{\partial s^2}(S_n', D_n') = \frac{\partial^2 \overline{U}}{\partial s^2}(S_{n-1}, D_{n-1}) + \frac{\partial^3 \overline{U}}{\partial s^3}(S_n'', D_n'')\Delta S_n' + \frac{\partial^3 \overline{U}}{\partial D \partial s^2}(S_n'', D_n'')\Delta D_n',$$

where (S_n'', D_n'') is a point strictly between (S_{n-1}, D_{n-1}) and (S_n', D_n'), and $\Delta S_n'$ and $\Delta D_n'$ satisfy $|\Delta S_n'| \leq |\Delta S_n|$ and $|\Delta D_n'| \leq |\Delta D_n|$. Plugging this equation and the heat equation, (2.24), into (2.28), we obtain

$$\Delta \overline{U}(S_n, D_n) = \frac{\partial \overline{U}}{\partial s}(S_{n-1}, D_{n-1})\Delta S_n + \frac{\partial \overline{U}}{\partial D}(S_{n-1}, D_{n-1})((\Delta S_n)^2 + \Delta D_n)$$

$$+ \frac{1}{2}\frac{\partial^3 \overline{U}}{\partial s^3}(S_n'', D_n'')\Delta S_n'(\Delta S_n)^2 + \frac{1}{2}\frac{\partial^3 \overline{U}}{\partial D \partial s^2}(S_n'', D_n'')\Delta D_n'(\Delta S_n)^2$$

$$+ \frac{\partial^2 \overline{U}}{\partial s \partial D}(S_n', D_n')\Delta S_n \Delta D_n + \frac{1}{2}\frac{\partial^2 \overline{U}}{\partial D^2}(S_n', D_n')(\Delta D_n)^2. \qquad (2.29)$$

We can now show that all the terms on the right-hand side of (2.29) except the first two are negligible.

All the partial derivatives involved in the remaining terms are bounded: the heat equation implies

$$\frac{\partial^3 \overline{U}}{\partial D \partial s^2} = \frac{\partial^3 \overline{U}}{\partial s^2 \partial D} = \frac{1}{2}\frac{\partial^4 \overline{U}}{\partial s^4}, \quad \frac{\partial^2 \overline{U}}{\partial s \partial D} = \frac{1}{2}\frac{\partial^3 \overline{U}}{\partial s^3}, \quad \frac{\partial^2 \overline{U}}{\partial D^2} = \frac{1}{2}\frac{\partial^3 \overline{U}}{\partial D \partial s^2} = \frac{1}{4}\frac{\partial^4 \overline{U}}{\partial s^4},$$

and $\partial^3\overline{U}/\partial s^3$ and $\partial^4\overline{U}/\partial s^4$, being averages of $U^{(3)}$ and $U^{(4)}$, are bounded. So the four terms will have at most the order of magnitude $O(\delta^3)$, and their total cumulative contribution will be at most $O(N\delta^3)$. The absolute difference between $\mathcal{M}_N$ and $U(S_N)$ is consequently bounded above by $cN\delta^3$ for some $c \geq 0$. If we choose δ to satisfy $\delta < \varepsilon^2/c^2$, then by (2.21) we will have $cN\delta^3 \leq c\delta^{1/2} < \varepsilon$.   □

The fact that $\overline{U}$ satisfies the heat equation is essential in the preceding proof. We cannot neglect the $(\Delta s)^2$ term in the Taylor expansion (2.26) simply because $(\Delta s)^2$ is so much smaller than $|\Delta s|$. The increment $\Delta s = y_n$ may change sign frequently, whereas $(\Delta s)^2$ never changes sign. Because the partial derivatives change sign relatively infrequently, this means that the successive values of the leading term, when summed, may tend to cancel each other out, while the successive values of the terms of smaller order tend to cumulate, possibly producing totals comparable to that of the lower-order leading term. We cannot ignore the term containing ΔD for a similar reason: the general tendency for this term is to be negative.

As its name suggests, the heat equation was first used to study the propagation of heat. In this context, D represents time, and the coefficient $1/2$ is replaced by an arbitrary constant. The value $\overline{U}(s, D)$ is the temperature at point s at time D, and the equation says that the increase in the temperature at a point is proportional to how much warmer adjacent points are. Figure 2.1 shows the resulting propagation of heat for one particular initial distribution $\overline{U}(s, 0)$. The propagation tends to smooth any distribution of heat, no matter how irregular, until it resembles the bell-shaped curve associated with the Gaussian distribution. Because the heat equation is also used to describe the diffusion of a substance in a fluid, it is often called the *diffusion equation*.

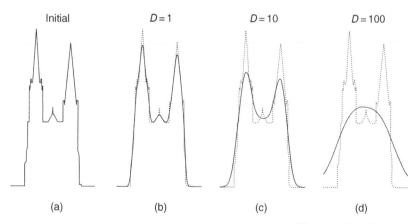

Figure 2.1   Heat propagation according to the equation $\partial\overline{U}/\partial D = \frac{1}{2}\partial^2\overline{U}/\partial s^2$. Part (a) shows an initial temperature distribution $U(s)$, or $\overline{U}(s, 0)$. Parts (b), (c), and (d) show $\overline{U}(s, D)$ as a function of s at times $D = 1, 10, 100$, respectively.

Our application is no exception to the rule that the variable whose first derivative appears in the equation represents time; D decreases with time.

De Moivre's Theorem as a Special Case

De Moivre's theorem for $p = 1/2$ is a special case of Proposition 2.10. To see this, first rescale the prices of the stock and the variance swap, multiplying the y_n by $1/\delta$ and the D_n by $1/\delta^2$. This eliminates δ, making D_0 the free parameter:

PARAMETER: $D_0 \in (1, \infty)$
   Skeptic announces $\mathcal{K}_0 \in \mathbb{R}$.
   FOR $n = 1, 2, \ldots$:
      Skeptic announces $M_n, V_n \in \mathbb{R}$.
      Reality announces $y_n \in [-1, 1]$.
      Forecaster announces $D_n \geq 0$ such that $|D_n - D_{n-1}| \leq 1$.
      $\mathcal{K}_n := \mathcal{K}_{n-1} + M_n y_n + V_n(y_n^2 + D_n - D_{n-1})$.
      If $D_n = 0$, $N := n$ and the game stops.
ADDITIONAL CONSTRAINT:
   Forecaster first announces $D_n := 0$ for some $n \leq D_0^{5/4}$.

Proposition 2.10 is not affected by the rescaling; for each $\varepsilon > 0$, there exists D_0 such that (2.22) holds. Because it is a statement about what Skeptic can accomplish no matter what his opponents do, the proposition also continues to hold if we fix N, set $D_0 := N$, constrain Forecaster to announce $D_n := N - n$ for $n = 1, \ldots, N$, and constrain Reality to choose y_n from the two-element set $\{-1, 1\}$ for $n = 1, \ldots, N$. Because $y_n^2 + D_n - D_{n-1}$ is then always equal to zero, we can remove Skeptic's move V_n, and the protocol takes on a more familiar shape:

Protocol 2.11
PARAMETER: $N \in \mathbb{N}$
   Skeptic announces $\mathcal{K}_0 \in \mathbb{R}$.
   FOR $n = 1, \ldots, N$:
      Skeptic announces $M_n \in \mathbb{R}$.
      Reality announces $y_n \in \{-1, 1\}$.
      $\mathcal{K}_n := \mathcal{K}_{n-1} + M_n y_n$.

In Protocol 2.11, where global upper probabilities are probabilities because Reality's choice is always binary, Proposition 2.10 says that for any $\varepsilon > 0$, there exists $N \in \mathbb{N}$ such that

$$\left| \mathbb{E}(U(N^{1/2} \, \bar{y}_N)) - \int_{-\infty}^{\infty} U(z) \mathcal{N}_{0,1}(dz) \right| < \varepsilon. \tag{2.30}$$

When we reparameterize replacing y_n by $2y_n - 1$ and letting Reality choose from $\{0, 1\}$ instead of $\{-1, 1\}$, we obtain Protocol 1.10 with $p = 1/2$, and Proposition 2.10

becomes the statement that for any $\varepsilon > 0$, there exists $N \in \mathbb{N}$ such that

$$\left| \mathbb{E}\left(U\left(2N^{1/2}\left(\bar{y}_N - \frac{1}{2} \right) \right) \right) - \int_{-\infty}^{\infty} U(z) \mathcal{N}_{0,1}(dz) \right| < \varepsilon.$$

This is equivalent to (2.3) with $p = 1/2$.

2.4 GLOBAL UPPER EXPECTED VALUES FOR BOUNDED FORECASTING

In Section 2.2, we generalized Bernoulli's theorem from the case where Reality has only a binary choice on each round, as in Protocol 1.10 or 2.11, to Protocol 2.1, where she chooses outcomes y_n from the interval $[-1, 1]$. In what sense can we similarly generalize De Moivre's theorem?

As shown in [349], we can use parabolic potential theory to find asymptotic global upper expected values in Protocol 2.1. Given a function U that satisfies certain regularity conditions, there is a unique continuous function $U^*(s, D)$, $s \in \mathbb{R}$ and $D > 0$, that approaches $U(s)$ as $D \to 0$, satisfies $U^*(s, D) \geq U(s)$ for all s and D, and satisfies the heat equation (2.24) for all s and D such that $U^*(s, D) > U(s)$. In Protocol 2.1,

$$\lim_{N \to \infty} \overline{\mathbb{E}}(U(N^{1/2} \bar{y}_N)) = U^*(0, 1). \tag{2.31}$$

In [349, section 6.3] we define $U^*(s, D)$ (called the *least superparabolic majorant*), prove (2.31) under the assumption that U is continuous and constant outside a finite interval, and discuss extensions to more general U.

We can gain insight into the behavior of the function U^* from its interpretation in terms of heat propagation. In this interpretation, we start at time $D = 0$ with a distribution of heat over the real line described by the function $U(s)$, and at every point s, there is a thermostat that switches on whenever this is necessary to keep the temperature from falling below $U(s)$. Whenever the temperature at s rises above $U(s)$, this local thermostat switches off and the evolution of the temperature is then governed by the heat equation. The function $U^*(s, D)$ gives the resulting temperature at s at time D. Figure 2.2 depicts the temperature distribution $s \mapsto U^*(s, D)$ for several values of D, starting with the same initial temperature distribution as in Figure 2.1, which depicts heat propagation with no heat sources or sinks.

Table 2.1 compares some asymptotic global upper probabilities in Protocol 2.1, obtained numerically from (2.31), with the corresponding asymptotic global probabilities in Protocol 2.11, given by De Moivre's theorem in the form (2.30), which must also be obtained numerically but are widely available from written sources and standard statistical software.

As Table 2.1 suggests, $\overline{\mathbb{P}}(|\bar{y}_N| \leq c) = 1$ in Protocol 2.1 for all $c > 0$. Skeptic cannot make money in this protocol if Reality always sets y_n equal to zero, and so he cannot

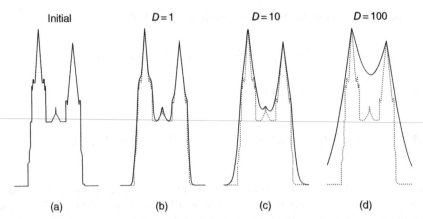

Figure 2.2 Heat propagation with thermostats set to keep the temperature at s from falling below $U(s)$. Part (a) shows the same initial temperature distribution U as in Figure 2.1. Parts (b), (c), and (d) show $\overline{U}(s, D)$ as a function of s at times $D = 1, 10, 100$, respectively.

Table 2.1 Some global upper probabilities for $|\bar{y}_N|$ in Protocols 2.1 and 2.11.

Event	Upper probability when $y_n \in [-1, 1]$	Probability when $y_n \in \{-1, 1\}$
$\|\bar{y}_N\| \leq N^{-1/2}$	1	0.683
$\|\bar{y}_N\| \leq 2N^{-1/2}$	1	0.955
$\|\bar{y}_N\| \leq 3N^{-1/2}$	1	0.997
$\|\bar{y}_N\| \geq N^{-1/2}$	0.629	0.317
$\|\bar{y}_N\| \geq 2N^{-1/2}$	0.091	0.045
$\|\bar{y}_N\| \geq 3N^{-1/2}$	0.005	0.003

force her to stay away from zero. He can only force her to stay close to zero. This picture, which is one-sided in comparison with De Moivre's theorem, arises as soon as Reality has more than a binary choice; we would obtain the same numbers in the middle column of Table 2.1 if Reality chose from the three-element set $\{-1, 0, 1\}$ instead of the interval $[-1, 1]$.

2.5 EXERCISES

Exercise 2.1 Verify the properties of the functional $\overline{\mathbb{E}}$ listed in Lemma 2.2, and list the corresponding properties of the functional $\underline{\mathbb{E}}$. ☐

Exercise 2.2 Show that for every bounded variable X in Protocol 2.1,

$$\overline{\mathbb{E}}(X) = \inf\{\alpha | \exists \mathcal{T} \in \mathbf{T}_0 : \mathcal{T}_N + \alpha \geq X\}$$

and

$$\underline{\mathbb{E}}(X) = \sup\{\alpha | \exists \mathcal{T} \in \mathbf{T}_0 : \mathcal{T}_N - \alpha \geq -X\},$$

where $\mathbf{T}_0$ is the set of supermartingales with initial value 0. □

Exercise 2.3 Consider the following protocol, which consists of a single round of Protocol 1.3.

> Skeptic announces $\mathcal{K}_0 \in \mathbb{R}$.
> Skeptic announces $M \in \mathbb{R}$.
> Reality announces $y \in [-1, 1]$.
> $\mathcal{K} := \mathcal{K}_0 + My$.

Here the sample space Ω is $[-1, 1]$.

1. Show that the variable Y given by $Y(y) := y$ satisfies $\overline{\mathbb{E}}(Y) = \underline{\mathbb{E}}(Y) = 0$.
2. Show that if $E = [1/2, 1]$, then $\overline{\mathbb{P}}(E) = 2/3$.
3. Find $\overline{\mathbb{P}}(E)$ for an arbitrary event E. □

Exercise 2.4 Prove the statement that follows (2.10): $\mathcal{T}_N \geq \mathbf{1}_E$ implies that $\mathcal{T}$ is nonnegative. □

Exercise 2.5 Given a situation s and a bounded below variable X in Protocol 2.1, set

$$\overline{\mathbb{E}}_s(X) := \inf\{\mathcal{T}(s) | \mathcal{T} \in \mathbf{T} \text{ and } \mathcal{T}_N \geq X\}.$$

Show that the process $s \mapsto \overline{\mathbb{E}}_s(X)$ is a supermartingale. □

Exercise 2.6 Suppose we stop play in Protocol 1.8 after N rounds. Then a strategy for Forecaster that takes account of Reality's previous moves while ignoring Skeptic's previous moves can be expressed as a function $\phi : \mathbb{S}_N \to [0, 1]$, where $\mathbb{S}_N$ is the set of all sequences of zeros and ones of length less than N. If we eliminate Forecaster by fixing such a strategy ϕ, we obtain the following protocol:

PARAMETERS: $N \in \mathbb{N}, \phi : \mathbb{S}_N \to [0, 1]$
> Skeptic announces $\mathcal{K}_0 \in \mathbb{R}$.
> FOR $n = 1, 2, \ldots$:
> > Skeptic announces $L_n \in \mathbb{R}$.
> > Reality announces $y_n \in \{0, 1\}$.
> > $\mathcal{K}_n := \mathcal{K}_{n-1} + L_n(y_n - \phi(y_1 \ldots y_{n-1}))$.

Generalize Proposition 2.5 by showing that every variable X in this protocol has a game-theoretic expected value $\mathbb{E}(X)$, and show that $\mathbb{E}(X)$ is also the classical expected value with respect to the probability distribution for $y_1 \ldots y_N$ defined by the strategy ϕ.   □

Exercise 2.7   Compare the two upper bounds on $\overline{\mathbb{P}}(|\bar{y}_N| \geq \varepsilon)$ discussed in Section 2.2, $1/\varepsilon^2 N$ and $2\exp(-\varepsilon^2 N/4)$.

1. How large must $\varepsilon^2 N$ be in order for each bound to be meaningful (i.e. < 1)?
2. What is the largest value of $\varepsilon^2 N$ for which $1/\varepsilon^2 N \leq 2\exp(-\varepsilon^2 N/4)$?
3. How do the two bounds compare when $\varepsilon^2 N = 20$?

We expect $\bar{y}_N$'s deviation from zero to have the order of magnitude $N^{-1/2}$ (see Section 2.3). If $\varepsilon \approx N^{-1/2}$, then $\varepsilon^2 N \approx 1$. So we are most interested in values of $\varepsilon^2 N$ within a few orders of magnitude of 1.   □

Exercise 2.8   From the inequalities (2.15) and (2.16) for Protocol 2.1, deduce corresponding inequalities for versions of Protocols 1.1, 1.8, and 1.10 in which play stops after a fixed number N of rounds. *Hint.* Reason as in the proof of Lemma 1.4, or else specialize Protocol 2.1 and then make a more direct translation. (In the case of Protocol 1.1, rescale Protocol 2.1 before specializing it; have Reality choose her moves from $[-2, 2]$ and modify the right-hand sides of (2.15) and (2.16) accordingly.)

1. Using (2.15), show that

$$\overline{\mathbb{P}}(|\bar{y}_N - \bar{m}_N| \geq \varepsilon) \leq \frac{4}{\varepsilon^2 N}$$

   in Protocol 1.1, and using (2.16), show that

$$\overline{\mathbb{P}}(|\bar{y}_N - \bar{m}_N| \geq \varepsilon) \leq 2\exp\left(-\frac{\varepsilon^2 N}{16}\right)$$

   in Protocol 1.1.
2. Deduce from the preceding statement that

$$\mathbb{P}(|\bar{y}_N - \bar{p}_N| \geq \varepsilon) \leq \frac{1}{\varepsilon^2 N}$$

   and

$$\mathbb{P}(|\bar{y}_N - \bar{p}_N| \geq \varepsilon) \leq 2\exp\left(-\frac{\varepsilon^2 N}{4}\right)$$

   in the finite-horizon version of Protocol 1.8.
3. Deduce that (2.19) and (2.20) hold in the finite-horizon version of Protocol 1.10.   □

Exercise 2.9 Show that, in the proof of Proposition 2.10, there is indeed no loss of generality in assuming that U is an infinitely differentiable function that is constant outside a finite interval. *Hint.* See, e.g. [5, Theorem 2.29]. □

Exercise 2.10

1. Deduce from (2.10) that if $\mathcal{T}$ is a nonnegative supermartingale, then, for $K > 0$,

$$\overline{\mathbb{P}}(\max_{1 \leq n \leq N} \mathcal{T}_n \geq K) \leq \frac{\mathcal{T}_0}{K}. \tag{2.32}$$

This is a version of *Ville's inequality* (see Exercise 8.8 for a more general version).

2. Deduce that, for $\varepsilon > 0$,

$$\overline{\mathbb{P}}(\max_{1 \leq n \leq N} |\bar{y}_n| \geq \varepsilon) \leq \frac{1}{\varepsilon^2 N}$$

in Protocol 2.1. □

Exercise 2.11

1. Show that

$$\lim_{N \to \infty} \overline{\mathbb{E}}(\bar{y}_N) = \lim_{N \to \infty} \underline{\mathbb{E}}(\bar{y}_N) = 0$$

holds in Protocol 2.1.

2. Extend this result by showing that

$$\lim_{N \to \infty} \overline{\mathbb{E}}(U(\bar{y}_N)) = \lim_{N \to \infty} \underline{\mathbb{E}}(U(\bar{y}_N)) = U(0)$$

holds for a wide class of functions U. □

Exercise 2.12 How large does N need to be in order for the upper and lower probabilities in Table 2.1 to be accurate? □

2.6 CONTEXT

Blaise Pascal (1623–1662), Pierre Fermat (1607–1665), and Christiaan Huygens (1629–1695)

The game-theoretic understanding of probability has been traced back to Pascal. In their 1654 correspondence, Pascal and Fermat used different methods to solve a problem concerning the division of stakes in a game of chance. Fermat used combinatorial principles to count the chances, whereas Pascal used backward recursion to find the value of a position in the game [123]. Both methods can be traced even further back, to authors writing centuries earlier [346]. Fermat's principles prefigured measure-theoretic probability. Pascal's principles, deepened by Huygens, prefigured game-theoretic probability. Our proof of Proposition 2.5 can be understood in terms of Pascal's and Huygens's principles [154, 331, 368].

See [130] for many additional references to the role of games of chance in the development of probability theory.

Expectation and Expected Value

The Latin version of Huygens's tract *De Ratiociniis in Ludo Aleae*, published in 1657, cast probability problems in terms of the value of a player's *expectatio* when he had various chances of winning various amounts [154]. From the time of De Moivre's *The Doctrine of Chances*, first published in 1718, and continuing into the twentieth century, authors writing about probability often asked about the *value* of a player's *expectation*. In German, beginning in the nineteenth century, *value of expectation* was sometimes *Erwartungswert*, and this was retranslated into English as *expected value*. These terms were originally used only for gambling problems, but by the middle of the twentieth century it became increasingly common for the mean value of any variable with a probability distribution to be called its *expected value* or *expectation*. See [345] for details.

In this book, the terms *expected value* and *expectation* come back to their game-theoretic roots. But Forecaster's betting offers often fall short of determining an expected value for a variable X; instead we obtain only an *upper expected value* $\overline{\mathbb{E}}(X)$ and a *lower expected value* $\underline{\mathbb{E}}(X)$. Moreover, the upper expectation $\overline{\mathbb{E}}$ is not determined by its upper probabilities for events (see Exercise 6.8). In the measure-theoretic framework, in contrast, the probabilities of events determine the expected values of random variables.

Jacob Bernoulli (1655–1705)

Bernoulli's *Ars Conjectandi* appeared, after his death, in 1713 [33]. Edith Sylla's translation of the book from Latin into English [34] includes a brief biography of Bernoulli.

Bernoulli stated and proved his theorem on the approximation of probability by frequency in Part IV of *Ars Conjectandi*. The theorem is usually stated by saying that if an event always happens with probability p, then the frequency with which it happens in a large number of independent trials is close to p with high probability. Bernoulli actually stated and proved a more concrete statement. He imagined drawing with replacement from an urn with a large but finite number of tokens, some white and some black, and he showed that if we know the total number of tokens and we want to estimate the number that are black and be exactly right with given odds, then we can find a number of draws that will allow us to accomplish this. Stephen Stigler translates Bernoulli's proof into modern notation [367, chapter 2].

Abraham De Moivre (1667–1754)

The oldest and simplest version of the central limit theorem, proven by De Moivre in 1733, is situated in the archetype of probability theory, the example of independent trials of an event that happens with probability p on each trial. Stigler gives a careful account of the differences between the way De Moivre thought about his approximations and the modern conception [367]. De Moivre's life and work have been studied in detail by David Bellhouse [26, 27].

We have described De Moivre's approximation (2.2) by saying that probability distribution of $\sum_{n=1}^{N} y_n$, which is discrete, can be approximated by the Gaussian distribution, which is continuous. De Moivre did not think in these terms; he did not yet have the concept of a continuous distribution of probability. Subsequent work by Laplace and Gauss led to the understanding that (2.2) is an approximation of a discrete distribution by a continuous one.

By the early twentieth century, there were many generalizations of De Moivre's theorem, all showing that the probability distribution for various averages can be approximated by the Gaussian distribution. In 1920 [312], Georg Pólya introduced the name *central limit theorem* for this family of theorems, on the grounds that they are central to probability theory. Since 1920, yet further central limit theorems have emerged. As Hans Fischer recently wrote in his comprehensive history of the central limit theorem [145, p. 1],

> These days the term "central limit theorem" is associated with a multitude of statements having to do with the convergence of probability distributions of functions of an increasing number of one- or multi-dimensional random variables or even more general random elements (with values in Banach spaces or more general spaces) to a normal distribution (or related distributions).

The result (2.31) stretches the notion of a central limit theorem even further, to the case where the objects are upper expectations rather than probability distributions.

Louis Bachelier (1870–1946)

Bachelier's doctoral thesis, *Théorie de Spéculation*, was the first mathematical analysis of Brownian motion, which he proposed as model for prices of financial securities, especially *rentes*, or perpetual annuities [190]. Bachelier called attention to connections with the heat equation, which intrigued his examiner Henri Poincaré. After noting that "the reasoning of Fourier is applicable almost without change to this problem, so different from that for which it was created," Poincaré expressed regret that Bachelier did not develop his idea further [20, chapter 2].

Our proof of Proposition 2.10 is modeled on the pioneering proof of the central limit theorem published by Lindeberg in 1920 [250] and 1922 [251]. Lindeberg's argument was taken up by Lévy, in his *Calcul des Probabilités* in 1925 [243, section 51], and it has been subsequently repeated in a number of textbooks, including those by Leo Breiman in 1968 [51, pp. 167–170] and Patrick Billingsley in 1968 [39, pp. 42–44]. Neither Lindeberg nor these subsequent authors made explicit reference to the heat equation, which tends to disappear from view when the error terms in the approximation (2.26) are treated rigorously. The underlying gambling ideas are also usually left implicit.

The Heat Equation

The heat equation was first successfully studied in the early nineteenth century by Joseph Fourier, who provided solutions in terms of trigonometric series for the case where initial conditions $U(s)$ are given on a finite interval in $[s_1, s_2]$, and boundary conditions $\overline{U}(s_1, D)$ and $\overline{U}(s_2, D)$ are also provided. Laplace, in 1809, was the first to provide the solution (2.23) for the case where $U(s)$ is given for all s and hence no boundary conditions are needed [171].

It is easy to verify informally that the function $\overline{U}(s, D)$ given by (2.23) satisfies the heat equation. Indeed, for a small positive constant δ,

$$\frac{\partial \overline{U}}{\partial D}(s, D)\delta \approx \overline{U}(s, D + \delta) - \overline{U}(s, D) = \int_{-\infty}^{\infty} U(s + z)\mathcal{N}_{0, D+\delta}(dz) - \overline{U}(s, D)$$

$$= \int_{-\infty}^{\infty} \overline{U}(s + z, D)\mathcal{N}_{0, \delta}(dz) - \overline{U}(s, D)$$

$$\approx \int_{-\infty}^{\infty} \left(\overline{U}(s, D) + \frac{\partial \overline{U}}{\partial s}(s, D)z + \frac{1}{2}\frac{\partial^2 \overline{U}}{\partial s^2}(s, D)z^2 \right) \mathcal{N}_{0, \delta}(dz) - \overline{U}(s, D)$$

$$= \frac{1}{2}\frac{\partial^2 \overline{U}}{\partial s^2}(s, D)\delta.$$

A rigorous proof requires some regularity conditions on U. For example, if U is a Borel function such that

$$\lim_{s \to \infty} |U(s)|e^{-\varepsilon s^2} = 0 \tag{2.33}$$

for any $\varepsilon > 0$ (this holds if $|U(s)|$ grows at most polynomially fast as $s \to \infty$), then a proof can be obtained by differentiating $\overline{U}(s, D)$, in the form

$$\frac{1}{\sqrt{2\pi D}} \int_{-\infty}^{\infty} U(z)e^{-(z-s)^2/2D} \, dz,$$

under the integral sign; this is authorized by Leibniz's differentiation rule for integrals [47, 61]. Other proofs are given in standard references on partial differential equations (e.g. [61]). It is easy to see that (2.23) converges to $U(s_0)$ as $D \to 0$ and $s \to s_0$ for any real number s_0, provided that U is continuous and (2.33) holds. This means that $\overline{U}$ solves the heat equation for the initial conditions given by U.

The heat equation is often interpreted in terms of Brownian motion. Under appropriate regularity conditions, the function $\overline{U}$ can be given by

$$\overline{U}(s, D) = \mathbb{E}(U(s + B_D)),$$

and the function U^* used in (2.31) can be given by

$$U^*(s, D) = \sup_\tau \mathbb{E}(U(s + B_\tau)), \tag{2.34}$$

where B is a standard Brownian motion, τ ranges over stopping times taking values in the interval $[0, D]$, and $\mathbb{E}$ is mathematical expectation. (See the section about terminology and notation at the end of the book for the standard continuous-time measure-theoretic definition of a stopping time.)

3

Some Basic Supermartingales

In Section 1.4, we discussed two simple supermartingales. One was quadratic in Reality's moves; the other was exponential. This chapter studies some additional quadratic and exponential supermartingales, which appear in various guises in various protocols. They can be used directly and in the construction of more complex supermartingales. We will see some of them when we study the law of the iterated logarithm (Section 5.4), defensive forecasting (Lemma 12.10), and continuous-time probability (in the construction of the Itô integral in Chapter 14 and the construction of portfolios in Chapter 17).

Quadratic supermartingales can grow moderately fast and are often used to show that moderately large deviations from expectations (i.e. from prices given by Forecaster) are unlikely. Exponential supermartingales can grow much faster. They are sometimes competitive with quadratic supermartingales for moderately large deviations, and they can be used to show that very large deviations are very unlikely.

In Section 3.1, we study a classic quadratic martingale, which we call Kolmogorov's martingale, after Kolmogorov's early work on the law of large numbers for dependent outcomes [220, 221]. Then we turn to some exponential supermartingales, to which we attach the names of Catherine Doléans (Section 3.2), Wassily Hoeffding (Section 3.3), and Sergei Bernstein (Section 3.4). Doléans's supermartingale was implicit in our proof of Borel's law of large numbers in Chapter 1. Hoeffding's and Bernstein's supermartingales lead to interesting large-deviation

Game-Theoretic Foundations for Probability and Finance, First Edition. Glenn Shafer and Vladimir Vovk.
© 2019 John Wiley & Sons, Inc. Published 2019 by John Wiley & Sons, Inc.

inequalities. These supermartingales appear in many protocols, a supermartingale or martingale in one protocol also being one in any specialization (see Section 1.2).

3.1 KOLMOGOROV'S MARTINGALE

The following protocol loosens Protocol 1.3 by removing the constraint $y_n \in [-1, 1]$.

Protocol 3.1
    Skeptic announces $\mathcal{K}_0 \in \mathbb{R}$.
    FOR $n = 1, 2, \ldots$:
        Skeptic announces $M_n \in \mathbb{R}$.
        Reality announces $y_n \in \mathbb{R}$.
        $\mathcal{K}_n := \mathcal{K}_{n-1} + M_n y_n$.

Proposition 3.2 *The process $\mathcal{T}$ given by*

$$\mathcal{T}_n := \left(\sum_{i=1}^{n} y_i \right)^2 - \sum_{i=1}^{n} y_i^2 \tag{3.1}$$

is a martingale in Protocol 3.1.

Proof The process $\mathcal{T}$'s gain on round n is

$$\mathcal{T}_n - \mathcal{T}_{n-1} = \left(\sum_{i=1}^{n} y_i \right)^2 - \left(\sum_{i=1}^{n-1} y_i \right)^2 - y_n^2 = 2 \left(\sum_{i=1}^{n-1} y_i \right) y_n. \tag{3.2}$$

Because this has the form $M_n y_n$, $\mathcal{T}$ is a capital process for a strategy for Skeptic that begins with $\mathcal{K}_0 := 0$. By symmetry, $-\mathcal{T}$ is as well.   □

We call (3.1) *Kolmogorov's martingale* (see Section 3.6). The quadratic supermartingale in Protocol 1.3 defined by (1.20) is obtained from it by Skeptic's giving up capital $1 - y_n^2$ on each round.

3.2 DOLÉANS'S SUPERMARTINGALE

The process

$$\exp \left(\kappa \sum_{i=1}^{n} y_i - \kappa^2 \sum_{i=1}^{n} y_i^2 \right) \tag{3.3}$$

played a central role in Chapter 1 and will reappear in various forms in later chapters. According to Exercise 1.10, it is a supermartingale in Protocol 1.3, where Reality is constrained to choose $y_n \in [-1, 1]$, provided that $\kappa \in (0, 1/2]$. This implies that the process $\mathcal{T}$ given by

$$\mathcal{T}_n := \exp\left(\sum_{i=1}^{n} y_i - \sum_{i=1}^{n} y_i^2\right) \tag{3.4}$$

is a supermartingale when Reality is constrained instead to choose $y_n \in [-1/2, 1/2]$. We call it *Doléans's supermartingale*. As we now show, it is still a supermartingale when the constraint $y_n \in [-1/2, 1/2]$ is replaced with $y_n \in [-1/2, \infty)$.

Protocol 3.3
    Skeptic announces $\mathcal{K}_0 \in \mathbb{R}$.
    FOR $n = 1, 2, \ldots$:
        Skeptic announces $M_n \in \mathbb{R}$.
        Reality announces $y_n \in [-0.5, \infty)$.
        $\mathcal{K}_n := \mathcal{K}_{n-1} + M_n y_n$.

Proposition 3.4 *The process $\mathcal{T}$ given by (3.4) is a supermartingale in Protocol 3.3.*

Proof It suffices to prove that Skeptic can multiply his capital by at least $\exp(y_n - y_n^2)$ on round n. This means finding $M \in \mathbb{R}$ such that

$$\mathcal{K}_{n-1} \exp(y - y^2) - \mathcal{K}_{n-1} \leq My$$

for all $y \in [-1/2, \infty)$. But this inequality holds when $M := \mathcal{K}_{n-1}$, because

$$\exp(y - y^2) - 1 \leq y$$

or

$$y - y^2 \leq \ln(1 + y) \tag{3.5}$$

for all $y \in [-1/2, \infty)$. To verify that (3.5) holds, notice that it holds with equality when $y = 0$, that the derivative of its left-hand side never exceeds the derivative of its right-hand side when $y > 0$, and that the opposite holds when $-1/2 \leq y < 0$. $\quad\square$

It can further be shown that (3.4) is still a supermartingale when the constraint $y_n \in [-1/2, \infty)$ in Protocol 3.3 is further relaxed to $y_n \in [-0.683, \infty)$ (see Exercise 3.2).

An interesting variation on Doléans's supermartingale has been considered by Shin-ichiro Takazawa [373], Theorem 3.1 (see Exercise 4.9).

Applying Proposition 3.4 to y_n and $-y_n$ and averaging the resulting supermartingales, we see that

$$\frac{1}{2}\left(\prod_{i=1}^{n} \exp(y_i - y_i^2) + \prod_{i=1}^{n} \exp(-y_i - y_i^2) \right) \tag{3.6}$$

is a supermartingale in the protocol obtained from Protocol 3.1 by adding the constraint $y_n \in [-1/2, 1/2]$.

3.3   HOEFFDING'S SUPERMARTINGALE

We now turn to an exponential supermartingale that implies a game-theoretic version of Hoeffding's inequality. We call it, naturally, *Hoeffding's supermartingale*.

Consider the following variation on Protocol 1.1, in which Forecaster specifies an interval from which Reality must choose her y_n.

Protocol 3.5
   Skeptic announces $\mathcal{K}_0 \in \mathbb{R}$.
   FOR $n = 1, 2, \ldots$:
       Forecaster announces $[a_n, b_n] \subseteq \mathbb{R}$ and $m_n \in [a_n, b_n]$.
       Skeptic announces $M_n \in \mathbb{R}$.
       Reality announces $y_n \in [a_n, b_n]$.
       $\mathcal{K}_n := \mathcal{K}_{n-1} + M_n(y_n - m_n)$.

Theorem 3.6   *For any $\kappa \in \mathbb{R}$, the process*

$$\prod_{i=1}^{n} \exp\left(\kappa(y_i - m_i) - \frac{\kappa^2}{8}(b_i - a_i)^2 \right) \tag{3.7}$$

is a supermartingale in Protocol 3.5.

Proof   Assume, without loss of generality, that Forecaster is constrained to choose each a_n and b_n so that $a_n < 0 < b_n$ and to choose $m_n := 0$ for all n. (Adding the same constant to a_n, b_n, and m_n will not change the gains $\mathcal{K}_n - \mathcal{K}_{n-1}$ available to Skeptic.)

It suffices to prove that Skeptic can multiply his capital by at least

$$\exp\left(\kappa y_n - \frac{\kappa^2}{8}(b_n - a_n)^2 \right)$$

on round n. This means finding $M \in \mathbb{R}$ such that

$$\mathcal{K}_{n-1} \exp\left(\kappa y - \frac{\kappa^2}{8}(b_n - a_n)^2 \right) - \mathcal{K}_{n-1} \leq My \tag{3.8}$$

for all $y \in [a_n, b_n]$. We will show that (3.8) holds when we set

$$M := \mathcal{K}_{n-1} \frac{e^{\kappa b_n} - e^{\kappa a_n}}{b_n - a_n} \exp\left(-\frac{\kappa^2}{8}(b_n - a_n)^2\right). \tag{3.9}$$

This means showing that

$$\exp\left(\kappa y - \frac{\kappa^2}{8}(b_n - a_n)^2\right) - 1 \le y \frac{e^{\kappa b_n} - e^{\kappa a_n}}{b_n - a_n} \exp\left(-\frac{\kappa^2}{8}(b_n - a_n)^2\right).$$

Simplifying by omitting the remaining subscripts, we see that we need to show that

$$\exp(\kappa y) \le \exp\left(\frac{\kappa^2}{8}(b - a)^2\right) + y \frac{e^{\kappa b} - e^{\kappa a}}{b - a},$$

when $a < 0 < b$ and $y \in [a, b]$. By the convexity of the function exp, it suffices to prove

$$\frac{y - a}{b - a} e^{\kappa b} + \frac{b - y}{b - a} e^{\kappa a} \le \exp\left(\frac{\kappa^2}{8}(b - a)^2\right) + y \frac{e^{\kappa b} - e^{\kappa a}}{b - a},$$

i.e.

$$\frac{b e^{\kappa a} - a e^{\kappa b}}{b - a} \le \exp\left(\frac{\kappa^2}{8}(b - a)^2\right),$$

i.e.

$$\ln(b e^{\kappa a} - a e^{\kappa b}) \le \frac{\kappa^2}{8}(b - a)^2 + \ln(b - a) \tag{3.10}$$

(the first ln is well defined since $b e^{\kappa a} > 0 > a e^{\kappa b}$). The derivative in κ of the left-hand side of (3.10) is

$$\frac{a b e^{\kappa a} - a b e^{\kappa b}}{b e^{\kappa a} - a e^{\kappa b}}$$

and the second derivative, after cancellations and regrouping, is

$$(b - a)^2 \frac{(b e^{\kappa a})(-a e^{\kappa b})}{(b e^{\kappa a} - a e^{\kappa b})^2}.$$

The last ratio is of the form $u(1 - u)$, where $0 < u < 1$. Hence, it does not exceed $1/4$, and the second derivative itself does not exceed $(b - a)^2/4$. Inequality (3.10) now follows from the second-order Taylor expansion of the left-hand side around $\kappa = 0$. $\qquad\square$

It is clear from the proof that we can generalize Theorem 3.6 by allowing the parameter κ in (3.7) to vary: given any sequence $\kappa_1, \kappa_2, \ldots$ in $\mathbb{R}$, the process

$$\prod_{i=1}^{n} \exp\left(\kappa_i(y_i - m_i) - \frac{\kappa_i^2}{8}(b_i - a_i)^2 \right) \tag{3.11}$$

is a supermartingale in Protocol 3.5.

Another variation on Theorem 3.6 arises if we constrain Skeptic by requiring all his moves to be nonnegative, thus obtaining the following protocol.

Protocol 3.7

    Skeptic announces $\mathcal{K}_0 \in \mathbb{R}$.
    FOR $n = 1, 2, \ldots$:
        Forecaster announces $[a_n, b_n] \subseteq \mathbb{R}$ and $m_n \in [a_n, b_n]$.
        Skeptic announces $M_n \geq 0$.
        Reality announces $y_n \in [a_n, b_n]$.
        $\mathcal{K}_n := \mathcal{K}_{n-1} + M_n(y_n - m_n)$.

The value of M given by (3.9) being nonnegative if κ is nonnegative, we see that (3.7) is a supermartingale in Protocol 3.7 if κ is nonnegative and that (3.11) is a supermartingale in Protocol 3.7 if all the κ_i are nonnegative.

If we replace the condition $M_n \geq 0$ in Protocol 3.7 with the condition $M_n \leq 0$, we similarly see that (3.7) is a supermartingale if κ is nonpositive and that (3.11) is a supermartingale if all the κ_i are nonpositive.

Hoeffding's Inequality

Suppose we further specialize Protocol 3.5 by requiring Forecaster to announce all the a_n and b_n at the outset. Then (3.7) is still a supermartingale, and we obtain the following game-theoretic version of an inequality proven by Hoeffding in measure-theoretic probability in 1963 [188].

Corollary 3.8 **(Hoeffding)** *Suppose $n \in \mathbb{N}$ and $\varepsilon > 0$, and suppose $a_1, b_1, \ldots, a_n, b_n$ are known to Skeptic at the outset of play. Then*

$$\overline{\mathbb{P}}\left(\frac{1}{n} \sum_{i=1}^{n} (y_i - m_i) \geq \varepsilon \right) \leq \exp\left(-\frac{2n^2\varepsilon^2}{\sum_{i=1}^{n}(b_i - a_i)^2} \right). \tag{3.12}$$

This is true both in Protocol 3.5 and in Protocol 3.7.

Proof Set

$$C := \sum_{i=1}^{n} (b_i - a_i)^2 \tag{3.13}$$

and notice that the supermartingale (3.7) begins at 1 and achieves

$$\prod_{i=1}^{n} \exp\left(\kappa(y_i - m_i) - \frac{\kappa^2}{8}(b_i - a_i)^2 \right) \geq \exp\left(\kappa n \varepsilon - \frac{\kappa^2}{8} C \right) \tag{3.14}$$

on the event

$$\frac{1}{n} \sum_{i=1}^{n} (y_i - m_i) \geq \varepsilon,$$

the inequality in (3.14) being satisfied for κ positive. The right-hand side of (3.14) attains its maximum at $\kappa := 4n\varepsilon/C$, which is positive, and the maximum is the inverse to (3.12). □

It is clear from the proof that it is sufficient for Skeptic to know C (defined by (3.13)) at the outset; he need not know the individual a_n and b_n until the nth round.

Suppose we further strengthen Skeptic in Protocol 3.5 by setting $m_n = 0$, $a_n = -1$, and $b_n = 1$ for all n, thus eliminating Forecaster and reducing the protocol to Protocol 1.3. Then (3.12) reduces to

$$\overline{\mathbb{P}}(\bar{y}_n \geq \varepsilon) \leq \exp\left(-\frac{\varepsilon^2 n}{2} \right). \tag{3.15}$$

By symmetry, we then obtain

$$\overline{\mathbb{P}}(|\bar{y}_n| \geq \varepsilon) \leq 2\exp\left(-\frac{\varepsilon^2 n}{2} \right). \tag{3.16}$$

This is an improvement on the upper bound $2\exp(-\varepsilon^2 n/4)$ that we derived in Lemma 2.8 from the exponential supermartingale (1.16). When $\varepsilon^2 n$ is greater than about 4.3, it is also an improvement on the upper bound $1/\varepsilon^2 n$, which we derived in Lemma 2.7 from the quadratic martingale (1.20) (see Figure 3.1).

It is also instructive to compare Hoeffding's bound on the upper probability of $\bar{y}_n \geq \varepsilon$ in Protocol 1.3 with the estimate given by the central limit theorem. Hoeffding's inequality for this event, (3.15), can be written

$$\overline{\mathbb{P}}\left(\bar{y}_n \geq \frac{c}{\sqrt{n}} \right) \leq \exp\left(-\frac{c^2}{2} \right), \tag{3.17}$$

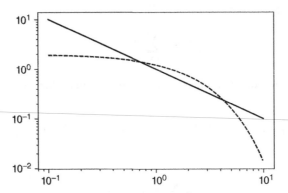

Figure 3.1   Log–log plot showing two upper bounds on $\overline{\mathbb{P}}(|\bar{y}_n| \geq \varepsilon)$ as functions of $x := \varepsilon^2 n$. The dashed upper bound, $2e^{-x/2}$, is given by Hoeffding's inequality (see Eq. (3.16)). The solid upper bound, $1/x$ (flat on the log scale), was derived from Kolmogorov's martingale (see (2.15)). When the two functions cross for a second time, approximately at $x = 4.3$, their value is only about 0.23. When $x = 20$, the upper bound $1/x$ has the more interesting value of 0.05, but the upper bound $2e^{-x/2}$ is vastly smaller: about 0.000 09.

where $c > 0$. Let us further restrict Reality by requiring that $y_n \in \{-1, 1\}$; this puts us in the framework of Protocol 2.11. By the central limit theorem, the asymptotic probability δ of the event in the parentheses on the left-hand side is determined by the equality $z_\delta = c$, where z_δ is the upper δ-quantile of the standard Gaussian distribution. It is easy to check (Exercise 3.5) that

$$\delta \leq e^{-c^2/2}, \tag{3.18}$$

when $c = z_\delta$. We can see that Hoeffding's inequality (3.17) is the combination of the asymptotic approximation given by the central limit theorem and the inequality (3.18). Asymptotically, Hoeffding's inequality is cruder, but it also works for a finite horizon.

The inequality (3.18) for $c = z_\delta$ can be rewritten as

$$z_\delta \leq \sqrt{2 \ln \frac{1}{\delta}}.$$

In this form, it is closely related to well-known approximations to z_δ in terms of $\sqrt{2 \ln(1/\delta)}$ due to Hastings [182] (and reproduced in [3], sections 26.2.22 and 26.2.23). It is easy to check that

$$z_\delta \sim \sqrt{2 \ln \frac{1}{\delta}} \tag{3.19}$$

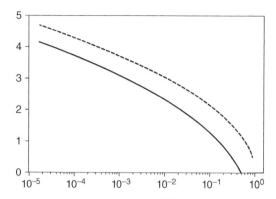

Figure 3.2 The functions z_δ (solid) and $\sqrt{2\ln(1/\delta)}$ (dashed) over a range of δ.

as $\delta \to 0$. Figure 3.2 compares the two sides of (3.19) over a wide range of δ. We can see that the result obtained by Hoeffding's inequality is competitive with the result obtained by the central limit theorem, even for moderate deviations.

3.4 BERNSTEIN'S SUPERMARTINGALE

We now turn to a protocol in which Skeptic can hedge against extreme negative values of y_n; our study of such protocols will be continued in Chapter 4.

Protocol 3.9
  Skeptic announces $\mathcal{K}_0 \in \mathbb{R}$.
  FOR $n = 1, 2, \ldots$:
    Forecaster announces numbers $c_n \geq 0$ and $v_n \geq 0$.
    Skeptic announces $M_n \geq 0$ and $V_n \geq 0$.
    Reality announces $y_n \leq c_n$.
    $\mathcal{K}_n := \mathcal{K}_{n-1} + M_n y_n + V_n(y_n^2 - v_n)$.

Theorem 3.10 *For any $\kappa \geq 0$, the process*

$$\prod_{i=1}^{n} \exp\left(\kappa y_i - \frac{e^{c_i \kappa} - 1 - c_i \kappa}{c_i^2} v_i \right) \tag{3.20}$$

is a supermartingale in Protocol 3.9.

We call (3.20) *Bernstein's supermartingale*.

Proof of Theorem 3.10 It suffices to prove that Skeptic can multiply his capital by at least

$$\exp\left(\kappa y_n - \frac{e^{c_n \kappa} - 1 - c_n \kappa}{c_n^2} v_n \right)$$

on round n. He can do so if there exist $M \geq 0$ and $V \geq 0$ such that

$$\mathcal{K}_{n-1} \exp\left(\kappa y - \frac{e^{c_n \kappa} - 1 - c_n \kappa}{c_n^2} v_n\right) - \mathcal{K}_{n-1} \leq My + V(y^2 - v_n)$$

for all $y \leq c_n$, and this is equivalent to the existence of $M \geq 0$ and $V \geq 0$ such that

$$\exp(\kappa y) - \exp\left(\frac{e^{c_n \kappa} - 1 - c_n \kappa}{c_n^2} v_n\right) \leq My + V(y^2 - v_n) \qquad (3.21)$$

for all $y \leq c_n$. To find such M and V, note first that the function $(e^u - 1 - u)/u^2$, defined by continuity as $1/2$ at $u = 0$, is increasing in u; see [152, Lemma 3.1] or [46, p. 35]. Thus,

$$\frac{e^{\kappa y} - 1 - \kappa y}{\kappa^2 y^2} \leq \frac{e^{\kappa c_n} - 1 - \kappa c_n}{\kappa^2 c_n^2}$$

or

$$\exp(\kappa y) \leq 1 + \kappa y + \frac{e^{c_n \kappa} - 1 - c_n \kappa}{c_n^2} y^2 \qquad (3.22)$$

when $y \leq c_n$. Then note that

$$\exp\left(\frac{e^{c_n \kappa} - 1 - c_n \kappa}{c_n^2} v_n\right) \geq 1 + \frac{e^{c_n \kappa} - 1 - c_n \kappa}{c_n^2} v_n. \qquad (3.23)$$

Subtracting (3.23) from (3.22), we obtain an inequality of the form (3.21).   □

Under reasonable assumptions, Bernstein's supermartingale tends to be more powerful (i.e. larger) than Hoeffding's supermartingale. To see this, let us strengthen Skeptic in the two protocols in ways that make them more similar:

- In Hoeffding's protocol (Protocol 3.5), require Forecaster to set $m_n := 0$ (with no loss of generality) and $[a_n, b_n] := [-c_n, c_n]$.
- In Bernstein's protocol (Protocol 3.9), require Forecaster to choose $y_n \in [-c_n, c_n]$ and $v_n \leq c_n^2$. (The second requirement is implicit in the first: even when $v_n = c_n^2$, Skeptic is not really being offered a hedge; he can only lose money by choosing V_n nonzero.)

Suppose further that the $c_n \kappa$ are small and that κ is positive. Then

$$\kappa y_n - \frac{e^{c_n \kappa} - 1 - c_n \kappa}{c_n^2} v_n \approx \kappa y_n - \frac{1}{2}\kappa^2 v_n^2 \geq \kappa y_n - \frac{1}{2}\kappa^2 c_n^2$$

$$= \kappa y_n - \frac{1}{8}\kappa^2 (b_n - a_n)^2,$$

and so (3.20) is greater than or equal to (3.7).

Bennett's and Bernstein's Inequalities

For some applications, it is convenient to specialize Protocol 3.9 by announcing all the v_n in advance and setting $c_n := 1$ for all n. This gives the following protocol.

Protocol 3.11
PARAMETERS: $v_1, v_2, \ldots \geq 0$.
   Skeptic announces $\mathcal{K}_0 \in \mathbb{R}$.
   FOR $n = 1, 2, \ldots$:
      Skeptic announces $M_n \geq 0$ and $V_n \geq 0$.
      Reality announces $y_n \leq 1$.
      $\mathcal{K}_n := \mathcal{K}_{n-1} + M_n y_n + V_n(y_n^2 - v_n)$.

In this specialization, Bernstein's supermartingale reduces to

$$\prod_{i=1}^{n} \exp(\kappa y_i - (e^\kappa - 1 - \kappa)v_i),$$

and we obtain the following game-theoretic analogs of inequalities due to George Bennett [28] and Sergei Bernstein [35, 36] as a corollary to Theorem 3.10.

Corollary 3.12 *Suppose $n \in \mathbb{N}$ and $\varepsilon > 0$. Set*

$$\sigma^2 := \frac{1}{n} \sum_{i=1}^{n} v_i;$$

we assume $\sigma^2 > 0$. Then the following inequalities hold in Protocol 3.11.

  1. Bennett's inequality:

$$\overline{\mathbb{P}}\left(\frac{1}{n}\sum_{i=1}^{n} y_i \geq \varepsilon\right) \leq \exp\left(-n\sigma^2 h\left(\frac{\varepsilon}{\sigma^2}\right)\right), \tag{3.24}$$

    where $h : (0, \infty) \to (0, \infty)$ is the function $h(u) := (1 + u)\ln(1 + u) - u$.
  2. Bernstein's inequality:

$$\overline{\mathbb{P}}\left(\frac{1}{n}\sum_{i=1}^{n} y_i \geq \varepsilon\right) \leq \exp\left(-\frac{n\varepsilon^2}{2\sigma^2 + \frac{2}{3}\varepsilon}\right). \tag{3.25}$$

Proof
  1. *Bennett's inequality.* The supermartingale of Theorem 3.10 starts from 1 and achieves

$$\prod_{i=1}^{n} \exp(\kappa y_i - (e^\kappa - 1 - \kappa)v_i) \geq \exp(\kappa n\varepsilon - (e^\kappa - 1 - \kappa)n\sigma^2) \tag{3.26}$$

on the event (3.24); remember that $\kappa \geq 0$. The right-hand side of (3.26) attains its maximum at $\kappa := \ln(1 + \varepsilon/\sigma^2) > 0$. The maximum is

$$\exp\left(n\sigma^2 h\left(\frac{\varepsilon}{\sigma^2}\right)\right).$$

2. *Bernstein's inequality*. It suffices to use the inequality

$$h(u) \geq \frac{u^2}{2 + \frac{2}{3}u} \tag{3.27}$$

(see Exercise 3.6). This inequality shows, in particular, that $h(u) > 0$ for any $u > 0$. □

As the proof of Corollary 3.12 makes clear, the inequalities (3.24) and (3.25) will still hold if we relax the assumption that Skeptic knows $v_1, v_2, \ldots$ in advance (i.e. they are parameters in the protocol) to the assumption that he knows σ^2 in advance. This is all he needs in order to choose the optimal value of κ.

Bennett's and Bernstein's inequalities also generalize immediately to the case where the bound $y_n \leq 1$ in the protocol is replaced with $y_n \leq c$, where $c > 0$. This is a mere rescaling, and the two inequalities become

$$\overline{\mathbb{P}}\left(\frac{1}{n}\sum_{i=1}^{n} y_i \geq \varepsilon\right) \leq \exp\left(-\frac{n\sigma^2}{c^2}h\left(\frac{c\varepsilon}{\sigma^2}\right)\right)$$

and

$$\overline{\mathbb{P}}\left(\frac{1}{n}\sum_{i=1}^{n} y_i \geq \varepsilon\right) \leq \exp\left(-\frac{n\varepsilon^2}{2\sigma^2 + \frac{2}{3}c\varepsilon}\right),$$

respectively.

3.5 EXERCISES

Exercise 3.1 Is

$$\prod_{i=1}^{n} \exp(|y_i| - y_i^2)$$

a supermartingale in the protocol obtained from Protocol 3.1 by adding the constraint $y_n \in [-1/2, 1/2]$? (This would simplify the supermartingale (3.6).) □

Exercise 3.2 Show that Proposition 3.4 will still be true if the constraint $y_n \in [-0.5, \infty)$ in Protocol 3.3 is replaced by $y_n \in [-0.683, \infty)$ but not if it is replaced by $[-0.684, \infty)$. □

Exercise 3.3 Show that

$$\prod_{i=1}^{n} \exp\left(y_i - \frac{y_i^2}{2} - |y_i|^3 \right)$$

is a supermartingale in the protocol obtained from Protocol 3.1 by adding the constraint $y_n \in [-0.8, 0.8]$. *Hint.* See [327, section 2]. □

Exercise 3.4 Verify formally the claim in the proof of Theorem 3.6 that no generality is lost by adding the constraint $m_n := 0$ for all n. In other words, show how a strategy for Skeptic that proves the theorem under the constraint can be adapted to prove the theorem without the constraint. □

Exercise 3.5 Check that the inequality (3.18) holds for $c := z_\delta$ and $\delta \in (0, 1/2]$. *Hint.* Use, e.g. [135, Lemma VII.1.2]. □

Exercise 3.6 Show that the function h given by $h(u) := (1 + u) \ln(1 + u) - u$ satisfies the inequality (3.27) (used in the proof of Bernstein's inequality in Corollary 3.12). □

Exercise 3.7 (research project). State and prove a game-theoretic version of McDiarmid's inequality (see [273]). □

Exercise 3.8 (research project). State and prove a game-theoretic version of Talagrand's inequality (see [378]). □

3.6 CONTEXT

Kolmogorov's Martingale

This martingale, (3.1), appeared in our proof of Lemma 2.7. This elementary proof echoes reasoning used by Pafnuty Chebyshev to prove classical versions of Bernoulli's theorem for independent random variables [67, 68]. In some of his earliest work in probability theory, Kolmogorov explored extending Chebyshev's reasoning to what we now call the martingale case, where the variables are dependent (see [220, formula (4)] and [221, formula (6)]).

The continuous-time counterpart of Kolmogorov's martingale is $X^2 - [X]$, where X is a martingale and $[X]$ is its quadratic variation, to be defined for continuous martingales X in a game-theoretic setting in Chapter 14. More generally, $XY - [X, Y]$ is a continuous martingale if X and Y are continuous martingales (cf. Lemma 14.14).

Hoeffding's Supermartingale

Theorem 3.6, concerning Hoeffding's supermartingale, was first stated and proven for the one-sided Protocol 3.7 by Gert de Cooman [89]. The two-sided version (Theorem 3.6 itself) had been proven earlier [400]; it is essentially the same as the inequality (4.16) in [188]. The measure-theoretic counterpart of Corollary 3.8 is sometimes called the Hoeffding–Azuma inequality, in honor of Kazuoki Azuma [17], but the martingale version was already stated by Hoeffding [188, end of section 2].

Bernstein's Supermartingale

The measure-theoretic version of Theorem 3.10 was first proven by David A. Freedman in 1975 [152, Corollary 1.4]. A sketch of the proof is given in [111, p. 130] and a complete proof is given in [46].

4

Kolmogorov's Law of Large Numbers

Borel's law of large numbers no longer holds if we modify Protocol 1.1 by allowing Reality to choose her moves from the whole range of real numbers. In this case Reality can turn any bet by Skeptic, no matter how small, into an arbitrarily large loss for Skeptic: as soon as Skeptic makes his first nonzero move, say M_n, Reality can make $y_n - m_n$ so far from zero in the direction opposite the sign of M_n that the negative change in Skeptic's capital, $M_n(y_n - m_n)$, will bankrupt him.

To put Skeptic back into the game, we must offer him additional bets that hedge against such extreme moves by Reality. The most natural hedge is one that sets a price v_n for the squared difference between m_n and y_n. When this hedge is available, Skeptic can choose as his gain function the sum of a multiple of $y_n - m_n$ and a multiple of $(y_n - m_n)^2 - v_n$. By choosing a positive multiple of $(y_n - m_n)^2 - v_n$, he can turn potential losses from extreme values of $y_n - m_n$ into gains.

In this chapter, we study a protocol that includes this hedge, and we construct a strategy for Skeptic that forces

$$\lim_{n\to\infty} \frac{1}{n} \sum_{i=1}^{n} (y_i - m_i) = 0 \tag{4.1}$$

if

$$\sum_{n=1}^{\infty} \frac{v_n}{n^2} < \infty. \tag{4.2}$$

Game-Theoretic Foundations for Probability and Finance, First Edition. Glenn Shafer and Vladimir Vovk.
© 2019 John Wiley & Sons, Inc. Published 2019 by John Wiley & Sons, Inc.

We also show that Reality can make (4.1) fail if the v_n do not satisfy (4.2). These results constitute a game-theoretic counterpart of the classical law of large numbers proven by Kolmogorov in 1930 [223].

We state the game-theoretic version of Kolmogorov's law more precisely in Section 4.1. The construction of Skeptic's strategy requires a game-theoretic version of Doob's convergence theorem, which we provide in Section 4.2. We then complete the proof of the game-theoretic version of Kolmogorov's law by constructing a strategy for Skeptic in Section 4.3 and a strategy for Reality in Section 4.4. In Section 4.5, we consider different ways of describing the game between Skeptic and Reality when one tries to make an event happen and the other tries to make it fail.

We will return to Skeptic's strategy in Section 9.3, where we show how the game-theoretic conclusion of Section 4.3 can be used to derive its measure-theoretic counterpart.

4.1   STATING KOLMOGOROV'S LAW

Here is a testing protocol in which Forecaster gives both a price m for Reality's move y and a price v for the hedge $(y - m)^2$.

Protocol 4.1
>   Skeptic announces $\mathcal{K}_0 \in \mathbb{R}$.
>   FOR $n = 1, 2, \ldots$:
>>   Forecaster announces $m_n \in \mathbb{R}$ and $v_n \geq 0$.
>>   Skeptic announces $M_n \in \mathbb{R}$ and $V_n \geq 0$.
>>   Reality announces $y_n \in \mathbb{R}$.
>> $\mathcal{K}_n := \mathcal{K}_{n-1} + M_n(y_n - m_n) + V_n((y_n - m_n)^2 - v_n).$ $\qquad(4.3)$

Much of the terminology we introduced in Chapter 1 carries over to this protocol:

- A condition on the moves by Skeptic's opponents is an *event*.
- A strategy for Skeptic *forces* an event E if it guarantees both (i) $\mathcal{K}_n \geq 0$ for all n and (ii) either $\mathcal{K}_n$ tends to infinity or E happens.
- A strategy for Skeptic *weakly forces* an event E if it guarantees both (i) $\mathcal{K}_n \geq 0$ for all n and (ii) either $\mathcal{K}_n$ is unbounded or E happens.

By the argument we made in Chapter 1 (Lemma 1.5), Skeptic has a strategy that forces E if and only if he has a strategy that weakly forces E. If he does have such strategies, we say that he *can force* E. We also say that E is *almost sure* or that it happens *almost surely*.

We also define a concept of forcing for Reality:

- A strategy for Reality *forces* an event E if it guarantees that either (i) $\mathcal{K}_n < 0$ for some n or (ii) $\mathcal{K}_n$ is bounded and E happens.

If Reality has a strategy that forces E, we say that she *can force E*. Notice that if either player can force E, then the other cannot force E^c. The following proposition is due to Kenshi Miyabe and Akimichi Takemura [285].

Proposition 4.2 *If Skeptic can force an event E in Protocol 4.1, then Reality can force E in Protocol 4.1.*

Proof Suppose Skeptic can force E, and choose a strategy ψ for Skeptic in Protocol 4.1 that guarantees both (i) $\mathcal{K}_n \geq 0$ for all n and (ii) either $\mathcal{K}_n \to \infty$ or E happens. Imagine an alternative Skeptic – call him Alt – who always plays the average of Skeptic's actual move and the move for Skeptic specified by ψ. Alt's capital will always be the average of Skeptic's and ψ's capital, both of which are nonnegative. Let Reality move on each round so that Alt has zero gain (see Exercise 4.7). When Reality does this, Alt's capital remains constant, equal to his initial capital, and thus both Skeptic's and ψ's capital stay bounded. This implies that $\mathcal{K}_n$ stays bounded and E happens. Therefore, this strategy for Reality forces E. □

Here is our game-theoretic counterpart of Kolmogorov's law:

Theorem 4.3 *In Protocol 4.1:*

1. *Skeptic can force*

$$\sum_{n=1}^{\infty} \frac{v_n}{n^2} < \infty \implies \frac{1}{n} \sum_{i=1}^{n} (y_i - m_i) \to 0. \tag{4.4}$$

2. *Reality can force*

$$\sum_{n=1}^{\infty} \frac{v_n}{n^2} = \infty \implies \frac{1}{n} \sum_{i=1}^{n} (y_i - m_i) \nrightarrow 0.$$

Proof This theorem is proven in Sections 4.1–4.4. The strategy for Skeptic that establishes Statement 1 is given in Section 4.3. The strategy for Reality that establishes Statement 2 is given in Section 4.4. □

To appreciate fully the content of Theorem 4.3, we must remember that a valid statement about what one player can achieve remains valid when we put additional constraints on the player's opponents. Three instances of this are salient for this theorem:

- Statements 1 and 2 both remain valid if we constrain Forecaster by fixing all his moves m_n and v_n at the outset. This brings the theorem closer to Kolmogorov's classical result.

- Statement 1 remains valid if we require $y_n, m_n \in [-1, 1]$ and $v_n = 4$ for all n. This effectively reduces the statement to Proposition 1.2. (With these constraints, Skeptic can only lose money by making V_n nonzero. So his ability to force an event is unaffected if we drop the term involving V_n from (4.3), thus reducing Protocol 4.1 to Protocol 1.1. Since (4.2) is satisfied when the v_n are all equal to 4, we can drop it as a condition in (4.4), making Statement 1 identical to Proposition 1.2.)

- Both statements remain valid if we require $v_n > 0$ for all n. Exercise 4.2 explores some consequences of this constraint.

Like our proof of the law of large numbers for bounded outcomes, our proof of Kolmogorov's law for unbounded outcomes will use a simplified protocol, in this case Protocol 4.4, in which Forecaster's m_n are all set to zero.

Protocol 4.4

    Skeptic announces $\mathcal{K}_0 \in \mathbb{R}$.

    FOR $n = 1, 2, \ldots$:

        Forecaster announces $v_n \geq 0$.

        Skeptic announces $M_n \in \mathbb{R}$ and $V_n \geq 0$.

        Reality announces $y_n \in \mathbb{R}$.

        $\mathcal{K}_n := \mathcal{K}_{n-1} + M_n y_n + V_n(y_n^2 - v_n)$.

In Section 4.3, we construct a strategy for Skeptic that forces

$$\sum_{n=1}^{\infty} \frac{v_n}{n^2} < \infty \implies \frac{1}{n} \sum_{i=1}^{n} y_i \to 0 \qquad (4.5)$$

in Protocol 4.4. In Section 4.4, we exhibit a strategy for Reality that forces

$$\sum_{n=1}^{\infty} \frac{v_n}{n^2} = \infty \implies \frac{1}{n} \sum_{i=1}^{n} y_i \nrightarrow 0 \qquad (4.6)$$

in Protocol 4.4.

    The simplification from Protocol 4.1 to Protocol 4.4 loses no generality. Skeptic can convert a strategy that forces his goal in Protocol 4.4 to one that forces his goal in Protocol 4.1 by responding to $y_1 m_1 v_1 \ldots y_n m_n v_n$ as he would to $(y_1 - m_1)v_1 \ldots (y_n - m_n)v_n$ in Protocol 4.4. Reality can convert a strategy that forces her goal in Protocol 4.4 to one that forces her goal in Protocol 4.1 by playing $y_n + m_n$ when her opponents' moves would have led her to play y_n in Protocol 4.4.

4.2 SUPERMARTINGALE CONVERGENCE THEOREM

To construct a strategy for Skeptic that forces (4.5) in Protocol 4.4, we need a game-theoretic version of Doob's convergence theorem. In order to state and prove it, we adapt and extend the terminology and notation developed for Protocol 1.3 in Section 1.2.

Situations and Paths

Situations and paths in Protocol 4.4 are sequences of elements of $[0, \infty) \times \mathbb{R}$. A *situation* is a finite sequence $(v_1, y_1) \ldots (v_n, y_n)$, representing initial moves by Forecaster and Reality. A *path* is an infinite sequence $(v_1, y_1)(v_2, y_2) \ldots$, representing a complete sequence of moves by Forecaster and Reality. As always, $\mathbb{S}$ is the set of all situations (the *situation space*), Ω is the set of all paths (the *sample space*), and $\square$ is the empty sequence, representing the initial situation. An *event* is a subset of Ω, and a *variable* is a real-valued function on Ω.

We define the notion of process for Protocol 4.4 just as we did for Protocol 1.3 in Section 1.2. A *process* is a real-valued function S on $\mathbb{S}$, sometimes represented as the sequence of variables $S_0, S_1, \ldots$, where $S_n : \omega \in \Omega \mapsto S(\omega^n) \in \mathbb{R}$. Our notion of a predictable process, however, looks slightly more complicated than that in Section 1.2. A *predictable process* $\mathcal{A}$ in Protocol 4.4 is a real-valued function on $\mathbb{S} \setminus \{\square\}$ such that when $s = (v_1, y_1) \ldots (v_n, y_n)$, the value $\mathcal{A}(s)$ depends only on $(v_1, y_1) \ldots v_n$ and not on y_n. As in Section 1.2, a predictable process $\mathcal{A}$ can be represented as the sequence of variables $\mathcal{A}_1, \mathcal{A}_2, \ldots$, where $\mathcal{A}_n : \omega \in \Omega \mapsto \mathcal{A}(\omega^n) \in \mathbb{R}$.

Supermartingales

A strategy ψ for Skeptic in Protocol 4.4 is a triplet $(\psi^{\text{stake}}, \psi^{\text{M}}, \psi^{\text{V}})$, where $\psi^{\text{stake}} \in \mathbb{R}$ and ψ^{M} and ψ^{V} are predictable processes. Here ψ^{stake} is the value ψ specifies for the initial capital $\mathcal{K}_0$, and $\psi^{\text{M}}(\omega^n)$ and $\psi^{\text{V}}(\omega^n)$ are the moves M_n and V_n that it specifies in the situation ω^n.

The set of strategies for Skeptic is a convex cone: (i) if $(\psi^{\text{stake}}, \psi^{\text{M}}, \psi^{\text{V}})$ is a strategy for Skeptic and $\beta \geq 0$, then $(\beta\psi^{\text{stake}}, \beta\psi^{\text{M}}, \beta\psi^{\text{V}})$ is a strategy for Skeptic and (ii) if $(\psi^{1,\text{stake}}, \psi^{1,\text{M}}, \psi^{1,\text{V}})$ and $(\psi^{2,\text{stake}}, \psi^{2,\text{M}}, \psi^{2,\text{V}})$ are strategies for Skeptic, then $(\psi^{1,\text{stake}} + \psi^{2,\text{stake}}, \psi^{1,\text{M}} + \psi^{2,\text{M}}, \psi^{1,\text{V}} + \psi^{2,\text{V}})$ is as well.

When Skeptic follows the strategy ψ, his capital process $\mathcal{K}^\psi$ is given by

$$\mathcal{K}_0^\psi := \psi^{\text{stake}}, \tag{4.7}$$

$$\mathcal{K}_n^\psi := \mathcal{K}_{n-1}^\psi + \psi_n^{\text{M}} y_n + \psi_n^{\text{V}}(y_n^2 - v_n). \tag{4.8}$$

Skeptic can force an event E if and only if Skeptic has a strategy ψ whose nonnegative capital process $\mathcal{K}^\psi$ tends to infinity on every path in E^c.

Because Skeptic must make his V_n nonnegative, the capital processes do not form a vector space: if $\mathcal{T}$ results from a strategy that does not always set $V_n := 0$, then $-\mathcal{T}$ is not a capital process. But it follows from (4.7) and (4.8) that the set of capital processes is a convex cone.

As in Section 1.4, we call a capital process in Protocol 4.4 or its slackening a *supermartingale*. The general understanding of supermartingales that we developed in Section 1.4 carries over to Protocol 4.4. Skeptic can force an event if and only if there is a nonnegative supermartingale that tends to infinity on paths where the event fails.

We call a process $\mathcal{U}$ a *semimartingale* if $\mathcal{U} = \mathcal{T} + \mathcal{A}$, where $\mathcal{T}$ is a supermartingale and $\mathcal{A} = \mathcal{A}_0, \mathcal{A}_1, \ldots$ is an increasing process such that $\mathcal{A}_0 = 0$ and $\mathcal{A}_1, \mathcal{A}_2, \ldots$ is a predictable process. We call the process $\mathcal{A}$ a *compensator* for $\mathcal{U}$. A semimartingale $\mathcal{U}$ is another kind of capital process for Skeptic, one that arises when $\mathcal{A}_n$ is a nonnegative side payment to Skeptic on round n that does not depend on Reality's move on round n.

Transforms

Given a predictable process $\mathcal{D}$ and a process $\mathcal{S}$, we write $\mathcal{D} \cdot \mathcal{S}$ for the process

$$(\mathcal{D} \cdot \mathcal{S})_n := \sum_{i=1}^{n} \mathcal{D}_i(\mathcal{S}_i - \mathcal{S}_{i-1}). \tag{4.9}$$

We call $\mathcal{D} \cdot \mathcal{S}$ the *transform of $\mathcal{S}$ by $\mathcal{D}$*. It is the discrete-time version of the Itô integral, defined in Section 14.3. We sometimes refer to $\Delta X_n := X_n - X_{n-1}$ as the *gain* of the process X on round $n \geq 1$; using this notation, (4.9) can be rewritten as $\Delta(\mathcal{D} \cdot \mathcal{S})_n = \mathcal{D}_n \Delta \mathcal{S}_n$.

When a predictable process $\mathcal{D}$ takes only the values 0 and 1, we call it a *sampler*. In this case, we say that the transform $\mathcal{D} \cdot \mathcal{S}$ *takes $\mathcal{S}$'s gains* on the rounds n when $\mathcal{D}_n = 1$ and *abstains* on the rounds n where $\mathcal{D}_n = 0$. We call a sequence of consecutive rounds when $\mathcal{D}$ has the value 1 an *on-period* and a sequence of consecutive rounds when $\mathcal{D}$ takes the value 0 an *off-period*.

Lemma 4.5 *Suppose $\mathcal{D}$ is a nonnegative predictable process and $\mathcal{S}$ is a process.*

1. *If $\mathcal{S}$ is a supermartingale, then $\mathcal{D} \cdot \mathcal{S}$ is a supermartingale.*
2. *If $\mathcal{S}$ is a semimartingale, then $\mathcal{D} \cdot \mathcal{S}$ is a semimartingale.*

Proof

1. A process S is a supermartingale if and only if there is a strategy ψ for Skeptic such that

$$S_n - S_{n-1} \leq \psi_n^{\mathrm{M}} y_n + \psi_n^{\mathrm{V}} (y_n^2 - v_n)$$

on every round n. If S satisfies this condition, then $\mathcal{D} \cdot S$ satisfies it with ψ^{M} replaced by $\mathcal{D}\psi^{\mathrm{M}}$ and ψ^{V} replaced by $\mathcal{D}\psi^{\mathrm{V}}$.

2. Similarly, S is a semimartingale if and only if there is a strategy ψ for Skeptic and a nonnegative increasing predictable process $\mathcal{A}$ such that

$$S_n - S_{n-1} \leq \psi_n^{\mathrm{M}} y_n + \psi_n^{\mathrm{V}} (y_n^2 - v_n) + \mathcal{A}_n - \mathcal{A}_{n-1}$$

on every round n, where $\mathcal{A}_0 := 0$. If S satisfies this condition, then $\mathcal{D} \cdot S$ satisfies it with ψ^{M} replaced by $\mathcal{D}\psi^{\mathrm{M}}$, ψ^{V} by $\mathcal{D}\psi^{\mathrm{V}}$, and $\mathcal{A}$ by the predictable process whose gain on round n is $\mathcal{D}_n(\mathcal{A}_n - \mathcal{A}_{n-1})$. $\qquad \square$

In general, we cannot necessarily form a convex combination of an arbitrary sequence $S^1, S^2, \dots$ of processes using an arbitrary sequence $\beta_1, \beta_2, \dots$ of nonnegative numbers adding to 1, because the infinite sum $\sum_{k=1}^{\infty} \beta_k S^k$ may fail to converge in $\mathbb{R}$; moreover, even when the sum is known to converge in $\mathbb{R}$, it might fail to be a supermartingale even where each S^k is a supermartingale, because the sum of the corresponding strategies for Skeptic might fail to converge in $\mathbb{R}$ (cf. Exercise 4.2). The exception noted in the following lemma is key to the generality of Doob's convergence theorem.

Lemma 4.6 *Suppose S is a process and $\mathcal{D}^1, \mathcal{D}^2, \dots$ are uniformly bounded predictable processes.[1] Suppose $\beta_1, \beta_2, \dots$ are nonnegative numbers adding to one. Then the convex combination $\sum_{k=1}^{\infty} \beta_k (\mathcal{D}^k \cdot S)$ exists and is equal to $\left(\sum_{k=1}^{\infty} \beta_k \mathcal{D}^k \right) \cdot S$.*

Proof Because the $\mathcal{D}^k$ are uniformly bounded, the convex combination $\sum_{k=1}^{\infty} \beta_k \mathcal{D}^k$ exists. The transform $\left(\sum_{k=1}^{\infty} \beta_k \mathcal{D}^k \right) \cdot S$ therefore exists, and as the following calculation verifies, it is equal to $\sum_{k=1}^{\infty} \beta_k (\mathcal{D}^k \cdot S)$ (in particular, the latter also exists). On

---

[1] Here *uniformly bounded* means that there exists $c \in \mathbb{R}$ such that $|\mathcal{D}^k| \leq c$ for all $k \in \mathbb{N}$. We are mainly interested in the case where the $\mathcal{D}^k$ are samplers (take only the values 0 and 1).

every round n,

$$
\left(\left(\sum_{k=1}^{\infty} \beta_k \mathcal{D}^k\right) \cdot S\right)_n = \sum_{i=1}^{n} \Delta\left(\left(\sum_{k=1}^{\infty} \beta_k \mathcal{D}^k\right) \cdot S\right)_i
$$

$$
= \sum_{i=1}^{n}\left(\sum_{k=1}^{\infty} \beta_k \mathcal{D}^k\right)_i \Delta S_i = \sum_{i=1}^{n}\left(\sum_{k=1}^{\infty} \beta_k \mathcal{D}_i^k\right) \Delta S_i
$$

$$
= \sum_{i=1}^{n} \sum_{k=1}^{\infty} \beta_k(\mathcal{D}_i^k \Delta S_i) = \sum_{i=1}^{n} \sum_{k=1}^{\infty} \beta_k \Delta(\mathcal{D}^k \cdot S)_i
$$

$$
= \sum_{k=1}^{\infty} \beta_k \sum_{i=1}^{n} \Delta(\mathcal{D}^k \cdot S)_i = \sum_{k=1}^{\infty} \beta_k(\mathcal{D}^k \cdot S)_n.
$$

□

Doob's Transform

Doob's celebrated argument for his convergence theorem constructs a transform that takes advantage of a process's oscillations to grow. The construction can be applied to any process S, but we will consider only the case where S is nonnegative.

Given a nonnegative process S, we begin the construction by defining a sampler $\mathcal{D}^{a,b}$ and hence a transform $\mathcal{D}^{a,b} \cdot S$ for every pair of positive rational numbers a and b such that $a < b$.[2] We define $\mathcal{D}^{a,b}$ by giving instructions for switching back and forth between taking S's gains and abstaining:

- Start taking S's gains the first time as soon as S is equal to a or less. (If $S_0 \le a$, then start on the first round; otherwise wait until S reaches a value equal to a or less; if it never does, never start.)
- Keep taking S's gains until S is equal to b or more; stop as soon as it does so. (If S never gets to b or more, never stop.)
- Repeat indefinitely: start again as soon as S is a or less, then stop as soon as it is b or more, etc.

As Figure 4.1 illustrates, $\mathcal{D}^{a,b} \cdot S$ takes S's gains during S's upcrossings across the interval $[a, b]$ and abstains on the downcrossings.

Now order all positive rational (a, b) with $a < b$ in a sequence $(a_1, b_1), (a_2, b_2), \ldots$ and define a predictable process $\mathcal{D}^S$ by, e.g.

$$
\mathcal{D}^S := \sum_{k=1}^{\infty} 2^{-k} \mathcal{D}^{a_k, b_k}. \tag{4.10}
$$

---

[2]The sampler $\mathcal{D}^{a,b}$ depends on S as well as on a and b, but we do not make this explicit in the notation.

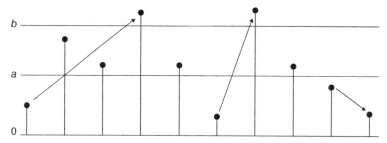

Figure 4.1 Evolution of a nonnegative process S and its transform $\mathcal{D}^{a,b} \cdot S$ over the first nine rounds. The heights of the dots are S's values starting from S_0. The sampler $\mathcal{D}^{a,b}$ divides the rounds $\{1, \dots, 9\}$ into three on-periods $\{1, 2, 3\}$, $\{6\}$, and $\{9\}$ and two off-periods $\{4, 5\}$ and $\{7, 8\}$. The transform $\mathcal{D}^{a,b} \cdot S$ takes S's gains in the on-periods, shown by the arrows. These gains are positive (at least $b - a$) except in the last period, where no more than a is lost.

If S oscillates indefinitely, not converging in $\overline{\mathbb{R}}$, then it will cross some rational interval $[a, b]$ an infinite number of times. As Figure 4.1 suggests, $\mathcal{D}^{a,b} \cdot S$ will therefore tend to infinity, and by Lemma 4.6 $\mathcal{D}^S \cdot S$ will as well. This leads to the following lemma.

Lemma 4.7 *Suppose S is a nonnegative process. Then $\mathcal{D}^S \cdot S$ is nonnegative and tends to infinity on every path on which S does not converge in $\overline{\mathbb{R}}$.*

Proof For a given path ω, set $\tau_0(\omega) := -1$, and for $k = 1, 2, \dots$, set

$$\sigma_k(\omega) := \min\{i > \tau_{k-1}(\omega) | S_i(\omega) \le a\}, \quad \tau_k(\omega) := \min\{i > \sigma_k(\omega) | S_i(\omega) \ge b\},$$

with $\min \emptyset := \infty$. Our sampler $\mathcal{D}^{a,b}$ is

$$\mathcal{D}_i^{a,b} := \begin{cases} 1 & \text{if } \exists k : \sigma_k < i \le \tau_k, \\ 0 & \text{otherwise,} \end{cases}$$

and on each round n we have

$$(\mathcal{D}^{a,b} \cdot S)_n = S_0 + \sum_{k=1}^{\infty}(S_{\tau_k \wedge n} - S_{\sigma_k \wedge n}), \tag{4.11}$$

where the zero terms are ignored (which makes the sum finite). All but possibly one of the terms in the sum in (4.11) are nonnegative. Indeed, S_0 is nonnegative, for all k such that $\tau_k \le n$ we have $S_{\tau_k \wedge n} - S_{\sigma_k \wedge n} = S_{\tau_k} - S_{\sigma_k} \ge b - a$, and for all k such that $\sigma_k \ge n$ we have $S_{\tau_k \wedge n} - S_{\sigma_k \wedge n} = 0$. The possible exception is the *current* k, i.e. k satisfying

$\sigma_k < n < \tau_k$ (such a k either does not exist or is unique). If k is the current one, then $S_{\tau_k \wedge n} - S_{\sigma_k \wedge n} = S_n - S_{\sigma_k}$ might be negative. But because $S_n \geq 0$ and $S_{\sigma_k} \leq a$, this one possible negative term in (4.11) is no less than $-a$.

From this analysis, we can conclude that $\mathcal{D}^{a,b} \cdot S$ is nonnegative. To see that

$$(\mathcal{D}^{a,b} \cdot S)_n \geq 0$$

in spite of the one possible negative term in (4.11), it suffices to consider three cases:

1. The sampler $\mathcal{D}$ is always on until round n. In this case, $\mathcal{D}^{a,b} \cdot S$ takes S's gains all the time up to n, and thus $(\mathcal{D}^{a,b} \cdot S)_n = S_n$.
2. The sampler is initially off, $\mathcal{D}_1^{a,b} = 0$. In this case, $S_0 > a$, and since the only possible negative term in (4.11) is no less than $-a$, the sum is positive.
3. The sampler is initially (but not always until round n) on. In this case, $\mathcal{D}^{a,b} \cdot S$ starts at S_0 but the first term in the sum (4.11) will be at least $b - S_0$. Again, since the only possible negative term in (4.11) is no less than $-a$, the sum is positive.

We can also conclude that if S is both a or less infinitely often and b or more infinitely often, then $\mathcal{D}^{a,b} \cdot S$ tends to infinity. This is because as n tends to infinity, the sum (4.11) will acquire an unbounded number of terms exceeding or equal to $b - a$, while never having more than one negative term, which is never less than $-a$.

By Lemma 4.6 and (4.10),

$$\mathcal{D}^S \cdot S = \sum_{k=1}^{\infty} 2^{-k} (\mathcal{D}^{a_k, b_k} \cdot S). \qquad (4.12)$$

Equation (4.12) puts us in a position to establish the lemma's two assertions:

1. Since the $\mathcal{D}^{a_k, b_k} \cdot S$ are all nonnegative, (4.12) implies that $\mathcal{D}^S \cdot S$ is nonnegative.
2. If S_n does not converge in $\overline{\mathbb{R}}$ as $n \to \infty$, there exist positive rational numbers $a < b$ such that S is both a or less infinitely often and b or greater infinitely often. In this case, $\mathcal{D}^{a,b} \cdot S$ tends to infinity, and therefore, by (4.12), $\mathcal{D}^S \cdot S$ also tends to infinity. □

Doob's convergence theorem takes many forms within measure-theoretic probability. One of the simplest in discrete time is the statement that a nonnegative measure-theoretic supermartingale converges in $\mathbb{R}$ almost surely in the measure-theoretic sense. Comparing Lemmas 4.5 and 4.7 and recalling that an event happens almost surely if there is a nonnegative supermartingale that tends to infinity when it does not happen, we see that Doob's construction implies the following analogous game-theoretic assertion for Protocol 4.4.

Lemma 4.8 *If $\mathcal{T}$ is a nonnegative supermartingale, then $\mathcal{T}_n$ converges in $\mathbb{R}$ almost surely.*

Proof Set

$$\mathcal{T}^* := \frac{1}{2}(\mathcal{T} + \mathcal{D}^{\mathcal{T}} \cdot \mathcal{T}). \tag{4.13}$$

Lemma 4.5 implies that $\mathcal{T}^*$ is a nonnegative supermartingale. Lemma 4.7 implies that it tends to infinity on any path where $\mathcal{T}_n$ does not converge in $\mathbb{R}$. So it witnesses $\mathcal{T}_n$'s almost sure convergence in $\mathbb{R}$. □

By considering samplers that stop a nonnegative semimartingale when its compensator gets too large, we also obtain the following more general result.

Lemma 4.9 *If $\mathcal{V}$ is a nonnegative semimartingale with $\mathcal{A}$ as a compensator, then $\mathcal{V}$ converges in $\mathbb{R}$ on almost all paths on which $\mathcal{A}_\infty := \lim_{n\to\infty}\mathcal{A}_n$ is finite.*

Proof Consider the supermartingale $\mathcal{T}$ defined by $\mathcal{T} := \mathcal{V} - \mathcal{A}$. For $k \in \mathbb{N}$, define a sampler $\mathcal{D}^k$ by $\mathcal{D}^k_n := 1_{\mathcal{A}_n \le k}$. Since $\mathcal{V}$ is nonnegative and $\mathcal{A}$ is increasing,

$$\mathcal{D}^k \cdot \mathcal{T} = \mathcal{D}^k \cdot (\mathcal{V} - \mathcal{A}) \ge -k,$$

and thus $\mathcal{D}^k \cdot \mathcal{T} + k$ is a nonnegative supermartingale. Now consider the predictable process

$$\mathcal{D}^\dagger = \sum_{k=1}^\infty 2^{-k}\mathcal{D}^k.$$

Using Lemma 4.6 and writing

$$\mathcal{D}^\dagger \cdot \mathcal{T} + 2 = \mathcal{D}^\dagger \cdot \mathcal{T} + \sum_{k=1}^\infty k2^{-k} = \sum_{k=1}^\infty 2^{-k}(\mathcal{D}^k \cdot \mathcal{T} + k), \tag{4.14}$$

we see that the supermartingale $\mathcal{D}^\dagger \cdot \mathcal{T} + 2$ is nonnegative.

Consider a path on which $\mathcal{A}_\infty < \infty$. By the definition of $\mathcal{D}^k$, we see that

- for $k \ge \mathcal{A}_\infty$, $(\mathcal{D}^k \cdot \mathcal{T})_n = \mathcal{T}_n$ for all $n \in \mathbb{N}$ and
- for $k < \mathcal{A}_\infty$, $(\mathcal{D}^k \cdot \mathcal{T})_n$ does not depend on n for n sufficiently large.

It follows from (4.14) that $(\mathcal{D}^\dagger \cdot \mathcal{T} + 2)_n$ converges in $\mathbb{R}$ as $n \to \infty$ if and only if $\mathcal{T}_n$ converges in $\mathbb{R}$: indeed, from some n on, $(\mathcal{D}^\dagger \cdot \mathcal{T} + 2)_n$ will be the convex mixture $\alpha c + (1 - \alpha)\mathcal{T}_n$ of a constant c and $\mathcal{T}_n$ with the same $\alpha \in [0, 1]$.

Now apply the construction in the proof of Lemma 4.8 to the nonnegative supermartingale $\mathcal{D}^\dagger \cdot \mathcal{T} + 2$. This yields a nonnegative supermartingale, say $\mathcal{V}^*$, such that $\mathcal{U}_n^*$ tends to infinity as $n \to \infty$ on every path on which $(\mathcal{D}^\dagger \cdot \mathcal{T} + 2)_n$ does not converge in $\mathbb{R}$. In particular, $\mathcal{U}_n^*$ tends to infinity on every path on which $\mathcal{A}_\infty < \infty$ and $\mathcal{U}_n$ does not converge in $\mathbb{R}$, because on such a path $\mathcal{T}_n$ does not converge in $\mathbb{R}$ and thus $(\mathcal{D}^\dagger \cdot \mathcal{T} + 2)_n$ does not converge in $\mathbb{R}$. $\square$

4.3 HOW SKEPTIC FORCES CONVERGENCE

We are now in a position to complete our proof of Statement 1 of Theorem 4.3 by showing that Skeptic can force (4.5) in Protocol 4.4.

Consider the supermartingale S given by

$$S_n := \sum_{i=1}^{n} \frac{y_i}{i}$$

and the nonnegative increasing predictable process $\mathcal{A}$ given by

$$\mathcal{A}_n := \sum_{i=1}^{n} \frac{v_i}{i^2}. \tag{4.15}$$

The difference

$$S_n^2 - \mathcal{A}_n = \left(\sum_{i=1}^{n} \frac{y_i}{i} \right)^2 - \sum_{i=1}^{n} \frac{v_i}{i^2}$$

$$= \sum_{i=1}^{n} M_i y_i + \sum_{i=1}^{n} V_i (y_i^2 - v_i), \tag{4.16}$$

where

$$M_i := \frac{2}{i} \left(\sum_{j=1}^{i-1} \frac{y_j}{j} \right) \quad \text{and} \quad V_i := \frac{1}{i^2},$$

is a supermartingale, and hence S^2 is a semimartingale with $\mathcal{A}$ as a compensator. Since $(S + 1)^2 - S^2 = 2S + 1$ is a supermartingale, it follows that $(S + 1)^2$ is also a semimartingale with $\mathcal{A}$ as a compensator. So by Lemma 4.9, both S^2 and $(S + 1)^2$ converge in $\mathbb{R}$ on almost all paths on which $\mathcal{A}_\infty$ is finite, and S, since it equals $((S + 1)^2 - S^2 - 1)/2$, also converges in $\mathbb{R}$ on almost all these paths.

Our conclusion that S converges in $\mathbb{R}$ on almost all paths where $\mathcal{A}_\infty$ is finite can alternatively be expressed by saying that Skeptic can force

$$\sum_{n=1}^{\infty} \frac{v_n}{n^2} < \infty \implies \sum_{n=1}^{\infty} \frac{y_n}{n} \text{exists and is finite.}$$

To complete the proof, we apply Kronecker's lemma (see, e.g. [358, Lemma IV.3.2]), which implies that

$$\sum_{n=1}^{\infty} \frac{y_n}{n} \text{ exists and is finite} \implies \lim_{n \to \infty} \frac{1}{n} \sum_{i=1}^{n} y_i = 0. \tag{4.17}$$

Implicit in the preceding proof and the proofs of the preceding lemmas is the construction of a strategy for Skeptic that produces a nonnegative supermartingale tending to infinity on all paths outside the event (4.5) (see Exercise 4.6 for details).

4.4 HOW REALITY FORCES DIVERGENCE

In this section, we prove Statement 2 of Theorem 4.3 by exhibiting a strategy for Reality that forces (4.6) in Protocol 4.4.

Let us assume, without loss of generality, that Skeptic sets $\mathcal{K}_0 := 1$. If Skeptic chooses some positive value greater than 1 for $\mathcal{K}_0$, the strategy for Reality we give will work when appropriately rescaled. If Skeptic makes $\mathcal{K}_0 \leq 1$, Reality can use any strategy that works for $\mathcal{K}_0 = 1$.

Define a strategy ρ for Reality as follows:

1. If $\mathcal{K}_{n-1} + V_n(n^2 - v_n) > 1$, set $y_n := 0$.
2. If $\mathcal{K}_{n-1} + V_n(n^2 - v_n) \leq 1$, and $M_n \leq 0$, set $y_n := n$.
3. If $\mathcal{K}_{n-1} + V_n(n^2 - v_n) \leq 1$, and $M_n > 0$, set $y_n := -n$.

Lemma 4.10 *The strategy ρ for Reality forces (4.6).*

Proof We must show that when Reality plays ρ, either (i) $\mathcal{K}_n < 0$ for some n or (ii) $\mathcal{K}_n$ is bounded and (4.6) happens.

Skeptic gains no advantage against ρ by making some of his M_n nonzero, because this only decreases his capital. So it suffices to show that ρ achieves its goal when Skeptic makes all his M_n zero. In this case,

$$\mathcal{K}_n = \mathcal{K}_{n-1} + V_n(y_n^2 - v_n),$$

and so ρ reduces to these two instructions:

1'. If $\mathcal{K}_{n-1} + V_n(n^2 - v_n) > 1$, set $y_n := 0$.
2'. If $\mathcal{K}_{n-1} + V_n(n^2 - v_n) \leq 1$, set $y_n := n$.

So Skeptic's capital is bounded by 1. It remains only to show that either $\mathcal{K}_n < 0$ for some n or (4.6) holds.

If the hypothesis of Instruction $2'$ is satisfied infinitely often, so that Reality sets $y_n = n$ infinitely often, then $\sum_{i=1}^{n} y_i/n \nrightarrow 0$, and hence (4.6) will hold.

If the hypothesis of Instruction $2'$ is not satisfied infinitely often, then there exists $N \in \mathbb{N}$ such that for all $n > N$,

(a) $y_n = 0$, and hence Skeptic loses the nonnegative amount $V_n v_n$ on the nth round,

(b) $\mathcal{K}_{n-1} + V_n(n^2 - v_n) > 1$, and hence $V_n > n^{-2}(1 - \mathcal{K}_{n-1})$, and hence, since Skeptic never increases his capital after N, $V_n > n^{-2}(1 - \mathcal{K}_N)$,

(c) and hence, by points (a) and (b) Skeptic loses at least $n^{-2}v_n(1 - \mathcal{K}_N)$ on the nth round, and more than this if $v_n > 0$.

Assuming that $\sum_n v_n n^{-2} = \infty$ (otherwise Reality's goal is achieved because (4.6) is satisfied), there will be arbitrarily large n for which $v_n > 0$, so that $\mathcal{K}_n$ is eventually less than 1 after N. So by taking N larger if need be, we can make sure that $1 - \mathcal{K}_N > 0$. It then follows from (c) and $\sum_n v_n n^{-2} = \infty$ that Skeptic's cumulative loss after N will be unbounded, eventually bringing Reality to her goal by making $\mathcal{K}_n < 0$.   □

4.5   FORCING GAMES

In this section, we use Martin's theorem on the determinacy of perfect-information games to gain more insight into the relationship between forcing by Skeptic and forcing by Reality. Martin's theorem says that in a two-player perfect-information game in which one of the players necessarily wins (i.e. ties are not permitted), one of the players has a winning strategy. There is only one qualification: if we call the players Player I and Player II, then the set of plays on which Player I wins (and hence also the set of plays on which Player II wins) must be quasi-Borel, to be defined shortly.[3] This is a very weak condition; as we will explain, the goals we set for our players in this book are always quasi-Borel.

To make a testing protocol into a game, we must add a rule for winning for Skeptic or Reality. Here are four ways of doing so for Protocol 4.1. In all four cases, the rule for winning involves an event E in the protocol, which is therefore a parameter of the game. Because a complete sequence of moves by Skeptic's opponents in the protocol has the form $m_1 v_1 y_1 m_2 v_2 y_2 \ldots$, the sample space is $\Omega := (\mathbb{R} \times [0, \infty) \times \mathbb{R})^\infty$.

Game 4.11
PARAMETER: *Event $E \subseteq \Omega$*
PROTOCOL: *Protocol 4.1*
RULE FOR WINNING: *Skeptic wins if both (i) $\mathcal{K}_n$ is always nonnegative and (ii) either E happens or $\mathcal{K}_n \to \infty$. Otherwise Reality wins.*

---

[3]Here a *play* is a complete sequence of moves by all players in the game. We distinguish a play from a *path*, which is a complete sequence of moves by Skeptic's opponents.

Game 4.12
PARAMETER: *Event $E \subseteq \Omega$*
PROTOCOL: *Protocol 4.1*
RULE FOR WINNING: *Skeptic wins if both (i) $\mathcal{K}_n$ is always nonnegative and (ii) either E happens or $\mathcal{K}_n$ is unbounded. Otherwise Reality wins.*

Game 4.13
PARAMETER: *Event $E \subseteq \Omega$*
PROTOCOL: *Protocol 4.1*
RULE FOR WINNING: *Reality wins if both (i) $\mathcal{K}_n$ is bounded above and (ii) either E happens or $\mathcal{K}_n$ is negative for some n. Otherwise Skeptic wins.*

Game 4.14
PARAMETER: *Event $E \subseteq \Omega$*
PROTOCOL: *Protocol 4.1*
RULE FOR WINNING: *Reality wins if both (i) $\mathcal{K}_n \nrightarrow \infty$ and (ii) either E happens or $\mathcal{K}_n$ is negative for some n. Otherwise Skeptic wins.*

In all four games, the rule for winning dictates that either Skeptic or Reality wins. Forecaster is not eligible to win, but there are many E for which neither Skeptic nor Reality has a winning strategy, and for these E Forecaster may be able to influence who wins (an example is the consequent of (4.4)).

By Section 4.1's definitions of forcing by Skeptic and Reality, Skeptic can force an event E if and only if he has a winning strategy in Game 4.11 with E as the value of its parameter, and Reality can force an event E if and only if she has a winning strategy in Game 4.12 with E^c as the value of its parameter. The following proposition gives other characterizations of enforceable quasi-Borel events.

Proposition 4.15   *If E is quasi-Borel, then:*

- *either Skeptic has a winning strategy in all four games with E, E, E^c, and E^c, respectively, as their parameter or he does not have a winning strategy in any of these four games;*
- *either Reality has a winning strategy in all four games with E, E, E^c, and E^c, respectively, as their parameter or she does not have a winning strategy in any of these four games.*

The proposition shows that the games are equivalent, as far as the existence of winning strategies for Skeptic and Reality is concerned. Games 4.11 and 4.14 involve the condition $\mathcal{K}_n \to \infty$ or its negation, whereas Games 4.12 and 4.13 replace $\mathcal{K}_n \to \infty$ by $\mathcal{K}_n$ being unbounded above (with "above" being optional in Game 4.12). We have seen earlier (Lemma 1.5) that Skeptic can turn $\mathcal{K}_n$ being unbounded (while nonnegative) into $\mathcal{K}_n \to \infty$, and the proposition extends this observation assuming

E is quasi-Borel. The pair of Games 4.11 and 4.12 is symmetric to the pair 4.13 and 4.14: the roles of Skeptic and Reality swap, together with swapping what we called their collateral duties in [349]; Skeptic's collateral duty is to keep $\mathcal{K}_n$ nonnegative, and Reality's collateral duty is to keep $\mathcal{K}_n$ bounded above (or to prevent it from tending to ∞).

To prove Proposition 4.15, we need to use some of the properties of the notion of a quasi-Borel set. The *quasi-Borel* subsets of the set of all paths in a game form the smallest class of such subsets that contains all open sets (the topology is generated by the $[s]$, s ranging over the finite sequences $m_1 v_1 M_1 V_1 y_1 \ldots m_n v_n M_n V_n y_n$ and $[s]$ being the set of the plays of the game starting by s) and is closed under (i) complementation, (ii) finite or countable union, and (iii) open-separated union.[4] It is easy to see that the winning sets in this book are also quasi-Borel. For example, consider Game 4.11 with E given by (4.4), and set

$$A := \{\forall n : \mathcal{K}_n \geq 0\} \quad \text{and} \quad B := \{\mathcal{K}_n \to \infty\}.$$

All three events E, A, and B can be expressed by means of events depending only on a finite number of the players' moves (such events are open and so quasi-Borel) using countable unions and intersections and therefore are quasi-Borel. So the event $A \cap (E \cup B)$, which is the winning set for Skeptic in Game 4.11, is also quasi-Borel.

Proof of Proposition 4.15 If Skeptic can guarantee his winning condition

$$\forall n : \mathcal{K}_n \geq 0 \quad \text{and} \quad (E \text{ or } \mathcal{K}_n \to \infty) \tag{4.18}$$

in Game 4.11 with parameter E, he can guarantee (using the same strategy) the winning condition

$$\forall n : \mathcal{K}_n \geq 0 \quad \text{and} \quad (E \text{ or } \sup_n \mathcal{K}_n = \infty) \tag{4.19}$$

in Game 4.12 with parameter E. This follows from (4.19) being a logical implication of (4.18).

If he can guarantee his winning condition (4.19) in Game 4.12 with parameter E, he can guarantee (using the same strategy) the winning condition

$$\sup_n \mathcal{K}_n = \infty \quad \text{or} \quad (E \text{ and } \forall n : \mathcal{K}_n \geq 0) \tag{4.20}$$

in Game 4.13 with parameter E^c. This follows from (4.20) being a logical implication of (4.19).

---

[4] We will not use the notion of open-separated union, but for completeness we define it: $E \subseteq \Gamma$ is the *open-separated union* of a family $\{F_j : j \in J\}$ of subsets of Γ if (i) $E = \cup_{j \in J} F_j$ and (ii) there are disjoint open sets D_j, $j \in J$, such that $F_j \subseteq D_j$ for each $j \in J$.

Now suppose he can guarantee his winning condition (4.20) in Game 4.13 with parameter E^c, and let us check that he can guarantee his winning condition

$$\mathcal{K}_n \to \infty \quad \text{or} \quad (E \text{ and } \forall n : \mathcal{K}_n \geq 0) \tag{4.21}$$

in Game 4.14 with parameter E^c. Fix a strategy for Skeptic guaranteeing (4.20). Because Reality can always make $\mathcal{K}_n$ remain negative (and so bounded above) if it ever becomes negative (see Exercise 4.7), this strategy for Skeptic guarantees $\forall n : \mathcal{K}_n \geq 0$. Since (4.19) is a logical implication of the conjunction of $\forall n : \mathcal{K}_n \geq 0$ and (4.20), this strategy for Skeptic guarantees (4.19). As in the proof of Lemma 1.5, Skeptic can make $\mathcal{K}_n$ being nonnegative and unbounded into $\mathcal{K}_n \to \infty$, so Skeptic has a strategy that guarantees (4.18). It remains to notice that (4.21) is a logical implication of (4.18).

If he can guarantee his winning condition (4.21) in Game 4.14 with parameter E^c, he can guarantee his winning condition (4.18) in Game 4.11 with parameter E: we showed this in the previous paragraph with (4.21) replaced by the weaker condition (4.20).

This completes the part of the proof concerning strategies for Skeptic; it is interesting that this part does not depend on the assumption of E being quasi-Borel.

Suppose Reality can guarantee her winning condition in Game 4.11 with parameter E, and let us show that she can guarantee her winning condition in Game 4.12 with parameter E. Arguing indirectly, suppose she cannot guarantee her winning condition in Game 4.12 with parameter E. By Martin's theorem, this means that Skeptic and Forecaster have a joint strategy that guarantees Skeptic's winning condition (4.19) in Game 4.12 with parameter E. As in the proof of Lemma 1.5, Skeptic can make $\mathcal{K}_n$ being nonnegative and unbounded into $\mathcal{K}_n \to \infty$; therefore, Skeptic and Forecaster have a joint strategy that guarantees Skeptic's winning condition (4.18) in Game 4.11 with parameter E. We have arrived at a contradiction.

Suppose Reality can guarantee her winning condition

$$\exists n : \mathcal{K}_n < 0 \quad \text{or} \quad (E^c \text{ and } \sup_n \mathcal{K}_n < \infty) \tag{4.22}$$

in Game 4.12 with parameter E, and let us show that she can guarantee her winning condition

$$\sup_n \mathcal{K}_n < \infty \quad \text{and} \quad (E^c \text{ or } \exists n : \mathcal{K}_n < 0) \tag{4.23}$$

in Game 4.13 with parameter E^c. Because Reality can always make $\mathcal{K}_n$ remain negative (and so bounded above) if it ever becomes negative (see Exercise 4.7), this strategy for Reality can be modified to guarantee, additionally, that $\mathcal{K}_n < 0$ from some n on whenever $\exists n : \mathcal{K}_n < 0$. Let us check that the modified strategy guarantees (4.23). Suppose Reality plays it. If the first disjunct in (4.22) is true, both conjuncts in (4.23) are true, the first one being true being ensured by our modification. And if the second disjunct in (4.22) is true, both conjuncts in (4.23) are obviously true as well.

Reality's winning condition (4.23) in Game 4.13 with parameter E^c logically implies her winning condition

$$\mathcal{K}_n \nrightarrow \infty \quad \text{and} \quad (E^c \text{ or } \exists n : \mathcal{K}_n < 0) \tag{4.24}$$

in Game 4.14 with parameter E^c.

Finally, Reality's winning condition (4.24) in Game 4.14 with parameter E^c logically implies her winning condition

$$\exists n : \mathcal{K}_n < 0 \quad \text{or} \quad (E^c \text{ and } \mathcal{K}_n \nrightarrow \infty)$$

in Game 4.11 with parameter E.

We can see that only one implication out of eight (Reality having a winning strategy in Game 4.11 implying her having winning strategy in Game 4.12) relies on the assumption that E is quasi-Borel. □

4.6   EXERCISES

Exercise 4.1   How are Skeptic's moves in Protocol 4.1 constrained when (as in Games 4.11–4.12) he must guarantee $\mathcal{K}_n \geq 0$ for all n regardless of how the other players move?

Fix $m \in \mathbb{R}$ and $v \geq 0$. For each $M \in \mathbb{R}$ and $V \geq 0$, let $g_{M,V}$ be the function given by

$$g_{M,V}(y) := M(y - m) + V((y - m)^2 - v)$$

for all $y \in \mathbb{R}$.

1. If $V = 0$ and $M \neq 0$, then $g_{M,V}$ is not bounded below. Verify that if $V > 0$, then $g_{M,V}$ is bounded below and attains its minimum value, $-M^2/4V - Vv$, at $y = m - M/2V$.

2. Fix $\mathcal{K} \geq 0$ and suppose $v = 0$. Verify that $\mathcal{K} + g_{M,V}(y) \geq 0$ for all $y \in \mathbb{R}$ if and only if $M^2 \leq 4V\mathcal{K}$.

3. Fix $\mathcal{K} \geq 0$ and suppose $v > 0$. Verify that $\mathcal{K} + g_{M,V}(y) \geq 0$ for all $y \in \mathbb{R}$ if and only if $M^2 \leq 4V(\mathcal{K} - Vv)$. (Notice that this requires $V \leq \mathcal{K}/v$.)

So Skeptic guarantees $\mathcal{K}_n \geq 0$ for all n if and only if he chooses $\mathcal{K}_0$ nonnegative and then, for $n = 1, 2, \ldots$, chooses M_n and V_n satisfying $M_n^2 \leq 4V_n(\mathcal{K}_{n-1} - V_n v_n)$. □

Exercise 4.2   This exercise uses the results of the previous exercise to study the effect of allowing $v_n = 0$ in Protocol 4.1.

For each $\mathcal{K} \geq 0$, set

$$A_{\mathcal{K}} := \{(M,V) | M \in \mathbb{R}, V \geq 0, \text{ and } \mathcal{K} + g_{M,V}(y) \geq 0 \text{ for all } y \in \mathbb{R}\}.$$

Thus

$$A_{\mathcal{K}} = \{(M,V) | M \in \mathbb{R}, V \geq 0, \text{ and } M^2 \leq 4V(\mathcal{K} - Vv)\}.$$

We see that $A_{\mathcal{K}}$ is a closed set in $\mathbb{R}^2$, which is unbounded when $v = 0$ but bounded when $v > 0$.

For each $\mathcal{K} \geq 0$, set

$$B_{\mathcal{K}} := \{h : \mathbb{R} \to \mathbb{R} | \exists (M,V) \in A_{\mathcal{K}} : -\mathcal{K} \leq h(y) \leq g_{M,V}(y) \text{ for all } y \in \mathbb{R}\}.$$

1. Show that $B_{\mathcal{K}}$ is closed under infinite convex combination if $v > 0$ but not if $v = 0$.

2. Suppose we modify Protocol 4.1 in two ways: (i) we allow Skeptic to substitute for the payoff $g_{M,V}$ any payoff h that is bounded above by $g_{M,V}$ and (ii) we require Forecaster to make all $v_n > 0$. Show that in this case, the set of strategies that begin with capital $\mathcal{K}_0 := 1$ and avoid bankruptcy is closed under infinite convex combination. Show that it is not closed under infinite convex combination if we make modification (i) but not modification (ii). □

Exercise 4.3 (loosening Protocol 4.1) Protocol 4.1 requires the v_n and the V_n to be nonnegative. Would it be interesting to drop these restrictions?

1. Show that if the protocol is modified to allow Forecaster to make v_n negative, then Skeptic can still force (4.4).

2. Show that if the protocol is modified to allow Skeptic to make V_n negative, then Theorem 4.3 still holds. □

Exercise 4.4 The following protocol differs from Protocol 4.1 only in that Skeptic is no longer allowed to make his M_n negative. This seems appropriate if Forecaster claims only that his forecasts m_n are good upper estimates: they do not systematically underestimate the y_n. Skeptic can test this claim by betting on each round that y_n will be greater than m_n.

Skeptic announces $\mathcal{K}_0 \in \mathbb{R}$.
FOR $n = 1, 2, \ldots$:
    Forecaster announces $m_n \in \mathbb{R}$ and $v_n \geq 0$.
    Skeptic announces $M_n \geq 0$ and $V_n \geq 0$.
    Reality announces $y_n \in \mathbb{R}$.
    $\mathcal{K}_n := \mathcal{K}_{n-1} + M_n(y_n - m_n) + V_n((y_n - m_n)^2 - v_n)$.

Show that Skeptic can force

$$\sum_{n=1}^{\infty} \frac{v_n}{n^2} < \infty \implies \limsup_{n \to \infty} \frac{1}{n} \sum_{i=1}^{n} (y_i - m_i) \le 0$$

in this protocol.   □

Exercise 4.5   Lemma 4.7 says nothing about paths on which S does converge in $\mathbb{R}$. Does $\mathcal{D}^S \cdot S$ necessarily converge in $\mathbb{R}$ on such a path? Prove that it does or give a counterexample.   □

Exercise 4.6   Describe explicitly a strategy for Skeptic that produces a nonnegative supermartingale tending to infinity on all paths outside the event (4.5). *Hint.* Such a strategy can be extracted from the proof in Section 4.3 and the proofs of Lemmas 4.7, 4.8, and 4.9. Namely, consider the strategy for Skeptic giving rise to the supermartingale (4.16) and its analogue for $(S + 1)^2 - A$. Apply to either of them the transformation given in the proof of Lemma 4.9 (which includes the transformation given in the proof of Lemma 4.8 as its last step). Average the resulting two strategies.   □

The following exercise was used in the proofs of Propositions 4.2 and 4.15. The two statements in it and their generalizations will be central to Chapter 12.

Exercise 4.7

1. Show that Reality can play in Protocol 4.1 in such a way that the sequence $\mathcal{K}_n$, $n = 0, 1, \ldots$, is decreasing. *Hint.* Reality can set $y_n := m_n$.

2. Show that Reality can play in Protocol 4.1 in such a way that $\mathcal{K}_n$ is constant.   □

Exercise 4.8   Show that the condition

$$\sum_{n=1}^{\infty} \frac{v_n}{g(A_n)} < \infty$$

in (4.25) follows from the condition $A_n \to \infty$ if the function g satisfies $\int_0^{\infty} dx/g(x) < \infty$. Show that

$$g(x) := 1 \vee (x \ln x \ln \ln x (\ln \ln \ln x)^2)$$

satisfies this condition. Compare the strength of the consequent in (4.25) with the consequent in the law of the iterated logarithm (see, e.g. (5.3)). *Hint.* These are Corollary 4.6 and Example 4.7 in [283].   □

Exercise 4.9   (Takazawa's supermartingale [373]) Show that the process

$$\prod_{i=1}^{n} \frac{\exp(y_i - y_i^2/2)}{1 + v_i/2}$$

is a supermartingale in Protocol 4.4. *Hint.* This is witnessed by this strategy for Skeptic:

$$M_n := \mathcal{K}_{n-1} \frac{\exp(-v_n/2)}{1 + v_n/2}, \quad V_n := \frac{1}{2}M_n.$$

For further details, see [373, Theorem 3.1]. □

Exercise 4.10 Show that the process

$$\prod_{i=1}^{n} \exp\ (y_i - y_i^2/2 - v_i/2)$$

is a supermartingale in Protocol 4.4. *Hint.* Deduce this from the previous exercise. Alternatively, follow [29, proof of Lemma 3.1]. □

4.7 CONTEXT

Doob's Convergence Theorem

Our proof of Lemma 4.7 uses Doob's upcrossing argument, which we will use again in Section 7.2 to establish a game-theoretic supermartingale convergence theorem that applies to all testing protocols. Doob introduced the argument in 1940 to establish the convergence of nonnegative measure-theoretic martingales [115, 255]. Ville had proven only that they are bounded with probability 1 [387].

Martingale Transforms

The term *martingale transform* has been used in measure-theoretic probability since the 1960s (see [59]).

Semimartingales

The term *semimartingale* is now used in measure-theoretic probability theory primarily in continuous time, where it names any process that can be decomposed into the sum of an adapted finite-variation process and a local martingale. According to Paul-André Meyer [281], this use of the term developed in the late 1960s in work by himself and Doléans.

Skeptic's Strategies

Our proof of Statement 1 of Theorem 4.3 follows Robert Liptser's proof of the measure-theoretic version [358, section VI.5]. In measure-theoretic probability it was first obtained by Yuan-Shih Chow in 1965 [77, Theorem 5a]. Chow's result will be deduced from Statement 1 of Theorem 4.3 in Section 9.3.

Japanese researchers have generalized Statement 1 in several directions. Kumon et al. [230, Theorem 3.1] show that if $y_n^2 - v_n$ is replaced by $|y_n|^{1+\epsilon} - v$ in Protocol 4.4, where ϵ and v are positive constants, then Skeptic can force $\bar{y}_n \to 0$. (In this setting Forecaster becomes irrelevant and can be removed from the protocol.) This ceases to be true when $\epsilon = 0$ [230, Proposition 2.1]. This result is then generalized to a game-theoretic version of the Marcinkiewicz–Zygmund law of large numbers [263], in which the rate of growth $o(n)$ of $\sum_{i=1}^{n} y_i$ is replaced by the stronger $o(n^{1/r})$ for $r \in (1, 2)$ (which requires use of stronger hedges than $|y_n|^{1+\epsilon} - v$).

Miyabe and Takemura have studied rates of convergence in the law of large numbers in Protocol 4.4 [283]. In their Corollary 4.5 they show that Skeptic can force

$$\left. \begin{array}{c} \mathcal{A}_n \to \infty \\ \sum_{n=1}^{\infty} v_n / g(\mathcal{A}_n) < \infty \end{array} \right\} \implies \frac{1}{\sqrt{g(\mathcal{A}_n)}} \sum_{i=1}^{n} y_i \to 0 \qquad (4.25)$$

(cf. (4.5)), where $\mathcal{A}_n$ is defined by (4.15) and $g : [0, \infty) \to (0, \infty)$ is a positive increasing function with $g(\infty) = \infty$ (see also Exercise 4.8). They extend these results to more general hedges than quadratic, developing the results of Kumon et al. [230].

Martin's Theorem

In a two-player perfect-information game that always has a winner, we expect one of the players to have a winning strategy. If the game has a finite horizon, Zermelo's theorem asserts without qualification that one of them has a winning strategy [434]. If the game can continue indefinitely, then Donald Martin's theorem on the determinacy of games asserts that one of them has a winning strategy under weak conditions on the event that determines the winner; it suffices, for example, that it be quasi-Borel [266–268]. All interesting strategies for Skeptic known to us are quasi-Borel.

When move spaces are countable, being quasi-Borel is the same as being Borel. Even in this special case, Martin's theorem was a product of long development. In 1953 [161], David Gale and Frank Stewart established it for a subclass of Borel sets known as Σ_1^0, and this result was extended to Σ_2^0 in 1955 [430], to Σ_3^0 in 1964 [94], to Σ_4^0 in 1972 [299], and finally to all Borel sets in 1975 [266]. In the countable case, Borel games form the largest natural class of two-person perfect information games whose determinacy can be proven in standard ZFC set theory [267]. Natural examples of two-player perfect-information games without a winning strategy are provided by Choquet games in general topology [76, chapter 7], [214, section 8.C].

Reality's Strategies

Kolmogorov's classical argument [223] shows that there exists a randomized strategy for Reality defeating Skeptic almost surely if his goal is to force the consequent of (4.4) when the antecedent of (4.4) is violated. It follows that Skeptic does not have a winning strategy, and by Martin's theorem this implies that Reality does have a winning strategy. This was the proof of Statement 2 of Theorem 4.3 used in [349].

In 2006, Takemura posed the problem of explicitly constructing pure strategies for Reality in problems of this type. This inspired the discovery of the strategy for Reality given in the proof of Statement 1 of Theorem 4.3 [411]. In articles published in 2012 and 2015, Miyabe and Takemura carried out an impressive program of studying pure strategies for Reality [283, 285]. Proposition 4.2 is from [283], but Miyabe and Takemura's result [283, Proposition 4.10] is stronger: they construct a strategy for Reality that forces (4.6) *strongly*, in that it does not allow Skeptic's capital to rise above its initial level.

A typical result obtained by Miyabe and Takemura [283, 285] is: in Protocol 4.1, Reality can strongly force

$$\sum_{n=1}^{\infty} \frac{v_n}{n^2} < \infty \iff \frac{1}{n} \sum_{i=1}^{n} (y_i - m_i) \to 0$$

[285, Theorem 4.1]. This strengthens Statement 2 of Theorem 4.3 (which holds with $\iff$ in place of $\implies$ in view of Proposition 4.2 and Statement 1 of Theorem 4.3). They obtain similar statements about Reality's strategies for other kinds of laws of large numbers in game-theoretic probability, such as Marcinkiewicz–Zygmund [230, Theorem 5.1] and (4.25). An important advance in [285] is Miyabe and Takemura's method of *derandomization* – a constructive procedure for transforming any randomized strategy for Reality that achieves its goal almost surely (as the strategy extracted from Kolmogorov's proof does) into a pure strategy.

5

The Law of the Iterated Logarithm

The law of the iterated logarithm concerns the rate and oscillation of the almost sure convergence guaranteed by laws of large numbers.

In this chapter, we prove a game-theoretic law of the iterated logarithm for a variant of Protocol 4.1, in which we proved Kolmogorov's law of large numbers. This game-theoretic law says that

$$\limsup_{n \to \infty} \frac{\sum_{i=1}^{n}(y_i - m_i)}{\sqrt{2A_n \ln \ln A_n}} = 1 \qquad (5.1)$$

almost surely provided that the $y_n - m_n$ do not grow too fast and that $A_n \to \infty$, where $A_n := \sum_{i=1}^{n} v_i$.

Roughly speaking, (5.1) says that $\sqrt{2A_n \ln \ln A_n}$ is almost surely the asymptotic least upper bound for the cumulative sum $\sum_{i=1}^{n}(y_i - m_i)$. By the symmetry between y_n and $-y_n$ in the protocol, it follows that $-\sqrt{2A_n \ln \ln A_n}$ is almost surely the asymptotic greatest lower bound for $\sum_{i=1}^{n}(y_i - m_i)$. We discuss only the upper bound, leaving corresponding observations about the lower bound to the reader.

We consider separately the validity and the sharpness of the upper bound. The bound is *valid* if (5.1) holds when = is replaced by ≤. This means that for any positive number ε, the ratio

$$\frac{\sum_{i=1}^{n}(y_i - m_i)}{\sqrt{2A_n \ln \ln A_n}} \qquad (5.2)$$

Game-Theoretic Foundations for Probability and Finance, First Edition. Glenn Shafer and Vladimir Vovk.
© 2019 John Wiley & Sons, Inc. Published 2019 by John Wiley & Sons, Inc.

almost surely eventually stays below $1 + \varepsilon$. The bound is *sharp* if the equality holds. This means that the bound is valid and that for any positive number ε, (5.2) almost surely comes within ε of 1 infinitely often.

Measure-theoretic statements of the law of the iterated logarithm usually give conditions on the $y_n - m_n$ and their variances that make the iterated-logarithm bound both valid and sharp. In the game-theoretic framework, in contrast, it is natural to note that Skeptic can force validity under weaker conditions than sharpness. To force sharpness as well as validity, he must have advance knowledge about how large $|y_n - m_n|$ can be.

The original law of the iterated logarithm, published by Khinchin in 1924 [217], considered only binary outcomes with constant probabilities. In the corresponding protocol, Protocol 1.10, Skeptic can force sharpness as well as validity (see Exercise 5.4). But in the more general protocols we have been studying, Protocols 1.1 and 4.1, the iterated-logarithm bound is only valid. To obtain sharpness, we must change the rules in Skeptic's favor, and these changes are not appropriate for all applications. When Reality is a financial market as in Chapter 16, for example, it may be reasonable to assume weak limits on price changes that make the bound valid but not the advance knowledge that can make it sharp.

In Section 5.1, we establish a theorem on validity for Protocol 4.1 and explore its consequences for some related protocols. In Section 5.2, we state a theorem that asserts sharpness as well as validity in a protocol in which Skeptic is stronger, Protocol 5.2. This theorem, Theorem 5.5, is a game-theoretic counterpart of Kolmogorov's 1929 generalization of Khinchin's result [222]. Its conclusions do not hold in Protocol 4.1, because Skeptic is too weak there. We refer the reader to [349, section 5.3] for a proof of the theorem, modeled after Kolmogorov's proof.

In Section 5.3, we discuss additional recent work on game-theoretic laws of the iterated logarithm. In Section 5.4, we discuss connections with large-deviation inequalities.

5.1 VALIDITY OF THE ITERATED-LOGARITHM BOUND

The following theorem makes precise the sense in which the iterated-logarithm bound is valid in Protocol 4.1. After proving this theorem, we consider its implications for some slightly different protocols.

Theorem 5.1 *Skeptic can force*

$$\left(A_n \to \infty \text{ and } |y_n - m_n| = o\left(\sqrt{\frac{A_n}{\ln \ln A_n}} \right) \right) \implies \limsup_{n \to \infty} \frac{\sum_{i=1}^{n} (y_i - m_i)}{\sqrt{2A_n \ln \ln A_n}} \leq 1 \tag{5.3}$$

in Protocol 4.1, where $A_n := \sum_{i=1}^{n} v_i$.

Proof We assume without loss of generality that $m_n = 0$ for all n, so that we are working in Protocol 4.4. The argument for Lemma 1.5 applies to this protocol: Skeptic can force any event that he can weakly force. So our goal is to construct a strategy for Skeptic that weakly forces

$$\left(A_n \to \infty \text{ and } |y_n| = o\left(\sqrt{\frac{A_n}{\ln \ln A_n}}\right)\right) \implies \limsup_{n \to \infty} \frac{\sum_{i=1}^{n} y_i}{\sqrt{2A_n \ln \ln A_n}} \le 1. \quad (5.4)$$

For each $\delta \in (0, 1)$ and $\kappa \in (0, 1)$, let $\psi^{\delta,\kappa}$ be the strategy for Skeptic that sets $\mathcal{K}_0 := 1$ and

$$M_n := \mathcal{K}_{n-1} \frac{\kappa}{1 + (1 + \delta)\kappa^2 v_n/2}, \quad V_n := \mathcal{K}_{n-1} \frac{(1 + \delta)\kappa^2/2}{1 + (1 + \delta)\kappa^2 v_n/2} \quad (5.5)$$

for $n \in \mathbb{N}$. Its capital process is the supermartingale $\mathcal{T}^{\delta,\kappa}$ given by $\mathcal{T}_0^{\delta,\kappa} = 1$ and[1]

$$\mathcal{T}_n^{\delta,\kappa} = \mathcal{T}_{n-1}^{\delta,\kappa} \frac{1 + \kappa y_n + (1 + \delta)\kappa^2 y_n^2/2}{1 + (1 + \delta)\kappa^2 v_n/2} \quad (5.6)$$

for $n = 1, 2, \ldots$. The constraints $\delta \in (0, 1)$ and $\kappa \in (0, 1)$ guarantee that $\mathcal{T}^{\delta,\kappa}$ is nonnegative.

For each $k \in \mathbb{N}$, set $\kappa(\delta, k) := \sqrt{2(1 + \delta)^{-k} \ln k}$. For each $\delta \in (0, 1)$, choose an integer K_δ large enough that

$$\kappa(\delta, k) \in (0, 1) \quad \text{for } k \ge K_\delta \quad (5.7)$$

and

$$\sum_{k=K_\delta}^{\infty} k^{-1-\delta} \le 1. \quad (5.8)$$

By (5.7), the definition and reasoning of the preceding paragraph apply when $k \ge K_\delta$: we have defined a strategy $\psi^{\delta,\kappa(\delta,k)}$ whose capital process is a nonnegative supermartingale $\mathcal{T}^{\delta,\kappa(\delta,k)}$. Equation (5.6) then implies that on any fixed round n and fixed path $\omega = v_1 y_1 v_2 y_2 \ldots$, the strategy's capital is bounded as a function of $\delta \in (0, 1)$ and $k \ge K_\delta$. This boundedness, together with (5.5), imply that on any fixed round and path

---

[1] The factor multiplying $\mathcal{T}_{n-1}^{\delta,\kappa}$ in (5.6) approximates $\exp(\kappa y_n - \kappa^2 v_n/2)$, when δ and κ are sufficiently small. If we could achieve this factor exactly, then the capital after the nth round would be $\exp\left(\kappa \sum_{i=1}^{n} y_i - \kappa^2 A_n/2\right)$. The value of κ that maximizes this capital is $\sum_{i=1}^{n} y_i/A_n$, which will repeatedly exceed $\sqrt{2 \ln \ln A_n/A_n}$ if the iterated-logarithm bound is violated. The basic idea of this proof is to average over many different values of κ, so as to take advantage of every violation (see (5.14)).

Skeptic's moves are bounded as functions of $\delta \in (0, 1)$ and $k \geq K_\delta$. The boundedness of these quantities with respect to k for given $\delta \in (0, 1)$ implies that we can define a strategy ψ^δ for Skeptic by

$$\psi^\delta := \sum_{k=K_\delta}^{\infty} k^{-1-\delta} \psi^{\delta, \kappa(\delta, k)}. \tag{5.9}$$

As a mixture of strategies with nonnegative capital processes, ψ^δ has a nonnegative capital process, say $\mathcal{T}^\delta$. By the definition of weakly forcing, ψ^δ weakly forces $\sup_n \mathcal{T}_n^\delta < \infty$. It follows that it also weakly forces

$$\sup_n \mathcal{T}_n^{\delta, \kappa(\delta, k)} = O(k^{1+\delta}), \quad k \geq K_\delta. \tag{5.10}$$

For a given $\delta \in (0, 1)$, consider a path ω on which both (5.10) and

$$A_n \to \infty \quad \text{and} \quad |y_n| = o\left(\sqrt{\frac{A_n}{\ln \ln A_n}}\right) \tag{5.11}$$

hold. Set

$$k = k(\delta, n) := \lfloor \log_{1+\delta} A_n \rfloor \quad \text{and} \quad \kappa_{\delta, n} := \kappa(\delta, k(\delta, n))$$

for n sufficiently large that these quantities are defined and $k(\delta, n) \geq K_\delta$; bear in mind that these quantities depend on ω even though the notation does not make this explicit. We obtain, as $n \to \infty$,

$$\ln \mathcal{T}_n^{\kappa_{\delta, n}} \leq (1 + \delta) \ln k(\delta, n) + O(1) \leq (1 + \delta) \ln \ln A_n + O(1).$$

By (5.6), we further obtain

$$\ln \prod_{i=1}^{n} \frac{1 + \kappa_{\delta, n} y_i + (1 + \delta) \kappa_{\delta, n}^2 y_i^2 / 2}{1 + (1 + \delta) \kappa_{\delta, n}^2 v_i / 2} \leq (1 + \delta) \ln \ln A_n + O(1), \tag{5.12}$$

that is,

$$\sum_{i=1}^{n} \ln \left(1 + \kappa_{\delta, n} y_i + (1 + \delta) \kappa_{\delta, n}^2 y_i^2 / 2\right)$$

$$\leq \sum_{i=1}^{n} \ln \left(1 + (1 + \delta) \kappa_{\delta, n}^2 v_i / 2\right) + (1 + \delta) \ln \ln A_n + O(1). \tag{5.13}$$

For n large enough,[2]

$$\frac{1}{1+\delta}\sqrt{\frac{2\ln\ln A_n}{A_n}} \le \kappa_{\delta,n} \le (1+\delta)\sqrt{\frac{2\ln\ln A_n}{A_n}}. \tag{5.14}$$

Together with (5.11), this gives

$$\lim_{n\to\infty}\sup_{i\le n}\kappa_{\delta,n}|y_i| = 0. \tag{5.15}$$

Because

$$\ln(1 + t + (1+\delta)t^2/2) \ge t$$

for t small enough in absolute value, (5.13) and (5.15) imply

$$\kappa_{\delta,n}\sum_{i=1}^{n} y_i \le \sum_{i=1}^{n} \ln(1 + (1+\delta)\kappa_{\delta,n}^2 v_i/2) + (1+\delta)\ln\ln A_n + O(1).$$

Because $\ln(1+t) \le t$, we further obtain

$$\sum_{i=1}^{n} y_i \le (1+\delta)\kappa_{\delta,n}A_n/2 + \frac{1+\delta}{\kappa_{\delta,n}}\ln\ln A_n + O\left(\frac{1}{\kappa_{\delta,n}}\right). \tag{5.16}$$

This and (5.14) imply that

$$\sum_{i=1}^{n} y_i \le (1+\delta)^2\sqrt{2A_n\ln\ln A_n} + O\left(\sqrt{\frac{A_n}{\ln\ln A_n}}\right) \tag{5.17}$$

on ω. So for each $\delta \in (0,1)$, the strategy ψ^δ weakly forces (5.17) or the failure of (5.11).

Because each strategy in the mixture (5.9) has unit initial capital, we know from (5.8) that the strategy ψ^δ has initial capital 1 or less. We also know that the moves specified by ψ^δ on each round and each path are bounded as functions of δ. So we can define a strategy ψ for Skeptic by

$$\psi := \sum_{j=1}^{\infty} 2^{-j}\psi^{1/j}.$$

The strategy ψ weakly forces (5.4).   □

---

[2]The definition of $\kappa_{\delta,n}$ is designed to achieve (5.14). This can be motivated by the argument in the preceding footnote or by noting that the sum of the first two addends on the right-hand side of (5.16) is minimized as a function of $\kappa_{\delta,n}$, when $\kappa_{\delta,n} = \sqrt{2\ln\ln A_n/A_n}$.

The following protocol modifies Protocol 4.1 by strengthening Skeptic in two ways (cf. Exercise 5.1):

1. It allows him to make V_n negative.
2. It gives him information about y_n before he makes his moves M_n and V_n: a bound c_n on how far y_n can be from m_n.

Protocol 5.2
Skeptic announces $\mathcal{K}_0 \in \mathbb{R}$.
FOR $n = 1, 2, \ldots$:
  Forecaster announces $m_n \in \mathbb{R}$, $c_n \geq 0$, and $v_n \geq 0$.
  Skeptic announces $M_n \in \mathbb{R}$ and $V_n \in \mathbb{R}$.
  Reality announces $y_n \in \mathbb{R}$ such that $|y_n - m_n| \leq c_n$.
  $\mathcal{K}_n := \mathcal{K}_{n-1} + M_n(y_n - m_n) + V_n((y_n - m_n)^2 - v_n)$.

Because he is stronger in Protocol 5.2 than in Protocol 4.1, Skeptic can force anything in Protocol 5.2 that he can force in Protocol 4.1. In particular, he can force (5.3). Since $|y_n - m_n| \leq c_n$ for all n, Skeptic can also force

$$\left(A_n \to \infty \text{ and } c_n = o\left(\sqrt{\frac{A_n}{\ln \ln A_n}} \right) \right) \implies \limsup_{n \to \infty} \frac{\sum_{i=1}^{n} (y_i - m_i)}{\sqrt{2A_n \ln \ln A_n}} \leq 1 \quad (5.18)$$

in Protocol 5.2. This formulation is of interest because, as we discuss in Section 5.2, Skeptic can also force the relationship with $\leq$ replaced by $=$; it is in this sense that the iterated-logarithm bound is sharp as well as valid in Protocol 5.2.

Skeptic can also force (5.18) in the following simpler protocol, in which Forecaster simply announces a bound c_n on $|y_n|$ instead of a price v_n for y_n^2.

Protocol 5.3
Skeptic announces $\mathcal{K}_0 \in \mathbb{R}$.
FOR $n = 1, 2, \ldots$:
  Forecaster announces $c_n \geq 0$ and $m_n \in [-c_n, c_n]$.
  Skeptic announces $M_n \in \mathbb{R}$.
  Reality announces $y_n \in [-c_n, c_n]$.
  $\mathcal{K}_n := \mathcal{K}_{n-1} + M_n(y_n - m_n)$.

Corollary 5.4   *Skeptic can force (5.18) in Protocol 5.3, where $A_n := \sum_{i=1}^{n} c_i^2$.*

Proof   Because both y_n and m_n are in $[-c_n, c_n]$, we have $y_n^2 - c_n^2 \leq m_n^2$, and hence,

$$M_n(y_n - m_n) + V_n((y_n - m_n)^2 - c_n^2) \leq (M_n - 2V_n m_n)(y_n - m_n)$$

for any $M_n \in \mathbb{R}$ and $V_n \geq 0$. It follows that Skeptic can make at least as much money from his opponents' moves m_n, c_n, and y_n in Protocol 5.3 as he can make from the same moves in Protocol 4.1, Forecaster's move c_n in Protocol 5.3 being relabeled v_n in Protocol 4.1. So he can force any event in Protocol 5.3 that he can force in Protocol 4.1. (An event in Protocol 5.3 is a condition on the path $m_1, c_1, y_1, \ldots$; the "same" event in Protocol 4.1 is the same condition on the path $m_1, v_1, y_1, \ldots$.) □

If we strengthen Skeptic in Protocol 5.3 by requiring Forecaster to set $c_n := 1$ for all n, then the protocol reduces to Protocol 1.1, which we used to study bounded outcomes in Chapter 1. So Corollary 5.4 implies that Skeptic can force

$$\limsup_{n \to \infty} \frac{\sum_{i=1}^{n}(y_i - m_i)}{\sqrt{2n \ln \ln n}} \leq 1 \tag{5.19}$$

in Protocol 1.1.

5.2 SHARPNESS OF THE ITERATED-LOGARITHM BOUND

As we just saw, Theorem 5.1 implies the validity of the iterated-logarithm bound in Protocol 5.2. The following theorem makes precise the sense in which the bound is sharp as well as valid in this protocol.

Theorem 5.5 *Skeptic can force*

$$\left(A_n \to \infty \text{ and } c_n = o\left(\sqrt{\frac{A_n}{\ln \ln A_n}} \right) \right) \implies \limsup_{n \to \infty} \frac{\sum_{i=1}^{n}(y_i - m_i)}{\sqrt{2A_n \ln \ln A_n}} = 1 \tag{5.20}$$

in Protocol 5.2, where $A_n := \sum_{i=1}^{n} v_i$.

This theorem is proven in [349, section 5.3]. The proof is modeled on Kolmogorov's 1929 proof [222] (see also [302, 303]).

The iterated-logarithm bound is not sharp in other protocols we have been considering: Protocols 1.1, 4.1, and 5.3. To formalize this point, let us define *forcing by Forecaster and Reality* just as we defined forcing by Reality in Section 4.1: a strategy for Forecaster and Reality forces an event E if it guarantees that either (i) $\mathcal{K}_n < 0$ for some n or (ii) $\mathcal{K}_n$ is bounded and E happens. When Forecaster and Reality can force an event E, Skeptic cannot force an event that contradicts E.

Proposition 5.6

1. In Protocol 1.1, Forecaster and Reality can force $y_n = m_n = 0$ for all n. Hence, Skeptic cannot force (5.19) with $\leq$ replaced by $=$.
2. In Protocol 4.1, Forecaster and Reality can force $\sum_{i=1}^{n} v_i = n$ and $y_n = m_n = 0$ for all n, even if we strengthen Skeptic by allowing negative V_n. Hence, Skeptic cannot force (5.3) with $\leq$ replaced by $=$ and $A_n := \sum_{i=1}^{n} v_i$.
3. In Protocol 5.3, Forecaster and Reality can force $\sum_{i=1}^{n} c_i^2 = n$ and $y_n = m_n = 0$ for all n, even if we strengthen Skeptic by allowing negative V_n. Hence, Skeptic cannot force (5.18) with $\leq$ replaced by $=$ and $A_n := \sum_{i=1}^{n} c_i^2$.

Proof
1. When Forecaster and Reality choose $y_n = m_n = 0$ for all n, Skeptic's capital never changes, and hence, Forecaster and Reality have forced the event.
2. Forecaster and Reality force the event by always choosing $y_n = m_n = 0$ and $v_n = 1$, except that if Skeptic makes V_n negative, Reality makes y_n so large that Skeptic's capital becomes negative.
3. Forecaster and Reality choose $y_n = m_n = 0$ and $c_n = 1$ for all n, with the same exception for negative V_n.   □

5.3   ADDITIONAL RECENT GAME-THEORETIC RESULTS

In this section, we review additional recent game-theoretic versions of the law of the iterated logarithm. For brevity, we generally report only simple special cases of results stated by their authors in stronger forms.

Our game-theoretic version of Kolmogorov's law of the iterated logarithm, Theorem 5.5, requires the deviation $|y_n - m_n|$ to be bounded by c_n. Miyabe and Takemura replace this somewhat awkward hard constraint by a softer one, adding another hedge to y_n^2 [284]. Here, we state their protocol as Protocol 5.7 and their result, [284, Theorem 1.4 and Example 2.4], as Theorem 5.8.

Protocol 5.7
PARAMETER: $\alpha \in (2, 3]$
    Skeptic announces $\mathcal{K}_0 \in \mathbb{R}$.
    FOR $n = 1, 2, \ldots$:
        Forecaster announces $m_n \in \mathbb{R}$, $v_n \geq 0$, and $w_n \geq 0$.
        Skeptic announces $M_n \in \mathbb{R}$, $V_n \in \mathbb{R}$, and $W_n \geq 0$.
        Reality announces $y_n \in \mathbb{R}$.
        $\mathcal{K}_n := \mathcal{K}_{n-1} + M_n(y_n - m_n) + V_n((y_n - m_n)^2 - v_n)$
                $+ W_n(|y_n - m_n|^{\alpha} - w_n)$.

Theorem 5.8 *Skeptic can force*

$$\left(A_n \to \infty \text{ and } \sum_n w_n \left(\frac{\ln \ln A_n}{A_n}\right)^{\alpha/2} < \infty\right) \implies \limsup_{n \to \infty} \frac{\sum_{i=1}^{n}(y_i - m_i)}{\sqrt{2A_n \ln \ln A_n}} = 1$$

in Protocol 5.7, where $A_n := \sum_{i=1}^{n} v_i$.

Protocol 5.7 requires that Skeptic choose $W_n \geq 0$; we can say that w_n is only an upper price for $|y_n - m_n|^\alpha$. We will use the condition $W_n \geq 0$ in the next paragraph, but it is not needed for Theorem 5.8. If we drop it, as Miyabe and Takemura do in [284], Skeptic will nevertheless choose $W_n \geq 0$ in order to avoid the bankruptcy that Reality could impose by choosing y_n very large in absolute value.

Theorem 5.8 implies the slightly weakened version of Theorem 5.5 in which the second condition

$$c_n = o\left(\sqrt{\frac{A_n}{\ln \ln A_n}}\right)$$

in (5.20), equivalent to

$$c_n^\alpha \left(\frac{\ln \ln A_n}{A_n}\right)^{\alpha/2} = o(1),$$

is replaced by

$$\sum_n c_n^\alpha \left(\frac{\ln \ln A_n}{A_n}\right)^{\alpha/2} < \infty.$$

Protocol 5.7 allows only values in the interval $(2, 3]$ for its parameter α. The value $\alpha = 2$ is excluded, as it would be useless: we already have $(y_n - m_n)^2$ as a hedge. But we can make α as close to 2 as we wish. Miyabe and Takemura show that we can get even a little closer to 2 than this by replacing $|y_n - m_n|^\alpha$ in the protocol with $h(y_n - m_n)$ for $h(x) := (1 + |x|)^2 \ln^2(1 + |x|) - x^2$, so that $h(x) < |x|^\alpha$ for any $\alpha > 2$ and sufficiently large $|x|$ [284, Example 2.5].

Takazawa [374, section 4] considers a simplified protocol, Protocol 5.9, in which the two hedges $(y_n - m_n)^2$ and $|y_n - m_n|^\alpha$ are replaced by just one hedge, $h(y_n - m_n)$ for $h(x) := e^{x^2/2} - 1$. His result is given as Theorem 5.10; it only establishes validity.

Protocol 5.9
PARAMETER: hedge h
    Skeptic announces $\mathcal{K}_0 \in \mathbb{R}$.
    FOR $n = 1, 2, \ldots$:
        Forecaster announces $m_n \in \mathbb{R}$ and $w_n \geq 0$.
        Skeptic announces $M_n \in \mathbb{R}$ and $W_n \geq 0$.
        Reality announces $y_n \in \mathbb{R}$.
        $\mathcal{K}_n := \mathcal{K}_{n-1} + M_n(y_n - m_n) + W_n(h(y_n - m_n) - w_n)$.

Theorem 5.10   *Skeptic can force*

$$B_n \to \infty \implies \limsup_{n\to\infty} \frac{\sum_{i=1}^{n}(y_i - m_i)}{\sqrt{2B_n \ln \ln B_n}} = 1$$

in Protocol 5.9 with $h(y) := e^{y^2/2} - 1$, *where* $B_n := 2\sum_{i=1}^{n} w_i$.

As our argument in previous sections shows, there can be no sharpness result for Takazawa's protocol 5.9 (Reality can set $y_n := m_n$ with impunity), but we can ask whether the constant 2 in the definition of B_n in Theorem 5.10 is optimal. A simple measure-theoretic argument shows that it is: see Exercise 5.5.

To conclude, we state a very precise version of the law of the iterated logarithm due to Sasai et al. [327] (with [326] as predecessor). Their very simple protocol is stated here as Protocol 5.11; without loss of generality, it could be simplified further by setting $m_n := 0$.

Protocol 5.11
> Skeptic announces $\mathcal{K}_0 \in \mathbb{R}$.
> FOR $n = 1, 2, \ldots$:
>> Forecaster announces $m_n \in \mathbb{R}$ and $c_n \geq 0$.
>> Skeptic announces $M_n \in \mathbb{R}$.
>> Reality announces $y_n \in [m_n - c_n, m_n + c_n]$.
>> $\mathcal{K}_n := \mathcal{K}_{n-1} + M_n(y_n - m_n)$.

We do not ask Forecaster to price the accuracy of m_n as a forecast of y_n. Instead, we set $A_n := \sum_{i=1}^{n}(y_i - m_i)^2$, measuring the empirical accuracy of m_n. It is shown in [327] that

$$\left(A_n \to \infty \text{ and } c_n = O\left(\sqrt{\frac{A_n}{(\ln \ln A_n)^3}}\right)\right) \implies \limsup_{n\to\infty} \frac{\sum_{i=1}^{n}(y_i - m_i)}{\sqrt{2A_n \ln \ln A_n}} = 1, \quad (5.21)$$

a weakening of (5.20), holds almost surely in Protocol 5.11. The cube is essential (see [134]). The statement (5.21) is an example of a *self-normalizing* law of the iterated logarithm. It normalizes the sum $\sum_{i=1}^{n}(y_i - m_i)$ using a function of Reality's moves rather than a function of Forecaster's moves.

Sasai et al. make the denominator in the consequent of (5.21) much more precise, in the spirit of Ivan Petrovsky's version of the law of the iterated logarithm (see Section 5.6). Let us say that a positive increasing function ψ is of *upper class* (in Protocol 5.11) if the consequent of (5.21) can be replaced by

$$\sum_{i=1}^{n}(y_i - m_i) \leq \sqrt{A_n}\psi(A_n) \text{ from some } n \text{ on.} \quad (5.22)$$

Similarly, such a function ψ is of *lower class* if the consequent of (5.21) can be replaced by

$$\sum_{i=1}^{n}(y_i - m_i) > \sqrt{A_n}\psi(A_n) \text{ for infinitely many } n. \tag{5.23}$$

Sasai et al. show that ψ is of upper class if and only if the integral

$$\int_1^\infty \psi(\lambda)\exp(-\psi(\lambda)^2/2)\frac{d\lambda}{\lambda} \tag{5.24}$$

is finite; and if the integral is infinite, ψ is of lower class. An example of a function of upper class is

$$\psi(\lambda) := \sqrt{2}(\ln \ln \lambda)^{-1/2}\left(\ln \ln \lambda + \frac{3}{4}\ln_3\lambda + \frac{1}{2}\ln_4\lambda\right.$$
$$\left. +\cdots+ \frac{1}{2}\ln_{k-1}\lambda + \left(\frac{1}{2}+\varepsilon\right)\ln_k\lambda\right)$$

for $\varepsilon > 0$ and $k \geq 4$, where $\ln_i$ stands for $\ln\cdots\ln$ (i times); the function ψ belongs to the lower class if $\varepsilon \leq 0$ [129].

A Law of the Iterated Logarithm for Finance

A fascinating one-sided law of the iterated logarithm was established by Sato et al. [328]. They study the following protocol.

Protocol 5.12
  $\mathcal{K}_0 := 1.$
  FOR $n = 1, 2, \ldots$:
    Skeptic announces $M_n \in \mathbb{R}$.
    Reality announces $y_n \in [-1, \infty)$.
    $\mathcal{K}_n := \mathcal{K}_{n-1} + M_n y_n.$

As usual, Skeptic must keep his capital nonnegative. So he always chooses $M_n \geq 0$; otherwise, Reality bankrupts him by choosing sufficiently large y_n. We take y_n to be the return of a financial security, or a portfolio of securities, over the nth round; the condition $y_n \in [-1, \infty)$ then says that the price of the security, or the value of the portfolio, cannot become negative. Skeptic starts with the initial capital 1, and M_n is the position he takes in the security at the beginning of the nth round; $\mathcal{K}_n$ is his capital at the end of the nth round.

Set $S_n := \sum_{i=1}^{n} y_i$. Let b_n be an increasing sequence of positive numbers such that $b_n \to \infty$. Sato et al. [328, Proposition 6.1] show that:

- If $\sum_n 1/b_n = \infty$, Skeptic can force $S_n/b_n \to 0$.
- If $\sum_n 1/b_n < \infty$, Reality can force $S_n/b_n \nrightarrow 0$.

The corresponding strategies for Skeptic and Reality are simple and explicit. This result looks very different from the law of the iterated logarithm, and this motivates considering self-normalizing results.

Set $A_n := \sum_{i=1}^{n} y_i^2$. The functions of upper and lower class are defined as before, by (5.22) and (5.23) (with m_i ignored), respectively. It is still true that a positive increasing function ψ is in the upper class if and only if the integral (5.24) is infinite [328, Corollary 5.5].

As we will see in Part IV (see the discussion at its beginning), interpreting results in Protocol 5.12, including the two we have just stated, is not completely straightforward. One possibility is to assume that y_n are returns of a financial market index, such as S&P 500, that is assumed to be an efficient numeraire in the sense explained in Chapter 16.

5.4 CONNECTIONS WITH LARGE DEVIATION INEQUALITIES

In measure-theoretic probability, the law of the iterated logarithm is usually deduced from large deviation inequalities. The corresponding game-theoretic approach would use large deviation supermartingales such as Doléans's, Hoeffding's, and Bernstein's, introduced in Chapter 3. But to obtain Theorem 5.1 (validity for Kolmogorov's law of the iterated logarithm), we used instead a specially crafted large-deviation supermartingale (5.6).

Proceeding in exactly the same way but using Hoeffding's supermartingale (3.7), we obtain a direct proof of Corollary 5.4. Indeed, for A_n as defined in Corollary 5.4 and T_n Hoeffding's supermartingale

$$T_n^{\delta,\kappa} = T_n^{\kappa} := \prod_{i=1}^{n} \exp\left(\kappa y_i - \frac{\kappa^2}{2} c_i^2 \right),$$

(5.10) with the same values of k and $\kappa_{\delta,n}$ gives (5.17).

Analogously, using Doléan's supermartingale (3.3) will give a self-normalizing law of the iterated logarithm. The constants, however, will not be optimal when specialized to the coin tossing.

5.5 EXERCISES

Exercise 5.1 As we discussed when introducing it, Protocol 5.2 strengthens Protocol 4.1 in two ways. Show that it is useless to allow Skeptic to make V_n negative without giving him information about y_n before he makes his moves M_n and V_n. Namely, consider Protocol 4.1 with "$V_n \geq 0$" replaced by "$V_n \in \mathbb{R}$" and show that any strategy ψ for Skeptic ensuring $\mathcal{K}^\psi \geq 0$ satisfies $\psi^V \geq 0$. *Hint.* See the proof of Statements 2 and 3 in Proposition 5.6.   □

Exercise 5.2 Consider the following binary version of Protocol 4.1.

> Skeptic announces $\mathcal{K}_0 \in \mathbb{R}$.
> FOR $n = 1, 2, \ldots$:
>   Forecaster announces $m_n \in \mathbb{R}$ and $v_n \geq 0$.
>   Skeptic announces $M_n \in \mathbb{R}$.
>   Reality announces $y_n \in \{m_n - \sqrt{v_n}, m_n + \sqrt{v_n}\}$.
> $\mathcal{K}_n := \mathcal{K}_{n-1} + M_n(y_n - m_n)$.

Notice that this protocol is a special case of Protocol 4.1 in which Reality is constrained to choosing y_n from the two-element set $\{m_n - \sqrt{v_n}, m_n + \sqrt{v_n}\}$, which makes the addend $V_n((y_n - m_n)^2 - v_n)$ in (4.3) equal to 0 and, therefore, Skeptic's choice of V_n irrelevant.

1. Show that

$$\exp\left(\kappa \sum_{i=1}^{n} y_i - \frac{1}{2}\kappa^2 A_n \right) \tag{5.25}$$

   is a supermartingale in this protocol (cf. footnote 1 in Section 5.1).
2. Replacing the supermartingale $\mathcal{T}^{\delta,\kappa}$ (see (5.6)) by (5.25), obtain

$$\kappa \sum_{i=1}^{n} y_i - \frac{1}{2}\kappa^2 A_n \leq (1 + \delta)\ln\ln A_n + O(1)$$

   in place of (5.12).
3. Proceeding as in the rest of the proof of Theorem 5.1, show that Skeptic can force

$$A_n \to \infty \Longrightarrow \limsup_{n\to\infty} \frac{\sum\limits_{i=1}^{n}(y_i - m_i)}{\sqrt{2A_n \ln\ln A_n}} \leq 1$$

(Marcinkiewicz, [436, pp. 27–28]).   □

Exercise 5.3 Consider the following testing protocol, in which, intuitively, Forecaster outputs m_n such that Reality's move y_n is symmetrically distributed around m_n, in the sense that the expected value of $f(y_n - m_n)$ is zero for any function f that is *odd* in the sense that $f(x) + f(-x) = 0$ for any $x \in \mathbb{R}$.

> Skeptic announces $\mathcal{K}_0 \in \mathbb{R}$.
> FOR $n = 1, 2, \ldots$:
>   Forecaster announces $m_n \in \mathbb{R}$.
>   Skeptic announces odd $f_n : \mathbb{R} \to \mathbb{R}$.
>   Reality announces $y_n \in \mathbb{R}$.
> $\mathcal{K}_n := \mathcal{K}_{n-1} + f_n(y_n - m_n)$.

(Notice that Skeptic does not need the permission to bet on $y_n - m_n$, since he can do it anyway using the oddness of the function $x \mapsto Mx$.) Using the result of Exercise 5.2 (or otherwise), show that Skeptic can force

$$A_n \to \infty \Longrightarrow \limsup_{n\to\infty} \frac{\sum_{i=1}^{n}(y_i - m_i)}{\sqrt{2A_n \ln \ln A_n}} \le 1,$$

where $A_n := \sum_{i=1}^{n} y_i^2$ (in measure-theoretic probability, this self-normalizing law of the iterated logarithm was obtained by Marcinkiewicz [436, p. 28]).   □

Exercise 5.4 Prove the game-theoretic form of the law of the iterated logarithm for binary outcomes and a constant probability: Skeptic can force

$$\limsup_{n\to\infty} \frac{\sum_{i=1}^{n} y_i - np}{\sqrt{2p(1 - p)n \ln \ln n}} = 1 \tag{5.26}$$

in Protocol 1.10.   □

Exercise 5.5 Show that constant 2 in the definitions of B_n in Theorem 5.10 is optimal (cannot be replaced by a smaller one). *Hint.* Notice that

$$h(y) = e^{y^2/2} - 1 \sim y^2/2 \text{ as } y \to 0,$$

and so for small y_n, the role of w_n is similar to that of $v_n/2$ in Protocol 5.2. Apply the measure-theoretic law of the iterated logarithm to the strategy $m_n := 0$ for Forecaster and the randomized strategy for Reality $y_n := \pm\varepsilon$, where ε is a small positive constant and the sign of y_n is chosen randomly. Connections between game-theoretic and measure-theoretic probability will be explored further in Chapter 9.   □

5.6 CONTEXT

The Development of Khinchin's Result

In his 1909 exposition of his law of large numbers [44], Borel gave an upper bound on the rate of convergence. This bound was far from sharp; it was of order $\sqrt{n} \ln n$ rather than $\sqrt{n \ln \ln n}$. He was not trying to give the tightest possible bound, and his method of proof – approximating the probability of his bound being violated for each n and then concluding that this would happen for only finitely many n because the sum of the probabilities is finite – readily yields tighter bounds.

In a two-page account of Borel's law in the chapter on measure theory at the end of his first book on set theory, in 1914 [183], Felix Hausdorff gave a different (and more complete) proof than Borel but announced a bound on the rate of convergence, $n^{1/2+\varepsilon}$ for any $\varepsilon > 0$, that is actually weaker than Borel's bound. Perhaps because of the shortcomings of Borel's proof, Hausdorff's contribution has been seen as an important step on the way from Borel to Khinchin. Feller [135, p. 196] states that Khinchin's law gave an answer to a problem "treated in a series of papers initiated by" Hausdorff and Hardy and Littlewood. Yuan-Shih Chow and Henry Teicher [78, p. 368] state that Kolmogorov's version of the law "was the culmination of a series of strides by mathematicians of the caliber of Hausdorff, Hardy, Littlewood, and Khintchine." Similar statements can be found in [134, 166, 237].

G.H. Hardy and John Littlewood touched on the rate of convergence in Borel's law in a lengthy article on number theory in 1914 [179, p. 190]. Their method was essentially the same as Borel's, but they found the tightest asymptotic bound the method will yield: with probability 1, the deviation of $\sum_{i=1}^{n} y_i$ from zero is, in their words, "not of order exceeding $\sqrt{n \ln n}$." They also made a start, as neither Borel nor Hausdorff had done, on investigating how large a deviation can be counted on; they showed that with probability 1, "the deviation, in both directions, is sometimes of order exceeding $\sqrt{n}$."

The big step after Borel was Khinchin's alone. Borel, Hausdorff, and Hardy and Littlewood had all followed the same strategy: they estimated the probability of a large deviation for $\sum_{i=1}^{n} y_i$ separately for each n and then summed these probabilities. This is crude because the deviations are highly dependent for adjacent n. Khinchin estimated the probability of at least one deviation within increasingly wide ranges of n. This much more difficult calculation, which Khinchin made in 1922 and published in 1924 [217], yielded an asymptotic bound on the rate convergence in Borel's law that was sharp as well as valid. Like his predecessors, Khinchin considered independent tosses of a coin with any probability p for heads; he showed that (5.26) will hold almost surely, where $\sum_{i=1}^{n} y_i$ is the number of heads.

Subsequent Contributions by Kolmogorov and Others

In 1929 [222], Kolmogorov generalized Khinchin's law from coin tossing to independent random variables $y_1, y_2, \ldots$ with means $m_1, m_2, \ldots$ and variances $v_1, v_2, \ldots$; he showed that (5.1), with $A_n := \sum_{i=1}^{n} v_i$, holds almost surely in the measure-theoretic sense if A_n tends to infinity and $y_n - m_n$ stays within bounds that do not grow too fast in absolute value. In 1970 [370], William Stout generalized Kolmogorov's result to the martingale setting, where the m_n and v_n are conditional means and variances relative to a filtration to which $y_1, y_2, \ldots$ is adapted. In Section 9.3, we explain a method by which Stout's theorem can be derived from our game-theoretic result.

Kolmogorov's condition

$$|y_n - m_n| = o\left(\sqrt{\frac{A_n}{\ln \ln A_n}}\right)$$

(see, e.g. (5.3)) was shown to be optimal for sharpness by Jósef Marcinkiewicz and Antoni Zygmund in 1937 [262, Lemma 10]. Zygmund later cited a remark by Marcinkiewicz that if the y_i are known to take only two possible values, a_i or $-a_i$, then Kolmogorov's condition is not needed for validity ([435, p. 38], [436]). A game-theoretic version of Marcinkiewicz's remark is the subject of Exercise 5.2.

In 1941 [181], Philip Hartman and Avi Wintner obtained another law of the iterated logarithm for independent identically distributed random variables. Although they used Kolmogorov's theorem as the starting point of their proof, their result does not follow simply from Kolmogorov's theorem or from Stout's. Nor does it have a simple game-theoretic counterpart. A game could no doubt be constructed on the basis of their proof, but it would price so many functions of the y_n that it would hold little interest. In general, we consider a game-theoretic generalization of a measure-theoretic result interesting only if it manages to reduce significantly the number of prices that are assumed.

Petrovsky's Definitive Form

The definitive form of the law of the iterated logarithm for coin tossing in terms of upper and lower classes was obtained by Ivan Petrovsky in 1935 [304] as a by-product of his study of the heat equation. Kolmogorov described Petrovsky's result and the main ideas behind its proof in (at least) two letters to Lévy [22, pp. 119–120], who included it in his book [245, p. 266] with a credit to Kolmogorov but without mentioning Petrovsky. Petrovsky briefly mentioned but did not emphasize connections with the law of the iterated logarithm (referring and expressing his gratitude for useful advice to Khinchin rather than to Kolmogorov). More explicit proofs were published by Ville in 1939 [387] (the validity part) and Paul Erdős in 1942 [129] (the sharpness part); Feller's [134] research leading to stronger versions of Petrovsky's law of the iterated logarithm was independent of Erdős's [129]. As we already mentioned, Ville's proof was very game-theoretic in spirit.

Sasai et al. [326, 327] call this family of results the Erdős–Feller–Kolmogorov–Petrowsky law of the iterated logarithm, but our attribution to Petrovsky alone in Section 5.3 appears to be justified from a historical point of view.

Game-Theoretic Versions

Section 5.1's proof of validity extends the proof in [102], which was adapted from [390]. The basic idea goes back to Ville's 1939 book [387]. The proof of sharpness first appeared in [349].

Kolmogorov's Nonasymptotic Formulation

The number n must be cosmically large in order for $\ln \ln n$ to be moderately large. Perhaps for this reason, the law of the iterated logarithm is usually stated in asymptotic terms and treated as a purely theoretical result, not of practical interest. The law is equivalent, however, to the existence of values of N such that certain bounds on probabilities hold for all n greater than N, and so we can restate it without using the notion of a limit. Kolmogorov gave such a nonasymptotic formulation in 1929 [222]. For details in the case of coin tossing, where the game-theoretic and measure-theoretic formulations are identical, see [349, section 5.8].

Part II

Abstract Theory in Discrete Time

In Part I's protocols, the players' move spaces were fully specified. In the abstract theory that we now study, they are not. Just as measure-theoretic probability studies classes of probability spaces satisfying various conditions, we study protocols in which the move spaces depend on parameters whose values are not specified. Concrete protocols, in which the players' move spaces are fully specified, are instantiations of these abstract protocols.

To make the abstract theory as simple as possible, we extend the choices available to Skeptic in Part I's concrete protocols. When Skeptic is offered a particular payoff (by Forecaster or by the protocol), we assume that he is also offered any smaller payoff, as in the slackenings we considered in Section 1.4.

In Chapter 6, we consider axioms for the offers to Skeptic on a single round. We express these axioms in two ways: as conditions on the set of payoffs offered to Skeptic (we call this set an *offer*) and as conditions on prices at which Skeptic can buy any payoff he wants (we call the pricing functional an *upper expectation*).

Given the flexibility of game theory and the historical richness of probability theory, we do not expect to capture all the possibilities for game-theoretic probability in one axiomatic system. Systems both weaker and stronger than the one developed in Chapter 6 may be of interest. This system does, however, cover nearly all the discrete-time examples considered in this book, and it yields a surprising range of results often thought to require measure-theoretic probability's stronger assumptions. The axiom that might be most naturally omitted is the continuity axiom, which

---

Game-Theoretic Foundations for Probability and Finance, First Edition. Glenn Shafer and Vladimir Vovk.
© 2019 John Wiley & Sons, Inc. Published 2019 by John Wiley & Sons, Inc.

implies countable subadditivity for upper expectations. The principal results of our abstract theory hold even when the continuity axiom is not assumed (see Section 6.5).

In Chapter 7, we consider discrete-time protocols in which the offers on each round satisfy Chapter 6's axioms, and we define upper and lower expected values of global functions of the moves by Forecaster and Reality. We prove Doob's convergence theorem and state Lindeberg's theorem in this context.

The main protocol considered in Chapter 7 is very simple: the same offer is made to Skeptic on every round, and hence Forecaster is absent. It turns out, however, that most other protocols that interest us can be embedded in this simple protocol, because the player who determines the outcomes can play the role of Forecaster as well as the role of Reality.

In Chapter 8, we derive a game-theoretic version of Lévy's zero-one law, one of the most fundamental results of abstract discrete-time probability theory. In Lévy's original formulation, this law says that when a random variable Z depends on a sequence $Z_1, Z_2, \ldots$ of random variables, its expected value given $Z_1, \ldots, Z_n$ converges almost surely toward its actual value as $n \to \infty$. When the random variable is the indicator variable for an event, this actual value is either 0 or 1, hence the name. Using Lévy's zero-one law, we show that the functional that gives global upper expected values satisfies the same axioms that are satisfied by the local upper expectations on each round.

In Chapter 9, we look at several ways in which game-theoretic and measure-theoretic probability are related. Assuming that the outcome space on individual rounds is finite and that the set of upper expectations available to Forecaster is parameterized by a compact metrizable space, we show that global game-theoretic upper probabilities and upper expected values, which are infima over supermartingales, are also suprema over probability measures on the canonical measurable space. In the case of noncanonical probability spaces, we show that standard measure-theoretic results can be obtained from the corresponding game-theoretic results by translating game-theoretic martingales into measure-theoretic martingales.

The abstract theory developed in these chapters permits infinite payoffs to Skeptic, even on a single round. This level of generality is mathematically convenient, and it is even helpful as an idealization in some of the applications that we consider in Part III. It requires, however, that we occasionally pay attention to the conventions of arithmetic in the extended real numbers: $\infty + (-\infty) = \infty - \infty = \infty$ and $0 \times \infty = 0 \times (-\infty) = 0$ (see the section on terminology and notation at the end of the book).

6

Betting on a Single Outcome

In the abstract testing protocols that we study in this part of the book, Reality chooses her move y from a set $\mathcal{Y}$, and Skeptic chooses a payoff that depends on y. The payoff is allowed to take infinite as well as finite values; it is a function from $\mathcal{Y}$ to the extended real numbers $\overline{\mathbb{R}} := [-\infty, \infty]$. We call any function from $\mathcal{Y}$ to $\overline{\mathbb{R}}$ an *extended variable*.

In this chapter, we study axioms for Skeptic's betting opportunities on each round. We consider two equivalent ways of specifying these opportunities:

1. Offer Skeptic certain extended variables for free. We call the set of extended variables offered, say $\mathbf{G}$, an *offer*. If Skeptic chooses $g \in \mathbf{G}$, he receives $g(y)$.

2. Set a price $\overline{\mathbf{E}}(f)$ for each extended variable f. If Skeptic chooses the extended variable f, he pays $\overline{\mathbf{E}}(f)$ and receives $f(y)$ in return. We call the functional $\overline{\mathbf{E}}$ an *upper expectation*.

We emphasize upper expectations, because they generalize the concept of expectation familiar from classical and measure-theoretic probability. On the other hand, the concept of an offer relates more directly to Part I's protocols.

In Section 6.1, we state axioms that characterize upper expectations and derive additional properties from these axioms. We also study the dual concept of lower expectation. We already encountered these functionals (or rather their values, which we call *upper and lower expected values*) in Part I, beginning in Section 2.1.

Game-Theoretic Foundations for Probability and Finance, First Edition. Glenn Shafer and Vladimir Vovk.
© 2019 John Wiley & Sons, Inc. Published 2019 by John Wiley & Sons, Inc.

Table 6.1 Notation and terminology for upper expectations.

Chapter 6 Abstract	Chapters 7 and 8 Local	Global
Outcome space $\mathcal{Y}$	Reality's move space $\mathcal{Y}$	Sample space Ω
Outcome $y \in \mathcal{Y}$	Reality's move $y \in \mathcal{Y}$	Path $\omega \in \Omega$
Event $E \subseteq \mathcal{Y}$	Local event $E \subseteq \mathcal{Y}$	Global event $E \subseteq \Omega$
Upper probability $\overline{\mathbf{P}}(E)$	Upper probability $\overline{\mathbf{P}}(E)$	Upper probability $\overline{\mathbb{P}}(E)$
Extended variable $f : \mathcal{Y} \to \overline{\mathbb{R}}$	Local variable $f : \mathcal{Y} \to \overline{\mathbb{R}}$	Global variable $X : \Omega \to \overline{\mathbb{R}}$
Upper expected value $\overline{\mathbf{E}}(f)$	Upper expected value $\overline{\mathbf{E}}(f)$	Upper expected value $\overline{\mathbb{E}}(X)$
Upper expectation $\overline{\mathbf{E}}$	Local upper expectation $\overline{\mathbf{E}}$	Global upper expectation $\overline{\mathbb{E}}$

Note the distinction between $\overline{\mathbf{E}}$, which we use in the general theory of upper expectations and for local upper expectations on individual rounds, and $\overline{\mathbb{E}}$, which we use for the global upper expectation. The local variables and global variables considered in Chapters 7 and 8 are extended variables (see Section 6.7).

In Section 6.2, we study upper and lower probabilities, which we also encountered in Part I. We relate the game-theoretic concept of *almost sure*, defined in Part I without reference to upper or lower probability, to the concept of a *null event*, an event that has upper probability 0.

In Section 6.3, we consider upper expectations with smaller domains. We call a functional $\overline{\mathbf{E}} : D^{\mathcal{Y}} \to \overline{\mathbb{R}}$, where $D \subseteq \overline{\mathbb{R}}$, a *D-upper expectation* if it is the restriction to $D^{\mathcal{Y}}$ of an upper expectation on $\mathcal{Y}$. We axiomatize D-upper expectations for $D = \mathbb{R}$ and $D = [0, \infty]$. The first case is relevant to examples, such as those in Part I, where the payoffs offered to Skeptic do not take infinite values. The second case is helpful when we want to incorporate into the protocol a prohibition on Skeptic risking bankruptcy.

In Section 6.4, we state axioms for offers and define a one-to-one correspondence between offers and upper expectations such that when $\mathbf{G}$ and $\overline{\mathbf{E}}$ are corresponding offer and upper expectation, $g \in \mathbf{G}$ if and only if $\overline{\mathbf{E}}(g) \leq 0$. This gives us two equivalent ways of stating testing protocols.

Section 6.5 focuses on the least essential of our axioms for upper expectations, the continuity axiom. We did not use it in Part I, and it is not essential for our abstract theory, but it holds in most applications and is mathematically convenient, simplifying our terminology and sometimes our reasoning. By assuming continuity for mathematical convenience, we are following the example of Kolmogorov and other twentieth-century probabilists (see the further discussion in Section 6.7).

In Chapters 7 and 8, we study abstract testing protocols in which the offers made to Skeptic on each round satisfy our axioms. As we will see in Section 8.2, the upper expected values of functions on the sample space then also satisfy these axioms. So whereas we consider offers and upper expectations abstractly in this chapter, we consider them locally and globally in Chapters 7 and 8. These three points of view produce some diversity in terminology and notation, summarized in Table 6.1.

6.1 UPPER AND LOWER EXPECTATIONS

Given a nonempty set $\mathcal{Y}$, we call a functional $\overline{\mathbf{E}} : \overline{\mathbb{R}}^{\mathcal{Y}} \to \overline{\mathbb{R}}$ an *upper expectation on* $\mathcal{Y}$ if it satisfies these five axioms:

Axiom E1. If $f_1, f_2 \in \overline{\mathbb{R}}^{\mathcal{Y}}$, then $\overline{\mathbf{E}}(f_1 + f_2) \leq \overline{\mathbf{E}}(f_1) + \overline{\mathbf{E}}(f_2)$.
Axiom E2. If $f \in \overline{\mathbb{R}}^{\mathcal{Y}}$ and $c \in (0, \infty)$, then $\overline{\mathbf{E}}(cf) = c\overline{\mathbf{E}}(f)$.
Axiom E3. If $f_1, f_2 \in \overline{\mathbb{R}}^{\mathcal{Y}}$ and $f_1 \leq f_2$, then $\overline{\mathbf{E}}(f_1) \leq \overline{\mathbf{E}}(f_2)$.
Axiom E4. For each $c \in \mathbb{R}$, $\overline{\mathbf{E}}(c) = c$.
Axiom E5. If $f_1 \leq f_2 \leq \cdots \in [0, \infty]^{\mathcal{Y}}$, then $\overline{\mathbf{E}}(\lim_{k \to \infty} f_k) = \lim_{k \to \infty} \overline{\mathbf{E}}(f_k)$.

We call Axiom E5 the *continuity axiom*. We call $\overline{\mathbf{E}}(f)$ f's *upper expected value*.

Example 6.1 **(supremum)** If $\mathcal{Y}$ is a nonempty set, then the functional $\overline{\mathbf{E}} : \overline{\mathbb{R}}^{\mathcal{Y}} \to \overline{\mathbb{R}}$ given by

$$\overline{\mathbf{E}}(f) := \sup_{y \in \mathcal{Y}} f(y)$$

is an upper expectation on $\mathcal{Y}$. ☐

Example 6.2 **(upper integral)** Suppose $(\mathcal{Y}, \mathcal{F}, P)$ is a probability space (see the section on terminology and notation at the end of the book for a definition of probability space). Let $\mathcal{E}$ be the set of extended random variables f for which the expected value $P(f)$ exists. For each $f \in \overline{\mathbb{R}}^{\mathcal{Y}}$, set

$$\overline{\mathbf{E}}(f) := \inf\{P(g)|g \in \mathcal{E} \text{ and } g \geq f\}. \tag{6.1}$$

When $f \in \mathcal{E}$, $\overline{\mathbf{E}}(f) = P(f)$. In general, $\overline{\mathbf{E}}(f)$ is f's *upper integral* [40, 358]. As the reader may verify, $\overline{\mathbf{E}}$ is an upper expectation on $\mathcal{Y}$ (Exercise 6.12). ☐

Example 6.3 (Protocol 1.1) For each $m \in [-1, 1]$, define $\overline{\mathbf{E}}^m : \overline{\mathbb{R}}^{[-1,1]} \to \overline{\mathbb{R}}$ by

$$\overline{\mathbf{E}}^m(f) := \inf\{\alpha \in \mathbb{R}|\exists M \in \mathbb{R} \; \forall y \in [-1, 1] : f(y) \leq M(y - m) + \alpha\}. \tag{6.2}$$

In particular, $\overline{\mathbf{E}}^{-1}(f) := f(-1)$ and $\overline{\mathbf{E}}^{1}(f) := f(1)$. Each of these functionals is an upper expectation on $[-1, 1]$. In Protocol 1.1, Skeptic chooses $M \in \mathbb{R}$ and receives $M(y - m)$. Chapter 1's results, which state that Skeptic can achieve certain goals in the protocol, remain true if we instead allow him to choose any $f \in \overline{\mathbb{R}}^{[-1,1]}$ such that $\overline{\mathbf{E}}^{m}(f) \leq 0$ and receive $f(y)$ (see Exercise 6.13). □

Proposition 6.4 *Suppose* $\overline{\mathbf{E}}$ *is an upper expectation on* $\mathcal{Y}$. *Then the following statements hold.*

1. *Translation. If* $f \in \overline{\mathbb{R}}^{\mathcal{Y}}$ *and* $c \in \mathbb{R}$, *then*

$$\overline{\mathbf{E}}(f + c) = \overline{\mathbf{E}}(f) + c. \tag{6.3}$$

2. *Bounds. If* $f \in \overline{\mathbb{R}}^{\mathcal{Y}}$, *then*

$$\inf f \leq \overline{\mathbf{E}}(f) \leq \sup f. \tag{6.4}$$

3. *Countable subadditivity. If* $f_1, f_2, \ldots \in [0, \infty]^{\mathcal{Y}}$, *then*

$$\overline{\mathbf{E}}\left(\sum_{k=1}^{\infty} f_k\right) \leq \sum_{k=1}^{\infty} \overline{\mathbf{E}}(f_k).$$

4. *Strong coherence. If* $f \in \overline{\mathbb{R}}^{\mathcal{Y}}$, *then*

$$f > 0 \Longrightarrow \overline{\mathbf{E}}(f) > 0. \tag{6.5}$$

5. *Generalization of Axiom E5. If* $f_1 \leq f_2 \leq \cdots \in \overline{\mathbb{R}}^{\mathcal{Y}}$, *and* f_1 *is bounded below,* *then* $\overline{\mathbf{E}}(\lim_{k\to\infty} f_k) = \lim_{k\to\infty} \overline{\mathbf{E}}(f_k)$.

Proof
1. To see that (6.3) holds, note that

$$\overline{\mathbf{E}}(f + c) \overset{\mathrm{E1}}{\leq} \overline{\mathbf{E}}(f) + \overline{\mathbf{E}}(c) \overset{\mathrm{E4}}{=} \overline{\mathbf{E}}(f) + c$$

and

$$\overline{\mathbf{E}}(f) \overset{\mathrm{E1}}{\leq} \overline{\mathbf{E}}(f + c) + \overline{\mathbf{E}}(-c) \overset{\mathrm{E4}}{=} \overline{\mathbf{E}}(f + c) - c.$$

(Above each relation we have named the axiom that justifies it.)
2. The relation (6.4) follows from Axioms E3 and E4.

3. Applying Axiom E5 to the variables $h_1 \leq h_2 \leq \cdots$, where $h_k := \sum_{i=1}^{k} f_i$, and using Axiom E1, we obtain

$$\overline{\mathbf{E}}\left(\sum_{i=1}^{\infty} f_i\right) = \lim_{k \to \infty} \overline{\mathbf{E}}\left(\sum_{i=1}^{k} f_i\right) \leq \lim_{k \to \infty} \sum_{i=1}^{k} \overline{\mathbf{E}}(f_i) = \sum_{i=1}^{\infty} \overline{\mathbf{E}}(f_i).$$

4. We prove the contrapositive of (6.5). Suppose $\overline{\mathbf{E}}(f) = 0$. Then $\overline{\mathbf{E}}(\sum_{k=1}^{\infty} f) \leq 0$ by countable subadditivity. Since $\overline{\mathbf{E}}(\infty) = \infty$ by Axioms E3 and E4, this implies that $\sum_{k=1}^{\infty} f$ is not identically equal to ∞ and hence that $f > 0$ is false.

5. This follows directly from Axiom E5 and Statement 1 of this proposition.

$\square$

In our protocols, Skeptic is presented with an upper expectation $\overline{\mathbf{E}}$ on $\mathcal{Y}$ and allowed to buy any $f \in \overline{\mathbb{R}}^{\mathcal{Y}}$ at the price $\overline{\mathbf{E}}(f)$. Because buying f for $\overline{\mathbf{E}}(f)$ is the same as selling $-f$ for $-\overline{\mathbf{E}}(f)$, we can equivalently say that Skeptic is allowed to sell any $f \in \overline{\mathbb{R}}^{\mathcal{Y}}$ at the price $-\overline{\mathbf{E}}(-f)$. With this in mind, we associate with $\overline{\mathbf{E}}$ the functional $\underline{\mathbf{E}}$ given by $\underline{\mathbf{E}}(f) := -\overline{\mathbf{E}}(-f)$ for all $f \in \overline{\mathbb{R}}^{\mathcal{Y}}$. We call $\underline{\mathbf{E}}$ the *lower expectation* corresponding to $\overline{\mathbf{E}}$, and we call $\underline{\mathbf{E}}(f)$ the *lower expected value* of f.

Proposition 6.5 *If $\overline{\mathbf{E}}$ is an upper expectation on $\mathcal{Y}$, then $\underline{\mathbf{E}}(f) \leq \overline{\mathbf{E}}(f)$ for all $f \in \overline{\mathbb{R}}^{\mathcal{Y}}$.*

Proof By Axioms E1, E3, and E4,

$$\overline{\mathbf{E}}(f) + \overline{\mathbf{E}}(-f) \overset{E1}{\geq} \overline{\mathbf{E}}(f - f) \overset{E3}{\geq} \overline{\mathbf{E}}(0) \overset{E4}{=} 0,$$

and hence $\underline{\mathbf{E}}(f) \leq \overline{\mathbf{E}}(f)$. (If $f(y)$ is ∞ or $-\infty$, then $(f - f)(y) = \infty$, and so in general we can say only that $f - f \geq 0$, not that $f - f = 0$.) $\square$

When $\underline{\mathbf{E}}(f)$ and $\overline{\mathbf{E}}(f)$ are equal, we call their common value f's (game-theoretic) *expected value*, and we write $\mathbf{E}(f)$ for this common value.

6.2 UPPER AND LOWER PROBABILITIES

We call a subset E of $\mathcal{Y}$ an *event*. When working with an upper expectation $\overline{\mathbf{E}}$ on $\mathcal{Y}$, we write $\overline{\mathbf{P}}(E)$ for $\overline{\mathbf{E}}(1_E)$. We call $\overline{\mathbf{P}}(E)$ the *upper probability* of E. Similarly, we write $\underline{\mathbf{P}}(E)$ for $\underline{\mathbf{E}}(1_E)$ and call it the *lower probability* of E.

It follows immediately from the properties of upper and lower expectation that

$$0 \leq \underline{\mathbf{P}}(E) \leq \overline{\mathbf{P}}(E) \leq 1.$$

When $\underline{P}(E) = \overline{P}(E)$, we designate their common value by $P(E)$, call $P(E)$ the (game-theoretic) *probability* of E, and say that E is *probabilized*. When $\underline{P}(E) < \overline{P}(E)$, we say that E is *unprobabilized*. When $\underline{P}(E) = 0$ and $\overline{P}(E) = 1$, we say that E is *fully unprobabilized*.

Exercise 6.7 lists some of the properties of upper and lower probabilities that follow from Axioms E1–E5. Notice in particular the property of countable subadditivity:

$$\overline{P}\left(\bigcup_{k=1}^{\infty} E_k\right) \le \sum_{k=1}^{\infty} \overline{P}(E_k) \qquad (6.6)$$

for any events $E_1, E_2, \ldots$.

In measure-theoretic probability, the probability measure, which gives probabilities for events, is a fundamental object. The expected values of random variables, being Lebesgue integrals with respect to the probability measure, carry no additional information, and hence the expectation operator is often considered less fundamental than the probability measure. In contrast, an upper expectation may not be fully determined by its upper and lower probabilities and hence must be considered more fundamental (see Exercise 6.8). For this reason, we give axioms for upper expectations but not for upper probabilities.

Markov's Inequality

A small value of $\overline{P}(E)$ has the same significance as we gave to small upper probabilities in Chapter 2: Skeptic can bet in such a way that what he risks is multiplied by the large factor $1/\overline{P}(E)$ if E happens. If Skeptic does bet in this way, and E does happen, then Skeptic has grounds to reject the hypothesis represented by the prices, and the smallness of $\overline{P}(E)$ measures the strength of these grounds.

The following proposition helps us compare this testing interpretation of small upper probabilities with the more general testing interpretation of upper expected values.

Proposition 6.6 (**Markov's inequality**) *Suppose $\overline{E}$ is an upper expectation on $\mathcal{Y}$, $f \in \overline{\mathbb{R}}^{\mathcal{Y}}, f \ge 0, \overline{E}(f) > 0$, and $K \in (0, \infty)$. Then*

$$\overline{P}(f \ge K\overline{E}(f)) \le 1/K.$$

Proof By Axioms E2 and E3,

$$\overline{P}(f \ge K\overline{E}(f)) = \overline{E}(1_{f \ge K\overline{E}(f)}) \le \overline{E}\left(\frac{f}{K\overline{E}(f)}\right) = \frac{1}{K}.$$

□

Consider two ways Skeptic can test the upper expectation $\overline{\mathbf{E}}$:

1. He selects a nonnegative $f \in \overline{\mathbb{R}}^{\mathcal{Y}}$ such that $\overline{\mathbf{E}}(f) > 0$ and a large positive number K. He pays $\overline{\mathbf{P}}(f \geq K\overline{\mathbf{E}}(f))$ for the indicator variable $\mathbf{1}_{f \geq K\overline{\mathbf{E}}(f)}$. If $f \geq K\overline{\mathbf{E}}(f)$ happens, he has multiplied the capital he risked by the factor $1/\overline{\mathbf{P}}(f \geq K\overline{\mathbf{E}}(f))$. By Markov's inequality, this is at least K.

2. He selects a nonnegative $f \in \overline{\mathbb{R}}^{\mathcal{Y}}$ such that $\overline{\mathbf{E}}(f) > 0$. He pays $\overline{\mathbf{E}}(f)$ for f. Whatever the outcome y is, he multiplies the capital he risked by $f(y)/\overline{\mathbf{E}}(f)$.

The second method has the advantage that K need not be selected in advance. Skeptic can wait to see how large the ratio $f(y)/\overline{\mathbf{E}}(f)$ is and then claim the full force of this ratio, however large it is, as evidence against $\overline{\mathbf{E}}$.

Almost Sure and Null Events with Respect to an Upper Expectation

We say that a subset E of $\mathcal{Y}$ is *almost sure* or happens *almost surely* with respect to an upper expectation $\overline{\mathbf{E}}$ on $\mathcal{Y}$ if there exists $f \in [0, \infty]^{\mathcal{Y}}$ such that $\overline{\mathbf{E}}(f) = 1$ and $f(y) = \infty$ for all $y \in E^{\mathrm{c}}$. In Chapter 8, we will learn how this concept relates to Part I's concept of almost sure with respect to a protocol (see Proposition 8.4).

We say that E is *null* with respect to $\overline{\mathbf{E}}$ if $\overline{\mathbf{P}}(E) = 0$.

Proposition 6.7 *Suppose $\overline{\mathbf{E}}$ is an upper expectation on $\mathcal{Y}$ and $E \subseteq \mathcal{Y}$.*

1. *E is almost sure with respect to $\overline{\mathbf{E}}$ if and only if E^{c} is null with respect to $\overline{\mathbf{E}}$.*
2. *E is almost sure with respect to $\overline{\mathbf{E}}$ if and only if there exists $f \in [0, \infty]^{\mathcal{Y}}$ such that $\overline{\mathbf{E}}(f) = 0$ and $f(y) = \infty$ for all $y \in E^{\mathrm{c}}$.*

Proof Both statements are established by the following argument.

1. Suppose E is almost sure with respect to $\overline{\mathbf{E}}$. Suppose $\overline{\mathbf{E}}(f) = 1$ and $f(y) = \infty$ for all $y \in E^{\mathrm{c}}$. For every $\varepsilon > 0$, εf satisfies $\overline{\mathbf{E}}(\varepsilon f) = \varepsilon$ and $\varepsilon f(y) = \infty$ for all $y \in E^{\mathrm{c}}$. Since $\mathbf{1}_{E^{\mathrm{c}}} \leq \varepsilon f$, $\overline{\mathbf{P}}(E^{\mathrm{c}}) \leq \varepsilon$. This being true for all $\varepsilon > 0$, it follows that $\overline{\mathbf{P}}(E^{\mathrm{c}}) = 0$. So E^{c} is null with respect to $\overline{\mathbf{E}}$.

2. Suppose E^{c} is null with respect to $\overline{\mathbf{E}}$. By Axiom E2, $\overline{\mathbf{E}}(k\mathbf{1}_{E^{\mathrm{c}}}) = 0$ for all $k \in \mathbb{N}$, and so by Axiom E5, $\overline{\mathbf{E}}(\infty\mathbf{1}_{E^{\mathrm{c}}}) = 0$. Setting $f := \infty\mathbf{1}_{E^{\mathrm{c}}}$ and $f' := f + 1$, we see that $\overline{\mathbf{E}}(f) = 0$, $\overline{\mathbf{E}}(f') = 1$, and $f(y)$ and $f'(y)$ are both infinite for all $y \in E^{\mathrm{c}}$. □

By the countable subadditivity of upper probability, (6.6), the union of a countable number of null events is null. Hence the intersection of a countable number of almost sure events is almost sure.

6.3   UPPER EXPECTATIONS WITH SMALLER DOMAINS

Given a nonempty set $\mathcal{Y}$ and nonempty $D \subseteq \overline{\mathbb{R}}$, we call a functional $\overline{\mathbf{E}} : D^{\mathcal{Y}} \to \overline{\mathbb{R}}$ a *D-upper expectation on* $\mathcal{Y}$ if there exists an upper expectation $\overline{\mathbf{E}}^*$ on $\mathcal{Y}$ such that $\overline{\mathbf{E}} = \overline{\mathbf{E}}^* |_{D^{\mathcal{Y}}}$. We are particularly interested in $\mathbb{R}$-upper expectations and $[0, \infty]$-upper expectations. The examples in Part I can be understood in terms of $\mathbb{R}$-upper expectations, because they involve only finite payoffs to Skeptic. In Part III, we will sometimes build the requirement that Skeptic avoid bankruptcy into our protocols by using $[0, \infty]$-upper expectations. The two following propositions show how $\mathbb{R}$-upper expectations and $[0, \infty]$-upper expectations can be axiomatized directly.

Proposition 6.8   *A functional* $\overline{\mathbf{E}} : \mathbb{R}^{\mathcal{Y}} \to \overline{\mathbb{R}}$, *where* $\mathcal{Y}$ *is a nonempty set, is an* $\mathbb{R}$-*upper expectation on* $\mathcal{Y}$ *if and only if it satisfies the following axioms:*

Axiom E1$^{\mathbb{R}}$.   *If* $f_1, f_2 \in \mathbb{R}^{\mathcal{Y}}$, *then* $\overline{\mathbf{E}}(f_1 + f_2) \le \overline{\mathbf{E}}(f_1) + \overline{\mathbf{E}}(f_2)$.
Axiom E2$^{\mathbb{R}}$.   *If* $f \in \mathbb{R}^{\mathcal{Y}}$ *and* $c \in (0, \infty)$, *then* $\overline{\mathbf{E}}(cf) = c\overline{\mathbf{E}}(f)$.
Axiom E3$^{\mathbb{R}}$.   *If* $f_1, f_2 \in \mathbb{R}^{\mathcal{Y}}$ *and* $f_1 \le f_2$, *then* $\overline{\mathbf{E}}(f_1) \le \overline{\mathbf{E}}(f_2)$.
Axiom E4$^{\mathbb{R}}$.   *For each* $c \in \mathbb{R}$, $\overline{\mathbf{E}}(c) = c$.
Axiom E5$^{\mathbb{R}}$.   *If* $f_1 \le f_2 \le \cdots$ *and* $\lim_{k \to \infty} f_k$ *are in* $[0, \infty)^{\mathcal{Y}}$, *then*

$$\overline{\mathbf{E}}(\lim_{k \to \infty} f_k) = \lim_{k \to \infty} \overline{\mathbf{E}}(f_k).$$

Proof   It follows immediately from Axioms E1–E5 that an $\mathbb{R}$-upper expectation on $\mathcal{Y}$ satisfies Axioms E1$^{\mathbb{R}}$–E5$^{\mathbb{R}}$.

If a functional $\overline{\mathbf{E}} : \mathbb{R}^{\mathcal{Y}} \to \overline{\mathbb{R}}$, satisfies Axioms E1$^{\mathbb{R}}$–E5$^{\mathbb{R}}$, then we can define an upper expectation $\overline{\mathbf{E}}^*$ on $\mathcal{Y}$ such that $\overline{\mathbf{E}} = \overline{\mathbf{E}}^* |_{\mathbb{R}^{\mathcal{Y}}}$ in two steps. First

$$\overline{\mathbf{E}}'(f) := \sup\{\overline{\mathbf{E}}(h) | h \in \mathbb{R}^{\mathcal{Y}} \text{ and } h \le f\}$$

for all $f \in (-\infty, \infty]^{\mathcal{Y}}$, and then

$$\overline{\mathbf{E}}^*(f) := \inf\{\overline{\mathbf{E}}'(h) | h \in (-\infty, \infty]^{\mathcal{Y}} \text{ and } h \ge f\}$$

for all $f \in \overline{\mathbb{R}}^{\mathcal{Y}}$.

It is evident that $\overline{\mathbf{E}}^* |_{\mathbb{R}^{\mathcal{Y}}} = \overline{\mathbf{E}}$. Let us check that $\overline{\mathbf{E}}^*$ satisfies Axiom E1 and Axiom E5, leaving it to the reader to the check the other three axioms.

To check Axiom E1 we first show that $\overline{\mathbf{E}}'(f_1 + f_2) \le \overline{\mathbf{E}}'(f_1) + \overline{\mathbf{E}}'(f_2)$ when $f_1, f_2 \in (-\infty, \infty]^{\mathcal{Y}}$. Choose $h \in \mathbb{R}^{\mathcal{Y}}$ such that $h \le f_1 + f_2$; we must show that $\overline{\mathbf{E}}(h) \le \overline{\mathbf{E}}'(f_1) + \overline{\mathbf{E}}'(f_2)$. By the axiom of choice, we can define $h_1, h_2 \in \mathbb{R}^{\mathcal{Y}}$ such that $h =$

$h_1 + h_2, h_1 \leq f_1$, and $h_2 \leq f_2$: namely, for each $y \in \mathcal{Y}$, we choose $z_1, z_2 \in \mathbb{R}$ such that $h(y) = z_1 + z_2, z_1 \leq f_1(y)$, and $z_2 \leq f_2(y)$, and we set $h_1(y) := z_1$ and $h_2(y) := z_2$. So

$$\overline{\mathbf{E}}(h) \leq \overline{\mathbf{E}}(h_1) + \overline{\mathbf{E}}(h_2) \leq \overline{\mathbf{E}}'(f_1) + \overline{\mathbf{E}}'(f_2).$$

Now suppose $f_1, f_2 \in \overline{\mathbb{R}}^{\mathcal{Y}}$. For an arbitrarily small $\varepsilon > 0$, choose $h_1, h_2 \in (-\infty, \infty]^{\mathcal{Y}}$ such that $h_1 \geq f_1, h_2 \geq f_2, \overline{\mathbf{E}}'(h_1) \leq \overline{\mathbf{E}}^*(f_1) + \varepsilon$, and $\overline{\mathbf{E}}'(h_2) \leq \overline{\mathbf{E}}^*(f_2) + \varepsilon$. We then have

$$\overline{\mathbf{E}}^*(f_1 + f_2) \leq \overline{\mathbf{E}}'(h_1 + h_2) \leq \overline{\mathbf{E}}'(h_1) + \overline{\mathbf{E}}'(h_2) \leq \overline{\mathbf{E}}^*(f_1) + \overline{\mathbf{E}}^*(f_2) + 2\varepsilon.$$

Axiom E5 concerns $f_1 \leq f_2 \leq \cdots \in [0, \infty]^{\mathcal{Y}}$, for which there is no difference between $\overline{\mathbf{E}}^*$ and $\overline{\mathbf{E}}'$. We need to show that $\overline{\mathbf{E}}'(\lim_k f_k) \leq \lim_k \overline{\mathbf{E}}'(f_k)$. It suffices to show that $\overline{\mathbf{E}}(h) \leq \lim_k \overline{\mathbf{E}}'(f_k)$ for $h \in \mathbb{R}^{\mathcal{Y}}$ satisfying $h \leq \lim_k f_k$. Fix such an h and assume, without loss of generality, that $h \geq 0$. Since $(f_k \wedge h) \uparrow h$, we have

$$\overline{\mathbf{E}}(h) = \lim_k \overline{\mathbf{E}}(f_k \wedge h) \leq \lim_k \overline{\mathbf{E}}'(f_k).$$

$\square$

Proposition 6.9 *A functional* $\overline{\mathbf{E}} : [0, \infty]^{\mathcal{Y}} \to \overline{\mathbb{R}}$, *where* $\mathcal{Y}$ *is a nonempty set, is a* $[0, \infty]$-*upper expectation on* $\mathcal{Y}$ *if and only if it satisfies the following axioms:*

Axiom E1$^{[0,\infty]}$. *If* $f_1, f_2 \in [0, \infty]^{\mathcal{Y}}$, *then* $\overline{\mathbf{E}}(f_1 + f_2) \leq \overline{\mathbf{E}}(f_1) + \overline{\mathbf{E}}(f_2)$.
Axiom E2$^{[0,\infty]}$. *If* $f \in [0, \infty]^{\mathcal{Y}}$ *and* $c \in (0, \infty)$, *then* $\overline{\mathbf{E}}(cf) = c\overline{\mathbf{E}}(f)$.
Axiom E3$^{[0,\infty]}$. *If* $f_1, f_2 \in [0, \infty]^{\mathcal{Y}}$ *and* $f_1 \leq f_2$, *then* $\overline{\mathbf{E}}(f_1) \leq \overline{\mathbf{E}}(f_2)$.
Axiom E4$^{[0,\infty]}$. *For each* $c \in [0, \infty)$, $\overline{\mathbf{E}}(c) = c$.
Axiom E5$^{[0,\infty]}$. *If* $f_1 \leq f_2 \leq \cdots$ *are in* $[0, \infty]^{\mathcal{Y}}$, *then*

$$\overline{\mathbf{E}}\left(\lim_{k \to \infty} f_k\right) = \lim_{k \to \infty} \overline{\mathbf{E}}(f_k).$$

Axiom E6$^{[0,\infty]}$. *If* $f \in [0, \infty]^{\mathcal{Y}}$ *and* $c \in (0, \infty)$, *then* $\overline{\mathbf{E}}(f + c) = \overline{\mathbf{E}}(f) + c$.

Proof It follows immediately from Axioms E1–E5 and Statement 1 of Proposition 6.4 that a $[0, \infty]$-upper expectation on $\mathcal{Y}$ satisfies Axioms E1$^{[0,\infty]}$–E6$^{[0,\infty]}$.

To show that a functional $\overline{\mathbf{E}} : [0, \infty]^{\mathcal{Y}} \to \overline{\mathbb{R}}$ that satisfies Axioms E1$^{[0,\infty]}$–E6$^{[0,\infty]}$ is a $[0, \infty]$-upper expectation on $\mathcal{Y}$, it suffices to show that we can define an upper expectation on $\mathcal{Y}$, say $\overline{\mathbf{E}}^*$, such that $\overline{\mathbf{E}} = \overline{\mathbf{E}}^*|_{[0,\infty]^{\mathcal{Y}}}$. We can do this in two steps. First, set

$$\overline{\mathbf{E}}'(f) := \overline{\mathbf{E}}(f - \inf f) + \inf f, \tag{6.7}$$

for all $f \in \overline{\mathbb{R}}^{y}$ that are bounded below, and then set

$$\overline{\mathbf{E}}^{*}(f) := \lim_{a \to -\infty} \overline{\mathbf{E}}'(\max(f, a)) \tag{6.8}$$

for all $f \in \overline{\mathbb{R}}^{y}$. Notice that, by Axiom E6$^{[0,\infty]}$, we can replace $\inf f$ by any $a < \inf f$ in the definition (6.7).

Using Axiom E6$^{[0,\infty]}$, we see that $\overline{\mathbf{E}} = \overline{\mathbf{E}}^{*}|_{[0,\infty]^{y}}$.

Now we check that $\overline{\mathbf{E}}^{*}$ satisfies Axioms E1–E5.

- Axiom E5 follows from $\overline{\mathbf{E}} = \overline{\mathbf{E}}^{*}|_{[0,\infty]^{y}}$, as the variables referenced by this axiom are in $[0, \infty]^{y}$.
- Axiom E4 follows immediately from the definitions.
- To check Axiom E3, first consider variables $f_1 \le f_2 \in \overline{\mathbb{R}}^{y}$ that are bounded below. Using Axiom E3$^{[0,\infty]}$, we obtain

$$\overline{\mathbf{E}}'(f_1) = \overline{\mathbf{E}}(f_1 - \inf f_1) + \inf f_1 \le \overline{\mathbf{E}}(f_2 - \inf f_1) + \inf f_1 = \overline{\mathbf{E}}'(f_2).$$

  We then see from (6.8) that Axiom E3 holds for $\overline{\mathbf{E}}^{*}$.
- Axiom E2 follows from the definitions and Axiom E2$^{[0,\infty]}$: when $c \in (0, \infty)$,

$$\overline{\mathbf{E}}^{*}(cf) = \lim_{a \to -\infty} \overline{\mathbf{E}}'(\max(cf, a)) = \lim_{a \to -\infty} \overline{\mathbf{E}}'(\max(cf, ca))$$

$$= \lim_{a \to -\infty} (\overline{\mathbf{E}}(\max(cf, ca) - ca) \mid ca)$$

$$= c \lim_{a \to -\infty} (\overline{\mathbf{E}}(\max(f, a) - a) + a)$$

$$= c \lim_{a \to -\infty} \overline{\mathbf{E}}'(\max(f, a)) = c\overline{\mathbf{E}}^{*}(f).$$

- To check Axiom E1, we use Axiom E1$^{[0,\infty]}$ to obtain

$$\overline{\mathbf{E}}'(f_1 + f_2) = \overline{\mathbf{E}}(f_1 + f_2 - \inf f_1 - \inf f_2) + \inf f_1 + \inf f_2$$

$$\le \overline{\mathbf{E}}(f_1 - \inf f_1) + \overline{\mathbf{E}}(f_2 - \inf f_2) + \inf f_1 + \inf f_2$$

$$= \overline{\mathbf{E}}'(f_1) + \overline{\mathbf{E}}'(f_2)$$

for bounded-below $f_1, f_2 \in \overline{\mathbb{R}}^{y}$. So if $f_1, f_2 \in \overline{\mathbb{R}}^{y}$, then

$$\overline{\mathbf{E}}^{*}(f_1 + f_2) = \lim_{a \to -\infty} \overline{\mathbf{E}}'(\max(f_1 + f_2, a)) = \lim_{a \to -\infty} \overline{\mathbf{E}}'(\max(f_1 + f_2, 2a))$$

$$\le \lim_{a \to -\infty} \overline{\mathbf{E}}'(\max(f_1, a) + \max(f_2, a))$$

$$\leq \lim_{a \to -\infty} \left(\overline{\mathbf{E}}'(\max(f_1, a)) + \overline{\mathbf{E}}'(\max(f_2, a)) \right)$$

$$= \lim_{a \to -\infty} \overline{\mathbf{E}}'(\max(f_1, a)) + \lim_{a \to -\infty} \overline{\mathbf{E}}'(\max(f_2, a))$$

$$= \overline{\mathbf{E}}^*(f_1) + \overline{\mathbf{E}}^*(f_2).$$

$\square$

6.4   OFFERS

Suppose $\mathcal{Y}$ is a nonempty set. Let us call a nonempty subset **G** of $\overline{\mathbb{R}}^{\mathcal{Y}}$ an *offer* on $\mathcal{Y}$ if it satisfies these six axioms:

Axiom G0. If $g \in \overline{\mathbb{R}}^{\mathcal{Y}}$ and $g - \varepsilon \in \mathbf{G}$ for every $\varepsilon \in (0, \infty)$, then $g \in \mathbf{G}$.

Axiom G1. If $g_1, g_2 \in \mathbf{G}$, then $g_1 + g_2 \in \mathbf{G}$.

Axiom G2. If $c \in [0, \infty)$ and $g \in \mathbf{G}$, then $cg \in \mathbf{G}$.

Axiom G3. If $g_1 \in \overline{\mathbb{R}}^{\mathcal{Y}}$, $g_2 \in \mathbf{G}$, and $g_1 \leq g_2$, then $g_1 \in \mathbf{G}$.

Axiom G4. If $g \in \mathbf{G}$, then $\inf g \leq 0$.

Axiom G5. If $g_1 \leq g_2 \leq \cdots$ are all in **G**, and g_1 is bounded below, then $\lim_{k \to \infty} g_k \in \mathbf{G}$.

Because **G** is nonempty, Axiom G2 with $c = 0$ implies $0 \in \mathbf{G}$. Were we to replace $[0, \infty)$ with $(0, \infty)$ in Axiom G2, the axioms would not guarantee $0 \in \mathbf{G}$. Instead, they would be satisfied by the one-element set containing the constant $-\infty$.

Axioms G1–G4 are a natural mathematical representation of the picture in which Skeptic is offered a collection of gambles with the option of taking as many and as much as he wants. Axiom G1 says Skeptic can combine offered gambles. Axiom G2 adds that he can take any multiple and any fraction of an offered gamble. Axiom G3 says Skeptic can accept less than an offered gamble pays, even turning its gain or loss into an infinite loss. Axiom G4 is a relatively weak version of the principle called *coherence* (in Bayesian statistics) or *no arbitrage* (in finance theory). It implies that there is no positive number α such that Skeptic is offered α for sure.

Axioms G0 and G5 can be thought of as regularity conditions; they are needed to make the notion of an offer equivalent to the notion of an upper expectation. For g that are bounded below, Axiom G0 obviously follows from Axiom G5 (see also Exercise 6.10).

Equivalence Between Offers and Upper Expectations

The following proposition describes the natural one-to-one correspondence between offers on $\mathcal{Y}$ and upper expectations on $\mathcal{Y}$.

Proposition 6.10

1. *If* **G** *is an offer on* $\mathcal{Y}$, *then the functional* $\overline{\mathbf{E}}_{\mathbf{G}} : \overline{\mathbb{R}}^{\mathcal{Y}} \to \overline{\mathbb{R}}$ *given by*

$$\overline{\mathbf{E}}_{\mathbf{G}}(f) := \inf\{\alpha \in \mathbb{R} | f - \alpha \in \mathbf{G}\} \tag{6.9}$$

   is an upper expectation on $\mathcal{Y}$.

2. *If* $\overline{\mathbf{E}}$ *is an upper expectation on* $\mathcal{Y}$, *then the set of variables* $\mathbf{G}_{\overline{\mathbf{E}}}$ *given by*

$$\mathbf{G}_{\overline{\mathbf{E}}} := \{g \in \overline{\mathbb{R}}^{\mathcal{Y}} | \overline{\mathbf{E}}(g) \le 0\} \tag{6.10}$$

   is an offer on $\mathcal{Y}$.

3. *The relations (6.9) and (6.10) define a bijection:* $\overline{\mathbf{E}} = \overline{\mathbf{E}}_{\mathbf{G}}$ *if and only if* $\mathbf{G} = \mathbf{G}_{\overline{\mathbf{E}}}$.

Proof

1. Let us show that $\overline{\mathbf{E}}_{\mathbf{G}}$ is an upper expectation when **G** is an offer. Axiom E1 follows from Axiom G1: if $f_1 - \alpha_1 \in \mathbf{G}$ and $f_2 - \alpha_2 \in \mathbf{G}$, then $(f_1 + f_2) - (\alpha_1 + \alpha_2) \in \mathbf{G}$ by Axiom G1, and hence

$$\inf\{\alpha \in \mathbb{R} | (f_1 + f_2) - \alpha \in \mathbf{G}\} \le \inf\{\alpha_1 \in \mathbb{R} | f_1 - \alpha_1 \in \mathbf{G}\} +$$
$$\inf\{\alpha_2 \in \mathbb{R} | f_2 - \alpha_2 \in \mathbf{G}\}.$$

   Axiom E2 follows from Axiom G2: since $cf - \alpha \in \mathbf{G}$ if and only if $f - \alpha/c \in \mathbf{G}$ by Axiom G2,

$$\overline{\mathbf{E}}_{\mathbf{G}}(cf) = \inf\{\alpha \in \mathbb{R} | f - \alpha/c \in \mathbf{G}\}$$
$$= c \inf\{\beta \in \mathbb{R} | f - \beta \in \mathbf{G}\} = c\overline{\mathbf{E}}_{\mathbf{G}}(f).$$

   We obtain Axiom E3 immediately from Axiom G3, which guarantees that if $f_1 \le f_2$ and $f_2 - \alpha \in \mathbf{G}$, then $f_1 - \alpha \in \mathbf{G}$. We obtain Axiom E4 from Axiom G4 and the fact that $0 \in \mathbf{G}$. To show that $\overline{\mathbf{E}}_{\mathbf{G}}$ satisfies Axiom E5, suppose $f_1 \le f_2 \le \cdots \in [0, \infty]^{\mathcal{Y}}$. By Statement 2 of Proposition 6.4 (which does not depend on Axiom E3), $\overline{\mathbf{E}}_{\mathbf{G}}(f_k) \ge 0$ for all k. By Axiom E3, $\overline{\mathbf{E}}_{\mathbf{G}}(\lim_{k \to \infty} f_k) \ge \lim_{k \to \infty} \overline{\mathbf{E}}_{\mathbf{G}}(f_k)$. So we only need to show that

$$\overline{\mathbf{E}}_{\mathbf{G}}\left(\lim_{k \to \infty} f_k\right) \le \lim_{k \to \infty} \overline{\mathbf{E}}_{\mathbf{G}}(f_k). \tag{6.11}$$

   Set $f := \lim_{k \to \infty} f_k$ and $c := \lim_{k \to \infty} \overline{\mathbf{E}}_{\mathbf{G}}(f_k)$. If $c = \infty$, then (6.11) certainly holds. Otherwise, $0 \le c < \infty$, and for all k, $\overline{\mathbf{E}}_{\mathbf{G}}(f_k) \le c$, whence

$$\overline{\mathbf{E}}_{\mathbf{G}}(f_k - c) \le \overline{\mathbf{E}}_{\mathbf{G}}(f_k - \overline{\mathbf{E}}_{\mathbf{G}}(f_k)) = 0$$

   and so $f_k - c \in \mathbf{G}$. Because $f_1 - c \le f_2 - c \le \cdots$ and $f_1 - c$ is bounded below, it follows from Axiom G5 that $f - c \in \mathbf{G}$. Thus $\overline{\mathbf{E}}_{\mathbf{G}}(f - c) \le 0$ or $\overline{\mathbf{E}}_{\mathbf{G}}(f) \le c$.

2. Now let us show that $\mathbf{G}_{\overline{\mathbf{E}}}$ is an offer when $\overline{\mathbf{E}}$ is an upper expectation. Axiom G0 follows from (6.10) and (6.3). Axiom G1 follows from Axiom E1, Axiom G2 for $c > 0$ from Axiom E2, Axiom G2 for $c = 0$ from Axiom E4, and Axiom G3 from Axiom E3. We obtain Axiom G4 from Axioms E3 and E4: if $\inf g > 0$, then $\overline{\mathbf{E}}(\inf g) > 0$ by Axiom E4, and hence $\overline{\mathbf{E}}(g) > 0$ by Axiom E3, so that $g \notin \mathbf{G}_{\overline{\mathbf{E}}}$. We obtain Axiom G5 from Axiom E5: if g is the limit of the sequence $g_1 \leq g_2 \leq \cdots \in \mathbf{G}_{\overline{\mathbf{E}}}$, then because $\overline{\mathbf{E}}(g_k) \leq 0$ for $k \in \mathbb{N}$, Statement 5 of Proposition 6.4 implies that $\overline{\mathbf{E}}(g) \leq 0$ and hence that $g \in \mathbf{G}_{\overline{\mathbf{E}}}$.

3a. Suppose $\overline{\mathbf{E}} = \overline{\mathbf{E}}_{\mathbf{G}}$. Let us show that $\mathbf{G} = \mathbf{G}_{\overline{\mathbf{E}}}$, or

$$\mathbf{G} = \{g \in \overline{\mathbb{R}}^{\mathcal{Y}} \mid \inf\{\alpha \in \mathbb{R} \mid g - \alpha \in \mathbf{G}\} \leq 0\}.$$

Clearly $g \in \mathbf{G}$ implies $\inf\{\alpha \in \mathbb{R} \mid g - \alpha \in \mathbf{G}\} \leq 0$. If $\inf\{\alpha \in \mathbb{R} \mid g - \alpha \in \mathbf{G}\} \leq 0$, on the other hand, Axioms G0 and G3 imply that $g \in \mathbf{G}$.

3b. Suppose $\mathbf{G} = \mathbf{G}_{\overline{\mathbf{E}}}$. Let us show that $\overline{\mathbf{E}} = \overline{\mathbf{E}}_{\mathbf{G}}$, or

$$\overline{\mathbf{E}}(f) = \inf\{\alpha \in \mathbb{R} \mid \overline{\mathbf{E}}(f - \alpha) \leq 0\}.$$

This reduces to $\overline{\mathbf{E}}(f) = \inf\{\alpha \in \mathbb{R} \mid \overline{\mathbf{E}}(f) \leq \alpha\}$ by (6.3).

□

When $\overline{\mathbf{E}}$ and $\mathbf{G}$ are corresponding offer and upper expectation, we refer to $\overline{\mathbf{E}}$ as $\mathbf{G}$'s *upper expectation* and to $\mathbf{G}$ as $\overline{\mathbf{E}}$'s *offer*. The upper expectations for the offers in Examples 6.1–6.3 are discussed in Exercises 6.11–6.13.

Here are two further insights gained by considering the relation between an offer $\mathbf{G}$ and its upper expectation $\overline{\mathbf{E}}$.

- By Axiom G4, $\inf g > 0$ implies $g \notin \mathbf{G}$. By (6.5), which depends on the continuity axiom, we can make the stronger assertion that $g > 0$ implies $g \notin \mathbf{G}$.
- When $\mathbf{G}$ is offered to Skeptic and $f - \alpha \in \mathbf{G}$, α being a real number, we have said that Skeptic can buy f for α. If $\overline{\mathbf{E}}(f)$ is not $-\infty$, then $f - \overline{\mathbf{E}}(f) \in \mathbf{G}$ by (6.10) and (6.3), and hence the infimum in (6.9) is attained by $\alpha = \overline{\mathbf{E}}(f)$. So when $\overline{\mathbf{E}}(f)$ is not $-\infty$, it is the lowest price at which Skeptic can buy f.

A version of Proposition 6.10 also holds for $\mathbb{R}$-upper expectations; the $\mathbb{R}$-offer corresponding to an $\mathbb{R}$-upper expectation $\overline{\mathbf{E}}$ on $\mathcal{Y}$ is of course $\{g \in \mathbb{R}^{\mathcal{Y}} \mid \overline{\mathbf{E}}(g) \leq 0\}$ (Exercise 6.17).

Abstract Testing Protocols

The notions of upper expectation and offer allow us to formulate the following abstract testing protocols.

Protocol 6.11

PARAMETER: Nonempty set $\mathcal{Y}$
    Skeptic announces $\mathcal{K}_0 \in \overline{\mathbb{R}}$.
    FOR $n = 1, 2, \ldots$:
        Forecaster announces an offer $\mathbf{G}_n$ on $\mathcal{Y}$.
        Skeptic announces $g_n \in \mathbf{G}_n$.
        Reality announces $y_n \in \mathcal{Y}$.
        $\mathcal{K}_n := \mathcal{K}_{n-1} + g_n(y_n)$.

Protocol 6.12

PARAMETER: Nonempty set $\mathcal{Y}$
    Skeptic announces $\mathcal{K}_0 \in \overline{\mathbb{R}}$.
    FOR $n = 1, 2, \ldots$:
        Forecaster announces an upper expectation $\overline{\mathbf{E}}_n$ on $\mathcal{Y}$.
        Skeptic announces $f_n \in \overline{\mathbb{R}}^{\mathcal{Y}}$ such that $\overline{\mathbf{E}}_n(f_n) \leq \mathcal{K}_{n-1}$.
        Reality announces $y_n \in \mathcal{Y}$.
        $\mathcal{K}_n := f_n(y_n)$.

These two protocols give Skeptic the same betting opportunities and hence afford him the same possible capital processes. Indeed, if $\mathbf{G}_n$ is an offer, $\overline{\mathbf{E}}_n$ is $\mathbf{G}_n$'s upper expectation, and $\mathcal{K}_{n-1}$ is Skeptic's current capital, then offering him the new capital

$$\mathcal{K}_{n-1} + g(y_n) \text{ for any } g \in \mathbf{G}_n$$

is the same as offering him the new capital

$$\mathcal{K}_{n-1} + g(y_n) \text{ for any } g \in \overline{\mathbb{R}}^{\mathcal{Y}} \text{ such that } \overline{\mathbf{E}}_n(g) \leq 0$$

or the new capital

$$\mathcal{K}_{n-1} + g(y_n) \text{ for any } g \in \overline{\mathbb{R}}^{\mathcal{Y}} \text{ such that } \overline{\mathbf{E}}_n(\mathcal{K}_{n-1} + g) \leq \mathcal{K}_{n-1}$$

or the new capital

$$f(y_n) \text{ for any } f \in \overline{\mathbb{R}}^{\mathcal{Y}} \text{ such that } \overline{\mathbf{E}}_n(f) \leq \mathcal{K}_{n-1}.$$

The inequality in Protocol 6.12 allows Skeptic to waste money. Though sometimes mathematically or computationally convenient, such wastefulness is never conceptually necessary. If $f_n \in \overline{\mathbb{R}}^{\mathcal{Y}}$ and $\overline{\mathbf{E}}(f_n) < \mathcal{K}_{n-1}$, then Skeptic can add the constant $\mathcal{K}_{n-1} - \overline{\mathbf{E}}(f_n)$ to the value for $\mathcal{K}_n$ he obtains by playing f_n by playing $h_n := f_n + \mathcal{K}_{n-1} - \overline{\mathbf{E}}(f_n)$ instead. When he plays h_n instead of f_n, we say that he is *playing f_n while conserving capital*. More generally, we say that Skeptic is *conserving capital* whenever his actual move f_n satisfies $\overline{\mathbf{E}}(f_n) = \mathcal{K}_{n-1}$. One way he can conserve capital is to choose f_n equal to the constant $\mathcal{K}_{n-1}$. In this case, we say that he is *not betting*.

We call Protocols 6.11 and 6.12 *abstract* because they have unspecified parameters. Protocols that have no parameters, so that the players' move spaces are all fully specified, are called *concrete*. A concrete protocol is an *instantiation* of an abstract protocol when it results from giving specific values to the parameters in the abstract protocol. When we constrain the values of the parameters in an abstract protocol without necessarily specifying them fully, or when we otherwise constrain Skeptic's opponents, perhaps in a way that depends on the outcomes of previous rounds, we say that we have *specialized* the protocol. This is consistent with the notions of instantiation and specialization for concrete testing protocols introduced in Part I (Sections 1.2 and 2.2).

The concrete testing protocols that we studied in Part I, in their slackened form and sometimes with minor adjustments, are obtained from specializations of Protocol 6.11 that are further modified by constraining Skeptic to choose only real-valued g_n that do not take the values ∞ or $-\infty$ (see Exercises 6.13 and 6.14). In effect, these protocols use $\mathbb{R}$-offers rather than offers. By Proposition 6.8, $\mathbb{R}$-offers can be enlarged to offers. The enlargement can only strengthen Skeptic by giving him more choices, and so it does not affect Part I's demonstrations that Skeptic can achieve certain goals. So Part I's results can be considered part of the theory of upper expectations.

In the remainder of this book, we use upper expectations rather than offers, and our testing protocols are specializations and variations on Protocol 6.12. Skeptic sometimes chooses from $\mathbb{R}^{\mathcal{Y}}$ or $[0, \infty]^{\mathcal{Y}}$ rather than from $\overline{\mathbb{R}}^{\mathcal{Y}}$. Since upper expectations on these domains can always be enlarged to upper expectations on $\overline{\mathbb{R}}^{\mathcal{Y}}$, we nevertheless say simply that Forecaster announces an upper expectation.

In Part III, we will frequently use testing protocols in which Forecaster announces a probability measure P_n on the nth round and Skeptic is allowed to choose an extended variable f_n such that $P_n(f_n)$, f_n's expected value with respect to P_n, does not exceed $\mathcal{K}_{n-1}$. By Example 6.2, the slackening of such a protocol is equivalent to a specialization of Protocol 6.12 in which P_n's upper integral is used as the upper expectation.

6.5 DROPPING THE CONTINUITY AXIOM

The axiom of continuity slightly simplifies the theory of upper expectations and makes it more easily accessible for readers accustomed to measure-theoretic probability. But we did not use this axiom in Part I, and our most important abstract discrete-time results do not depend on it. Moreover, there are interesting theories, including probabilistic number theory and the theory of imprecise probabilities, that use functionals that satisfy Axioms E1–E4 without satisfying the continuity axiom.

Given a nonempty set $\mathcal{Y}$, let us call a functional $\overline{\mathbf{E}} : \overline{\mathbb{R}}^{\mathcal{Y}} \to \overline{\mathbb{R}}$ a *broad-sense upper expectation* on $\mathcal{Y}$ if it satisfies Axioms E1–E4, and let us call a nonempty subset of $\overline{\mathbb{R}}^{\mathcal{Y}}$

a *broad-sense offer* on $\mathcal{Y}$ if it satisfies Axioms G0–G4. The mappings (6.10) and (6.9) define a bijection when they are applied to broad-sense offers and broad-sense upper expectations.

The payoffs available to Skeptic in the testing protocols of Part I constitute broad-sense offers that either qualify as offers by satisfying Axiom G5 or can be enlarged so that they do (see Exercises 6.13, 6.14, and 6.17).

A simple example of a broad-sense upper expectation that is not an upper expectation is the functional limsup on $\overline{\mathbb{R}}^{\mathbb{N}}$ (see Exercise 6.21). Another well-known example is provided by probabilistic number theory.

Example 6.13   (Probabilistic number theory) Probabilistic number theory studies the behavior of certain functions on the natural numbers. A standard approach considers the distribution of a function's values on $\{1, \ldots, N\}$ and investigates whether it weakly converges to a probability measure as $N \to \infty$. Because such limiting probability measures exist for many functions, this brings the resources of measure-theoretic probability into number theory. An older and more elementary picture, which considers limits of averages without using measure theory, can be given a game-theoretic interpretation.

In the older picture, a subset E of $\mathbb{N}$ is said to have a *natural density* equal to

$$\lim_{n \to \infty} \frac{1}{n} |\{1, \ldots, n\} \cap E| \tag{6.12}$$

if this limit exists. Well-known examples of events with natural densities include the set of prime numbers, which has natural density 0, and the set of numbers not divisible by a square m^2 for $m \in \{2, 3, \ldots\}$, which has natural density $6/\pi^2$. Natural densities are a feature of the broad-sense upper expectation $\overline{\mathbf{E}}$ on $\mathbb{N}$ given by

$$\overline{\mathbf{E}}(f) := \limsup_{n \to \infty} \frac{1}{n} \sum_{k=1}^{n} f(k). \tag{6.13}$$

An event $E \subseteq \mathbb{N}$ is probabilized by $\overline{\mathbf{E}}$ if and only if the natural density (6.12) exists, and in this case this limit is its game-theoretic probability.   □

Results in this chapter that hold for all broad-sense upper expectations include Markov's inequality, the inequalities $\underline{\mathbf{E}}(f) \le \overline{\mathbf{E}}(f)$ and $\inf f \le \overline{\mathbf{E}}(f) \le \sup f$, and the equality $\overline{\mathbf{E}}(f + c) = \overline{\mathbf{E}}(f) + c$ for every $c \in \mathbb{R}$ (see Exercise 6.19). Doob's convergence theorem, which we prove for upper expectations in Chapter 7, and Lévy's zero-one law, which we prove for upper expectations in Chapter 8, also hold for broad-sense upper expectations.

The most important properties of upper expectations that may fail for broad-sense upper expectations are countable subadditivity and strong coherence (Statements 3 and 4 of Proposition 6.4). Because countable subadditivity may fail, a countable union of null events is not necessarily null (Exercise 6.21). Nor is it always true that the intersection of countably many almost sure events is almost sure. Proposition 6.7 does not hold; E being almost sure implies that E^c is null but is not equivalent to E^c being null. Whereas E being almost sure means that there exists a nonnegative extended variable with unit upper expected value that takes the value infinity on E^c, E^c being null means only that for every $C > 0$ there exists a nonnegative extended variable with unit upper expected value that equals or exceeds C on E^c (see Exercise 6.22).

6.6 EXERCISES

Exercise 6.1   Show that if $\overline{\mathbf{E}}$ is an upper expectation on $\mathcal{Y}$ and f is a bounded-below extended variable on $\mathcal{Y}$ satisfying $\overline{\mathbf{E}}(f) < \infty$, then $\overline{\mathbf{P}}(\{y \in \mathcal{Y} \,|\, f(y) = \infty\}) = 0$.   □

Exercise 6.2   Show that if $\mathcal{Y}$ is a finite nonempty set and $\overline{\mathbf{E}} : \overline{\mathbb{R}}^{\mathcal{Y}} \to \overline{\mathbb{R}}$ satisfies Axioms E1–E4, then $\overline{\mathbf{E}}$ satisfies the conclusion of Axiom E5 when $\lim_{k\to\infty} f_k < \infty$. *Hint.* Use Statement 2 of Proposition 6.4.   □

Exercise 6.3   Suppose $\overline{\mathbf{E}}$ is an upper expectation on $\mathcal{Y}$, $A \subseteq \mathcal{Y}$, and $\overline{\mathbf{P}}(A) = \underline{\mathbf{P}}(A) > 0$. Define a functional $\overline{\mathbf{E}}^* : \overline{\mathbb{R}}^A \to \overline{\mathbb{R}}$ by $\overline{\mathbf{E}}^*(f) = \overline{\mathbf{E}}(f^*)/\overline{\mathbf{P}}(A)$, where

$$f^*(y) := \begin{cases} f(y) & \text{if } y \in A, \\ 0 & \text{if } y \notin A. \end{cases}$$

Show that $\overline{\mathbf{E}}^*$ is an upper expectation on A.   □

Exercise 6.4   Give formulas for the lower expectations in Examples 6.1–6.3.   □

Exercise 6.5

1. Show that a functional $\underline{\mathbf{E}} : \overline{\mathbb{R}}^{\mathcal{Y}} \to \overline{\mathbb{R}}$ is the lower expectation for some upper expectation on $\mathcal{Y}$ if and only if it satisfies the following conditions.

    LE1. If $f_1, f_2 \in \overline{\mathbb{R}}^{\mathcal{Y}}$, then $\underline{\mathbf{E}}(f_1 +^* f_2) \geq \underline{\mathbf{E}}(f_1) +^* \underline{\mathbf{E}}(f_2)$.

    LE2. If $f \in \overline{\mathbb{R}}^{\mathcal{Y}}$ and $c \in (0, \infty)$, then $\underline{\mathbf{E}}(cf) = c\underline{\mathbf{E}}(f)$.

    LE3. If $f_1, f_2 \in \overline{\mathbb{R}}^{\mathcal{Y}}$ and $f_1 \leq f_2$, then $\underline{\mathbf{E}}(f_1) \leq \underline{\mathbf{E}}(f_2)$.

    LE4. For each $c \in \mathbb{R}$, $\underline{\mathbf{E}}(c) = c$.

    LE5. If $f_1 \geq f_2 \geq \cdots$ is a decreasing sequence of variables in $\overline{\mathbb{R}}^{\mathcal{Y}}$, and f_1 bounded above, then $\underline{\mathbf{E}}(\lim_{k\to\infty} f_k) = \lim_{k\to\infty} \underline{\mathbf{E}}(f_k)$.

    Here the operation $+^*$ follows our usual rules for $+$, except that $\infty +^* (-\infty) = -\infty$.

2. Suppose $\overline{\mathbf{E}}$ is an $\mathbb{R}$-upper expectation on $\mathcal{Y}$. Let $\mathcal{E}$ be the set of variables with finite expected values: $\mathcal{E} := \{f \in \mathbb{R}^{\mathcal{Y}} \,|\, -\infty < \underline{\mathbf{E}}(f) = \overline{\mathbf{E}}(f) < \infty\}$. Show that $\mathcal{E}$ is a vector space.   □

Exercise 6.6   Verify that the game-theoretic expected values associated with an upper expectation on $\mathcal{Y}$ (see the end of Section 6.1) have the following properties.

1. If f_1 and f_2 take values in $\mathbb{R}$ and have finite expected values, then $f_1 + f_2$ has an expected value and $\mathbf{E}(f_1 + f_2) = \mathbf{E}(f_1) + \mathbf{E}(f_2)$.

2. If f has an expected value and $c \in \mathbb{R}$, then cf has the expected value $\mathbf{E}(cf) = c\mathbf{E}(f)$.

3. If $f_1 \leq f_2$ have expected values, then $\mathbf{E}(f_1) \leq \mathbf{E}(f_2)$.

4. For each $c \in \mathbb{R}$, $\mathbf{E}(c) = c$.

5. If $f_1 \leq f_2 \leq \cdots \in [0, \infty]^{\mathcal{Y}}$ have expected values, then $\lim_{k \to \infty} f_k$ also has an expected value and $\mathbf{E}(\lim_{k \to \infty} f_k) = \lim_{k \to \infty} \mathbf{E}(f_k)$.

Compare these statements with Whittle's axioms for expectation [427, p. 15].   □

Exercise 6.7   Verify the following properties of the upper and lower probabilities derived from an upper expectation on $\mathcal{Y}$.

1. $\underline{\mathbf{P}}(E) = 1 - \overline{\mathbf{P}}(E^c)$.

2. If $E_1 \subseteq E_2$, then $\overline{\mathbf{P}}(E_1) \leq \overline{\mathbf{P}}(E_2)$ and $\underline{\mathbf{P}}(E_1) \leq \underline{\mathbf{P}}(E_2)$.

3. $\overline{\mathbf{P}}(E_1 \cup E_2) \leq \overline{\mathbf{P}}(E_1) + \overline{\mathbf{P}}(E_2)$.

4. $\overline{\mathbf{P}}\left(\bigcup_{k=1}^{\infty} E_k\right) \leq \sum_{k=1}^{\infty} \overline{\mathbf{P}}(E_k)$.

5. $\underline{\mathbf{P}}(\emptyset) = \overline{\mathbf{P}}(\emptyset) = 0$ and $\underline{\mathbf{P}}(\mathcal{Y}) = \overline{\mathbf{P}}(\mathcal{Y}) = 1$.   □

Exercise 6.8   (Walley [426]). Let $\mathcal{Y} := \{a, b, c\}$ and consider the probability measures on $\mathcal{Y}$ defined by

$$P_1\{a\} = 2/3, \qquad P_1\{b\} = 1/3, \qquad P_1\{c\} = 0,$$
$$P_2\{a\} = 1/3, \qquad P_2\{b\} = 0, \qquad P_2\{c\} = 2/3,$$
$$P_3\{a\} = 2/3, \qquad P_3\{b\} = 0, \qquad P_3\{c\} = 1/3.$$

For each extended variable f on $\mathcal{Y}$ set

$$\overline{\mathbb{E}}_1(f) := \max(P_1(f), P_2(f)),$$
$$\overline{\mathbb{E}}_2(f) := \max(P_1(f), P_2(f), P_3(f)).$$

Show that

- $\overline{\mathbb{E}}_1$ and $\overline{\mathbb{E}}_2$ are upper expectations,
- the corresponding upper probabilities coincide, $\overline{\mathbb{P}}_1 = \overline{\mathbb{P}}_2$,
- $\overline{\mathbb{E}}_1 \neq \overline{\mathbb{E}}_2$.

Hint. An example of f such that $\overline{\mathbb{E}}_1(f) \neq \overline{\mathbb{E}}_2(f)$ is given by $f(a) := 1, f(b) := -1$, and $f(c) := 0$ (see also [426, section 2.7.3]).   □

Exercise 6.9 (upper expectations with bounded-below support).

1. Given a nonempty set $\mathcal{Y}$, let $\mathbb{G}_y$ be the set of all bounded-below variables on $\mathcal{Y}$. Give axioms that characterize the functionals that are restrictions to $\mathbb{G}_y$ of upper expectations on $\mathcal{Y}$.

2. Let us say that an upper expectation $\overline{\mathbf{E}}$ on $\mathcal{Y}$ *has bounded-below support* if

$$\overline{\mathbf{E}}(f) := \lim_{a \to -\infty} \overline{\mathbf{E}}(\max(f, a))$$

   for all $f \in \overline{\mathbb{R}}^{\mathcal{Y}}$. Give an example of an upper expectation that does not have bounded-below support. *Hint*. See Exercise 6.15. □

Exercise 6.10 Show that, in the presence of the other axioms, Axiom G0 is equivalent to **G** being closed in the uniform metric. □

Exercise 6.11 As noted in Example 6.1, the supremum is an upper expectation. What is the corresponding offer? □

Exercise 6.12 Consider Example 6.2.

1. Show that in a probability space $(\mathcal{Y}, \mathcal{F}, P)$, the set

$$\mathbf{G} := \{g \in \overline{\mathbb{R}}^{\mathcal{Y}} | \exists f \in \mathcal{E} : g \le f \text{ and } P(f) \le 0\}$$

   is an offer, where $\mathcal{E}$ is the set of extended random variables that have expected values. *Hint*. To establish Axiom G5, choose $f_k \in \mathcal{E}$ such that $g_k \le f_k$ and $P(f_k) \le 0$, and consider $h_k := \inf_{n \ge k} f_n$.

2. Given a variable $f \in \overline{\mathbb{R}}^{\mathcal{Y}}$, let $\overline{\mathbf{E}}(f)$ be f's upper integral (6.1) with respect to P. Show that $f \in \mathbf{G}$ if and only if $\overline{\mathbf{E}}(f) \le 0$ and thus that $\overline{\mathbf{E}}$ is **G**'s upper expectation. □

Exercise 6.13 Consider Example 6.3. For all $m \in [-1, 1]$, set

$$\mathbf{G}^m := \{g \in \overline{\mathbb{R}}^{[-1,1]} | \exists M \in \overline{\mathbb{R}} \, \forall y \in [-1, 1] : g(y) \le M(y - m)\}$$

and

$$\mathbf{F}^m := \{g \in \mathbb{R}^{[-1,1]} | \exists M \in \mathbb{R} \, \forall y \in [-1, 1] : g(y) \le M(y - m)\}.$$

1. Verify that each $\mathbf{G}^m$ is an offer and that $\overline{\mathbf{E}}^m$ is $\mathbf{G}^m$'s upper expectation.

2. Verify that $\mathbf{F}^{m_n}$ is the set of moves available to Skeptic on the nth round of Protocol 1.1's slackening.

3. Verify that the functional $\overline{\mathbf{E}} : \mathbb{R}^{\mathcal{Y}} \to \overline{\mathbb{R}}$ given by $\overline{\mathbf{E}}(f) := \inf\{\alpha \in \mathbb{R} | f - \alpha \in \mathbf{F}^{-1}\}$ satisfies Axioms E1$^{\mathbb{R}}$–E4$^{\mathbb{R}}$ but not Axiom E5$^{\mathbb{R}}$.

4. In Chapter 1, we showed that Skeptic can achieve certain goals in Protocol 1.1. If we allow him to choose his nth move from $\mathbf{G}^{m_n}$ instead of the smaller set $\mathbf{F}^{m_n}$, he can achieve slightly more. Show, in fact, that he can force $m_n = -1 \implies y_n = -1$. Discuss. □

Exercise 6.14 On the nth round of the slackening of Protocol 4.1, Skeptic chooses a variable from

$$\{g \in \mathbb{R}^{\mathbb{R}} | \exists M \in \mathbb{R} \; \exists V \in [0, \infty) \; \forall y \in \mathbb{R} : g(y) \leq M(y - m_n) + V((y - m_n)^2 - v_n)\},$$

where $m_n \in \mathbb{R}$ and $v_n \in [0, \infty)$, and his capital becomes $\mathcal{K}_n := \mathcal{K}_{n-1} + g(y_n)$. Analyze this in terms of upper expectations and offers as Exercise 6.13 does for Protocol 1.1. (The case $v_n = 0$ involves complications similar to those involving $m_n = -1$ or $m_n = 1$ in Protocol 1.1.) $\square$

Exercise 6.15 Show that $\mathbf{G} := \{g \in \overline{\mathbb{R}}^{\mathbb{N}} | g \leq 0\} \cup \{g \in \overline{\mathbb{R}}^{\mathbb{N}} | \lim_{n \to \infty} g(n) = -\infty\}$ is an offer. Give an example of $g_1 \leq g_2 \leq \cdots \in \mathbf{G}$ such that $\lim_{k \to \infty} g_k \notin \mathbf{G}$. $\square$

Exercise 6.16 Suppose $(\mathbf{G}_\gamma)_{\gamma \in \Gamma}$ is a family of offers on $\mathcal{Y}$. Verify the following:

1. The set $\mathbf{G} := \bigcap_{\gamma \in \Gamma} \mathbf{G}_\gamma$ is an offer on $\mathcal{Y}$.

2. $\overline{\mathbf{E}}_{\mathbf{G}} = \sup_{\gamma \in \Gamma} \overline{\mathbf{E}}_{\mathbf{G}_\gamma}$. $\square$

Exercise 6.17 Give axioms for $\mathbb{R}$-offers. Adapt the proof of Proposition 6.10 to show that $\mathbb{R}$-offers and $\mathbb{R}$-upper expectations are in one-to-one correspondence, and adapt the discussion of the equivalence of Protocols 6.12 and 6.11 to the case where $\mathbb{R}$-offers and $\mathbb{R}$-upper expectations are used. $\square$

Exercise 6.18 Suppose $\overline{\mathbf{E}}$ is a broad-sense upper expectation on $\mathcal{Y}$. Show that if $f_1, f_2 \in \overline{\mathbb{R}}^{\mathcal{Y}}$ and $f_1 \leq f_2$ is almost sure with respect to $\overline{\mathbf{E}}$, then $\overline{\mathbf{E}}(f_1) \leq \overline{\mathbf{E}}(f_2)$. $\square$

Exercise 6.19 Suppose $\overline{\mathbf{E}}$ is a broad-sense upper expectation on $\mathcal{Y}$.

1. Show that for every $f \in \overline{\mathbb{R}}^{\mathcal{Y}}$, $\underline{\mathbf{E}}(f) \leq \overline{\mathbf{E}}(f)$, $\inf f \leq \overline{\mathbf{E}}(f) \leq \sup f$, and $\overline{\mathbf{E}}(f + c) = \overline{\mathbf{E}}(f) + c$ for every $c \in \mathbb{R}$.

2. Show that if $\overline{\mathbf{P}}(E) = 0$ and $c \in (0, \infty)$, then $\overline{\mathbf{E}}(f + c\mathbf{1}_E) = \overline{\mathbf{E}}(f)$.

3. Verify that $\overline{\mathbf{E}}$ satisfies Markov's inequality (Proposition 6.6).

4. Use Markov's inequality to show that if $\overline{\mathbf{E}}(f) < \infty$, then $\overline{\mathbf{P}}(f = \infty) = 0$. $\square$

Exercise 6.20 Suppose $\overline{\mathbf{E}}$ is a broad-sense upper expectation on $\mathcal{Y}$ that is countably subadditive. Use Markov's inequality (see Exercise 6.19) to show that $\overline{\mathbf{E}}$ satisfies the *first Borel–Cantelli lemma*: If $\sum_{k=1}^{\infty} \overline{\mathbf{P}}(E_k) < \infty$, then $\overline{\mathbf{P}}(\text{infinitely many } E_k \text{ happen}) = 0$. $\square$

Exercise 6.21

1. Verify that lim sup is a broad-sense upper expectation on $\mathbb{N}$ but not an upper expectation. *Hint.* Show that the property (6.6) of countable subadditivity is violated for $E_k := \{k\}$.

2. Verify that the functional given by (6.13) is a broad-sense upper expectation on $\mathbb{N}$ but not an upper expectation. $\square$

Exercise 6.22 Give an example of a broad-sense upper expectation on $[0, 1]$ and an event $E \subseteq [0, 1]$ such that $\overline{\mathbf{P}}(E) = 0$ but E is not null. $\square$

6.7 CONTEXT

Upper Expectation

Axioms for upper expectations in game-theoretic probability were studied previously in [351, 418]. These functionals have also arisen in a variety of other contexts and have been studied under a variety of names.

- *Outer probability contents.* Concepts of inner and outer content (*äussere und innere Inhalt* in German) were developed in the nineteenth century and influenced the later development of the concepts of inner and outer measure [184, section 3.1]. In his thorough study of the problem of extending finitely and countably additive probability measures from marginals [189], Jørgen Hoffmann-Jørgensen called broad-sense upper expectations *outer probability contents*.

- *Coherent risk measures.* Another context in which upper expectations have been studied is that of *coherent risk measures*, which were first studied axiomatically by Artzner et al. in 1999 [14]. In 2006, Andrzej Ruszczyński and Alexander Shapiro extended the concept to infinite-dimensional spaces and connected it to conjugate duality for convex optimization [324] (see also [147] and [148, chapter 4]).

- *Imprecise probabilities.* Many aspects of the theory of offers and upper expectations have been studied in the theory of imprecise probabilities [15, 381]. Peter Walley initiated this theory in the 1980s [426], drawing on earlier work by Peter Williams [428, 429].

  Walley emphasized the viewpoint of a player he called You. Like our Forecaster, You identifies a set of variables or *gambles*. But whereas Forecaster merely announces gambles from which Skeptic can choose and does not necessarily serve as Skeptic's counterparty, You's gambles are gambles he is inclined to make. Walley gave axioms for such a *set of desirable gambles*. If **G** is a broad-sense offer, then the set **H** := $\{-g|g \in \mathbf{G}, g$ is bounded$\}$ satisfies Walley's axioms, and any set satisfying his axioms is of this form, except that Axiom G0 may not be required and the gamble 0 may be omitted [426, p. 60]. Nearly equivalent axioms given by Troffaes and de Cooman do require that 0 be in the set [381, p. 30]. The minus sign enters because You is Skeptic's counterparty; when Skeptic gets g, You gets $-g$.

  Upper and lower expectations are called *upper* and *lower previsions* in the theory of imprecise probabilities. Because the theory takes You's viewpoint, it emphasizes lower rather than upper previsions. As noted in Section 6.1, buying X for α is the same as selling $-X$ for $-\alpha$. But because most people buy more often than they sell, ordinary language is more developed for buying than for selling, and we tend to develop theories in terms of buying prices. As we have learned, Skeptic's buying prices are given by the upper functional. You being Skeptic's counterparty, his buying prices are given by the lower functional.

Countable Additivity and Subadditivity

In measure-theoretic probability, where we begin by assuming finite additivity, the further assumption of continuity is equivalent to countable additivity. The use of countable additivity in probability theory goes back to Jacob Bernoulli in the seventeenth century (see the further discussion in Section 7.8), but it was adopted explicitly as an axiom for probability only in the twentieth century, as a result of the success of Lebesgue integration as a mathematical tool [350]. Fréchet, Kolmogorov, and others responsible for its adoption saw probability as a way of describing the empirical world, but they emphasized that they were assuming countable additivity (or continuity) for mathematical convenience, not because it had any direct empirical meaning. Kolmogorov wrote as follows concerning his axiom of continuity:

> Since the new axiom is meaningful only for infinite probability fields, it would hardly be possible to explain its empirical significance .... In a description of any actually observable random process, we can obtain only finite probability fields. Infinite probability fields appear only as idealized models of real random processes.
>
> The German original [224, p. 14]: Da das neue Axiom nur für unendlichen Wahrscheinlichkeitsfelder wesentlich ist, währe es kaum möglich, seine empirische Bedeutung zu erklären .... Bei einer Beschreibung irgendwelcher wirklich beobachtbarer zufälliger Prozesse kann man nur endliche Wahrscheinlichkeitsfelder erhalten. Unendlich Wahrscheinlichkeitsfelder erscheinen nur als idealisierte Schemata reeller zufälliger Prozesse.

Fréchet compared the introduction of countable additivity to the introduction of irrational numbers [151, p. 14].

There has been continued interest, however, in probability measures that are only finitely additive. There are two somewhat distinct motivations for this interest.

1. On the one hand, some proponents of the subjective interpretation of probability have argued that finite additivity is a reasonable property to demand of subjective probabilities but countable additivity is not. Bruno de Finetti is well known for this stance [206, 317].

2. Some authors, while studying gambling and dynamic programming, have found that insisting on countable additivity introduces unmanageable technicalities. Lester Dubins and Leonard Savage's *How to Gamble if You Must* [119] is the classic example.

When we weaken additivity (the rule $P(f_1 + f_2) = P(f_1) + P(f_2)$ for probability measures) to subadditivity (the rule $\overline{\mathbf{E}}(f_1 + f_2) \leq \overline{\mathbf{E}}(f_1) + \overline{\mathbf{E}}(f_2)$ for upper expectations) or superadditivity (the rule $\underline{\mathbf{E}}(f_1 + f_2) \geq \underline{\mathbf{E}}(f_1) + \underline{\mathbf{E}}(f_2)$ for lower expectations), the question remains whether we should extend the condition to countable sums, directly or by adopting a continuity axiom. Although he argued in favor of countable superadditivity for lower previsions, Walley did not adopt it as a basic axiom, and there is no consensus on the question among proponents of imprecise probability. Some, echoing de Finetti, see such a condition as an unjustified restriction when lower previsions are interpreted as measures of belief.

Markov's Inequality

The game-theoretic version of this inequality is a direct generalization of an inequality for classical probabilities credited to Andrei Markov. It appeared in Markov's probability textbook, first published in Russian in 1900 [264]. The proof given in Section 6.2 is essentially Markov's proof.

Probabilistic Number Theory

The use of probabilistic language to describe asymptotic frequencies of properties of the natural numbers goes back to the nineteenth century. For example, the assertion that two randomly chosen natural numbers will be relatively prime with probability $6/\pi^2$ was published by Ernesto Cesàro in the 1880s [65, 66] and has been attributed to lectures delivered decades earlier by Dirichlet and Chebyshev. Some authors, including Sergei Bernstein, opposed such formulations on the grounds that they cannot be given any direct empirical meaning. We cannot draw a natural number at random in the way that we might draw a ball at random from an urn. But beginning in the 1930s, after Kolmogorov established an axiomatic basis for probability theory independent of any empirical interpretation, the use of probability in the theory of numbers flourished. The monograph by Jonas Kubilius, translated from Russian into English in 1964, was an influential milestone, and P. D. T. A. Elliot used *Probabilistic Number Theory* as the title of a two-volume treatise in 1979 and 1980 [125, 126]. Sergei Abramovich and Yakov Nikitin [2] have surveyed the many rediscoveries of Cesàro's probability for two natural numbers being relatively prime. Elliott surveys the twentieth-century work on probabilistic number theory up to 1980 in his introduction [125]. For more recent expositions, see [365, 380].

Borel–Cantelli Lemma

This lemma, first stated by Borel in 1909 [44], has two parts: (i) if the probabilities for a countable sequence of events add to less than infinity, then almost surely only finitely many of them will happen, and (ii) if the probabilities for a countable sequence of independent events add to infinity, then almost surely infinitely many of them will happen. Borel failed to note the need for the condition of independence in the second part, and Cantelli has been credited for correcting the omission. A game-theoretic version of the first part of the lemma was published by Kumon et al. in 2007 [230]; another is provided by Exercise 6.20. A game-theoretic version of the second part is given in Exercise 8.9.

Abstract Testing Protocols

In this chapter, we study supermartingales and global upper expected values in abstract testing protocols. We begin with the following protocol.

Protocol 7.1
PARAMETERS: Nonempty set $\mathcal{Y}$; upper expectation $\overline{\mathbf{E}}$ on $\mathcal{Y}$
    Skeptic announces $\mathcal{K}_0 \in \mathbb{R}$.   (7.1)
    FOR $n = 1, 2, \ldots$:
        Skeptic announces $f_n \in \overline{\mathbb{R}}^{\mathcal{Y}}$ such that $\overline{\mathbf{E}}(f_n) \le \mathcal{K}_{n-1}$.   (7.2)
        World announces $y_n \in \mathcal{Y}$.
        $\mathcal{K}_n := f_n(y_n)$.   (7.3)

Protocol 7.1 specializes Protocol 6.12, in which $\mathcal{Y}$ is the only parameter and Forecaster can choose a different upper expectation on each round. The specialization fixes a single upper expectation $\overline{\mathbf{E}}$ and eliminates Forecaster by constraining him to use $\overline{\mathbf{E}}$ on every round. We have changed Reality's name to World. At first glance, Protocol 7.1 seems too special to serve as the setting for a widely applicable theory. But most of the results we obtain for it will extend directly to other testing protocols.

We study Protocol 7.1 in Sections 7.1–7.4. Section 7.1 reviews terminology and notation. Section 7.2 proves Doob's convergence theorem. Section 7.3 defines the upper expected value $\overline{\mathbb{E}}_s(X)$ of a global variable X in a situation s and studies its properties. We see in particular that $\overline{\mathbb{E}}_s(X)$ becomes a supermartingale when we fix

Game-Theoretic Foundations for Probability and Finance, First Edition. Glenn Shafer and Vladimir Vovk.
© 2019 John Wiley & Sons, Inc. Published 2019 by John Wiley & Sons, Inc.

X and allow s to vary. Section 7.4 illustrates Protocol 7.1's potential by stating an abstract game-theoretic version of Lindeberg's theorem, which was proven in [349].

Section 7.5 extends the results for Protocol 7.1 to testing protocols in which the upper expectation may vary from round to round, depending perhaps on the outcomes of previous rounds and decisions by Forecaster. As we will see, play in these seemingly more general protocols can be represented as play in Protocol 7.1. When Skeptic plays against Forecaster and Reality, World represents them both. Forecaster's move on a given round is revealed to Skeptic as part of World's subsequent move, and hence Skeptic's move, rather than being a response to Forecaster's move, is a strategy for responding to that move.

In Section 7.6 we discuss how the results of Part I can be expressed in Protocol 7.1 and other abstract testing protocols.

7.1 TERMINOLOGY AND NOTATION

We use terminology and notation along the lines laid out in Sections 1.2 and 4.2. See also the definitions and notation for sequences given in the section on terminology and notation at the end of the book and the distinction between *local* and *global* objects summarized in Table 6.1.

- A *situation* is a finite sequence of elements of $\mathcal{Y}$. We write $\mathbb{S}$ for the set of all situations and call it the *situation space*. The empty sequence $\square$ is the *initial situation*.
- A *path* is an infinite sequence of elements of $\mathcal{Y}$. We write Ω for the set of all paths and call it the *sample space*. For each $s \in \mathbb{S}$, Ω_s is the set of all paths that pass through s. In particular, $\Omega_\square = \Omega$.
- A *local event* is a subset of $\mathcal{Y}$; a *global event* is a subset of Ω.
- A *local variable* is an extended real-valued function on $\mathcal{Y}$; a *global variable* is an extended real-valued function on Ω.
- A *process* is an extended real-valued function on $\mathbb{S}$. We identify a process S with the sequence of global variables $S_0, S_1, \ldots$, where $S_n(\omega) := S(\omega^n)$.
- A strategy ψ for Skeptic is a pair $(\psi^{\text{stake}}, \psi^{\text{bet}})$, where $\psi^{\text{stake}} \in \overline{\mathbb{R}}$ and $\psi^{\text{bet}} : \mathbb{S} \setminus \{\square\} \to \overline{\mathbb{R}}^{\mathcal{Y}}$ is *predictable* in the sense that $\psi^{\text{bet}}(y_1 y_2 \ldots y_n)$ depends only on $y_1, \ldots, y_{n-1}$. The number ψ^{stake} specifies Skeptic's initial capital, and $\psi^{\text{bet}}(y_1 y_2 \ldots y_n)$ is the move for Skeptic specified by the strategy on the nth round when World's previous moves are $y_1, \ldots, y_{n-1}$. We sometimes represent ψ^{bet} as a sequence $\psi_1^{\text{bet}}, \psi_2^{\text{bet}}, \ldots$, where $\psi_n^{\text{bet}} : \omega \in \Omega \mapsto \psi^{\text{bet}}(\omega^n) \in \overline{\mathbb{R}}^{\mathcal{Y}}$.

7.2 SUPERMARTINGALES

As usual, we call the capital process resulting from a strategy for Skeptic a *supermartingale*. If (7.2) always holds with equality when Skeptic plays the strategy ψ,

Table 7.1 Three sets of processes in Protocol 7.1.

Set	Elements of the set
T	Supermartingales
L	Supermartingales that converge in $\overline{\mathbb{R}}$
M	Martingales

Note that $\mathbf{L} \subseteq \mathbf{T}$ and $\mathbf{M} \subseteq \mathbf{T}$.

then we call ψ and its supermartingale *efficient*. An important difference from the supermartingales of Part I is that a supermartingale $\mathcal{T}$ can now take the values $\pm\infty$; in particular, $\mathcal{T}_0 = \infty$ is allowed.

Just as in Part I's testing protocols, we say that a global event E happens *almost surely* if there exists a nonnegative supermartingale that begins with a finite value and tends to infinity on every path not in E. (See Sections 1.2 and 1.4; beginning with a finite value is automatic in Part I.) We sometimes abbreviate *almost surely* to a.s.

From (7.1) and (7.3) in Protocol 7.1 we see that the supermartingale $\mathcal{T}$ for the strategy $(\psi^{\mathrm{stake}}, \psi^{\mathrm{bet}})$ is given by

$$\mathcal{T}(\square) = \psi^{\mathrm{stake}} \tag{7.4}$$

and

$$\mathcal{T}(y_1 \dots y_n) = \psi^{\mathrm{bet}}(y_1 \dots y_n)(y_n) \tag{7.5}$$

for all noninitial situations $y_1 \dots y_n$. The relations (7.4) and (7.5) show that the correspondence between strategies for Skeptic and their supermartingales is a bijection. We sometimes take advantage of this to simplify our language; instead of saying that a strategy instructs Skeptic to bet a certain way, we simply say that the supermartingale bets that way.

As in Chapter 1, we call a supermartingale $\mathcal{M}$ a *martingale* if $-\mathcal{M}$ is also a supermartingale. As in Chapters 2 and 4, we write $\mathbf{T}$ for the set of supermartingales. We further write $\mathbf{M}$ for the set of martingales and $\mathbf{L}$ for the subset of $\mathbf{T}$ consisting of supermartingales that converge in $\overline{\mathbb{R}}$ on every path (see Table 7.1).

Properties of Supermartingales

The following nearly obvious proposition is a convenient characterization of game-theoretic supermartingales.

Proposition 7.2 *A process $\mathcal{T}$ is a supermartingale if and only if*

$$\overline{\mathbf{E}}(\mathcal{T}(s \cdot)) \leq \mathcal{T}(s) \tag{7.6}$$

for every $s \in \mathbb{S}$. A supermartingale $\mathcal{T}$ is efficient if and only if (7.6) holds with equality for all $s \in \mathbb{S}$.

Proof Lines (7.1) and (7.2) in Protocol 7.1 give necessary and sufficient conditions for a pair $(\psi^{\text{stake}}, \psi^{\text{bet}})$ to qualify as a strategy for Skeptic:

- $\psi^{\text{stake}} \in \overline{\mathbb{R}}$, and
- ψ^{bet} is a mapping from $\mathbb{S} \setminus \{\square\}$ to $\overline{\mathbb{R}}^{\mathcal{Y}}$ satisfying[1]

$$\overline{\mathbf{E}}(\psi^{\text{bet}}(y')) \le \psi^{\text{stake}} \tag{7.7}$$

and

$$\overline{\mathbf{E}}(\psi^{\text{bet}}(syy')) \le \psi^{\text{bet}}(sy')(y) \tag{7.8}$$

for all $s \in \mathbb{S}$ and all $y, y' \in \mathcal{Y}$.

Using (7.4) and (7.5), we can restate (7.7) and (7.8) as conditions on ψ's supermartingale $\mathcal{T}$: $\mathcal{T}(\square) \in \overline{\mathbb{R}}$,

$$\overline{\mathbf{E}}(\mathcal{T}(\square \cdot)) \le \mathcal{T}(\square) \tag{7.9}$$

and

$$\overline{\mathbf{E}}(\mathcal{T}(sy \cdot)) \le \mathcal{T}(sy) \tag{7.10}$$

for every $s \in \mathbb{S}$ and $y \in \mathcal{Y}$. The condition $\mathcal{T}(\square) \in \overline{\mathbb{R}}$ adds nothing to the assumption that $\mathcal{T}$ is a process. The inequalities (7.9) and (7.10) say that (7.6) holds for all $s \in \mathbb{S}$.

The supermartingale $\mathcal{T}$ is efficient if and only if (7.2) holds with equality in all situations, and this is equivalent to the inequalities (7.9) and (7.10) always holding with equality – i.e. to (7.6) always holding with equality. □

The following lemma compares a supermartingale's value in a situation s with the values it may take subsequently. Being a statement about all paths through s, not a statement about almost all paths, it has no counterpart in measure-theoretic probability.

Lemma 7.3 *If $\mathcal{T}$ is a supermartingale, then*

$$\inf_{\omega \in \Omega_s} \limsup_{n \to \infty} \mathcal{T}_n(\omega) \le \mathcal{T}(s) \tag{7.11}$$

for every $s \in \mathbb{S}$.

---

[1] Recall that, for any $t \in \mathbb{S}$, $\psi^{\text{bet}}(ty')$ does not depend on $y' \in \mathcal{Y}$.

Proof It suffices to show that if $r \in \mathbb{R}$ and $\mathcal{T}(s) < r$, then there exists $\omega \in \Omega_s$ such that $\lim \sup_{n \to \infty} \mathcal{T}(\omega^n) \le r$. The existence of such an ω follows by induction once we show the existence of an $y \in \mathcal{Y}$ such that $\mathcal{T}(sy) < r$. But this follows from the relation $\inf f \le \overline{\mathbf{E}}(f)$ and Proposition 7.2: $\inf_{y \in \mathcal{Y}} \mathcal{T}(sy) \le \overline{\mathbf{E}}(\mathcal{T}(s\cdot)) \le \mathcal{T}(s)$. □

Lemma 7.3 is quite useful. It implies the much weaker but still useful result that a supermartingale $\mathcal{T}$ satisfies $\inf_{\omega \in \Omega_s} \lim \inf_n \mathcal{T}_n(\omega) \le \mathcal{T}(s)$ for every $s \in \mathbb{S}$. This in turn has the simple but important consequence that a supermartingale $\mathcal{T}$ is bounded below by a constant c if $\lim \inf_n \mathcal{T}_n$ is.

The following lemma lists some properties of the sets **T** and **L**.

Lemma 7.4

1. *The sets* **T** *and* **L** *are convex cones.*
2. *If* $\mathcal{T}^\alpha \in \mathbf{T}$ *for every α in a nonempty set A, then* $\inf_{\alpha \in A} \mathcal{T}^\alpha \in \mathbf{T}$.
3. *If* $\mathcal{T}^1 \le \mathcal{T}^2 \le \cdots \in \mathbf{T}$ *and all $\mathcal{T}^k$ are bounded below, then* $\lim_{k \to \infty} \mathcal{T}^k \in \mathbf{T}$.
4. *If* $\mathcal{T}^1, \mathcal{T}^2, \ldots$ *are nonnegative supermartingales, then* $\sum_{k=1}^\infty \mathcal{T}^k$ *is a nonnegative supermartingale.*

Proof
1. If $c_1, c_2 \in [0, \infty)$ and $\mathcal{T}^1, \mathcal{T}^2 \in \mathbf{T}$, then

$$\overline{\mathbf{E}}(c_1 \mathcal{T}^1(s \cdot) + c_2 \mathcal{T}^2(s \cdot)) \le c_1 \overline{\mathbf{E}}(\mathcal{T}^1(s \cdot)) + c_2 \overline{\mathbf{E}}(\mathcal{T}^2(s \cdot))$$

by Axioms E1 and E2, and hence $c_1 \mathcal{T}^1 + c_2 \mathcal{T}^2 \in \mathbf{T}$ by Proposition 7.2. So **T** is a convex cone. Because the property of having a limit is preserved by convex combination, **L** is also a convex cone.

2. By Axiom E3 and Proposition 7.2,

$$\overline{\mathbf{E}} \left(\inf_{\alpha \in A} \mathcal{T}^\alpha(s \cdot) \right) \le \inf_{\alpha \in A} \overline{\mathbf{E}}(\mathcal{T}^\alpha(s \cdot)) \le \inf_{\alpha \in A} \mathcal{T}^\alpha(s)$$

for all $s \in \mathbb{S}$. So $\inf_{\alpha \in A} \mathcal{T}^\alpha$ is a supermartingale by Proposition 7.2.

3. By Axiom E5 and Proposition 7.2,

$$\overline{\mathbf{E}} \left(\lim_{k \to \infty} \mathcal{T}^k(s \cdot) \right) = \lim_{k \to \infty} \overline{\mathbf{E}} \left(\mathcal{T}^k(s \cdot) \right) \le \lim_{k \to \infty} \mathcal{T}^k(s)$$

for all $s \in \mathbb{S}$. So $\lim_{k \to \infty} \mathcal{T}^k$ is a supermartingale by Proposition 7.2.

4. Statement 4 follows from Statements 1 and 3. □

Although $\mathcal{T}^1 + \mathcal{T}^2$ is a supermartingale whenever $\mathcal{T}^1$ and $\mathcal{T}^2$ are super-martingales, efficiency for $\mathcal{T}^1$ and $\mathcal{T}^2$ does not imply efficiency for $\mathcal{T}^1 + \mathcal{T}^2$. Indeed, if the corresponding strategies ψ^1 and ψ^2 satisfy the strict inequality $\overline{\mathbf{E}}(\psi^{1,\text{bet}}(s) + \psi^{2,\text{bet}}(s)) < \overline{\mathbf{E}}(\psi^{1,\text{bet}}(s)) + \overline{\mathbf{E}}(\psi^{2,\text{bet}}(s))$ for some noninitial situation s, as permitted by Axiom E1, then the inequality in (7.2) will be strict for $\psi^1 + \psi^2$'s moves in those s.

Doob's Convergence Theorem

Doob's convergence theorem, which we stated for Protocol 4.4 as Lemma 4.8, asserts that if $\mathcal{T} \in \mathbf{T}$ is nonnegative and $\mathcal{T}_0 < \infty$, then $\mathcal{T}$ converges in $\mathbb{R}$ almost surely. The following theorem asserts this and a little more for Protocol 7.1.

Theorem 7.5 *If $\mathcal{T} \in \mathbf{T}$ is nonnegative, then there exists nonnegative $\mathcal{L} \in \mathbf{L}$ that satisfies $\mathcal{L}_0 = \mathcal{T}_0$ and tends to infinity on every path on which $\mathcal{T}$ does not converge in $\mathbb{R}$.*

Proof Assume, without loss of generality, that $\mathcal{T}_0 < \infty$. Let $\psi = (\psi^{\text{stake}}, \psi^{\text{bet}})$ be the strategy that produces $\mathcal{T}$. Let $[a_k, b_k]$, $k = 1, 2, \ldots$, be an enumeration of all closed intervals such that a_k and b_k are rational and $0 < a_k < b_k < \infty$.

For each $k \in \mathbb{N}$, we define a strategy ψ^k with a nonnegative supermartingale $\mathcal{T}^k$ by setting $\psi^{k,\text{stake}} := \psi^{\text{stake}} = \mathcal{T}_0$ and defining $\psi^{k,\text{bet}}$ by the following algorithm.

Start If $\mathcal{T}_0 \leq a_k$, go to On Mode for instructions. Otherwise, go to Off Mode for instructions.

On Mode While conserving your capital (see Section 6.4), bet on the current round as instructed by ψ^{bet} and continue doing this in subsequent rounds until you (i.e. Skeptic) first reach a round n such that

$$\mathcal{T}_n \geq b_k. \tag{7.12}$$

If and when you first reach a round n satisfying (7.12), go to Off Mode for instructions on what to do starting from round $n + 1$. If you never reach a round n satisfying (7.12), just continue following ψ^{bet} conserving your capital.

Off Mode Do not bet on the current round and continue not betting until you first reach a round n such that

$$\mathcal{T}_n \leq a_k. \tag{7.13}$$

If and when you reach a round n satisfying (7.13), go to On Mode for instructions on what to do on round $n + 1$. If you never reach a round n satisfying (7.13), just continue not betting.

To show that this algorithm defines a strategy for Skeptic and that its supermartingale is nonnegative, we must check that the algorithm has enough capital on each round to make its bet and be sure that its capital will still be nonnegative on the next round. It suffices to check that it has at least as much capital as $\mathcal{T}$ on rounds in On Mode, where it makes $\mathcal{T}$'s bet. (In Off Mode, it keeps its capital constant and therefore nonnegative by not betting.) During an On period the algorithm's capital will move up or down just like $\mathcal{T}$'s, so it suffices to check that it has as much capital as $\mathcal{T}$ at the beginning of each On period (i.e. at the beginning of the first round of each On period). This is straightforward: it has the same capital as $\mathcal{T}$ at the beginning of round 1 (which may or may not begin an On period), and at the beginning of each subsequent On period (if there are any), it has kept its capital constant during the preceding Off period, during which $\mathcal{T}$ went down.

In addition to showing that ψ^k is a strategy and $\mathcal{T}^k$ is nonnegative, this argument shows that $\mathcal{T}_n^k \geq \mathcal{T}_n$ for every round n that is in an On period or ends an Off period. A refinement of the argument shows that for any round n that is in an On period or ends an Off period,

$$\mathcal{T}_n^k - \mathcal{T}_n \geq \mathcal{T}_N^k - \mathcal{T}_N \tag{7.14}$$

for all $N \leq n$. Again, it suffices to check that this inequality holds when round $n + 1$ starts an On period, because then the difference between $\mathcal{T}^k$ and $\mathcal{T}$ will not change until after the beginning of the next Off period (if there is one). The inequality obviously holds if $n = 0$. It holds if n ends an Off period because then $\mathcal{T}^k$ has been constant since the beginning of the first round of this Off period, whereas $\mathcal{T}$ is smaller than it was then and in any situation in the interim.

It follows from (7.14) that

$$\inf_{n \geq N} (\mathcal{T}_n^k - \mathcal{T}_N^k) \geq \inf_{n \geq N} (\mathcal{T}_n - \mathcal{T}_N) \tag{7.15}$$

for every nonnegative integer N. To check this, it suffices to check

$$\mathcal{T}_m^k - \mathcal{T}_N^k \geq \inf_{n \geq N} (\mathcal{T}_n - \mathcal{T}_N) \tag{7.16}$$

for any $m \geq N$ such that m is in an Off period that it does not end. There are two possibilities:

- If N is in the same Off period, the left-hand side of (7.16) is 0, whereas the right-hand side is nonpositive;
- If not, let m' be the round that ends the On period that precedes m. Then $\mathcal{T}_m^k = \mathcal{T}_{m'}^k$, and so (7.16) follows from (7.14).

Since the $\mathcal{T}^k$ are nonnegative, the sum

$$\mathcal{T}' := \sum_{k=1}^{\infty} 2^{-k} \mathcal{T}^k \tag{7.17}$$

exists. By Statement 4 of Lemma 7.4, $\mathcal{T}' \in \mathbf{T}$. Its initial value, $\mathcal{T}'_0$, is equal to $\mathcal{T}_0$. From (7.15) and (7.17), we obtain

$$\inf_{n \geq N} (\mathcal{T}'_n - \mathcal{T}'_N) \geq \inf_{n \geq N} (\mathcal{T}_n - \mathcal{T}_N) \tag{7.18}$$

for every nonnegative integer N.

The asymptotic behavior of the nonnegative supermartingale $\mathcal{T}'$ depends on that of $\mathcal{T}$ as follows:

1. If $\mathcal{T}_n$ converges in $\mathbb{R}$ as $n \to \infty$, then $\mathcal{T}'_n$ converges in $\overline{\mathbb{R}}$. This follows from (7.18); see Exercise 7.2.
2. If $\mathcal{T}_n \to \infty$, then $\mathcal{T}'_n$ may or may not converge in $\overline{\mathbb{R}}$.
3. If $\mathcal{T}_n$ does not converge in $\overline{\mathbb{R}}$, then $\mathcal{T}'_n \to \infty$, because there exists $k \in \mathbb{N}$ such that $\mathcal{T}_n$ passes upward across $[a_k, b_k]$ an infinite number of times as $n \to \infty$, increasing by at least $b_k - a_k$ each time and therefore tending to infinity.

Let $\mathcal{L} := (\mathcal{T} + \mathcal{T}')/2$ be the nonnegative supermartingale obtained by averaging $\mathcal{T}$ and $\mathcal{T}'$. Then $\mathcal{L}_0 = \mathcal{T}_0$. To see that $\mathcal{L}$ also satisfies the other conditions in the statement of the theorem ($\mathcal{L} \in \mathbf{L}$ and $\mathcal{L}_n \to \infty$ on any path on which $\mathcal{T}_n$ fails to converge in $\mathbb{R}$), consider again $\mathcal{T}$'s three possible asymptotic behaviors:

1. If $\mathcal{T}_n$ converges in $\mathbb{R}$, then $\mathcal{T}'_n$ converges in $\overline{\mathbb{R}}$, and hence the average $\mathcal{L}_n$ converges in $\overline{\mathbb{R}}$.
2. If $\mathcal{T}_n$ tends to infinity, then $\mathcal{L}_n$, its average with a nonnegative process, also tends to infinity.
3. If $\mathcal{T}_n$ does not converge in $\overline{\mathbb{R}}$, then $\mathcal{T}'_n$ tends to infinity, and so $\mathcal{L}_n$ does as well. $\qquad\square$

7.3   GLOBAL UPPER EXPECTED VALUES

Given a global variable X in Protocol 7.1, we set

$$\overline{\mathbb{E}}(X) := \inf\{\mathcal{T}_0 | \mathcal{T} \in \mathbf{T}, \inf \mathcal{T} > -\infty, \liminf_{n \to \infty} \mathcal{T}_n \geq X\}, \tag{7.19}$$

and we call $\overline{\mathbb{E}}(X)$ the *upper expected value of* X. More generally, given a global variable X and a situation s, we set

$$\overline{\mathbb{E}}_s(X) := \inf\{\mathcal{T}(s) | \mathcal{T} \in \mathbf{T}, \inf \mathcal{T} > -\infty, \text{ and } \forall \omega \in \Omega_s : \liminf_{n \to \infty} \mathcal{T}_n(\omega) \geq X(\omega)\}, \tag{7.20}$$

and we call $\overline{\mathbb{E}}_s(X)$ the *upper expected value of X in s*. This is a generalization because $\overline{\mathbb{E}}(X) = \overline{\mathbb{E}}_\square(X)$. It is a small generalization, as $\overline{\mathbb{E}}_s(X) = \overline{\mathbb{E}}(X_{s+})$, where $X_{s+} : \omega \in \Omega \mapsto X(s\omega) \in \mathbb{R}$. When studying the functional $\overline{\mathbb{E}}_s$, we can often assume without loss of generality that $s = \square$.

In a situation where the value of a bounded below global variable depends only on World's next move, the global variable's upper expected value is given by the local upper expectation. This is spelled out in the following lemma.

Lemma 7.6 *Suppose s is a situation, f is a bounded below local variable, and X is a global variable satisfying $X(\omega) = f(\omega_{|s|+1})$ for all $\omega \in \Omega_s$. Then $\overline{\mathbb{E}}_s(X) = \overline{\mathbb{E}}(f)$.*

Proof Suppose $\mathcal{T} \in \mathbf{T}$ and $\lim \inf_n \mathcal{T}_n(\omega) \geq X(\omega)$ for all $\omega \in \Omega_s$. Define a local variable g by $g(y) := \mathcal{T}(sy)$ for all $y \in \mathcal{Y}$. By Lemma 7.3,

$$g(y) = \mathcal{T}(sy) \geq \inf_{\omega \in \Omega_{sy}} \limsup_n \mathcal{T}_n(\omega) \geq \inf_{\omega \in \Omega_{sy}} X(\omega) = f(y).$$

So $f \leq g$ and hence $\overline{\mathbb{E}}(f) \leq \overline{\mathbb{E}}(g)$. The strategy that produces $\mathcal{T}$ chooses g on round $|s| + 1$ in situation s, and so it follows from (7.2) that $\overline{\mathbb{E}}(g) \leq \mathcal{T}(s)$. Hence $\overline{\mathbb{E}}(f) \leq \mathcal{T}(s)$. It follows from (7.20) that $\overline{\mathbb{E}}(f) \leq \overline{\mathbb{E}}_s(X)$.

Now let $\mathcal{T}$ be the supermartingale produced by the strategy that sets $\mathcal{K}_0 := \overline{\mathbb{E}}(f)$, plays f in the situation s, and plays the constant equal to its current capital in all other situations. Then $\mathcal{T}(s) = \overline{\mathbb{E}}(f)$ and $\lim \inf_n \mathcal{T}_n(\omega) = f(\omega_{|s|+1}) = X(\omega)$ for all $\omega \in \Omega_s$. It follows from (7.20) that $\overline{\mathbb{E}}_s(X) \leq \overline{\mathbb{E}}(f)$. □

Alternative Definitions of Upper Expected Value

Equation (7.19) says that if Skeptic begins with slightly more than $\overline{\mathbb{E}}(X)$, he can make sure that his capital eventually stays at or above X no matter what path World follows. This is consistent with the definition of upper expected value we gave in Chapter 1 for protocols with finite horizons (see Eq. (2.4)). But we might prefer to generalize from finite to infinite horizons in some different way. We might ask how much initial capital Skeptic needs to make his capital repeatedly come arbitrarily close to X, or how much he needs to make his capital converge to a limit that equals or exceeds X. As the following proposition shows, $\overline{\mathbb{E}}(X)$, as we have defined it, is the answer to these questions as well.

Proposition 7.7 *If X is a bounded below global variable and $s \in \mathbb{S}$, then*

$$\overline{\mathbb{E}}_s(X) = \inf\{\mathcal{T}(s)|\mathcal{T} \in \mathbf{T} \text{ and } \forall \omega \in \Omega_s : \limsup_{n \to \infty} \mathcal{T}_n(\omega) \geq X(\omega)\}$$

$$= \inf\{\mathcal{L}(s)|\mathcal{L} \in \mathbf{L} \text{ and } \forall \omega \in \Omega_s : \lim_{n \to \infty} \mathcal{L}_n(\omega) \geq X(\omega)\}.$$

Proof Assuming without loss of generality that $s = \square$, we need to show that $c_1 = c_2 = c_3$, where

$$c_1 := \inf\{\mathcal{L}_0 | \mathcal{L} \in \mathbf{L}, \lim_{n\to\infty} \mathcal{L}_n \geq X\},$$

$$c_2 := \inf\{\mathcal{T}_0 | \mathcal{T} \in \mathbf{T}, \liminf_{n\to\infty} \mathcal{T}_n \geq X\},$$

$$c_3 := \inf\{\mathcal{T}_0 | \mathcal{T} \in \mathbf{T}, \limsup_{n\to\infty} \mathcal{T}_n \geq X\}.$$

Since

$$\{\mathcal{L} \in \mathbf{L} | \lim_n \mathcal{L}_n \geq X\} \subseteq \{\mathcal{T} \in \mathbf{T} | \liminf_n \mathcal{T}_n \geq X\} \subseteq \{\mathcal{T} \in \mathbf{T} | \limsup_n \mathcal{T}_n \geq X\},$$

$c_1 \geq c_2 \geq c_3$. So we only need to show that $c_3 \geq c_1$. It suffices, given $\varepsilon \in (0, 1)$ and $\mathcal{T} \in \mathbf{T}$ satisfying $\limsup_n \mathcal{T}_n \geq X$, to construct a supermartingale $\mathcal{L} \in \mathbf{L}$ satisfying

$$\mathcal{L}_0 \leq \mathcal{T}_0 + \varepsilon \tag{7.21}$$

and

$$\lim_n \mathcal{L}_n \geq X. \tag{7.22}$$

By Lemma 7.3, $\mathcal{T}$ is bounded below.

If $\mathcal{T}_0 = \infty$, let $\mathcal{L}$ be the supermartingale identically equal to ∞. If $\mathcal{T}_0 < \infty$, define a positive supermartingale B satisfying $B_0 = 1$ by setting

$$B := \frac{\mathcal{T} - C}{\mathcal{T}_0 - C},$$

where C is a constant satisfying $C < \inf \mathcal{T}$. On each path ω, B has the same asymptotic behavior as $\mathcal{T}$: either B and $\mathcal{T}$ both converge in $\mathbb{R}$ on ω or both tend to infinity on ω or both do neither (i.e. do not converge in $\overline{\mathbb{R}}$).

Construct a nonnegative supermartingale B' from B as in the proof of Theorem 7.5 and set $\mathcal{L} := \mathcal{T} + \varepsilon B'$. Because $\mathcal{T}$ has the same asymptotic behavior as B, we conclude as in the proof of Theorem 7.5 that $\mathcal{L} \in \mathbf{L}$. We have (7.21) because $B'_0 = 1$, and we have (7.22) because $\limsup_n \mathcal{T}_n \geq X$ and B' is nonnegative. $\square$

On the other hand, we do not get an equivalent definition if we replace lim inf by sup in (7.19) (see Exercise 7.3).

The Supermartingale of Upper Expected Values

Consider the process obtained from $\overline{\mathbb{E}}_s(X)$ by fixing X. We write $\overline{\mathbb{E}}_0(X), \overline{\mathbb{E}}_1(X), \dots$ for this process; these are the variables given by

$$\overline{\mathbb{E}}_n(X)(\omega) := \overline{\mathbb{E}}_{\omega^n}(X) \qquad (7.23)$$

for $n = 0, 1, \ldots$ and $\omega \in \Omega$.

Proposition 7.8 *For each global variable* X, $\overline{\mathbb{E}}_0(X), \overline{\mathbb{E}}_1(X), \ldots$ *is a supermartingale.*

Proof For simplicity, write S for the process $\overline{\mathbb{E}}_0(X), \overline{\mathbb{F}}_1(X), \ldots$. Suppose $S(s) < r$. By (7.20), there exists a bounded below supermartingale $\mathcal{T}$ such that $\mathcal{T}(s) < r$ and $\lim\inf_n \mathcal{T}_n \geq X$. Again using (7.20), but with sy in the place of s, we see that $S(sy) \leq \mathcal{T}(sy)$ for all $y \in \mathcal{Y}$. So by Axiom E3 and Proposition 7.2,

$$\overline{\mathbf{E}}(S(s \cdot)) \leq \overline{\mathbf{E}}(\mathcal{T}(s \cdot)) \leq \mathcal{T}(s) < r.$$

Since this holds for all $r > S(s)$, we conclude that

$$\overline{\mathbf{E}}(S(s \cdot)) \leq S(s),$$

and hence, by Proposition 7.2, that S is a supermartingale.   □

7.4 LINDEBERG'S CENTRAL LIMIT THEOREM FOR MARTINGALES

To illustrate the scope of Protocol 7.1, we now use it to state an abstract game-theoretic version of Lindeberg's theorem. We omit the proof, which uses the same ideas as the game-theoretic proof of De Moivre's theorem in Chapter 2 and is given with slightly different terminology in [349, pp. 153–158].

Given a real-valued martingale $\mathcal{M}$ and a positive real number δ, define local variables $\Delta_{y_1 \ldots y_{i-1}}$ and processes $\overline{\mathcal{V}}$, $\underline{\mathcal{V}}$, and $\mathcal{L}^\delta$ by

$$\Delta_{y_1 \ldots y_{i-1}}(y) := \mathcal{M}(y_1 \ldots y_{i-1}y) - \mathcal{M}(y_1 \ldots y_{i-1}),$$

$$\overline{\mathcal{V}}_n := \sum_{i=1}^n \overline{\mathbf{E}}\left(\Delta^2_{y_1 \ldots y_{i-1}}\right), \qquad \underline{\mathcal{V}}_n := \sum_{i=1}^n \underline{\mathbf{E}}\left(\Delta^2_{y_1 \ldots y_{i-1}}\right),$$

$$\mathcal{L}^\delta_n := \sum_{i=1}^n \overline{\mathbf{E}}\left(\Delta^2_{y_1 \ldots y_{i-1}} \mathbf{1}_{|\Delta_{y_1 \ldots y_{i-1}}| \geq \delta}\right).$$

The variables $\overline{\mathcal{V}}_n$ and $\underline{\mathcal{V}}_n$ measure $\mathcal{M}$'s cumulative pathwise variance in the first n rounds, while $\mathcal{L}^\delta_n$ measures the portion of that variance contributed by jumps of size δ or more.

A mapping $N : \Omega \to \mathbb{N}$ is a *finite stopping time* if $N(\omega) = N(\omega')$ whenever $\omega^{N(\omega)} = \omega'^{N(\omega)}$. We write $\underline{\mathbb{E}}(X)$ for $-\overline{\mathbb{E}}(-X)$ and $\underline{\mathbb{P}}(E)$ for $\underline{\mathbb{E}}(\mathbf{1}_E)$.

Theorem 7.9   (Lindeberg's theorem) *For every bounded continuous function U and every $\varepsilon > 0$, there exists $\delta > 0$, not depending on $\underline{\mathcal{Y}}$ or $\overline{\mathrm{E}}$, such that if $\mathcal{M}$ is a real-valued martingale in Protocol 7.1, the processes $\overline{\mathcal{V}}$, $\underline{\mathcal{V}}$, and $\mathcal{L}^{\delta}$ are defined for $\mathcal{M}$ as above, N is a finite stopping time, and*

$$\underline{\mathbb{P}}\left(1 - \delta \le \underline{\mathcal{V}}_N \le \overline{\mathcal{V}}_N \le 1 + \delta \text{ and } \mathcal{L}_N^{\delta} \le \delta\right) > 1 - \delta,$$

then

$$\left| \underline{\mathbb{E}}\left(U(\mathcal{M}_N)\right) - \int_{-\infty}^{\infty} U(z)\, \mathcal{N}_{0,1}(dz) \right| < \varepsilon \tag{7.24}$$

and

$$\left| \overline{\mathbb{E}}(U(\mathcal{M}_N)) - \int_{-\infty}^{\infty} U(z)\, \mathcal{N}_{0,1}(dz) \right| < \varepsilon. \tag{7.25}$$

When ε is small, the inequalities (7.24) and (7.25) say that $\underline{\mathbb{E}}\left(U(\mathcal{M}_N)\right)$ and $\overline{\mathbb{E}}(U(\mathcal{M}_N))$ are nearly equal to each other and to the Gaussian expected value $\int_{-\infty}^{\infty} U(z)\, \mathcal{N}_{0,1}(dz)$, so that $U(\mathcal{M}_N)$ has an approximate global expected value equal to the Gaussian expected value. The theorem says that this will be true when $\mathcal{M}$ is likely to have a cumulative pathwise variance about equal to 1, very little of which is contributed by large jumps.

7.5   GENERAL ABSTRACT TESTING PROTOCOLS

In this section, we extend what we have learned about Protocol 7.1 to testing protocols in general. We proceed in two steps. First we generalize to an abstract testing protocol in which World remains Skeptic's only opponent, but the local upper expectation may vary from round to round, possibly depending on World's previous moves. Then we generalize further, to an abstract testing protocol in which World is replaced by Forecaster and Reality.

Varying the Local Upper Expectation

In order to permit World's move space to vary from round to round, perhaps depending on World's own previous moves, we now generalize Protocol 7.1's concept of a situation space. We say that a set $\mathbb{S}$ of finite sequences of elements of a nonempty set $\mathcal{Y}$ is a *situation space* on $\mathcal{Y}$ if

- the empty sequence $\square$ is in $\mathbb{S}$,
- if s is in $\mathbb{S}$, then every prefix of s is in $\mathbb{S}$, and

- for each $s \in \mathbb{S}$, the set

$$\mathcal{Y}_s := \{y \in \mathcal{Y}|sy \in \mathbb{S}\} \tag{7.26}$$

is nonempty.

The condition that (7.26) be nonempty is required only because we have chosen to study a protocol with an infinite horizon.

The following abstract protocol has as parameters a situation space in the sense just defined, together with an upper expectation for each situation. As the name *situation space* suggests, $\mathbb{S}$ is the set of all the possible initial sequences of moves by Skeptic's opponent World in this protocol.

Protocol 7.10

PARAMETERS: Situation space $\mathbb{S}$ on $\mathcal{Y}$; upper expectation $\overline{\mathbf{E}}_s$ on $\mathcal{Y}_s$ for each $s \in \mathbb{S}$

Skeptic announces $\mathcal{K}_0 \in \overline{\mathbb{R}}$.

FOR $n = 1, 2, \ldots$:

Skeptic announces $f_n \in \overline{\mathbb{R}}^{\mathcal{Y}_{y_1 \ldots y_{n-1}}}$ such that $\overline{\mathbf{E}}_{y_1 \ldots y_{n-1}}(f_n) \le \mathcal{K}_{n-1}$.

World announces $y_n \in \mathcal{Y}_{y_1 \ldots y_{n-1}}$.

$\mathcal{K}_n := f_n(y_n)$.

The *sample space*, the set of all possible complete sequences of moves by World, is the set $\Omega := \{\omega \in \mathcal{Y}^{\infty} | \forall n \in \mathbb{N} : \omega^n \in \mathbb{S}\}$.

The assumption that play never stops does not limit the generality of Protocol 7.10 with respect to the results of this book. Play effectively stops after a given situation s if $\mathcal{Y}_t$ is a one-element set for every subsequent situation t. World then has no further choices to make and Skeptic's only choices are whether to give up money and how much.

Protocol 7.1 is a specialization of Protocol 7.10 – the specialization in which World's move space and the local upper expectation are constant. We can use essentially the same terminology and notation in Protocol 7.10 as in Protocol 7.1, and when we do so, the results and proofs of Sections 7.2–7.3 remain valid. Only the most obvious editing is needed to take account of the dependence of World's move space and the local upper expectation on the situation: add the appropriate subscript (usually s or $\square$) to $\mathcal{Y}$ and $\overline{\mathbf{E}}$ throughout. This means in particular that $\psi_n^{\text{bet}}(\omega)$ is an element of $\overline{\mathbb{R}}^{\mathcal{Y}_{\omega^{n-1}}}$ rather than $\overline{\mathbb{R}}^{\mathcal{Y}}$.

Instead of reviewing Sections 7.2–7.3 in detail to confirm that their arguments carry over to Protocol 7.10, we can make a general argument that embeds play in an instantiation of Protocol 7.10 (call it the *original protocol*) in an *enlarged protocol* that instantiates Protocol 7.1.

Assume, without loss of generality, that the $\mathcal{Y}_s$ in the original protocol are all disjoint and that $\mathcal{Y} = \cup_s \mathcal{Y}_s$. Define an upper expectation $\overline{\mathbf{E}} : \overline{\mathbb{R}}^{\mathcal{Y}} \to \overline{\mathbb{R}}$ by

$$\overline{\mathbf{E}}(f) := \sup_{s \in \mathbb{S}} \overline{\mathbf{E}}_s(f|_{\mathcal{Y}_s}) \tag{7.27}$$

(see Exercise 7.8). We take as the enlarged protocol the instantiation of Protocol 7.1 with parameters $\mathcal{Y}$ and $\overline{\mathbf{E}}$. Its sample space is $\mathcal{Y}^\infty$, and its situation space is $\mathcal{Y}^*$. We continue to write Ω and $\mathbb{S}$ for the original protocol's sample space and situation space, respectively.

The move spaces $\mathcal{Y}_s$ being subsets of $\mathcal{Y}$, and the sample space Ω being a subset of $\mathcal{Y}^\infty$, all the moves and paths available to World in the original protocol are also available to him in the enlarged protocol. Our definition of $\overline{\mathbf{E}}$, (7.27), implies that if World, moving in the enlarged protocol, sticks with moves and hence paths available to him in the original protocol, then Skeptic can also play in essentially the same way as in the original protocol. The following lemma spells this out, exhibiting a many-to-one correspondence between supermartingales in the enlarged protocol and those in the original protocol.

Lemma 7.11

1. *If $\mathcal{T}$ is a supermartingale in the enlarged protocol, then $\mathcal{T}|_\mathbb{S}$ is a supermartingale in the original protocol.*

2. *If $\mathcal{T}$ is a supermartingale in the original protocol, then the process $\mathcal{T}^\uparrow$ in the enlarged protocol defined by $\mathcal{T}_n^\uparrow(\omega) := \mathcal{T}_m(\omega)$, where m is the largest integer less than or equal to n such that $\omega^m \in \mathbb{S}$, is a supermartingale in the enlarged protocol. (This supermartingale replicates $\mathcal{T}$ so long as World stays in situations in $\mathbb{S}$ and stops changing as soon as World moves into a situation not in $\mathbb{S}$.)*

Proof
1. Skeptic produces the capital process $\mathcal{T}|_\mathbb{S}$ in the original protocol by starting with $\mathcal{T}_0$ and making the same bet as $\mathcal{T}$ in every $s \in \mathbb{S}$. More precisely, he moves $f|_{\mathcal{Y}_s}$ in s when $\mathcal{T}$ moves f in s.

2. Skeptic produces the capital process $\mathcal{T}^\uparrow$ in the enlarged protocol by starting with $\mathcal{T}_0$, making the same bet as $\mathcal{T}$ makes so long as he is in a situation $s \in \mathbb{S}$ (more precisely, making the move $f^\uparrow$ where f is $\mathcal{T}$'s move and $f^\uparrow(y)$ is equal to $f(y)$ for $y \in \mathcal{Y}_s$ and to $\mathcal{T}(s)$ for $y \in \mathcal{Y} \setminus \mathcal{Y}_s$), and then not betting as soon as he is in a situation in $\mathcal{Y}^* \setminus \mathbb{S}$.   □

The correspondences described in Lemma 7.11 preserve the properties we have been studying. If a supermartingale $\mathcal{T}$ in the enlarged protocol is nonnegative or always converges in $\mathbb{R}$ or $\overline{\mathbb{R}}$, then $\mathcal{T}|_\mathbb{S}$, which begins with the same capital as $\mathcal{T}$, has the same property. If a supermartingale $\mathcal{T}$ in the original protocol is nonnegative or converges in $\mathbb{R}$ or $\overline{\mathbb{R}}$, then $\mathcal{T}^\uparrow$, which begins with the same capital as $\mathcal{T}$, has the same property. This enables us to use these correspondences to generalize various results from Protocol 7.1 to 7.10.

Consider, for example, Doob's convergence theorem (Theorem 7.5). If $\mathcal{T}$ is a nonnegative supermartingale in the original protocol, then $\mathcal{T}^\uparrow$ is a nonnegative supermartingale in the enlarged protocol. The theorem says that there exists in the enlarged protocol a nonnegative supermartingale $\mathcal{L}$ that always converges in $\overline{\mathbb{R}}$ and tends to infinity on every path on which $\mathcal{T}^\uparrow$ does not converge in $\mathbb{R}$, and this implies that $\mathcal{L}|_\mathbb{S}$ is a nonnegative supermartingale in the original protocol that always converges in $\overline{\mathbb{R}}$ and tends to infinity on paths where $\mathcal{T}$ does not converge in $\mathbb{R}$.

In order to derive results concerning the upper expected value of a global variable X in the original protocol, we define a counterpart $X^\uparrow$ in the enlarged protocol by

$$X^\uparrow(\omega) := \begin{cases} X(\omega) & \text{if } \omega \in \Omega, \\ -\infty & \text{otherwise.} \end{cases}$$

A supermartingale in the original protocol reaches or exceeds X if and only if the corresponding supermartingale in the enlarged protocol (which begins with the same capital) reaches or exceeds $X^\uparrow$:

- A supermartingale $\mathcal{T}$ in the enlarged protocol will asymptotically reach or exceed $X^\uparrow$ in any given sense (lim inf, lim sup, lim, or sup) if and only if $\mathcal{T}|_\mathbb{S}$ reaches or exceeds X in this same sense.
- A supermartingale $\mathcal{T}$ in the original protocol will asymptotically reach or exceed X in any given sense if and only if $\mathcal{T}^\uparrow$ reaches or exceeds $X^\uparrow$ in this same sense.

It follows that X's upper expected value in the original protocol is the same as $X^\uparrow$'s in the enlarged protocol, and hence that Proposition 7.7 and our other results concerning global upper expected value in Protocol 7.1 generalize to Protocol 7.10.

Bringing Back Forecaster

As noted in the introduction to this chapter, World can represent both Forecaster and Reality when we are considering what Skeptic can achieve playing against them. What a strategy for Skeptic can achieve as a function of their moves does not change if he reveals at the beginning of each round how the strategy tells him to move as a function of Forecaster's move on that round. In this case, Forecaster effectively moves after Skeptic rather than before, and we can pair his move with Reality's, calling it a single move by World.

We now formalize this point by using an abstract testing protocol that pits Skeptic against Forecaster and Reality. We show that any instantiation of this protocol can be translated into an instantiation of the protocol where Skeptic faces only World (Protocol 7.10). By verifying that Skeptic has the same strategies with the same capital processes when World makes the moves as when Forecaster and Reality make the same moves, we will confirm that our results about what Skeptic can accomplish in the World protocol also hold in the Forecaster/Reality protocol.

We begin with the concept of a *family of upper expectations* $(\mathcal{Y}_\theta, \overline{\mathbf{E}}_\theta)_{\theta \in \Theta}$. Here Θ and the $\mathcal{Y}_\theta$ are nonempty sets, and $\overline{\mathbf{E}}_\theta$ is an upper expectation on $\mathcal{Y}_\theta$. A *situation space for a family of upper expectations* $(\mathcal{Y}_\theta, \overline{\mathbf{E}}_\theta)_{\theta \in \Theta}$ is a set $\mathbb{S}$ of finite sequences $s = (\theta_1, y_1) \ldots (\theta_n, y_n)$, where $n \in \{0, 1, \ldots\}$, $\theta_i \in \Theta$, and $y_i \in \mathcal{Y}_{\theta_i}$, that satisfies these conditions:

- the empty sequence $\square$ is in $\mathbb{S}$,
- if $s = (\theta_1, y_1) \ldots (\theta_n, y_n)$ is in $\mathbb{S}$, then every prefix $(\theta_1, y_1) \ldots (\theta_i, y_i)$, where $i \leq n$, is in $\mathbb{S}$,
- if $s \in \mathbb{S}$, then the set

$$\Theta_s := \{\theta \in \Theta | \exists y : s\theta y \in \mathbb{S}\}$$

  is nonempty, and
- if $s \in \mathbb{S}$ and $\theta \in \Theta_s$, then $s\theta y \in \mathbb{S}$ if and only if $y \in \mathcal{Y}_\theta$.

Such a set $\mathbb{S}$ is a situation space on the set $\mathcal{Y} := \{\theta y | \theta \in \Theta \text{ and } y \in \mathcal{Y}_\theta\}$ in the sense defined at the beginning of this section.

The following abstract testing protocol uses a situation space for a family of upper expectations as a parameter and gives rules of play that make it the situation space in our usual sense – the set of all possible initial sequences of moves by Skeptic's opponents.

Protocol 7.12

PARAMETERS: Situation space $\mathbb{S}$ for family of upper expectations $(\mathcal{Y}_\theta, \overline{\mathbf{E}}_\theta)_{\theta \in \Theta}$

> Skeptic announces $\mathcal{K}_0 \in \overline{\mathbb{R}}$.
> FOR $n = 1, 2, \ldots$:
>> Forecaster announces $\theta_n \in \Theta_{\theta_1 y_1 \ldots \theta_{n-1} y_{n-1}}$.
>> Skeptic announces $f_n \in \overline{\mathbb{R}}^{\mathcal{Y}_{\theta_n}}$ such that $\overline{\mathbf{E}}_{\theta_n}(f_n) \leq \mathcal{K}_{n-1}$.
>> Reality announces $y_n \in \mathcal{Y}_{\theta_n}$.
>> $\mathcal{K}_n := f_n(y_n)$.

The *sample space*, the set of all possible complete sequences of moves by Skeptic's opponents, is the set Ω of infinite sequences $\omega = (\theta_1, y_1)(\theta_2, y_2) \ldots$ all of whose prefixes $\omega^n = (\theta_1, y_1) \ldots (\theta_n, y_n)$ are in $\mathbb{S}$. We may omit the parentheses as well as the commas when representing an element of $\mathbb{S}$ or Ω.

Protocol 7.12 reveals the full scope of the notion of a testing protocol. Formally, a *testing protocol* is any protocol that can be obtained from Protocol 7.12 by the operations of instantiation and specialization. We will sometimes also apply the name to more complicated protocols obtained when we add other players with other goals.

It is evident that Protocol 7.10 is a specialization of Protocol 7.12. To reduce Protocol 7.12 to Protocol 7.10, we constrain Forecaster to setting $\theta_n := y_1 \ldots y_{n-1}$ on round n. Forecaster and the θs no longer having a role, we remove Forecaster from the protocol and the θs from the notation.

But as we have already asserted and will now demonstrate, we can also go in the other direction. For every instantiation of Protocol 7.12 there is a corresponding instantiation of Protocol 7.10 that has the same situation and sample spaces and the same supermartingales.

Instantiate Protocol 7.12 by fixing the parameters $(\mathcal{Y}_\theta, \overline{\mathbf{E}}_\theta)_{\theta \in \Theta}$ and $\mathbb{S}$ (and hence also the derived parameters $(\Theta_s)_{s \in \mathbb{S}}$ and Ω). Assume, without loss of generality, that all the Θ_s are disjoint. Then define a corresponding instantiation of Protocol 7.10 by using the same situation space $\mathbb{S}$ and hence the same sample space Ω and setting

$$\overline{\mathbf{E}}_s(f) := \sup_{\theta \in \Theta_s} \overline{\mathbf{E}}_\theta(f(\theta \cdot))$$

for all $s \in \mathbb{S}$ and $f \in \overline{\mathbb{R}}^{\mathcal{Y}_s}$, where, per (7.26), $\mathcal{Y}_s = \{\theta y | s\theta y \in \mathbb{S}\}$. The reader may easily verify that $\overline{\mathbf{E}}_s$ is an upper expectation on $\mathcal{Y}_s$ (cf. Eq. (7.27)).

To a sequence of moves $\omega = \theta_1 y_1 \theta_2 y_2 \ldots$ by Forecaster and Reality in the instantiation of Protocol 7.12 corresponds the same sequence of moves by World in the instantiation of Protocol 7.10. This is certainly a one-to-one correspondence. There is also a one-to-one correspondence between strategies for Skeptic in the two instantiations. To the strategy for Skeptic in Protocol 7.10 that prescribes the move $f_n \in \overline{\mathbb{R}}^{\mathcal{Y}_{\theta_1 y_1 \ldots \theta_{n-1} y_{n-1}}}$ when World has moved $\theta_1 y_1 \ldots \theta_{n-1} y_{n-1}$ corresponds the strategy in Protocol 7.12 that prescribes the move $f_n(\theta_n \cdot) \in \overline{\mathbb{R}}^{\mathcal{Y}_{\theta_n}}$ when Forecaster and Reality have moved $\theta_1 y_1 \ldots \theta_n$. The strategy in the one instantiation produces the same supermartingale on $\mathbb{S}$ as the corresponding strategy in the other instantiation. Because Skeptic has the same supermartingales in the two instantiations, the results we have verified for Protocol 7.10 are also valid for Protocol 7.12.

7.6 MAKING THE RESULTS OF PART I ABSTRACT

In Part I, we made our results as concrete as possible by framing them as direct assertions about moves by Skeptic's opponents. Chapter 4's law of large numbers, for example, was an assertion about Reality's moves $y_1 y_2 \ldots$ and Forecaster's moves $m_1 v_1 m_2 v_2 \ldots$. In most applications, however, the individual, system, or process that plays Reality's role is not merely and directly producing a sequence of numbers $y_1 y_2 \ldots$. Rather, each y_n summarizes some aspect of what the system is doing. The theory or process that provides the m_n and the v_n may also be doing more. This suggests a more abstract formulation, which uses an abstract protocol such as Protocol 7.12 and treats the y_n as local variables.

Recall that when the upper and lower expected values of a local variable f with respect to an upper expectation are equal, we write $\mathbf{E}(f)$ for their common value and call it f's expected value (Section 6.1). When both f and $(f - \mathbf{E}(f))^2$ have expected values, let us similarly write $\mathbf{V}(f)$ for $\mathbf{E}((f - \mathbf{E}(f))^2)$ and call it f's *variance*. Then we can state an abstract version of Statement 1 of Theorem 4.3 as follows.

Corollary 7.13 *Consider an instantiation of Protocol 7.12 and a family $(h_\theta)_{\theta \in \Theta}$ of real-valued local variables $h_\theta : \mathcal{Y}_\theta \to \mathbb{R}$. Suppose that for each $\theta \in \Theta$ the local variable h_θ has an expected value $\mathbf{E}_\theta(h_\theta)$ and a variance $\mathbf{V}_\theta(h_\theta)$. Then*

$$\sum_{n=1}^{\infty} \frac{\mathbf{V}_{\theta_n}(h_{\theta_n})}{n^2} < \infty \implies \frac{1}{n} \sum_{i=1}^{n} (h_{\theta_i}(y_i) - \mathbf{E}_{\theta_i}(h_{\theta_i})) \to 0 \quad \text{a.s.}$$

We can prove Corollary 7.13 by repeating the proof of Statement 1 of Theorem 4.3, with appropriate changes in notation. Or we can observe that the strategy for Skeptic in Protocol 4.1 that proves Statement 1 of Theorem 4.3 maps to a strategy in Protocol 7.12 that proves Corollary 7.13. Skeptic's move (M_n, V_n) on round n of Protocol 4.1 maps to the move

$$\mathcal{K}_{n-1} + M_n(h_{\theta_n} - \mathbf{E}_{\theta_n}(h_{\theta_n})) + V_n((h_{\theta_n} - \mathbf{E}_{\theta_n}(h_{\theta_n}))^2 - \mathbf{V}_{\theta_n}(h_{\theta_n}))$$

in Protocol 7.12. Write $\mathcal{T}$ and $\mathcal{T}^\dagger$ for the resulting nonnegative supermartingales in Protocols 4.1 and 7.12, respectively. Since $\mathcal{T}^\dagger$ is the same function of $h_{\theta_1} h_{\theta_2} \ldots$ as $\mathcal{T}$ is of $y_1 y_2 \ldots$, it will tend to infinity on a sequence of values for $h_{\theta_1} h_{\theta_2} \ldots$ if and only if $\mathcal{T}$ tends to infinity for the same sequence of values for $y_1 y_2 \ldots$. So Corollary 7.13 follows from our conclusions about $\mathcal{T}$ in Chapter 4.

Similarly, the proofs of Theorems 5.1 and 5.5 imply the following abstract versions.

Corollary 7.14 *Consider again an instantiation of Protocol 7.12 and a family $(h_\theta)_{\theta \in \Theta}$ of real-valued local variables $h_\theta : \mathcal{Y}_\theta \to \mathbb{R}$. Suppose again that for each $\theta \in \Theta$ the local variable h_θ has an expected value $\mathbf{E}_\theta(h_\theta)$ and a variance $\mathbf{V}_\theta(h_\theta)$. Then*

$$\left(A_n \to \infty \ \& \ |h_{\theta_n}(y_n) - \mathbf{E}_{\theta_n}(h_{\theta_n})| = o\left(\sqrt{\frac{A_n}{\ln \ln A_n}} \right) \right)$$
$$\implies \limsup_{n \to \infty} \frac{\sum_{i=1}^{n} (h_{\theta_i}(y_i) - \mathbf{E}_{\theta_i}(h_{\theta_i}))}{\sqrt{2A_n \ln \ln A_n}} \leq 1$$

almost surely, where $A_n := \sum_{i=1}^{n} \mathbf{V}_{\theta_i}(h_{\theta_i})$.

Suppose further that for each $\theta \in \Theta$, there exists a constant $c_\theta > 0$ such that $|h_\theta - \mathbf{E}_\theta(h_\theta)| \leq c_\theta$. Then

$$\left(A_n \to \infty \ \& \ c_{\theta_n} = o\left(\sqrt{\frac{A_n}{\ln \ln A_n}} \right) \right) \implies \limsup_{n \to \infty} \frac{\sum_{i=1}^{n} (h_{\theta_i}(y_i) - \mathbf{E}_{\theta_i}(h_{\theta_i}))}{\sqrt{2A_n \ln \ln A_n}} = 1$$

almost surely.

7.7 EXERCISES

Exercise 7.1 Suppose P is a probability measure on a measurable set $\mathcal{Y}$, and consider the following protocol:

> Skeptic announces $\mathcal{K}_0 \in \overline{\mathbb{R}}$.
> FOR $n = 1, 2, \ldots$:
>> Skeptic announces $f_n \in \overline{\mathbb{R}}^{\mathcal{Y}}$ such that $P(f_n)$ exists and $P(f_n) \leq \mathcal{K}_{n-1}$.
>> World announces $y_n \in \mathcal{Y}$.
>> $\mathcal{K}_n := f_n(y_n)$.

Verify that the slackening of this protocol is a specialization of Protocol 7.1. *Hint.* See Exercise 6.12. □

Exercise 7.2 Suppose $a_1, a_2, \ldots \in [0, \infty]$.

1. If $a_1, a_2, \ldots$ converges in $\mathbb{R}$, then $\lim_{N \to \infty} \inf_{n \geq N}(a_n - a_N) = 0$.
2. If $\lim \inf_{N \to \infty} \inf_{n \geq N}(a_n - a_N) \geq 0$, then $a_1, a_2, \ldots$ converges in $\overline{\mathbb{R}}$. □

Exercise 7.3 Consider the slackening of Protocol 1.10 with $p = 1/2$. This is the instantiation of Protocol 7.1 with $\mathcal{Y} = \{0, 1\}$ and $\overline{\mathbb{E}}(f) = (f(0) + f(1))/2$.

1. Let N be the number of 1s at the beginning of a path ω. That is to say,

$$N(\omega) := \sup\{n | \omega_1 = \cdots = \omega_n = 1\},$$

with $\sup \emptyset := 0$. Show that N has a geometric distribution by verifying that $\mathbb{E}(N) = 1$ and, more generally, that

$$\mathbb{E}(f(N)) = \sum_{n=0}^{\infty} \frac{f(n)}{2^{n+1}} \tag{7.28}$$

when f is a nonnegative function.

2. Show that for each $\varepsilon \in (0, 1]$ there exists a bounded nonnegative global variable X such that $\overline{\mathbb{E}}(X) = 1$ and $\overline{\mathbb{E}}^*(X) = \varepsilon$, where

$$\overline{\mathbb{E}}^*(X) := \inf\left\{ \mathcal{T}_0 | \mathcal{T} \in \mathbf{T}, \ \mathcal{T} \geq 0, \ \sup_n \mathcal{T}_n \geq X \right\}.$$

This shows that lim inf, lim sup, and lim cannot be replaced in the definition of global upper expected value by sup (cf. Eq. (7.19) and Proposition 7.7). *Hint.* Consider the unbounded global variable $Y := 2^N \varepsilon$. Show that $\overline{\mathbb{E}}^*(Y) = \varepsilon$ and $\overline{\mathbb{E}}(Y) = \infty$. Show that there exists $a \geq \varepsilon$ such that $\overline{\mathbb{E}}(\min(Y, a)) = 1$ (the function $a \mapsto \overline{\mathbb{E}}(\min(Y, a))$ is continuous), and set $X := \min(Y, a)$.

3. Does this remain true when $\varepsilon = 0$? □

Exercise 7.4   Consider the special case of Protocol 7.1 with $\mathcal{Y} = \{0, 1\}$ and $\overline{\mathbf{E}}(f) = (f(0) + f(1))/2$ for all $f \in \overline{\mathbb{R}}^{\mathcal{Y}}$.

1. Show that a nonnegative supermartingale with initial value 0 in this protocol must be identically equal to 0.

2. Let E be the set of all $\omega \in \Omega$ containing only finitely many 1s. Show that $\overline{\mathbb{P}}(E) = 0$. *Hint 1.* One way to do this is to note that $\overline{\mathbb{P}}(E) = L(E)$ for all Borel sets in $\{0, 1\}^\infty$, where L is the uniform probability measure on $\{0, 1\}^\infty$ equipped with the Borel σ-algebra (this is a special case of Theorem 9.3). *Hint 2.* Consider the strategy for Skeptic that multiplies his capital by $1 \pm \varepsilon$ at each step, with $\varepsilon \in (0, 1)$ and the capital going up when Reality chooses 0.

3. Show that the infimum in (7.19) is not attained when $s := \square$ and $X := \mathbf{1}_E$.

4. Show that in this case there is no smallest element in the class of all supermartingales $\mathcal{T}$ satisfying $\liminf_n \mathcal{T}_n \geq X$. $\square$

Exercise 7.5   The offer corresponding to the upper expectation $\overline{\mathbb{E}}$ in Protocol 7.1 is the set

$$\{X \in \overline{\mathbb{R}}^{\mathcal{Y}} | \overline{\mathbb{E}}(X) \leq 0\}$$

$$= \{X \in \overline{\mathbb{R}}^{\mathcal{Y}} | \forall \varepsilon > 0 \, \exists \mathcal{T} \in \mathbf{T} : \inf \mathcal{T} > -\infty, \mathcal{T}_0 \leq \varepsilon, \liminf_n \mathcal{T}_n \geq X\}$$

$$= \{X \in \overline{\mathbb{R}}^{\mathcal{Y}} | \forall \varepsilon > 0 \, \exists \mathcal{T} \in \mathbf{T} : \inf \mathcal{T} > -\infty, \mathcal{T}_0 = 0, \liminf_n \mathcal{T}_n \geq X - \varepsilon\}. \quad (7.29)$$

1. Show by example that the set (7.29) is not necessarily equal to

$$\{X \in \overline{\mathbb{R}}^{\mathcal{Y}} | \exists \mathcal{T} \in \mathbf{T} : \inf \mathcal{T} > -\infty, \mathcal{T}_0 = 0, \liminf_n \mathcal{T}_n \geq X\}. \quad (7.30)$$

Hint. See Exercise 7.4.

2. Show that the set (7.30) satisfies Axioms G1–G4 but may fail to satisfy Axioms G0 and G5. $\square$

Exercise 7.6   A set σ of situations in a testing protocol is a *cut* if each path passes through exactly one of the situations in σ. If σ is a cut and ω is a path, we write $\sigma(\omega)$ for this unique situation.

1. Show that the function $\omega \in \Omega \mapsto |\sigma(\omega)|$ is a finite stopping time.

2. Show that each finite stopping time can be obtained this way. $\square$

Exercise 7.7   Suppose $\mathcal{Y}$ is a nonempty set.

1. Show that if Ω is the sample space for a situation space $\mathbb{S}$ on $\mathcal{Y}$, as defined in Section 7.5, then

$$\mathbb{S} = \{\omega^n | \omega \in \Omega, n \in \{0, 1, \dots\}\}. \quad (7.31)$$

2. Show that $\Omega \subseteq \mathcal{Y}^\infty$ is the sample space for a situation space on $\mathcal{Y}$ if and only if it is nonempty and closed in the topology generated by $\{\Omega_s | s \in \mathcal{Y}^*\}$ as its base.

3. Conclude that instead of taking the notion of a situation space as our starting point, we could define the notion of a sample space on $\mathcal{Y}$ by condition 2 and then define the notion of a situation space on $\mathcal{Y}$ by (7.31).   □

Exercise 7.8   Show that the functional defined by (7.27) is an upper expectation. □

Exercise 7.9   (tower rule) Recall from Exercise 7.6 the definition of a cut. Given a global variable X and a cut σ, let $\overline{\mathbb{E}}_\sigma(X)$ denote the global variable given by $\overline{\mathbb{E}}_\sigma(X)(\omega) := \overline{\mathbb{E}}_{\sigma(\omega)}(X)$. Show that $\overline{\mathbb{E}}(X) = \overline{\mathbb{E}}(\overline{\mathbb{E}}_\sigma(X))$.   □

Exercise 7.10   (supermartingale of upper expected values) Recall the definition (7.23) of the supermartingale of upper expected values. In the terminology of Exercise 7.9, $\overline{\mathbb{E}}_n$ is $\overline{\mathbb{E}}_\sigma$ when σ is the cut consisting of all situations of length n.

1. Suppose the global variable X is bounded and depends only on World's move on round n. In other words, there is a bounded local variable f such that $X(\omega) = f(\omega_n)$ for all $\omega \in \Omega$. Show that $\overline{\mathbb{E}}(X) = \overline{\mathbf{E}}(f)$.

2. Show that a bounded process $\mathcal{T}_0, \mathcal{T}_1, \ldots$ is a supermartingale if and only if $\overline{\mathbb{E}}_{n-1}(\mathcal{T}_n) \le \mathcal{T}_{n-1}$ for all $n \in \mathbb{N}$. *Hint.* Apply Lemma 7.6 to $\mathcal{T}_n - \mathcal{T}_{n-1}$ and each situation of length $n - 1$ and use Lemma 7.3.   □

Exercise 7.11   (causal independence) Show that if X and Y are nonnegative global variables in a testing protocol such that $X(\omega)$ depends only on ω_j and $Y(\omega)$ only on ω_k, where $j < k$, then $\overline{\mathbb{E}}(XY) = \overline{\mathbb{E}}(X)\overline{\mathbb{E}}(Y)$.   □

7.8 CONTEXT

The results of this chapter and the next chapter were first reported in 2012 [351].

Dropping the Continuity Axiom

Most of the results of this chapter, including Theorem 7.5 and Proposition 7.7, still hold when the local variables are priced by broad-sense upper expectations that do not necessarily satisfy the continuity axiom. The proofs we have given for those two results use the continuity axiom only to verify that the infinite convex combination (7.17) is a supermartingale. As noted in Statement 4 of Lemma 7.4, the continuity axiom implies that any countable sum of nonnegative supermartingales is a nonnegative supermartingale. The particular infinite convex combination $\mathcal{T}'$ given by (7.17) being a nonnegative supermartingale can be demonstrated, however, without using the continuity axiom. One such demonstration is given in [351, p. 12]; it notices that $\mathcal{T}'$'s increment is always a multiple of $\mathcal{T}$'s and uses Proposition 7.2 to deduce that $\mathcal{T}'$ is a supermartingale. (See also the proof for Protocol 4.4 in Section 4.2.) Another way of demonstrating that $\mathcal{T}'$ is a supermartingale is to delay the start of betting for the component strategies, starting ψ^k only on the kth round of play. This makes the sums $\sum_{k=1}^{\infty} \psi^{k,\text{bet}}(s)$ and $\sum_{k=1}^{\infty} \mathcal{T}^k(s)$ finite for each situation s.

Causal Independence

The notion of causal independence suggested by Exercise 7.11 was developed in [340] and [342]. As shown there, it generalizes to several different game-theoretic notions of conditional independence, which can be used to deepen standard accounts of probabilistic causality. See also the related theory of event spaces [341, 347].

8

Zero-One Laws

If an event E depends on a sequence $Z_1, Z_2, \ldots$ of random variables, and $\mathbb{P}_n(E)$ is its probability after $Z_1, \ldots, Z_n$ are known, then, almost surely, $\mathbb{P}_n(E)$ tends to 1 if E happens and to 0 if not. More generally, if an integrable random variable X is a function of $Z_1, Z_2, \ldots$, and $\mathbb{E}_n(X)$ is its expected value after $Z_1, \ldots, Z_n$ are known, then $\mathbb{E}_n(X)$ tends almost surely to X. This is Lévy's classical zero-one law, roughly as he stated it in 1937 [245, section 41]. The relation between Lévy's formulation and later measure-theoretic formulations is discussed in Section 8.7.

In the preceding chapter, we learned how to define the global upper expected value $\overline{\mathbb{E}}_s(X)$ for a global variable X and a situation s in a testing protocol. We also learned that $\overline{\mathbb{E}}_0(X), \overline{\mathbb{E}}_1(X), \ldots$ is a supermartingale, where $\overline{\mathbb{E}}_n(X)(\omega) := \overline{\mathbb{E}}_{\omega^n}(X)$. In this chapter, we prove a game-theoretic version of Lévy's law: if X is bounded, then $\liminf_{n \to \infty} \overline{\mathbb{E}}_n(X) \geq X$ and $\limsup_{n \to \infty} \underline{\mathbb{E}}_n(X) \leq X$ almost surely in the game-theoretic sense. This game-theoretic version of Lévy's law plays as important a role in game-theoretic probability as the measure-theoretic version plays in measure-theoretic probability. As we will see in Chapter 9, it also helps clarify the relation between game-theoretic and measure-theoretic probability.

In Section 8.1, we state and prove the game-theoretic version of Lévy's law for Protocol 7.1, where Skeptic plays against World and the local upper expectation is the same on every round. By the arguments we gave in Section 7.5, the law also holds in testing protocols where the local upper expectation may vary.

Game-Theoretic Foundations for Probability and Finance, First Edition. Glenn Shafer and Vladimir Vovk.
© 2019 John Wiley & Sons, Inc. Published 2019 by John Wiley & Sons, Inc.

In Section 8.2, we use the game-theoretic law to show that the global upper expected values in Protocol 7.1 form an upper expectation – the protocol's *global upper expectation*. This fact gives us a new way of thinking about almost sure events in testing protocols.

In Section 8.3, we consider what Lévy's law tells us about global upper probabilities, and in Section 8.4, we give conditions on a global variable X under which the game-theoretic version of Lévy's law reduces to the classical version: along almost every path, X has expected values that converge to X.

In Section 8.5, we use Lévy's law to derive game-theoretic versions of several other zero-one laws: Kolmogorov's zero-one law, the ergodicity of Bernoulli shifts, and Pál Bártfai and Pál Révész's approximate zero-one law.

8.1 LÉVY'S ZERO-ONE LAW

In this section, we prove the game-theoretic version of Lévy's law and consider what it tells us about another way of defining global upper expected value.

Theorem 8.1   *If X is a bounded-below global variable in Protocol 7.1, then*

$$\liminf_{n\to\infty} \overline{\mathbb{E}}_n(X) \geq X \quad \text{a.s.} \tag{8.1}$$

Proof Assume, without loss of generality, that X is bounded below by a positive constant c. (By Statement 1 of Proposition 6.4, adding a constant to X will add the same constant to $\liminf_n \overline{\mathbb{E}}_n(X)$.)

By the definition of upper expected value, we can choose for each situation $s \in \mathbb{S}$ a supermartingale $\mathcal{U}^s$ such that

$$\mathcal{U}^s(s) < \overline{\mathbb{E}}_s(X) + 2^{-|s|} \tag{8.2}$$

and

$$\liminf_{n\to\infty} \mathcal{U}^s_n(\omega) \geq X(\omega) \quad \text{for all } \omega \in \Omega_s. \tag{8.3}$$

We may choose $\mathcal{U}^s$ so that $\mathcal{U}^s(t) = \mathcal{U}^s(s)$ for situations t that do not follow s, and it then follows from (8.3) that $\mathcal{U}^s$ is bounded below by c. (See the remarks following Lemma 7.3.)

Let $[a_k, b_k]$, $k \in \mathbb{N}$, be an enumeration of all intervals with $0 < a_k < b_k < \infty$ and both endpoints rational. For each $k \in \mathbb{N}$, define a strategy ψ^k and its supermartingale $\mathcal{T}^k$ by setting $\psi^{k,\text{stake}} := 1$ and defining $\psi^{k,\text{bet}}$ by the following algorithm.

Start Go to Off Mode for instructions.

Off Mode Do not bet on the current round and continue not betting until you first reach a round n such that

$$n > \max(k, -\log(b_k - a_k)) \quad \text{and} \quad \overline{\mathbb{E}}_n(X) < a_k, \tag{8.4}$$

where log is the binary logarithm. Then go to On Mode for instructions. If you never reach a round n satisfying (8.4), just continue not betting.

On Mode On the round when you enter this mode, say n, bet following $\mathcal{U}^{\cdot \omega^n}$ (ω being the path in Ω you are taking), scaled to your current capital $\mathcal{K}_n$ (i.e. multiply $\mathcal{U}^{\cdot \omega^n}$'s move by $\mathcal{K}_n/\mathcal{U}_n^{\cdot \omega^n}$). Continue following this scaled version of $\mathcal{U}^{\cdot \omega^n}$ on subsequent rounds i so long as $\mathcal{U}_i^{\cdot \omega^n} \le b_k$. (Note that $\mathcal{U}_n^{\cdot \omega^n} \le b_k$ by (8.2) and (8.4).) If and when you reach a round i such that $\mathcal{U}_i^{\cdot \omega^n} > b_k$, go to Off Mode for instructions.

The supermartingale $\mathcal{T}^k$ is positive, because all the $\mathcal{U}^s$ are positive. Moreover, $\mathcal{T}^k$ tends to infinity whenever

$$\liminf_{n \to \infty} \overline{\mathbb{E}}_n(X) < a_k < b_k < X. \tag{8.5}$$

Indeed, by (8.4) it leaves every Off period that it enters along such a path, and by (8.3) it leaves every On period it enters; thus ψ^k passes through an infinite number of On periods on such a path. During each On period, the positive supermartingale $\mathcal{U}^{\cdot \omega^n}$, where n is the initial round in the period and ω is the path taken, increases from $\mathcal{U}_n^{\cdot \omega^n}$, which satisfies

$$\mathcal{U}_n^{\cdot \omega^n} < \overline{\mathbb{E}}_n(X)(\omega) + 2^{-n} < a_k + 2^{-n} < b_k,$$

to more than b_k. This implies that $\mathcal{U}^{\cdot \omega^n}$ and hence also $\mathcal{T}^k$ is multiplied by more than

$$\frac{b_k}{a_k + 2^{-n}},$$

which is greater than 1 and grows with n.

Now define a strategy ψ by $\psi := \sum_{k=1}^{\infty} 2^{-k} \psi^k$. This means that

$$\psi^{\text{stake}} = \sum_{k=1}^{\infty} 2^{-k} \psi^{k,\text{stake}} = 1$$

and

$$\psi_n^{\text{bet}} = \sum_{k=1}^{\infty} 2^{-k} \psi_n^{k,\text{bet}} \tag{8.6}$$

for all $n \in \mathbb{N}$. To see that the infinite sum on the right-hand side of (8.6) is well defined, note that the variable $\psi_n^{k,\text{bet}}(\omega)$ is a positive constant when $k \geq n$ (by (8.4), ψ^k does not bet in this case); thus only a finite number of the variables in the infinite sum can take negative values, and these are all bounded below.

Designate the supermartingale produced by ψ by $\mathcal{T}$. Then $\mathcal{T}_0 = 1$, $\mathcal{T} > 0$ (because the capital for each ψ^k is always positive), and $\mathcal{T}_n \to \infty$ when

$$\liminf_{n \to \infty} \overline{\mathbb{E}}_n(X) < X,$$

because in this case there is at least one $k \in \mathbb{N}$ for which (8.5) holds and hence $\mathcal{T}_n^k$ tends to infinity. This proves (8.1).   □

In Section 7.3, we considered several equivalent ways of defining the global upper expected value $\overline{\mathbb{E}}_s(X)$. Here is another.

Proposition 8.2   *Suppose X is a global variable and $s \in \mathbb{S}$. Then*

$$\overline{\mathbb{E}}_s(X) = \inf\{\mathcal{T}(s) | \mathcal{T} \in \mathbf{T}, \inf \mathcal{T} > -\infty, \liminf_n \mathcal{T}_n \geq X \text{ a.s. in } s\}. \tag{8.7}$$

Proof Assume without loss of generality that $s = \square$. Then our task is to show that

$$\inf\{\mathcal{T}_0 | \mathcal{T} \in \mathbf{T}, \inf \mathcal{T} > -\infty, \liminf_n \mathcal{T}_n \geq X \text{ a.s.}\}$$

$$= \inf\{\mathcal{T}_0 | \mathcal{T} \in \mathbf{T}, \inf \mathcal{T} > -\infty, \liminf_n \mathcal{T}_n \geq X\}.$$

The first infimum is certainly no greater than the second. On the other hand, for every $\varepsilon > 0$ and every $\mathcal{T} \in \mathbf{T}$ such that $\liminf_n \mathcal{T}_n \geq X$ a.s., there exists a nonnegative supermartingale $\mathcal{S}$ satisfying $\mathcal{S}_0 = \varepsilon$ and tending to infinity when $\liminf_n \mathcal{T}_n < X$. Because $\mathcal{T} + \mathcal{S} \in \mathbf{T}$, $\liminf_n(\mathcal{T}_n + \mathcal{S}_n) \geq X$, and $\mathcal{T}_0 + \mathcal{S}_0 = \mathcal{T}_0 + \varepsilon$, this implies that the first infimum is no less than the second.   □

If X is bounded below, then by Theorem 8.1 the supermartingale $\overline{\mathbb{E}}_n(X)$, $n = 0, 1, \ldots$, attains the infimum in (8.7) for every situation s. The infimum in (7.19) is not attained in general (see Exercise 7.4).

8.2 GLOBAL UPPER EXPECTATION

We now show that the functional $\overline{\mathbb{E}}_s$ obtained from $\overline{\mathbb{E}}_s(X)$ by fixing s and allowing X to vary is an upper expectation. We prove this for Protocol 7.1, but by the arguments given in Section 7.5, the result holds generally in testing protocols.

Proposition 8.3   *For each situation s, $\overline{\mathbb{E}}_s : X \in \overline{\mathbb{R}}^\Omega \mapsto \overline{\mathbb{E}}_s(X) \in \overline{\mathbb{R}}$ is an upper expectation on Ω.*

Proof Without loss of generality, set $s := \square$ and write $\overline{\mathbb{E}}$ for $\overline{\mathbb{E}}_\square$. So our task is to prove that $\overline{\mathbb{E}}$ is an upper expectation. Because (i) $\liminf_n(\mathcal{T}_n^1 + \mathcal{T}_n^2) \geq \liminf_n \mathcal{T}_n^1 + \liminf_n \mathcal{T}_n^2$ and $\liminf_n c\mathcal{T}_n = c\liminf_n \mathcal{T}_n$ for any bounded below processes $\mathcal{T}^1, \mathcal{T}^2$, $\mathcal{T}$ and any $c \geq 0$ and (ii) $\mathbf{T}$ is a convex cone (Statement 1 of Lemma 7.4), it follows directly from (7.19) that $\overline{\mathbb{E}}$ satisfies Axioms E1 and E2. Axiom E3 also follows directly from (7.19). Axiom E4 follows from Lemma 7.3.

To establish Axiom E5, consider $X_1 \leq X_2 \leq \cdots \in [0, \infty]^\Omega$ and set $X := \lim_{k\to\infty} X_k = \sup_k X_k$. Axiom E3 implies that $\overline{\mathbb{E}}(X) \geq \lim_{k\to\infty} \overline{\mathbb{E}}(X_k)$. So we only need to show that $\overline{\mathbb{E}}(X) \leq \lim_{k\to\infty} \overline{\mathbb{E}}(X_k)$. By Proposition 7.8 and Theorem 8.1, for each X_k the process $\mathcal{T}^k(s) := \overline{\mathbb{E}}_s(X_k)$ is a nonnegative supermartingale such that $\liminf_n \mathcal{T}_n^k \geq X_k$ a.s. The sequence $\mathcal{T}^k$ is increasing, $\mathcal{T}^1 \leq \mathcal{T}^2 \leq \cdots$, so the limit $\mathcal{T} := \lim_{k\to\infty} \mathcal{T}^k = \sup_k \mathcal{T}^k$ exists and is a nonnegative supermartingale (Statement 3 of Lemma 7.4) such that $\mathcal{T}_0 = \lim_{k\to\infty} \overline{\mathbb{E}}(X_k)$ and $\liminf_n \mathcal{T}_n \geq X$ almost surely. We can get rid of "almost surely" by adding εS to $\mathcal{T}$, where $\varepsilon > 0$ can be arbitrarily small and S is a nonnegative supermartingale that starts at $S_0 < \infty$ and satisfies $\lim_n S_n(\omega) = \infty$ for all $\omega \in \Omega$ violating $\liminf_n S_n(\omega) \geq X(\omega)$. $\square$

We call $\overline{\mathbb{E}}_s$ the protocol's *global upper expectation in s*. We write $\underline{\mathbb{E}}_s$ for the corresponding lower expectation (see Section 6.1). Thus $\underline{\mathbb{E}}_s(X) \leq \overline{\mathbb{E}}_s(X)$ (Proposition 6.5). We write $\overline{\mathbb{E}}$ for $\overline{\mathbb{E}}_\square$ and $\underline{\mathbb{E}}$ for $\underline{\mathbb{E}}_\square$.

Almost Sure Global Events

We now have two game-theoretic concepts of *almost sure* for a testing protocol. By the definition we stated in Section 7.2 and have used so far, a global event E is almost sure *in a testing protocol* if there exists a nonnegative supermartingale that begins with a finite value and tends to infinity on all paths not in E. But we now have a global upper expectation $\overline{\mathbb{E}}$ for the protocol, and according to the definition in Section 6.2, an event E is almost sure *with respect to an upper expectation* $\overline{\mathbb{E}}$ if there exists a nonnegative extended variable X such that $\overline{\mathbb{E}}(X) = 1$ and $X(\omega) = \infty$ for all $\omega \notin E$.

The following proposition confirms that these two concepts are equivalent.

Proposition 8.4 *A global event is almost sure in Protocol 7.1 if and only if it is almost sure with respect to $\overline{\mathbb{E}}$.*

Proof Suppose E is almost sure in Protocol 7.1. By choosing a nonnegative supermartingale that is finite at $\square$ and infinite on all paths not in E and then scaling if necessary, we obtain a nonnegative supermartingale $\mathcal{T}$ satisfying $\mathcal{T}_0 = 1$ and $\lim_n \mathcal{T}_n(\omega) = \infty$ for all $\omega \notin E$. Set $X := \liminf_n \mathcal{T}_n$. Then X and $\overline{\mathbb{E}}(X)$ are nonnegative and $X(\omega) = \infty$ for all $\omega \notin E$. Adding a positive constant if necessary, we obtain a nonnegative global variable X with $\overline{\mathbb{E}}(X) = 1$ and $X(\omega) = \infty$ for all $\omega \notin E$. So E is almost sure with respect to $\overline{\mathbb{E}}$.

Now suppose E is almost sure with respect to $\overline{\mathbb{E}}$. Choose a nonnegative global variable X such that $\overline{\mathbb{E}}(X) = 1$ and $X(\omega) = \infty$ for all $\omega \notin E$. By the definition of upper expected value, there exists a supermartingale $\mathcal{T}$ that begins with a value arbitrarily close to 1 and hence finite and satisfies $\lim \inf_n \mathcal{T}_n \geq X$. By Lemma 7.3, $\mathcal{T}$ is nonnegative. So E is almost sure in the protocol. $\qquad\square$

8.3 GLOBAL UPPER AND LOWER PROBABILITIES

In this section, we consider what Lévy's law says about global upper probability, and we show that lim inf can be replaced by sup in its definition.

We write $\overline{\mathbb{P}}_s(E)$ and $\underline{\mathbb{P}}_s(E)$ for the upper and lower probabilities given by the upper expectation $\overline{\mathbb{E}}_s$, and we write $\overline{\mathbb{P}}_n(E)$ and $\underline{\mathbb{P}}_n(E)$ for the variables $\overline{\mathbb{E}}_n(\mathbf{1}_E)$ and $\underline{\mathbb{E}}_n(\mathbf{1}_E)$. Recall the properties of upper and lower probabilities discussed in Section 6.2.

When Do Global Upper Probabilities Converge?

When we take the global variable X in (8.1) to be the indicator $\mathbf{1}_E$ for an event E, we obtain

$$\lim_{n\to\infty} \inf \overline{\mathbb{P}}_n(E) \geq \mathbf{1}_E \quad \text{a.s.} \tag{8.8}$$

Upper probabilities being bounded above by 1, we obtain the following corollary.

Corollary 8.5 *For any global event E, $\lim_{n\to\infty} \overline{\mathbb{P}}_{\omega^n}(E) = 1$ for almost all $\omega \in E$.*

Because $\underline{\mathbb{P}}_s(E^c) = 1 - \overline{\mathbb{P}}_s(E)$, it is also true that $\lim_{n\to\infty} \underline{\mathbb{P}}_{\omega^n}(E^c) = 0$ for almost all $\omega \in E$. But as the following example shows, we cannot replace $\geq$ in (8.8) by $=$.

Example 8.6 Consider the instantiation of Protocol 7.1 in which $\mathcal{Y} = \{0, 1\}$ and $\overline{\mathbb{E}}$ is the supremum. Let E be the subset of Ω consisting of paths that contain only finitely many 1s. Then $\overline{\mathbb{P}}_s(E) = 1$ for all $s \in \mathbb{S}$, and hence $\lim_{n\to\infty} \overline{\mathbb{P}}_n(E) = 1$ holds on all paths, including those not in E. $\qquad\square$

Another Way of Defining Global Upper Probability

By Proposition 7.7, we have three ways to express $\overline{\mathbb{P}}_s(E)$ as an infimum over supermartingales:

$$\overline{\mathbb{P}}_s(E) = \inf\{\mathcal{T}(s) | \mathcal{T} \in \mathbf{T}, \mathcal{T} \geq 0, \forall \omega \in E \cap \Omega_s : \lim_n \inf \mathcal{T}_n(\omega) \geq 1\}$$

$$= \inf\{\mathcal{T}(s) | \mathcal{T} \in \mathbf{T}, \mathcal{T} \geq 0, \forall \omega \in E \cap \Omega_s : \lim_n \sup \mathcal{T}_n(\omega) \geq 1\}$$

$$= \inf\{\mathcal{L}(s) | \mathcal{L} \in \mathbf{L}, \mathcal{L} \geq 0, \forall \omega \in E \cap \Omega_s : \lim_n \mathcal{L}_n(\omega) \geq 1\}. \tag{8.9}$$

As the following lemma shows, we can also substitute sup for lim inf in (8.9).

Lemma 8.7

$$\overline{\mathbb{P}}_s(E) = \inf\{\mathcal{T}(s)|\mathcal{T} \in \mathbf{T}, \mathcal{T} \geq 0, \forall \omega \in E \cap \Omega_s : \sup_n \mathcal{T}_n(\omega) \geq 1\}. \qquad (8.10)$$

Proof The value of the right-hand side of (8.10) does not change if we replace $\sup_n \mathcal{T}_n(\omega) \geq 1$ with $\sup_n \mathcal{T}_n(\omega) > 1$. Suppose Skeptic has a strategy ψ that produces a nonnegative supermartingale $\mathcal{T}$ satisfying $\sup_n \mathcal{T}_n > 1$ on $E \cap \Omega_s$. Then he can obtain a nonnegative supermartingale satisfying $\liminf_n \mathcal{T}_n > 1$ on $E \cap \Omega_s$ by starting to play ψ and stopping betting as soon his capital exceeds 1. □

On the other hand, we cannot substitute inf for lim inf in (8.9) (see Exercise 8.3).

8.4 GLOBAL EXPECTED VALUES AND PROBABILITIES

In this section, we look at conditions under which a global variable's upper and lower expected values merge into a global expected value, so that our game-theoretic version of Lévy's law looks like Lévy's classical statement of it.

When $\underline{\mathbb{E}}_s(X) = \overline{\mathbb{E}}_s(X)$, we write $\mathbb{E}_s(X)$ for the common value, and we call it X's *expected value in s*. Similarly, when $\overline{\mathbb{P}}_s(E) = \underline{\mathbb{P}}_s(E)$, we write $\mathbb{P}_s(E)$ for their common value and call it E's probability in s. Usually we omit explicit reference to the initial situation; when X has an expected value in □, we say simply that X has an expected value and write $\mathbb{E}(X)$ instead of $\mathbb{E}_\square(X)$.

Lemma 8.8 *Suppose a bounded global variable X has an expected value. Then for almost all ω, X has an expected value in ω^n for every $n \in \mathbb{N}$.*

Proof First we show that for fixed $\delta > 0$, $\overline{\mathbb{E}}_{\omega^n}(X) - \underline{\mathbb{E}}_{\omega^n}(X) \leq \delta$ for all n for almost all ω. Choose $\varepsilon > 0$ and supermartingales $\mathcal{T}^1$ and $\mathcal{T}^2$ such that

$$\mathcal{T}_0^1 < \overline{\mathbb{E}}(X) + \varepsilon/2 \quad \text{and} \quad \mathcal{T}_0^2 < \overline{\mathbb{E}}(-X) + \varepsilon/2 \qquad (8.11)$$

and

$$\liminf_n \mathcal{T}_n^1 \geq X \quad \text{and} \quad \liminf_n \mathcal{T}_n^2 \geq -X. \qquad (8.12)$$

Set $\mathcal{T} := \mathcal{T}^1 + \mathcal{T}^2$. By assumption, $\overline{\mathbb{E}}(X) = \underline{\mathbb{E}}(X) \in \mathbb{R}$, so that $\overline{\mathbb{E}}(X) + \overline{\mathbb{E}}(-X) = 0$. So the supermartingale $\mathcal{T}$ satisfies $\mathcal{T}_0 < \varepsilon$ by (8.11). It satisfies $\liminf_n \mathcal{T}_n \geq 0$ by (8.12), and it is therefore nonnegative. Now let E be the event that $\overline{\mathbb{E}}_{\omega^n}(X)$ and $\underline{\mathbb{E}}_{\omega^n}(X)$ differ by more than δ for some n — i.e. the set of ω satisfying

$$\exists n : \overline{\mathbb{E}}_{\omega^n}(X) + \overline{\mathbb{E}}_{\omega^n}(-X) > \delta.$$

By (8.12) and the definition of upper expected value in ω^n,

$$T_n^1(\omega) \geq \overline{\mathbb{E}}_{\omega^n}(X) \quad \text{and} \quad T_n^2(\omega) \geq \overline{\mathbb{E}}_{\omega^n}(-X).$$

So $T_n(\omega) > \delta$ and hence $\sup_n T_n(\omega) > \delta$ for all $\omega \in E$. Thus T/δ is a nonnegative supermartingale with initial value less than ε/δ and supremum exceeding 1 for all $\omega \in E$. It follows by Lemma 8.7 that the upper probability of E is less than ε/δ. Since ε may be arbitrarily small, this shows that E has upper probability 0; it is a null event.

Letting δ range over the strictly positive rational numbers and recalling that the union of a countable number of null events is null, we conclude that $\underline{\mathbb{E}}_{\omega^n}(X) = \overline{\mathbb{E}}_{\omega^n}(X)$ for all n almost surely. $\qquad \square$

The following corollary of Theorem 8.1 resembles Lévy's original statement of his zero-one law.

Corollary 8.9   *Suppose X is a bounded global variable and has an expected value. Then*

$$\lim_{n \to \infty} \mathbb{E}_n(X) = X \quad \text{a.s.}$$

(When we assert that a relation involving terms that are not always defined holds almost surely, we mean that, almost surely, they are defined and the relation is true. By Lemma 8.8, $\mathbb{E}_n(X)$ exists for all n almost surely under the hypotheses of this corollary.)

Proof By Theorem 8.1,

$$\liminf_{n \to \infty} \overline{\mathbb{E}}_{\omega^n}(X) \geq X(\omega) \tag{8.13}$$

and

$$\liminf_{n \to \infty} \overline{\mathbb{E}}_{\omega^n}(-X) \geq -X(\omega) \tag{8.14}$$

for almost all $\omega \in \Omega$. By Lemma 8.8, $\overline{\mathbb{E}}_{\omega^n}(X) = \underline{\mathbb{E}}_{\omega^n}(X) = \mathbb{E}_{\omega^n}(X)$ for almost all $\omega \in \Omega$. Substituting $\mathbb{E}_{\omega^n}(X)$ for $\overline{\mathbb{E}}_{\omega^n}(X)$ in (8.13) and $-\mathbb{E}_{\omega^n}(X)$ for $\overline{\mathbb{E}}_{\omega^n}(-X)$ in (8.14), they become

$$\liminf_{n \to \infty} \mathbb{E}_{\omega^n}(X) \geq X(\omega)$$

and

$$\limsup_{n \to \infty} \mathbb{E}_{\omega^n}(X) \leq X(\omega),$$

respectively. $\qquad \square$

Specializing Corollary 8.9 to the case where $X = \mathbf{1}_E$ for some event E, we obtain the following.

Corollary 8.10 *Suppose the event E has a probability. Then*

$$\lim_{n \to \infty} \mathbb{P}_n(E) = \mathbf{1}_E \quad \text{a.s.}$$

In other words, $\mathbb{P}_{\omega^n}(E)$ converges almost surely to either 0 or 1 – 0 if ω is not in E and 1 if ω is in E.

8.5 OTHER ZERO-ONE LAWS

In this section, we derive game-theoretic versions of three additional zero-one laws: Kolmogorov's zero-one law, the ergodicity of Bernoulli shifts, and Bártfai and Révész's zero-one law. All three are corollaries of our game-theoretic version of Lévy's zero-one law (Theorem 8.1 generalized to all testing protocols by the arguments in Section 7.5).

Kolmogorov's Zero-One Law

The most natural setting for this zero-one law is the following specialization of Protocol 7.10, which differs from Protocol 7.1 only by allowing the local upper expectation on round n to vary with n.

Protocol 8.11
PARAMETERS: Upper expectations $(\mathcal{Y}_n, \overline{\mathbf{E}}_n)$, $n \in \mathbb{N}$
   Skeptic announces $\mathcal{K}_0 \in \overline{\mathbb{R}}$.
   FOR $n = 1, 2, \ldots$:
      Skeptic announces $f_n \in \overline{\mathbb{R}}^{\mathcal{Y}_n}$ such that $\overline{\mathbf{E}}_n(f_n) \leq \mathcal{K}_{n-1}$.
      World announces $y_n \in \mathcal{Y}_n$.
      $\mathcal{K}_n := f_n(y_n)$.

We say that two paths ω and ω' *agree from round N onward* if $\omega_n = \omega'_n$ for all $n \geq N$. We say that an event E *is determined by the rounds from N onward* if any two paths that agree from round N onward are either both in E or neither in E. Let $\mathcal{G}_N$ be the set of all events that are determined by the rounds from N onward. Set $\mathcal{G} := \cap_{N \in \mathbb{N}} \mathcal{G}_N$ and call the elements of $\mathcal{G}$ *tail events*. Thus an event E is a tail event if any path that agrees with a path in E from some round onward is also in E.

The following proposition can be considered a corollary of Theorem 8.1.

Corollary 8.12 *If E is a tail event in Protocol 8.11, then $\overline{\mathbb{P}}(E) \in \{0, 1\}$.*

Proof Suppose E is a tail event. First we show that $\overline{\mathbb{P}}_s(E)$ is the same for all situations of the same length N. Suppose s and t are two situations of length N, where $N \in \mathbb{N}$. Suppose $\mathcal{T} \in \mathbf{T}$ and $\liminf_{n \to \infty} \mathcal{T}_n(\omega) \geq \mathbf{1}_E(\omega)$ for all $\omega \in \Omega_s$. Then since $E \in \mathcal{G}_{N+1}$, there exists $\mathcal{T}' \in \mathbf{T}$ such that $\mathcal{T}'(t) = \mathcal{T}(s)$ and $\liminf_{n \to \infty} \mathcal{T}'_n(\omega) \geq \mathbf{1}_E(\omega)$ for all $\omega \in \Omega_t$. (Skeptic produces $\mathcal{T}'$ by starting in □ with capital $\mathcal{T}(s)$ and not betting except in t and situations following t, betting in tu for $u \in \mathbb{S}$ as he would in su to produce $\mathcal{T}$.) Hence $\overline{\mathbb{P}}_t(E) \leq \overline{\mathbb{P}}_s(E)$ and so, by symmetry, $\overline{\mathbb{P}}_s(E) = \overline{\mathbb{P}}_t(E)$.

It is obvious, and follows by Exercise 7.9, that $\overline{\mathbb{P}}(E)$ is equal to the common value of $\overline{\mathbb{P}}_s(E)$ for s of length N. Since this is true for all $N \in \mathbb{N}$, $\overline{\mathbb{P}}_s(E)$ is the same for all $s \in \mathbb{S}$, always equal to $\overline{\mathbb{P}}(E)$. To say it another way, the martingale $\overline{\mathbb{P}}_.(E)$ is a constant always equal to $\overline{\mathbb{P}}(E)$. By (8.8) this constant supermartingale $\overline{\mathbb{P}}(E)$ satisfies

$$\liminf_{n \to \infty} \overline{\mathbb{P}}(E) \geq \mathbf{1}_E \quad \text{a.s.} \tag{8.15}$$

If $\overline{\mathbb{P}}(E)$ is not equal to 1, then (8.15) implies that $\mathbf{1}_E = 0$ almost surely, in which case $\mathbb{P}(E)$ is equal to 0. □

Since complements of tail events are also tail events, we obtain the following corollary to Corollary 8.12.

Corollary 8.13 *If $E \subseteq \Omega$ is a tail event in Protocol 8.11, then E is almost sure, almost impossible, or fully unprobabilized.*

Ergodicity of Bernoulli Shifts

Turn again to Protocol 7.1, the simple Skeptic vs. World protocol where the same outcome space $\mathcal{Y}$ and the same local upper expectation $\overline{\mathbf{E}}$ is used on every round.

We write θ for the shift operator, which deletes the first element from a sequence in $\mathcal{Y}^\infty$:

$$\theta : y_1 y_2 y_3 \ldots \mapsto y_2 y_3 \ldots.$$

In accordance with standard terminology, we call an event E *invariant* if $E = \theta^{-1}E$. We call an event E *weakly invariant* if $\theta E \subseteq E$. Since $\theta\theta^{-1}E = E$ for any event E, an invariant event is weakly invariant.

Lemma 8.14 *An event E is invariant if and only if both E and E^c are weakly invariant.*

Proof If E is invariant, then E^c is also invariant: see Exercise 8.6.

Conversely, suppose E and E^c are weakly invariant but E is not invariant. Since E is weakly invariant but not invariant, there exists $\omega \in E$ such that $y\omega \notin E$ for some

$y \in \mathcal{Y}$ (see Exercise 8.6). And since E^c is weakly invariant, $y\omega \notin E$ implies $\omega \notin E$. This contradiction completes the proof. □

The following corollary asserts the ergodicity of Bernoulli shifts.

Corollary 8.15 *If E is a weakly invariant event in Protocol 7.1, then $\overline{\mathbb{P}}(E) \in \{0, 1\}$.*

Proof Suppose E is a weakly invariant event. First we show that for any situation s, $\overline{\mathbb{P}}_s(E) \leq \overline{\mathbb{P}}(E)$. Suppose $\mathcal{T} \in \mathbf{T}$ and $\lim \inf_n \mathcal{T}_n \geq \mathbf{1}_E$. Then since $\theta^{|s|} E \subseteq E$, there exists $\mathcal{T}' \in \mathbf{T}$ such that $\mathcal{T}'(s) = \mathcal{T}_0$ and $\lim \inf_n \mathcal{T}_n'(\omega) \geq \mathbf{1}_E(\omega)$ for all $\omega \in \Omega_s$. (Skeptic produces $\mathcal{T}'$ by starting with capital $\mathcal{T}_0$ in □ and not betting except in s and situations following s, betting in st for all $t \in \mathbb{S}$ as he would in t to produce $\mathcal{T}$.) Hence $\overline{\mathbb{P}}_s(E) \leq \overline{\mathbb{P}}(E)$.

The supermartingale $\overline{\mathbb{P}}_{\cdot}(E)$ satisfies (8.8). If $\overline{\mathbb{P}}(E)$ is not equal to 1, then (8.8) implies that $\mathbf{1}_E = 0$ a.s., in which case $\overline{\mathbb{P}}(E)$ is equal to 0. □

In view of Lemma 8.14 we obtain the following corollary to Corollary 8.15.

Corollary 8.16 *If E is an invariant event in Protocol 7.1, then E is almost sure, almost impossible, or fully unprobabilized.*

Since each invariant event is a tail event, Corollary 8.16 also follows from Corollary 8.13.

Bártfai and Révész's Zero-One Law

We now derive a game-theoretic analog of an approximate zero-one law that Bártfai and Révész proved for dependent random variables in 1967 [23]. A suitable general setting for this result is the specialization of Protocol 7.10 in which the outcome space (World's move space) depends only on the round of the game, but the upper expectation may depend more specifically on the situation. Let us designate the outcome space for the nth round by $\mathcal{Y}_n$ and the resulting situation space, which consists of the prefixes of sequences in $\prod_{n=1}^{\infty} \mathcal{Y}_n$, by $\mathbb{S}$. Then we require an upper expectation $\overline{\mathbf{E}}_s$ on $\mathcal{Y}_{|s|+1}$ for each $s \in \mathbb{S}$.

Protocol 8.17

PARAMETERS: Nonempty sets $\mathcal{Y}_n$, $n \in \mathbb{N}$; upper expectation $\overline{\mathbf{E}}_s$ on $\mathcal{Y}_{|s|+1}$, $s \in \mathbb{S}$
  Skeptic announces $\mathcal{K}_0 \in \overline{\mathbb{R}}$.
  FOR $n = 1, 2, \ldots$:
    Skeptic announces $f_n \in \overline{\mathbb{R}}^{\mathcal{Y}_n}$ such that $\overline{\mathbf{E}}_{y_1 \ldots y_{n-1}}(f_n) \leq \mathcal{K}_{n-1}$.
    World announces $y_n \in \mathcal{Y}_n$.
  $\mathcal{K}_n := f_n(y_n)$.

We define $\mathcal{G}_N$ for $N \in \mathbb{N}$ and the concept of *tail event* as in Protocol 8.11. Given $\delta \in [0, 1)$, we say that Protocol 8.17 is δ-*mixing* if there exists a function $a : \mathbb{N} \to \mathbb{N}$ such that, for all $n \in \mathbb{N}$ and $E \in \mathcal{G}_{n+a(n)}$,

$$\overline{\mathbb{P}}_n(E) - \overline{\mathbb{P}}(E) \leq \delta \quad \text{a.s.} \tag{8.16}$$

The following approximate zero-one law is a game-theoretic analog of the main result of [23].

Corollary 8.18   *Suppose Protocol 8.17 is δ-mixing, where $\delta \in [0, 1)$. Suppose E is a tail event. Then $\overline{\mathbb{P}}(E) \geq 1 - \delta$ or $\overline{\mathbb{P}}(E) = 0$.*

Proof Fix a tail event E. Since a countable union of null events is null, (8.16) holds for all n almost surely. So if $\overline{\mathbb{P}}(E) < 1 - \delta$, there exists $\varepsilon > 0$ such that $\overline{\mathbb{P}}(E) < 1 - \delta - \varepsilon$, and then

$$\forall n \in \mathbb{N} : \overline{\mathbb{P}}_n(E) \leq 1 - \varepsilon \quad \text{a.s.} \tag{8.17}$$

The supermartingale $\overline{\mathbb{P}}.(E)$ satisfies (8.8). When (8.17) holds, (8.8) is possible only if $\mathbf{1}_E = 0$ a.s. – i.e. $\overline{\mathbb{P}}(E) = 0$.   □

The following consequence of Corollary 8.18 can be compared with Corollary 1 in [23].

Corollary 8.19   *Suppose Protocol 8.17 is δ-mixing, where $\delta \in [0, 1/2)$. Suppose E is a tail event. Then $\underline{\mathbb{P}}(E) = 1$, $\overline{\mathbb{P}}(E) = 0$, or E is unprobabilized.*

Proof Apply Corollary 8.18 to the tail events E and E^c.   □

Given $\delta \in [0, 1)$, we say that Protocol 8.17 is *asymptotically δ-mixing* if (8.16) holds for each $n \in \mathbb{N}$ and each tail event E. The following corollary is similar to (but much simpler than) Theorems 2 and 3 in [23]. (In their measure-theoretic treatment, Bártfai and Révész do not use a notion of asymptotic δ-mixing; instead, they use notions intermediate between δ-mixing and asymptotic δ-mixing.)

Corollary 8.20   *In Protocol 8.17 and for $\delta \in [0, 1)$, the following two conditions are equivalent:*

1. *The protocol is asymptotically δ-mixing.*
2. *Every tail event E satisfies $\overline{\mathbb{P}}(E) = 0$ or $\overline{\mathbb{P}}(E) \geq 1 - \delta$.*

Proof The argument of Corollary 8.18 shows that the first condition implies the second. Let us now assume the second condition and deduce the first. Let $n \in \mathbb{N}$ and E be a tail event. If $\overline{\mathbb{P}}(E) = 0$, then we can prove that $\overline{\mathbb{P}}_n(E) = 0$ a.s. by an argument similar to the proof of Lemma 8.8, and so (8.16) holds. If $\overline{\mathbb{P}}(E) \geq 1 - \delta$, (8.16) is vacuous. □

8.6 EXERCISES

Exercise 8.1   Give an example of an instantiation of Protocol 7.1 with a global variable X that is not bounded below and for which it is not true that $\liminf_n \overline{\mathbb{E}}_n(X) \geq X$ almost surely. □

Exercise 8.2   Simplify the proof of Theorem 8.1 using Axiom E5. □

Exercise 8.3   Show that we cannot replace lim inf with inf in (8.9), because

$$\inf\{\mathcal{T}(s)|\mathcal{T} \in \mathbf{T}, \mathcal{T} \geq 0, \forall \omega \in E \cap \Omega_s : \inf_n \mathcal{T}_n(\omega) \geq 1\} = \begin{cases} 0 & \text{if } E \cap \Omega_s = \emptyset, \\ 1 & \text{otherwise.} \end{cases}$$
  □

Exercise 8.4   Consider a modification of Protocol 7.1 in which we assume that the parameter $\overline{\mathbf{E}}$ does not necessarily satisfy the continuity axiom; instead we assume that it is a broad-sense upper expectation (i.e. that it satisfies Axioms E1–E4) and that it is countably subadditive. Show that $\overline{\mathbb{E}}_s$ is then also countably subadditive, for each $s \in \mathbb{S}$. □

Exercise 8.5   The following example shows that the condition that $\overline{\mathbb{E}}(X) = \underline{\mathbb{E}}(X)$ be different from $-\infty$ and ∞ is essential in the statement of Lemma 8.8. Consider the coin-tossing protocol (Protocol 1.10). Let $f : \Omega \to [0, 1]$ be a function whose upper integral is 1 and lower integral is 0 with respect to the uniform probability measure. Show that the functions

$$X(\omega) := \begin{cases} f(\omega') & \text{if } \omega = 1\omega' \text{ starts from 1,} \\ \infty & \text{otherwise} \end{cases}$$

(for which $\overline{\mathbb{E}}(X) = \underline{\mathbb{E}}(X) = \infty$) and $-X$ can serve as counterexamples. □

Exercise 8.6   Show that $E \subseteq \mathcal{Y}^\infty$ being invariant can be equivalently expressed as follows: if $\omega \in E$ and $y \in \mathcal{Y}$, then both $\theta\omega \in E$ and $y\omega \in E$. This condition means that replacing any finite prefix of an element of E by any other sequence in $\mathcal{Y}^*$ does not lead outside E. □

Exercise 8.7
 1. Check that θ^{-1} commutes with complementation: for any $E \subseteq \mathcal{Y}^\infty$, $\theta^{-1}(E^c) = (\theta^{-1}E)^c$.

2. Conclude that E^c is invariant whenever E is invariant. Notice that this gives another demonstration of the part of Lemma 8.14 saying that E^c is weakly invariant if E is invariant. □

Exercise 8.8   (Ville's inequality) Show that if $\mathcal{T}$ is a nonnegative supermartingale, $\mathcal{T}_0 \in (0, \infty)$, and $K \in (0, \infty)$, then

$$\overline{\mathbb{P}} \left(\sup_n \mathcal{T}_n \geq K\mathcal{T}_0 \right) \leq 1/K. \tag{8.18}$$

This generalizes an inequality first proven by Jean Ville [387, p. 100]. We will explore its implications in Chapter 11. □

Exercise 8.9   (second Borel–Cantelli lemma) Suppose $A_1, A_2, \ldots$ are local events in Protocol 7.1 that satisfy $\sum_{k=1}^{\infty} \underline{\mathbf{P}}(A_k) = \infty$. Show that $\underline{\mathbb{P}}(y_k \in A_k$ for infinitely many $k) = 1$. □

8.7 CONTEXT

Lévy's Zero-One Law

Lévy published his measure-theoretic zero-one law in a 1935 article [244] and in his 1937 book on the addition of random variables [245]. In the book (p. 129), he states the law in terms of a property E that a sequence $X_1, X_2, \ldots$ of random variables might or might not have. He writes $\Pr.\{E\}$ for the initial probability of E, and $\Pr_n\{E\}$ for its probability after $X_1, \ldots, X_n$ are known. He remarks that if $\Pr.\{E\}$ is determined (i.e. if E is measurable), then the conditional probabilities $\Pr_n\{E\}$ are also determined. Then he states the law as follows (our translation from the French):

> Except in cases that have probability zero, if $\Pr.\{E\}$ is determined, then $\Pr_n\{E\}$ tends, as n tends to infinity, to one if the sequence $X_1, X_2, \ldots$ verifies the property E, and to zero in the contrary case.

As Lévy acknowledged, related results had been obtained earlier by Børge Jessen and Kolmogorov.

In the first paragraph of this chapter, we used $\mathbb{E}_n(X)$ and $\mathbb{P}_n(E)$ to state Lévy's theorem. Elsewhere in this book, however, we reserve these symbols for the game-theoretic concepts of probability and expected value and write $P(X)$ (respectively $P(E)$) for X's expected value (respectively E's probability) with respect to the probability measure P. See "Measure-Theoretic Probability and Statistics" in the section on terminology and notation at the end of the book.

Countable Additivity and Subadditivity

When World's move space $\mathcal{Y}$ in Protocol 7.1 is finite, the continuity axiom is nearly a theorem rather than an axiom. As noted in Exercise 6.2, Axioms E1–E4 imply the conclusion of the continuity axiom for $\overline{\mathbf{E}}$ when $\mathcal{Y}$ is finite and $\lim_{k\to\infty} f_k < \infty$. If $\overline{\mathbf{E}}$ does satisfy the continuity axiom, then $\overline{\mathbb{E}}$ as defined by (7.19) satisfies the continuity axiom by Proposition 8.3.

The picture is somewhat similar in measure-theoretic probability, where continuity is equivalent to countable additivity. We need not adopt countable additivity as an axiom for a finite probability space $(\mathcal{Y}, P)$; it is satisfied once we assume finite additivity. Countable additivity is still satisfied for a finite $\mathcal{Y}$ when we extend P to the algebra $\mathcal{F}$ of cylinder sets in $\mathcal{Y}^\infty$ by assigning each cylinder set its product measure, inasmuch as cylinder sets that are countable disjoint unions of cylinder sets have probability equal to the sum of the probabilities of the components. (We will discuss more general ways of assigning probabilities to the cylinder sets in Section 9.3.) Carathéodory's extension theorem [63] tells how to extend P further to a probability measure on a σ-algebra containing $\mathcal{F}$: define the outer measure P^* by

$$P^*(E) := \inf \left\{ \sum_{k=1}^{\infty} P(E_k) | E_1, E_2, \ldots \in \mathcal{F}, E \subseteq \bigcup_{k=1}^{\infty} E_k \right\}.$$

The sets E satisfying $P^*(F) = P^*(F \cap E) + P^*(F \cap E^c)$ for all $F \subseteq \mathcal{Y}^\infty$ form a σ-algebra, and P^*'s restriction to this σ-algebra is a probability measure (i.e. satisfies countable additivity). Here, as in our game-theoretic definitions, infinite operations enter in the extension from the local to the global. In fact, Carathéodory's extension can be thought of as a special case of the game-theoretic extension (see Exercise 9.1).

As Marie-France Bru and Bernard Bru have pointed out [54], countable additivity was first used in probability theory to obtain values of expectations in games that do not necessarily end in any given finite number of trials. Unless the rules of play are very simple, so that there are only a few positions where players can find themselves, problems of this type usually require the calculation of an infinite sum. Huygens, in a letter to Jan Hudde in 1665 [193, pp. 116–123], may have been the first to make such a calculation. Jacob Bernoulli may have been the first to publish one. To illustrate Bernoulli's approach, consider an infinite sequence of coin tosses with probability $1/2$ for heads (Protocol 1.10 with $p = 1/2$). What is the probability that the first head will appear on an odd-numbered toss? Or, to put it differently, if Peter and Paul alternate tossing, and Peter goes first, what is the probability that Peter will be the first to get a head? Bernoulli explained his method for solving problems of this type in his *Ars Conjectandi* [33, appendix to part I]. In our example (which is simpler than those Bernoulli discussed), we can add the probabilities for the first head coming on the first toss, the third toss, etc.:

$$\frac{1}{2} + \left(\frac{1}{2}\right)^3 + \left(\frac{1}{2}\right)^5 + \cdots = \frac{2}{3}.$$

To justify the infinite summation, Bernoulli imagined that each toss is made by a different player. Peter receives the gains of those who make the odd tosses, and Paul receives the gains of those who make the even tosses. In the preface to his *The Doctrine of Chances* [109], first published in 1718, De Moivre gave a different game-theoretic argument. He imagined that Peter and Paul cash out each time they have a turn. Peter's right to the first toss entitles him to half the stake. Instead of tossing the coin, he pockets this half and gives Paul his turn. Then Paul also cashes out, pocketing half the remaining stake, and so on. In the limit, Peter gets two-thirds of the stake, and Paul gets one-third.

As Bru and Bru point out, Bernoulli's way of putting the argument suggests an actual infinity; he supposes that there are infinitely many players. De Moivre's argument, on the other hand, uses only the notion of a limit as more and more trials are considered, a notion that was less controversial than actual infinity in the seventeenth and eighteenth centuries. The game-theoretic definitions of upper expected value and upper probability are akin to De Moivre's argument in this respect.

Other Measure-Theoretic Zero-One Laws

Beginning in his 1929 thesis, Jessen had constructed the Lebesgue integral on the infinite-dimensional space Q consisting of infinite sequences $x = (x_1, x_2, \ldots)$ of real numbers modulo 1. In a 1934 article [203, section 14], he showed that if f is an integrable function on this space, then

$$f(x) = \lim_{n \to \infty} \int_Q f(x_1, \ldots, x_n, \omega)\, d\omega$$

almost everywhere. Bernard Bru and Salah Eid have reviewed the correspondence between Jessen and Lévy concerning the relation between their results [55]. From the viewpoint of modern measure-theoretic probability, both Jessen's theorem and Lévy's zero-one law are martingale convergence theorems.

Kolmogorov's measure-theoretic zero-one law, which Lévy also cited, first appeared in an appendix to his *Grundbegriffe* in 1933 [224]. It says when $x_1, x_2, \ldots$ are independent random variables, and A is a tail event, the probability of A is either 0 or 1. Kolmogorov actually proved a more general statement: if f is a Baire function of any sequence of random variables $x_1, x_2, \ldots$, and if for any value of n, the conditional probability of $f(x_1, x_2, \ldots) = 0$ given values of the first n variables is always equal to its unconditional probability, then the unconditional probability is either 0 or 1.

For the measure-theoretic version of the result on Bernoulli shifts, see, e.g. [82, section 8.1, Theorem 1].

Game-Theoretic Zero-One Laws

The study of zero-one laws from the game-theoretic point of view was initiated by Takemura, who was also a coauthor of [375], which first established game-theoretic versions of Kolmogorov's zero-one law and the ergodicity of Bernoulli shifts, and [351], which first established the game-theoretic version of Lévy's zero-one law.

The Role of the Continuity Axiom

Our proof of Lévy's zero-one law, Theorem 8.1, does not rely on the continuity axiom. On the other hand, Lemma 8.8 relies on it very strongly.

9

Relation to Measure-Theoretic Probability

As we noted in the preface, the game-theoretic and measure-theoretic pictures are equivalent for coin tossing but generalize this core example in different directions. We have also noted in passing various other ways in which the two pictures overlap and complement each other. As we see in this chapter, we can connect them more systematically, recasting game-theoretic martingales and supermartingales as measure-theoretic martingales and supermartingales or vice versa. This can permit the ready derivation of many results in one picture from corresponding results in the other.

Measure-theoretic probability has two distinct ways of representing a sequence of outcomes. In one, the probability space is *canonical*, as it consists of the different possibilities for how the sequence can come out. In the other, successive outcomes are treated as random variables. The canonical representation comes closest to game-theoretic probability, at least as presented in this book in the case of discrete time. So we consider it first, in Sections 9.1 and 9.2. We consider the more general noncanonical representation in Section 9.3. In each case, we use a law of large numbers to illustrate how a game-theoretic result can be translated into a measure-theoretic result or vice versa.

Long before measure-theoretic probability was put in its modern form, mathematicians knew how to construct probabilities for a finite or infinite sequence of outcomes by combining probabilities for the first outcome with successive conditional probabilities. This idea was formalized within measure-theoretic probability using canonical

Game-Theoretic Foundations for Probability and Finance, First Edition. Glenn Shafer and Vladimir Vovk.
© 2019 John Wiley & Sons, Inc. Published 2019 by John Wiley & Sons, Inc.

spaces by Cassius T. Ionescu Tulcea in 1949 [196]. According to a theorem already proven by Ville in the 1930s, the Ionescu Tulcea probability measure agrees with the global upper expectation obtained by interpreting the probabilities and successive conditional probabilities as a strategy for Forecaster. In the general form in which we prove it in Section 9.1, Ville's theorem permits the derivation of many canonical discrete-time measure-theoretic results, including the classical limit theorems, from their game-theoretic counterparts. We illustrate this point using Kolmogorov's law of large numbers.

Ville's theorem can be thought of as a game-theoretic interpretation of classical probability. It shows how to embed in a canonical probability space any protocol where the outcome on each round has definite probabilities. It does not relate to protocols where Forecaster may announce upper expectations that are not probability measures. This is the topic of Section 9.2, where we consider only a finite outcome space and assume that the set of upper expectations available to Forecaster is parameterized by a compact metrizable space. Under these admittedly restrictive assumptions, we show that global game-theoretic upper probabilities and upper expected values, which are infima over supermartingales, are also suprema over probability measures. This result can be used to navigate between game-theoretic probability and canonical measure-theoretic probability. We illustrate the point by deriving a discrete variant of the game-theoretic version of Borel's law of large numbers that we established in Chapter 1 from a measure-theoretic statement of the law.

Most of this chapter's exercises are devoted to special cases of Section 9.2's very general Theorem 9.7. The first four special cases specialize Theorem 9.7 to one-round protocols (Exercises 9.3–9.6), whereas the simplest infinite-horizon special case (Exercises 9.7) corresponds to Ville's theorem.

In Section 9.3, we generalize Ville's theorem in a different way. Here we bring to the surface the key role played by martingales and supermartingales in both the game-theoretic and measure-theoretic pictures. As we learned in Chapter 8, game-theoretic probabilities are defined in terms of supermartingales. Doob's convergence theorem and related inequalities make the connection between martingales and probabilities measure-theoretic. So we can obtain measure-theoretic probabilities by translating game-theoretic martingales into measure-theoretic martingales. We illustrate this idea by deriving the measure-theoretic version of Kolmogorov's law of large numbers from our game-theoretic version.

We use standard results from measure-theoretic probability. For basic definitions, see the section on terminology and notation at the end of the book.

9.1 VILLE'S THEOREM

As we have seen, game-theoretic probability generalizes the classical picture of sequential betting in two ways. First, the betting opportunities on each round may be fewer; not every variable necessarily has a price at which it can be bought or sold.

Second, betting opportunities may be specified on each round by a player in the game rather than being specified in advance as a function of outcomes of previous rounds. Discrete-time measure-theoretic probability does not generalize the classical picture in these respects. So to derive measure-theoretic results from game-theoretic results, we first strip the game-theoretic picture of its extra generality: require Forecaster to announce complete probabilities for the outcome of each round, and then eliminate him by fixing his strategy.

Consider the following protocol, where Forecaster announces a probability measure rather than an upper expectation. Here $P(\mathcal{Y})$ is the set of all probability measures on the measurable space $(\mathcal{Y}, \mathcal{F})$, and $P(f)$ is f's expected value with respect to P. We follow the convention that an assertion about an expected value asserts in particular that it exists (see the section on terminology and notation at the end of the book). So the protocol requires Skeptic to choose an $f_n \in \overline{\mathbb{R}}^{\mathcal{Y}}$ that is measurable and has an expected value.

Protocol 9.1
PARAMETER: Measurable space $(\mathcal{Y}, \mathcal{F})$
    Skeptic announces $\mathcal{K}_0 \in \overline{\mathbb{R}}$.
    FOR $n = 1, 2, \ldots$:
        Forecaster announces $P_n \in P(\mathcal{Y})$.
        Skeptic announces $f_n \in \overline{\mathbb{R}}^{\mathcal{Y}}$ such that $P_n(f_n) \leq \mathcal{K}_{n-1}$.
        Reality announces $y_n \in \mathcal{Y}$.
        $\mathcal{K}_n := f_n(y_n)$.

The slackening of this protocol is the specialization of Protocol 7.12 in which Reality's move space $\mathcal{Y}$ is constant and Forecaster's upper expectation is always the upper integral with respect to a probability measure (see Exercise 7.1). So the theory of Chapters 7 and 8 applies.

Consider a strategy ϕ for Forecaster in Protocol 9.1 that takes into account only previous moves by Reality, ignoring previous moves by Skeptic and any other information. For each natural number n, let $\phi_n(y_1 y_2 \ldots)$ be the strategy's recommendation for P_n when Reality's previous moves are $y_1, \ldots, y_{n-1}$. We say that (ϕ_n) is a *probability forecasting system* if $\phi_n(\omega)(E)$ is a measurable function of ω for all n and all measurable $E \subseteq \mathcal{Y}$.

Ionescu Tulcea's theorem tells us that a probability forecasting system ϕ determines a probability measure on $\mathcal{Y}^\infty$ – the unique probability measure P_ϕ such that for any $n \in \mathbb{N}$ and for any bounded measurable function X on $\mathcal{Y}^\infty$ that depends on $\omega = y_1 y_2 \ldots$ only through $y_1 \ldots y_n$,

$$P_\phi(X) = \int \cdots \int X(\omega) \phi_n(\omega)(dy_n) \cdots \phi_1(\omega)(dy_1).$$

If each ϕ_n is constant – i.e. for each n there exists $P_n \in P(\mathcal{Y})$ such that $\phi_n(\omega) = P_n$ for all ω – then this Ionescu Tulcea probability measure is the product $P_1 \times P_2 \times \cdots$.

Requiring Forecaster to use a particular probability forecasting system ϕ and then removing him from the protocol, we obtain the following further specialization.

Protocol 9.2
PARAMETERS: Measurable space $(\mathcal{Y}, \mathcal{F})$, probability forecasting system ϕ
    Skeptic announces $\mathcal{K}_0 \in \overline{\mathbb{R}}$.
    FOR $n = 1, 2, \ldots$:
        Skeptic announces $f_n \in \overline{\mathbb{R}}^{\mathcal{Y}}$ such that $\phi_n(y_1, y_2, \ldots)(f_n) \leq \mathcal{K}_{n-1}$.
        Reality announces $y_n \in \mathcal{Y}$.
        $\mathcal{K}_n := f_n(y_n)$.

If $\mathcal{Y} = \{0, 1\}$ and $\phi_n(\omega)$ is the probability measure that gives probability p to 1, for all n and ω, then the protocol reduces to Protocol 1.10, the protocol for coin tossing with constant probability p.

Theorem 9.3   **(Ville's theorem)** *Every bounded measurable function X on $\mathcal{Y}^\infty$ has a game-theoretic expected value $\mathbb{E}(X)$ in Protocol 9.2, and $\mathbb{E}(X) = P_\phi(X)$.*

Proof Our goal is to prove that

$$\inf\{V_0 \mid \liminf_{n \to \infty} V_n \geq X\} \leq P_\phi(X) \tag{9.1}$$

and

$$P_\phi(X) \leq \inf\{V_0 \mid \liminf_{n \to \infty} V_n \geq X\}, \tag{9.2}$$

V ranging over game-theoretic supermartingales (i.e. capital processes in Protocol 9.2).

We will consider measure-theoretic supermartingales in the filtered probability space $(\mathcal{Y}^\infty, \mathcal{F}, (\mathcal{F}_n)_{n \in \mathbb{N}}, P_\phi)$, where $(\mathcal{F}_n)_{n \in \mathbb{N}}$ is the canonical filtration: $\mathcal{F}_n$ is the smallest σ-algebra making the mapping $\omega \in \mathcal{Y}^\infty \mapsto \omega^n$ measurable.

Let us first check (9.1). By Paul Lévy's measure-theoretic zero-one law [358, Theorem VII.4.3], there is a measure-theoretic martingale V that begins at $V_0 = P_\phi(X)$ and converges to X P_ϕ-a.s. Because V_n is a measure-theoretic martingale,

$$\phi_n(y_1, y_2, \ldots)(V_n) = V_{n-1}, \quad n \in \mathbb{N}, \tag{9.3}$$

almost surely with respect to P_ϕ. Let us change V so that it satisfies (9.3) always, not merely P_ϕ-a.s., and hence qualifies as a game-theoretic martingale. We do this by induction in n: V_0 is a constant, and after we have dropped the "P_ϕ-a.s." for $1, \ldots, n - 1$, we redefine V_n as

$$V_n := V_n \mathbf{1}_{E^c} + V_{n-1} \mathbf{1}_E,$$

where $E \in \mathcal{F}_{n-1}$ is the event $\phi_n(y_1, y_2, \ldots)(V_n) \neq V_{n-1}$. Because we have only changed V on a set of probability 0, it still converges to X P_ϕ-a.s. Now we use the fact that for any P_ϕ-null event, there is a measurable nonnegative game-theoretic martingale V' that starts from 1 and becomes infinitely rich on that event. Ville constructed such a martingale in his 1939 book [387, Theorem IV.1]; for a detailed self-contained proof, not relying on Ville's, see [349, section 8.5]. We choose such a V' for the P_ϕ-null event that V does not converge to X. Then $V + \epsilon V'$, where $\epsilon > 0$, is a measurable game-theoretic martingale that begins at $P_\phi + \epsilon$ and satisfies $\liminf_n (V + \epsilon V')_n \geq X$. Letting $\epsilon \to 0$ establishes (9.1).

Let us now derive the inequality (9.2), remembering that, according to our definition, game-theoretic supermartingales need not be measurable. We argue indirectly. Suppose (9.2) is violated and fix a game-theoretic supermartingale V such that $V_0 < P_\phi(X)$ and V superhedges X everywhere: $\liminf_{n\to\infty} V_n \geq X$. By the argument in the previous paragraph (applied to $-X$ in place of X), there is a measurable game-theoretic martingale V' such that $V_0 < V_0'$ and V' subhedges X everywhere, so that we have

$$\limsup_{n\to\infty} V_n'(\omega) \leq X(\omega) \leq \liminf_{n\to\infty} V_n(\omega)$$

for all ω. This leads to a contradiction. Let $\epsilon > 0$ be such that $V_0' - V_0 > \epsilon$. Choose y_1 such that $V'(y_1) - V(y_1) > \epsilon$. Given y_1, choose y_2 such that $V'(y_1 y_2) - V(y_1 y_2) > \epsilon$. Continue in this way getting an infinite sequence ω such that $V_n'(\omega) - V_n(\omega) > \epsilon$ for all n. Therefore, it is impossible that V superhedges X at ω and V' subhedges X at ω. $\qquad\square$

We can combine Theorem 9.3 with the game-theoretic version of a limit theorem to obtain a measure-theoretic version of the same theorem. The following corollary is an example; we combine Theorem 9.3 with Theorem 4.3 (Statement 1) and obtain a measure-theoretic martingale version of Kolmogorov's law of large numbers.

Corollary 9.4 *Suppose ϕ is a probability forecasting system such that the expected value and variance*

$$m_n := \int_{\mathbb{R}} y \phi_n(y_1, y_2, \ldots)(dy), \qquad v_n := \int_{\mathbb{R}} (y - m_n)^2 \phi_n(y_1, y_2, \ldots)(dy)$$

exist for all paths $y_1 y_2 \ldots$ and all n. Then

$$\sum_{n=1}^{\infty} \frac{v_n}{n^2} < \infty \implies \frac{1}{n} \sum_{i=1}^{n} (y_i - m_i) \to 0 \qquad (9.4)$$

almost surely with respect to the probability measure P_ϕ.

Proof According to Theorem 4.3 (Statement 1), (9.4) holds almost surely – i.e. has game-theoretic probability one – in Protocol 4.1. Because Protocol 9.2 with parameter $\mathcal{Y} = \mathbb{R}$ differs from Protocol 4.1 only by giving Skeptic more choices, it follows that (9.4) has game-theoretic probability 1 in it as well. By Theorem 9.3, (9.4) also has P_ϕ-probability 1. □

The notion of a probability forecasting system and that of a probability measure on a product space are extremely close, and the difference can be regarded as purely technical. The measure-theoretic framework *starts* with a probability measure P on $\mathcal{Y}^\infty$, and then the question is whether there exists a probability forecasting system ϕ such that $P_\phi = P$. Standard results on the existence of regular conditional probabilities [358, Theorem II.7.5] imply an affirmative answer when $\mathcal{Y}$ is a Borel space (in particular, when $\mathcal{Y} = \mathbb{R}$).

9.2 MEASURE-THEORETIC REPRESENTATION OF UPPER EXPECTATIONS

We now consider how measure-theoretic probability can be deployed to understand protocols in which the upper expectations announced on individual rounds are not necessarily probability measures. We continue to work in canonical spaces, but we now make two restrictive assumptions. We require the outcome space $\mathcal{Y}$ to be finite, and we require Forecaster to choose upper expectations from a parameterized family that satisfies strong topological conditions. Under these assumptions, we demonstrate a duality: the upper expected values that we obtain as infima over game-theoretic supermartingales are also suprema over probability measures. This duality allows us to derive game-theoretic results from measure-theoretic ones and vice versa. At the end of the section, we illustrate this by deriving a discrete variant of Chapter 1's game-theoretic version of Borel's law of large numbers from the measure-theoretic version.

We say that a probability measure P on a finite set $\mathcal{Y}$ *sharpens* an upper expectation $\overline{\mathbf{E}}$ on $\mathcal{Y}$ if $P(f) \leq \overline{\mathbf{E}}(f)$ for all $f \in \overline{\mathbb{R}}^{\mathcal{Y}}$. We write $P \preceq \overline{\mathbf{E}}$ when P sharpens $\overline{\mathbf{E}}$. The interpretation of sharpening depends on the interpretation we choose to give to probability measures on $\mathcal{Y}$; see the discussion in Section 9.5.

The outcome space $\mathcal{Y}$ being finite, we have natural topologies on $\overline{\mathbb{R}}^{\mathcal{Y}}$ and $\mathcal{Y}^\infty$. We write $\overline{\mathcal{E}}(\mathcal{Y})$ for the set of all upper expectations on $\mathcal{Y}$. The following protocols use a compact metrizable space Θ and a function $\overline{\mathbf{E}} : \theta \in \Theta \mapsto \overline{\mathbf{E}}_\theta \in \overline{\mathcal{E}}(\mathcal{Y})$ that is upper semicontinuous, in the sense that the subset $\{(\theta, g) | \overline{\mathbf{E}}_\theta(g) < c\}$ of $\Theta \times \overline{\mathbb{R}}^{\mathcal{Y}}$ is open for each $c \in \mathbb{R}$. Upper semicontinuity is natural in this context: $\overline{\mathbf{E}}_\theta(g) < c$ says that Skeptic can buy g at the price c, and upper semicontinuity ensures that he can do so as soon as θ and g are given with sufficient accuracy.

Consider the two following protocols.

Protocol 9.5

PARAMETERS: Finite nonempty set $\mathcal{Y}$, compact metrizable space Θ,
upper semicontinuous mapping $\overline{\mathbf{E}} : \theta \in \Theta \mapsto \overline{\mathbf{E}}_\theta \in \overline{\mathcal{E}}(\mathcal{Y})$
Skeptic announces $\mathcal{K}_0 \in \mathbb{R}$.
FOR $n = 1, 2, \ldots$:
    Forecaster announces $\theta_n \in \Theta$.
    Skeptic announces $f_n \in \overline{\mathbb{R}}^{\mathcal{Y}}$ such that $\overline{\mathbf{E}}_{\theta_n}(f_n) \leq \mathcal{K}_{n-1}$.
    Reality announces $y_n \in \mathcal{Y}$.
    Skeptic announces $\mathcal{K}_n \leq f_n(y_n)$.

Protocol 9.6

PARAMETERS: Finite nonempty set $\mathcal{Y}$, compact metrizable space Θ,
upper semicontinuous mapping $\overline{\mathbf{E}} : \theta \in \Theta \mapsto \overline{\mathbf{E}}_\theta \in \overline{\mathcal{E}}(\mathcal{Y})$
FOR $n = 1, 2, \ldots$:
    Forecaster announces $\theta_n \in \Theta$.
    Probability Forecaster announces $P_n \in \mathcal{P}(\mathcal{Y})$ such that $P_n \leq \overline{\mathbf{E}}_{\theta_n}$.
    Reality announces $y_n \in \mathcal{Y}$.

Protocol 9.5 is a testing protocol, a specialization of Protocol 7.12. It is adversarial in the sense that the basic forecasting game between Forecaster and Reality is complemented by Forecaster's adversary betting against his predictions. In this section, we will use the terminology, notation, and results of Chapters 7 and 8 with reference to this protocol. Thus our sample space is $\Omega := (\Theta \times \mathcal{Y})^\infty$, our set of situations is $\mathbb{S} := (\Theta \times \mathcal{Y})^*$, and by *supermartingale*, we mean a supermartingale in this protocol.

Protocol 9.6 may be called a *forecasting protocol*. It is cooperative, having no Skeptic and no betting; it merely expresses the idea that a second forecaster, who we call Probability Forecaster, sharpens Forecaster's upper expectation to a probability measure. We call a joint strategy for the two forecasters a *joint forecasting system*. If ϕ is a joint forecasting system, its recommendation $\phi_n(y_1 y_2 \ldots)$ for (θ_n, P_n) is an element of

$$\Psi := \{(\theta, P) \in \Theta \times \mathcal{P}(\mathcal{Y}) \mid P \leq \overline{\mathbf{E}}_\theta\};$$

remember that $\phi_n(y_1 y_2 \ldots) \in \Psi$ depends only on $y_1, \ldots, y_{n-1}$. A joint forecasting system ϕ can be split into two strategies: $\phi_n(\alpha) = (\phi_n^{\mathrm{F}}(\alpha), \phi_n^{\mathrm{P}}(\alpha))$, where ϕ^{F} is a strategy for Forecaster, and ϕ^{P} is a strategy for Probability Forecaster. Noting that ϕ^{P} is a probability forecasting system as defined in Section 9.1, we write $P_{\phi^{\mathrm{P}}}$ for the corresponding Ionescu Tulcea probability measure on $\mathcal{Y}^\infty$ and often abbreviate $P_{\phi^{\mathrm{P}}}$ to P_ϕ. For $\alpha = (y_1, y_2, \ldots) \in \mathcal{Y}^\infty$, we set

$$\alpha_\phi = \alpha_{\phi^{\mathrm{F}}} := (\phi_1^{\mathrm{F}}(\alpha), y_1, \phi_2^{\mathrm{F}}(\alpha), y_2, \ldots) \in \Omega,$$

using the notation $\alpha_{\phi^{\mathrm{F}}}$ when we wish to emphasize that α_ϕ depends on ϕ only via ϕ^{F}. We write $\overline{P}_{\phi^{\mathrm{P}}}(W)$ or $\overline{P}_\phi(W)$ for the upper integral of $W : \mathcal{Y}^\infty \to \overline{\mathbb{R}}$ with respect to $P_{\phi^{\mathrm{P}}}$ (i.e. the infimum of $P_\phi(Z)$ over measurable functions Z dominating W).

Given a joint forecasting system ϕ and an extended variable $X : \Omega \to \overline{\mathbb{R}}$, we set

$$\mathbb{E}^{\phi}(X) := \overline{P}_{\phi^P}(X(\cdot_{\phi^F})),$$

where $X(\cdot_{\phi^F}) : \alpha \in \mathcal{Y}^{\infty} \mapsto X(\alpha_{\phi^F}) \in \overline{\mathbb{R}}$. Here ϕ^F reduces X to a function of Reality's moves, and ϕ^P upgrades ϕ^F to a probability measure for Reality's moves. Now set

$$\mathbb{E}^{\text{meas}}(X) := \sup_{\phi} \mathbb{E}^{\phi}(X),$$

where ϕ ranges over all joint forecasting systems. Roughly speaking, $\mathbb{E}^{\text{meas}}(X)$ is the greatest expected value that can be attributed to X by consistent strategies for the two forecasters. In this sense, it is a reasonable price for a cautious seller to demand for X, and this makes it a measure-theoretic analog to the game-theoretic upper expected value $\overline{\mathbb{E}}(X)$.

The analogy extends to game-theoretic upper probabilities. Given an event E in Protocol 9.5, write $\mathbb{P}^{\text{meas}}(E)$ for $\mathbb{E}^{\text{meas}}(1_E)$. If E's happening indicates some kind of disagreement between Reality and Forecaster, and then a zero or very small value of $\mathbb{P}^{\text{meas}}(E)$ can be interpreted as a prediction that E will not happen: there are no strategies for the forecasters that give E an appreciable probability.

The following theorem, which is the main result of this section, supports the idea that $\mathbb{E}^{\text{meas}}(X)$ is a measure-theoretic interpretation of the game-theoretic upper expected value $\overline{\mathbb{E}}(X)$.

Theorem 9.7   *If X is a bounded Suslin global variable in Protocol 9.5, then $\overline{\mathbb{E}}(X) = \mathbb{E}^{\text{meas}}(X)$.*

The condition that X be Suslin is weak, because all Borel functions are Suslin. For a definition of Suslin functions, see Dellacherie [104, Definition I.4], and for a proof that all Borel functions are Suslin, see [104, Corollary I.6]; Dellacherie refers to Suslin functions as analytic (which is also a standard term).

Proof of Theorem 9.7

We begin by proving the inequality $\overline{\mathbb{E}}(X) \geq \mathbb{E}^{\text{meas}}(X)$, which holds whenever X is bounded below.

Proposition 9.8   *For all bounded below $X : \Omega \to \overline{\mathbb{R}}, \overline{\mathbb{E}}(X) \geq \mathbb{E}^{\text{meas}}(X)$.*

Let ϕ be a joint forecasting system. A ϕ-*supermartingale* is a function $V : \mathcal{Y}^* \to \overline{\mathbb{R}}$ satisfying

$$V_{n-1}(\alpha) \geq \phi_n^P(\alpha)(V(\alpha^{n-1}\cdot)) \tag{9.5}$$

for all $n \in \mathbb{N}$ and $\alpha \in \mathcal{Y}^\infty$ (with the usual convention $0(\pm\infty) := 0$). We say that V is a ϕ-*martingale* if (9.5) holds with $=$ in place of $\geq$.

If $\mathcal{T}$ is a supermartingale, the function $\mathcal{T}^\phi : \mathcal{Y}^* \to \overline{\mathbb{R}}$ defined by

$$\mathcal{T}_n^\phi(\alpha) := \mathcal{T}_n(\alpha_\phi), \quad n \in \mathbb{N}, \alpha \in \mathcal{Y}^\infty,$$

is a ϕ-supermartingale. Indeed, the definition of a game-theoretic supermartingale gives, for all $n \in \mathbb{N}$ and $\alpha \in \mathcal{Y}^\infty$,

$$\mathcal{T}_{n-1}^\phi(\alpha) = \mathcal{T}_{n-1}(\alpha_\phi) \geq \overline{\mathbf{E}}_{\phi_n^F(\alpha)} \left(\mathcal{T}(\alpha_\phi^{n-1}, \phi_n^F(\alpha), \cdot) \right)$$
$$= \overline{\mathbf{E}}_{\phi_n^F(\alpha)}(\mathcal{T}^\phi(\alpha^{n-1} \cdot)) \geq \phi_n^P(\alpha)(\mathcal{T}^\phi(\alpha^{n-1} \cdot)),$$

where we used the abbreviation $\alpha_\phi^k := (\alpha_\phi)^k$. Because $\mathcal{Y}$ is finite, the domain of $\mathcal{T}^\phi$ is discrete, and there are no measurability problems.

Proof of Proposition 9.8 It suffices to prove that $\mathbb{E}^\phi(X) \leq \mathcal{T}_0$ for any joint forecasting system ϕ and any supermartingale $\mathcal{T}$ satisfying $\liminf_n \mathcal{T}_n \geq X$. This follows from Fatou's lemma (applicable to random variables that are uniformly bounded below):

$$\mathbb{E}^\phi(X) = \overline{P}_{\phi^P}(X(\cdot_{\phi^F})) \leq \overline{P}_{\phi^P} \left(\liminf_{n \to \infty} \mathcal{T}_n(\cdot_{\phi^F}) \right) = P_{\phi^P} \left(\liminf_{n \to \infty} \mathcal{T}_n^\phi \right)$$
$$\leq \liminf_{n \to \infty} P_{\phi^P} \left(\mathcal{T}_n^\phi \right) \leq \liminf_{n \to \infty} \mathcal{T}_0^\phi = \mathcal{T}_0.$$

Notice that our application of Fatou's lemma (in the second inequality of this chain) is to Borel functions, since $\mathcal{T}_n^\phi(\alpha)$ depends only on the first n elements of α and each element takes only finitely many values. $\square$

Lemma 9.9 *For each* $\overline{\mathbf{E}} \in \overline{\mathcal{E}}(\mathcal{Y})$ *and* $f \in \overline{\mathbb{R}}^\mathcal{Y}$, $\overline{\mathbf{E}}(f) = \sup_{P \leq \overline{\mathbf{E}}} P(f)$. *Moreover, the supremum is attained for all* f.

Proof This well-known result is true in great generality and immediately follows from the Hahn–Banach theorem (see, e.g. [106, X.31b] for details). $\square$

Now we can prove a basic special case of Theorem 9.7.

Lemma 9.10 *Suppose* $X : \Omega \to \mathbb{R}$ *is a bounded upper semicontinuous function. Then* $\mathbb{E}^{meas}(X) = \overline{\mathbb{E}}(X)$.

Proof Let $X_1 \geq X_2 \geq \cdots$ be a decreasing uniformly bounded sequence of upper semicontinuous functions such that each $X_j = X_j(\omega)$ depends on ω only via ω^j and $X = \inf_{j \in \mathbb{N}} X_j$. (The existence of such a sequence follows, e.g. from each upper semicontinuous function on Ω being the limit of a decreasing sequence of continuous functions; see, e.g. [128, Problem 1.7.15(c)].) For each $j \in \mathbb{N}$, define a supermartingale W^j by setting

$$W^j_n := X_j, \quad n \geq j, \tag{9.6}$$

and then proceeding inductively as follows. If W^j_n is already defined for some $n \in \{j, j-1, \ldots, 1\}$, define W^j_{n-1} by

$$W^j_{n-1}(\omega) := \sup_{\theta \in \Theta} \overline{\mathbb{E}}_\theta \left(W^j(\omega^{n-1}, \theta, \cdot) \right), \quad \omega \in \Omega. \tag{9.7}$$

It is clear that $W^1 \geq W^2 \geq \cdots$.

Let us check that, for all j and n, $W^j_n(\omega)$ is upper semicontinuous as a function of ω^n. By (9.6) this is true for $n \geq j$. Suppose it is true for some $n \in \{j, j-1, \ldots, 2\}$, and let us prove that it is true for $n-1$ in place of n, using the inductive definition (9.7). Our assumption that $\overline{\mathbb{E}}.(\cdot)$ is upper semicontinuous and the finiteness of $\mathcal{Y}$ imply that $f(s, \theta) := \overline{\mathbb{E}}_\theta(W^j(s, \theta, \cdot))$ is upper semicontinuous as function of $\theta \in \Theta$ and $s \in (\Theta \times \mathcal{Y})^{n-1}$. (Indeed, if (s', θ') is sufficiently close to (s, θ), $W^j(s', \theta', y)$ will exceed $W^j(s, \theta, y)$ by an arbitrarily small amount for all $y \in \mathcal{Y}$, and so $\overline{\mathbb{E}}_{\theta'}(W^j(s', \theta', \cdot))$ will exceed $\overline{\mathbb{E}}_\theta(W^j(s, \theta, \cdot))$ by an arbitrarily small amount.) It is well known that $\sup_\theta f(s, \theta)$ is upper semicontinuous whenever f is upper semicontinuous and s and θ range over compact metrizable spaces (see, e.g. [104, Theorem I.2(d)]). Therefore, $W^j(s) = \sup_{\theta \in \Theta} f(s, \theta)$ is an upper semicontinuous function of $s \in (\Theta \times \mathcal{Y})^{n-1}$.

Since an upper semicontinuous function always attains its supremum over a compact set (this can be easily deduced from [128, Problem 3.12.23(g)]), the supremum in (9.7) is attained. For each $j \in \mathbb{N}$, we can now define a joint forecasting system $\phi^j = (\phi^{\mathrm{F}j}, \phi^{\mathrm{P}j})$ as follows. For each $n = 1, \ldots, j$ and $\alpha \in \mathcal{Y}^\infty$, choose $\phi^{\mathrm{F}j}_n(\alpha)$ and $\phi^{\mathrm{P}j}_n(\alpha)$ such that $\phi^{\mathrm{P}j}_n(\alpha) \leq \overline{\mathbb{E}}_{\phi^{\mathrm{F}j}_n(\alpha)}$ and

$$\phi^{\mathrm{P}j}_n(\alpha) \left(W^j \left(\alpha^{n-1}_{\phi^{\mathrm{F}j}}, \phi^{\mathrm{F}j}_n(\alpha), \cdot \right) \right) = \overline{\mathbb{E}}_{\phi^{\mathrm{F}j}_n(\alpha)} \left(W^j \left(\alpha^{n-1}_{\phi^{\mathrm{F}j}}, \phi^{\mathrm{F}j}_n(\alpha), \cdot \right) \right)$$

$$= \sup_{\theta \in \Theta} \overline{\mathbb{E}}_\theta \left(W^j \left(\alpha^{n-1}_{\phi^{\mathrm{F}j}}, \theta, \cdot \right) \right) = W^j_{n-1}(\alpha_{\phi^{\mathrm{F}j}}) \tag{9.8}$$

(this is an inductive definition; in particular, $W^j_{n-1}(\alpha_{\phi^{\mathrm{F}j}})$ is already defined at the time of defining $\phi^j_n(\alpha)$). The chain (9.8) should be read back-to-front. The last equality holds by our definition of W^j_{n-1}. The second equality defines $\phi^{\mathrm{F}j}_n(\alpha)$, and its existence follows from the supremum in (9.7) being attained. Finally, the first equality defines the probability measure $\phi^{\mathrm{P}j}_n(\alpha)$, and its existence follows from Lemma 9.9. For $n > j$, define ϕ^j_n arbitrarily, e.g. as the uniform probability measure. The important property of ϕ^j is that $(W^j)^{\phi^j}$ is a $\phi^{\mathrm{P}j}$-martingale and so $P_{\phi^{\mathrm{P}j}}(X_j(\cdot_{\phi^{\mathrm{F}j}})) = W^j_0$.

Since the set of all joint forecasting systems is compact in the product topology (this follows from the set Ψ being closed), the sequence ϕ^j has a convergent subsequence ϕ^{j_k}, $k \in \mathbb{N}$; let $\phi := \lim_{k \to \infty} \phi^{j_k}$. We assume, without loss of generality, $j_1 < j_2 < \cdots$. Set

$$c := \inf_j W_0^j = \lim_{j \to \infty} W_0^j. \tag{9.9}$$

Fix an arbitrarily small $\epsilon > 0$. Let us prove that $\mathbb{E}^\phi(X) \geq c - \epsilon$. Let $K \in \mathbb{N}$. The restriction of $P_{\phi^{P,j_k}}$ to $\mathcal{Y}^{j_K}$ (more formally, the probability measure assigning weight $P_{\phi^{P,j_k}}(\mathcal{Y}_s^\infty)$ to each singleton $\{s\}$, $s \in \mathcal{Y}^{j_K}$) comes within ϵ of the restriction of P_{ϕ^P} to $\mathcal{Y}^{j_K}$ in total variation distance from some k on; let the total variation distance be at most ϵ for all $k \geq K' \geq K$. Let $k \geq K'$. Since $\overline{P}_{\phi^{P,j_k}}(X_{j_k}(\cdot_{\phi^{F,j_k}})) \geq c$, it is also true that $\overline{P}_{\phi^{P,j_k}}(X_{j_K}(\cdot_{\phi^{F,j_k}})) \geq c$. So $\overline{P}_{\phi^P}(X_{j_K}(\cdot_{\phi^{F,j_k}})) \geq c - \epsilon$. By Fatou's lemma (applicable because the sequence X_j is uniformly bounded), we now obtain

$$\overline{P}_{\phi^P}(\limsup_k X_{j_K}(\cdot_{\phi^{F,j_k}})) \geq \limsup_{k \to \infty} \overline{P}_{\phi^P}(X_{j_K}(\cdot_{\phi^{F,j_k}})) \geq c - \epsilon. \tag{9.10}$$

On the other hand,

$$\limsup_k X_{j_K}(\cdot_{\phi^{F,j_k}}) \leq X_{j_K}(\cdot_{\phi^F}). \tag{9.11}$$

This follows immediately from the fact that $\phi^{j_k} \to \phi$ in the product topology and the function X_{j_K} is upper semicontinuous. From (9.10) and (9.11) we can see that

$$\forall K \in \mathbb{N} : \overline{P}_{\phi^P}(X_{j_K}(\cdot_{\phi^F})) \geq c - \epsilon.$$

This implies, again by Fatou's lemma, $\overline{P}_{\phi^P}(X(\cdot_{\phi^F})) \geq c - \epsilon$. Since this holds for all ϵ,

$$\mathbb{E}^\phi(X) = \overline{P}_{\phi^P}(X(\cdot_{\phi^F})) \geq c.$$

The rest of the proof is straightforward: since

$$\overline{\mathbb{E}}(X) \leq c \leq \mathbb{E}^\phi(X) \leq \mathbb{E}^{\mathrm{meas}}(X) \leq \overline{\mathbb{E}}(X)$$

(the last inequality following from Proposition 9.8), we have

$$\overline{\mathbb{E}}(X) = c = \mathbb{E}^\phi(X) = \mathbb{E}^{\mathrm{meas}}(X), \tag{9.12}$$

which achieves our goal. □

To complete the proof of Theorem 9.7, we extend Lemma 9.10 to the Suslin functions using Choquet's capacitability theorem. Remember that a $[0, \infty]$-valued function γ (such as $\overline{\mathbb{E}}$ or $\mathbb{E}^{\text{meas}}$) defined on the set of all $[0, \infty]$-valued functions on a topological space Z is a *capacity* if:

- for any $[0, \infty]$-valued functions f and g on Z,

$$f \leq g \implies \gamma(f) \leq \gamma(g); \tag{9.13}$$

- for any increasing sequence $f_1 \leq f_2 \leq \cdots$ of $[0, \infty]$-valued functions on Z,

$$\gamma(\sup_{k \in \mathbb{N}} f_k) = \lim_{k \to \infty} \gamma(f_k); \tag{9.14}$$

- for any decreasing sequence $f_1 \geq f_2 \geq \cdots$ of upper semicontinuous $[0, \infty]$-valued functions on Z,

$$\gamma(\inf_{k \in \mathbb{N}} f_k) = \lim_{k \to \infty} \gamma(f_k). \tag{9.15}$$

Choquet's capacitability theorem is stated below as Theorem 9.14. Since it only covers nonnegative functions, we will assume that the function X in Theorem 9.7 is also nonnegative; it is clear that this can be done without loss of generality. We will use the version of Choquet's theorem in which capacities are defined, by the same conditions, on the set of $[0, C]$-valued functions for a fixed $C \in (0, \infty)$; this version is correct since any capacity γ on the set of $[0, C]$-valued functions can be extended to a capacity γ' on the set of $[0, \infty]$-valued functions by $\gamma'(f) := \gamma(f \wedge C)$. Therefore, we will mainly consider functions (variables in our context) taking values in $[0, C]$.

Lemma 9.11 *The function $\overline{\mathbb{E}}$ is a capacity on $[0, C]^{\Omega}$.*

By Proposition 8.3, $\overline{\mathbb{E}}$ is an upper expectation and therefore satisfies (9.13) and (9.14). We can extract a proof of condition (9.15) from the proof of Lemma 9.10 as follows.

Lemma 9.12 *If $X_1 \geq X_2 \geq \cdots$ is a decreasing sequence of upper semicontinuous $[0, C]$-valued variables,*

$$\overline{\mathbb{E}}(\inf_{k \in \mathbb{N}} X_k) = \lim_{k \to \infty} \overline{\mathbb{E}}(X_k). \tag{9.16}$$

Proof In the proof of Lemma 9.10 we showed (in a different notation, with X in place of X_k) that each X_k can be represented in the form $X_k = \inf_{j \in \mathbb{N}} X_{k,j}$, where $X_{k,1} \geq X_{k,2} \geq \cdots$, and each $X_{k,j} = X_{k,j}(\omega)$ is upper semicontinuous and depends on ω only via ω^j. We will use the equality

$$\overline{\mathbb{E}}(X_k) = \lim_{j \to \infty} \overline{\mathbb{E}}(X_{k,j}). \tag{9.17}$$

This equality follows from

$$\overline{\mathbb{E}}(X_k) = c_k := \lim_{j\to\infty} W_0^{k,j} \geq \lim_{j\to\infty} \overline{\mathbb{E}}(X_{k,j}),$$

where we now write $W^{k,j}$ for W^j; the opposite inequality is obvious. Without loss of generality we will assume that $X_{1,j} \geq X_{2,j} \geq \cdots$ for all j. Then the function $X :=$ $\inf_{k\in\mathbb{N}} X_k$ can be represented as $X = \inf_{j\in\mathbb{N}} X_{j,j}$, and so (9.16) follows from

$$\overline{\mathbb{E}}(X) = \overline{\mathbb{E}}(\inf_{j\in\mathbb{N}} X_{j,j}) = \lim_{j\to\infty} \overline{\mathbb{E}}(X_{j,j}) = \lim_{k\to\infty} \lim_{j\to\infty} \overline{\mathbb{E}}(X_{k,j}) = \lim_{k\to\infty} \overline{\mathbb{E}}(X_k).$$

(The second equality in this chain follows from (9.12) and (9.9) applied to $X_{j,j}$ in place of that proof's X_j, and the last equality follows from (9.17).) ◻

Let us now check that the measure-theoretic definition also produces a capacity.

Lemma 9.13 *The function $\mathbb{E}^{meas}$ is a capacity on $[0, C]^\Omega$.*

Proof Property (9.13) is obvious for $\mathbb{E}^{meas}$. Property (9.15) follows from Lemma 9.10 and the validity of (9.15) for $\overline{\mathbb{E}}$.

Let us now check the remaining property (9.14), with $\mathbb{E}^{meas}$ as γ. Suppose there exists an increasing sequence $X_1 \leq X_2 \leq \cdots$ of $[0, C]$-valued variables such that

$$\mathbb{E}^{meas}(\sup_{k\in\mathbb{N}} X_k) > \lim_{k\to\infty} \mathbb{E}^{meas}(X_k).$$

Let ϕ be a joint forecasting system satisfying

$$\mathbb{E}^\phi(\sup_{k\in\mathbb{N}} X_k) > \lim_{k\to\infty} \mathbb{E}^{meas}(X_k).$$

Then ϕ will satisfy $\mathbb{E}^\phi(\sup_{k\in\mathbb{N}} X_k) > \lim_{k\to\infty} \mathbb{E}^\phi(X_k)$, which is equivalent to the obviously wrong $\overline{P}_\phi(\sup_{k\in\mathbb{N}} X_k(\cdot_\phi)) > \lim_{k\to\infty} \overline{P}_\phi X_k(\cdot_\phi)$. ◻

Here is Choquet's capacitability theorem as it is stated in, e.g. [104, Theorem II.5]. As we already mentioned, it remains true if $[0, \infty]$ is replaced by $[0, C]$.

Theorem 9.14 (Choquet's capacitability theorem) *If Z is a compact metrizable space, γ is a capacity on $[0, \infty]^Z$, and $X : Z \to [0, \infty]$ is a Suslin function,*

$$\gamma(X) = \sup\{\gamma(f) | f \text{ is upper semicontinuous and } f \leq X\}.$$

Proof of Theorem 9.7   Combining Choquet's capacitability theorem (applied to the compact metrizable space Ω) with Lemmas 9.10, 9.11, and 9.13, we obtain

$$\overline{\mathbb{E}}(X) = \sup_{f \leq X} \overline{\mathbb{E}}(f) = \sup_{f \leq X} \mathbb{E}^{\text{meas}}(f) = \mathbb{P}^{\text{meas}}(X),$$

f ranging over the nonnegative upper semicontinuous functions.   □

Deducing Game-Theoretic Results from Measure-Theoretic Results

When the outcome space $\mathcal{Y}$ is finite, Theorem 9.7 allows us to go two ways. We can use it to deduce measure-theoretic results from game-theoretic ones, just as we used Ville's theorem in Section 9.1, but we can also use it to deduce game-theoretic results from measure-theoretic ones, even for game-theoretic protocols that allow Forecaster to announce upper expectations that are not probability measures.

To illustrate this point, consider the following discrete version of Borel's law of large numbers, which is an analog and straightforward consequence of Corollary 9.4.

Corollary 9.15   *Suppose λ is a probability forecasting system in Protocol 9.1 with $\mathcal{Y} = [-1, 1]$, and write $m_n := \int_{-1}^{1} y\lambda(y_1, y_2, \ldots)(dy)$ for its expected value for y_n, $n \in \mathbb{N}$. Then*

$$\frac{1}{n}\sum_{i=1}^{n}(y_i - m_i) \to 0 \tag{9.18}$$

almost surely with respect to the Ionescu Tulcea probability measure P_λ.

Using this measure-theoretic result and Theorem 9.7, we now deduce the following game-theoretic counterpart, which is a discrete version of Theorem 1.2.

Corollary 9.16   *Suppose $\mathcal{Y}$ is a finite nonempty subset of $[-1, 1]$. Then the event (9.18) happens almost surely in the protocol obtained by replacing $y_n \in [-1, 1]$ with $y_n \in \mathcal{Y}$ in Protocol 1.1.*

Proof   Consider an instantiation of Protocol 9.6 in which $\Theta = [-1, 1]$ and $\overline{\mathbb{E}}_\theta$, for $\theta \in \Theta$, is the upper expectation on $\mathcal{Y}$ denoted by $\overline{\mathbb{E}}^\theta$ in (6.2). Let ϕ be a joint forecasting system in this protocol. Then it follows from Exercise 9.2 that, for all $n \in \mathbb{N}$ and $\alpha \in \mathcal{Y}^\infty$, $\phi_n^F(\alpha)$ is the expected value $\int_{-1}^{1} y\phi^P(\alpha)(dy)$ of $\phi^P(\alpha)$. So by Corollary 9.15, $P_\phi(E) = 0$, where E is the complement of (1.2). It follows by the definition of $\mathbb{E}^{\text{meas}}$ that $\mathbb{P}^{\text{meas}}(E) = 0$ and then by Theorem 9.7 that $\overline{\mathbb{P}}(E) = 0$.   □

9.3 EMBEDDING GAME-THEORETIC MARTINGALES IN PROBABILITY SPACES

Most of our game-theoretic theorems are based on the construction of game-theoretic martingales and supermartingales. A general recipe for translating such a game-theoretic theorem into an abstract (as opposed to canonical) probability space could involve three steps:

1. Observe that the game-theoretic theorem implies the existence of a game-theoretic supermartingale or martingale satisfying certain conditions.
2. Show that this game-theoretic supermartingale or martingale becomes a measure-theoretic supermartingale or martingale in a relevant abstract probability space when the moves by Skeptic's opponent are replaced by random variables.
3. Appeal to Doob's convergence theorem or a suitable inequality to obtain a theorem in measure-theoretic probability.

One apparent obstacle to this program is the absence of measurability conditions in the discrete-time game-theoretic picture. The mathematics of the preceding chapters did not require the assumption that variables and other objects be measurable, and so we did not impose that assumption. (The situation will be different in the continuous-time theory of Part IV. There we do impose measurability in order to avoid unacceptable consequences of the axiom of choice; see Section 13.5.) This obstacle is only apparent, however. The supermartingales and martingales that we have constructed in the preceding chapters are all measurable as functions of the moves of Skeptic's opponents. In fact, as discussed in Section 1.6, they are computable, a property stronger than measurability and one that can be far more important, because it can take on practical as well as theoretical significance.

As an example, consider Protocol 4.1, which we used in Chapter 4 for Kolmogorov's law of large numbers. In the notation of Chapter 4, a strategy ψ for Skeptic in this protocol is a triplet $(\psi^{\text{stake}}, \psi^{\text{M}}, \psi^{\text{V}})$, where ψ^{stake} specifies $\mathcal{K}_0$ and ψ^{M} and ψ^{V} are predictable processes specifying $M_n := \psi_n^{\text{M}}$ and $V_n := \psi_n^{\text{V}}$. If the functions ψ_n^{M} and ψ_n^{V} are all Borel measurable, then we say that the strategy is *Borel measurable*. Let us say that Skeptic *can Borel force* an event E if he has a Borel measurable strategy that forces E.

The strategy we constructed for Skeptic in Section 4.3, being the result of simple arithmetic and limiting processes, is obviously Borel measurable, and hence we may strengthen Statement 1 of Theorem 4.3 to the following:

Proposition 9.17 *Skeptic can Borel force*

$$\sum_{n=1}^{\infty} \frac{v_n}{n^2} < \infty \implies \lim_{n\to\infty} \frac{1}{n} \sum_{i=1}^{n} (y_i - m_i) = 0 \tag{9.19}$$

in Protocol 4.1.

Le $\mathcal{L}_0, \mathcal{L}_1, \ldots$ be the capital process from a Borel-measurable strategy ψ in Protocol 4.1. Then $\mathcal{L}_0 := \psi^{\text{stake}}$ is the initial capital and

$$\mathcal{L}_n(\omega) := \mathcal{L}_{n-1}(\omega) + \psi_n^M(\omega)(y_n - m_n) + \psi_n^V(\omega)((y_n - m_n)^2 - v_n),$$

$$\omega = (m_1, v_1, y_1, m_2, v_2, y_2, \ldots), \qquad (9.20)$$

is the capital at the end of the nth round of play. The functions $\mathcal{L}_n$ specified by (9.20) are very simple – certainly measurable – functions of the moves by Forecaster and Reality. So if all the m_n, v_n, and y_n are taken to be measurable functions on a measurable space $(\Omega, \mathcal{F})$, then $\mathcal{L}_0, \mathcal{L}_1, \ldots$ will also become measurable functions on $(\Omega, \mathcal{F})$. (Here $(\Omega, \mathcal{F})$ is an arbitrary measurable space. We are not using the symbol Ω to designate the sample space for a protocol, as we usually do.) In fact, if $(\mathcal{F}_n)_{n=0}^{\infty}$ is a filtration in $(\Omega, \mathcal{F})$, and the functions m_n, v_n, and y_n are measurable with respect to $\mathcal{F}_n$ for each n, then $\mathcal{L}_n$ will also be measurable with respect to $\mathcal{F}_n$ for each n.

The following corollary of Proposition 9.17 is the measure-theoretic form of Kolmogorov's law of large numbers.

Corollary 9.18   ([77], Theorem 5(a)) *If $y_1, y_2, \ldots$ is an adapted sequence of random variables in a filtered probability space $(\Omega, \mathcal{F}, (\mathcal{F}_n)_{n=0}^{\infty}, P)$, then*

$$\sum_{n=1}^{\infty} \frac{V_P(y_n | \mathcal{F}_{n-1})}{n^2} < \infty \implies \lim_{n \to \infty} \frac{1}{n} \sum_{i=1}^{n} (y_i - P(y_i | \mathcal{F}_{i-1})) = 0 \quad \text{a.s.} \qquad (9.21)$$

The conclusion of the corollary is true even if some or all of the y_n fail to have (finite) conditional expectations or variances on a set of $\omega \in \Omega$ of a positive measure. At such ω the conditional variance $V_P(y_n | \mathcal{F}_{n-1})$ does not exist or is infinite, and so the left-hand side of the implication in (9.21) fails and thus the implication itself happens (remember that $(A \implies B) = A^c \cup B$).

Proof of Corollary 9.18   According to Proposition 9.17, Skeptic has a Borel measurable strategy ψ in Protocol 4.1 for which the capital process $\mathcal{L}_0, \mathcal{L}_1, \ldots$, given by (9.20), is nonnegative no matter how Forecaster and Reality move and tends to infinity if (9.19) fails. Fix versions of the conditional expectations $P(y_n | \mathcal{F}_{n-1})$ and then of the conditional variances $V_P(y_n | \mathcal{F}_{n-1})$ (making sure the latter are nonnegative), and substitute them for m_n and v_n in (9.20). Similarly, substitute the random variable y_n for the move y_n in (9.20). As we explained in our comments following (9.20), the resulting function $\mathcal{L}_n$ on Ω is measurable with respect to $\mathcal{F}_n$. Similarly, ψ_n^M and ψ_n^V become functions on Ω measurable with respect to $\mathcal{F}_{n-1}$, and we can rewrite (9.20) in the form

$$\mathcal{L}_n := \mathcal{L}_{n-1} + \psi_n^M(y_n - P(y_n | \mathcal{F}_{n-1})) + \psi_n^V((y_n - P(y_n | \mathcal{F}_{n-1}))^2 - V_P(y_n | \mathcal{F}_{n-1})).$$

This implies that $P(\mathcal{L}_n | \mathcal{F}_{n-1}) = \mathcal{L}_{n-1}$. So $\mathcal{L}_0, \mathcal{L}_1, \dots$ is a measure-theoretic martingale. It is nonnegative and tends to infinity if (9.21) fails. But by Doob's convergence theorem [358, Theorem VII.4.1], a nonnegative measure-theoretic martingale tends to infinity with probability 0. So (9.21) happens almost surely.   □

9.4 EXERCISES

Exercise 9.1 Consider the specialization of Protocol 7.1 in which $\mathcal{Y}$ is finite and $\overline{\mathbf{E}}$ is given by $\overline{\mathbf{E}}(f) = P(f)$ for some probability measure P on $\mathcal{Y}$. This is also a specialization of Protocol 9.1, and according to Theorem 9.3, any global variable that depends on only a finite number of rounds of play has an expected value. Thus in particular any global event that depends only on a finite number of rounds of play has a probability. Let $\mathcal{F}$ denote the set of all global events that depend on only a finite number of rounds of play.

1. Show that the game-theoretic upper expected value of a bounded global variable is identical with its upper integral with respect to P^∞. *Hint.* Use Theorem 9.3.

2. Conclude that

$$\overline{\mathbb{P}}(E) = \inf \left\{ \sum_{k=1}^{\infty} \mathbb{P}(E_k) | E_1, E_2, \dots \in \mathcal{F}, E \subseteq \bigcup_{k=1}^{\infty} E_k \right\}$$

for every $E \subseteq \mathcal{Y}^\infty$. *Hint.* Use Carathéodory's extension theorem.   □

Exercise 9.2 Let P be a probability measure on $[-1, 1]$, and let $m \in [-1, 1]$. Show that $P \preceq \overline{\mathbf{E}}^m$, in the notation of Protocol 9.6 and Example 6.3, if and only if m is P's expected value: $m = \int_{-1}^{1} y P(dy)$.   □

The remaining exercises consider three ways of specializing Protocols 9.5 and 9.6 and the $2^3 - 1 = 7$ cases that arise when we specialize in one or more of these ways. The first way of specializing is to reduce the protocols to one round. (There are intermediate versions between the infinite-horizon versions in Section 9.2 and one-round versions, such as N-round versions for $N > 1$; see, e.g. [340, Propositions 12.6 and 12.7].) The second way is to make the protocols *strategic*: we fix a strategy for Forecaster. The third is to make them *precise*: each $\overline{\mathbf{E}}_\theta$ is an expectation (rather than upper expectation) operator.

Exercise 9.3 Consider the one-round, strategic, and precise versions of Protocols 9.5 and 9.6. What do the definitions of $\overline{\mathbb{E}}(X)$ and $\mathbb{E}^{\mathrm{meas}}(X)$ become, for an $\mathbb{R}$-valued X? (This is the simplest possible situation, the one of elementary probability theory.)   □

Exercise 9.4 Consider the one-round and strategic versions of Protocols 9.5 and 9.6. In Protocol 9.5, Forecaster outputs an upper expectation $\overline{\mathbf{E}}$, and we have $\overline{\mathbb{E}}(X) = \overline{\mathbf{E}}(X)$. In Protocol 9.6 Probability Forecaster can output any expectation P dominated by $\overline{\mathbf{E}}$. Therefore, $\mathbb{E}^{\mathrm{meas}}(X) = \sup_{P \preceq \overline{\mathbf{E}}} P(X)$, and the statement of Theorem 9.7 becomes a basic result of the theory of imprecise probabilities called the *lower envelope theorem* [426, Proposition 3.3.3] (in our setting, this becomes an upper envelope theorem). Check all details. □

Exercise 9.5 Consider the one-round and precise versions of Protocols 9.5 and 9.6. The indexing system $\overline{\mathbb{E}}_\theta$ now becomes a statistical model $(P_\theta)_{\theta \in \Theta}$. Show that in Protocol 9.5 we have $\overline{\mathbb{E}}(X) = \sup_{\theta \in \Theta} P_\theta(X)$. Also show directly that in Protocol 9.6 we have $\mathbb{E}^{\text{meas}}(X) = \sup_{\theta \in \Theta} P_\theta(X)$, which agrees with Theorem 9.7. If $X = \mathbf{1}_E$ for some set E, the value $\overline{\mathbb{P}}(E) = \mathbb{P}^{\text{meas}}(E)$ can be interpreted as the size of E regarded as a critical region for testing the composite hypothesis $(P_\theta \mid \theta \in \Theta)$. □

Exercise 9.6 Consider the one-round versions of Protocols 9.5 and 9.6. Show that in Protocol 9.5 we have $\overline{\mathbb{E}}(X) = \sup_{\theta \in \Theta} \overline{\mathbb{E}}_\theta(X)$. Show directly that in Protocol 9.6 we obtain the same expression for $\mathbb{E}^{\text{meas}}(X)$: $\mathbb{E}^{\text{meas}}(X) = \sup_{\theta \in \Theta} \overline{\mathbb{E}}_\theta(X)$. If $X = \mathbf{1}_E$ for some set E, the value $\overline{\mathbb{P}}(E) = \mathbb{P}^{\text{meas}}(E)$ can be interpreted as the size of E regarded as a critical region for testing the composite hypothesis $(P_\theta)_{\theta \in \Theta}$. This generalizes the *lower envelope theorem* (see Exercise 9.4) from the setting of imprecise probabilities to a statistical setting. □

Exercise 9.7 Consider the strategic and precise versions of Protocols 9.5 and 9.6. Now we are in the situation of Protocol 9.2. In Protocol 9.5, how would you characterize $\overline{\mathbb{E}}(X)$ in terms of Section 9.1? In Protocol 9.6, how would you characterize $\mathbb{E}^{\text{meas}}(X)$? Deduce Theorem 9.3 (whose conditions are strengthened to the conditions of Theorem 9.7) from Theorem 9.7. □

Exercise 9.8 Consider the strategic versions of Protocols 9.5 and 9.6. Can you simplify the general argument in the proof of Theorem 9.7 for this case? Now consider the finite-horizon version of Protocols 9.5 and 9.6 with horizon N (i.e. replace "FOR $n = 1, 2, \ldots$" by "FOR $n = 1, \ldots, N$"). Using backward induction and Exercise 9.6, show the counterpart of Theorem 9.7 for this case. □

Exercise 9.9 Consider the precise versions of Protocols 9.5 and 9.6. Now we are in the situation of Protocol 9.1. Specialize Theorem 9.7 (and all the definitions leading up to it) to this case. □

9.5 CONTEXT

Interpretation

The relation $P \preceq \overline{\mathbf{E}}$ between a probability measure P and an upper expectation $\overline{\mathbf{E}}$ on a set $\mathcal{Y}$, introduced in Section 9.2, can be understood in terms of the notion of an offer, discussed in Section 6.4. Indeed, $P \preceq \overline{\mathbf{E}}$ if and only if $\mathbf{G}_{\overline{\mathbf{E}}} \subseteq \mathbf{G}_P$, where $\mathbf{G}_{\overline{\mathbf{E}}}$ and $\mathbf{G}_P$ are the offers corresponding to $\overline{\mathbf{E}}$ and P, respectively. When $\mathcal{Y}$ is finite, as we assume in Section 9.2, the offer $\mathbf{G}_P$ corresponding to a probability measure P on $\mathcal{Y}$ is maximal; it cannot be enlarged to a different offer. Thus $P \preceq \overline{\mathbf{E}}$ means that P is a maximal enlargement of the betting offers made to Skeptic.

This picture lends itself to more than one substantive interpretation. In Walley's extension of de Finetti's interpretation of probability as personal belief, we think of $G_{\overline{E}}$ as a set of gambles that You is inclined to take, which You may enlarge on further reflection (see [426] and the discussion in Section 6.7). A larger set of gambles G_P, where P is a probability measure, is as far as this enlargement can go without losing coherence. The game-theoretic form of Cournot's principle, which says that a probabilistic theory makes predictions by asserting that given betting offers cannot be combined to multiply the capital risked by a large factor (see Section 10.2), gives another substantive interpretation: the enlargement from $G_{\overline{E}}$ to G_P means that we are making more predictions, again to a degree that is maximal.

There are other interpretations of probability measures that are not so easily extended to upper expectations that are not probability measures. For these interpretations, the P seems more basic than $\overline{E}$, and $P \le \overline{E}$ may have no meaning beyond its definition: $P(f) \le \overline{E}(f)$ for all f. For example, if $P(E)$ is interpreted as the frequency of the property E in a specified population, then $\overline{E}(1_E)$ is merely an upper bound on this frequency. It is not clear, however, why we should want or expect a system of upper bounds on frequencies to satisfy the axioms for an upper expectation.

Ionescu Tulcea's Theorem

Ionescu Tulcea's theorem is a result of a series of attempts to generalize the Kolmogorov extension theorem in Kolmogorov's *Grundbegriffe* [224, section III.4]. According to Kolmogorov's theorem, a compatible family of probability measures on the finite-dimensional measurable cylinder sets in $\mathbb{R}^I$, where the index set I may be uncountable, determines a probability measure on the whole of $\mathbb{R}^I$ with the marginals coinciding with that family. In 1938 Doob claimed that $\mathbb{R}$ can be replaced by any measurable space [114, Theorem 1.1], but he was wrong, as shown by Sparre Andersen and Jessen in 1948 [10]. Doob and Jessen planned to write an article correcting the error [10, p. 5], but they were preempted by Ionescu Tulcea. Ionescu Tulcea's 1949 article is in French [196]. For an account of his theorem in English (in the case of countable I, which is sufficient for our purposes), see [358, section II.9].

Measure-Theoretic vs. Game-Theoretic Probability in Mathematical Finance

The duality between measure-theoretic and game-theoretic probability has become a popular topic in mathematical finance, where game-theoretic upper expected value has been called minimal super-replicating cost. Key articles include [25,112,113]. They show the equality between the minimal super-replicating cost and the supremum of expected values over the martingale measures under various more or less restrictive conditions.

Choquet's Capacitability Theorem

Our main technical tool in this chapter is Gustave Choquet's capacitability theorem for functions. The standard reference for the topological form of Choquet's theorem for sets (as opposed to functions) is Choquet's 1954 article [74], which emphasized alternating and monotone capacities. The abstract form, which is very popular in the theory of stochastic processes, was established by Choquet in 1959 [75]. Choquet did not state his theorem for functions explicitly in these articles, but he sketched a wide generalization at the end of [75].

The key notion used in the statement of Choquet's capacitability theorem is that of a Suslin, or analytic, function or set. The realization that analytic sets are wider than Borel sets was due to Nikolai Luzin and his research student Mikhail Suslin. Suslin disproved Henri Lebesgue's claim that the projection of any Borel set in $\mathbb{R}^2$ onto its first coordinate is also Borel, and such projections came to be known as analytic sets. Luzin's role in Suslin's accomplishment was questioned when he came under attack by Soviet authorities in 1936 [107]. Analytic sets also arise in Exercise 13.1.

Part III

Applications in Discrete Time

Mathematical probability began as a theory of betting, and its applications to other topics, beginning in the eighteenth century, were developed in the language of betting. Annuities and insurance policies are bets, and error theory was developed in terms of betting. In the nineteenth century the discourse about betting largely gave way to a discourse about frequencies, and in the twentieth century it gave way to a more abstract language thought more mathematically rigorous. But as we will see in this part of the book, the rigor offered by testing protocols provides new opportunities to exploit insights that the notion of betting can bring to applications of probability to science, technology, and actual forecasting.

In many applications, events for which we can provide probabilistic predictions are interleaved in time with events, decisions, and other new information for which we cannot provide such predictions. When the theories and models used in these applications can be framed as forecasting strategies in testing protocols, their resistance to betting strategies becomes a general criterion for their reliability and validity.

In Chapter 10, we discuss in detail how statisticians can use testing protocols to study a probability model for color-blindness. We also look at testing protocols for least squares, for survival analysis, for John von Neumann's formalism for quantum mechanics, and for quantum computing. In the last section of the chapter, we use our criterion for validity to justify *Jeffreys's law*, the assertion that two competing probabilistic theories cannot both be valid unless they concur asymptotically.

Game-Theoretic Foundations for Probability and Finance, First Edition. Glenn Shafer and Vladimir Vovk.
© 2019 John Wiley & Sons, Inc. Published 2019 by John Wiley & Sons, Inc.

Chapter 11 considers a new issue raised by game-theoretic probability's dynamic notion of testing. In a testing protocol, Skeptic can stop betting at any point and count his success thus far as evidence against the validity of the forecasts. But he loses this evidence if he continues betting and loses all or much of what he had won. We show how to devise strategies to mitigate such loss of evidence, strategies that correspond to the purchase of a certain kind of lookback option for an investor in a financial market. Such strategies also give us ways to calibrate p-values in statistics to correct for the fact that they are not necessarily significance levels chosen in advance.

In Chapter 12, we turn from what Skeptic, the player who takes betting offers, can accomplish to what Forecaster, the player who announces those offers, can accomplish. We find that if he has the advantage of the feedback provided by a perfect-information testing protocol, Forecaster can often devise strategies that can pass Skeptic's tests. The theory developed in this chapter makes contact with theories of universal estimation and prediction that have been developed in mathematical statistics, machine learning, and the theory of algorithmic complexity.

10

Using Testing Protocols in Science and Technology

Probability has become ubiquitous in science and technology. Probabilities appear in scientific theories and in models used for control, prediction, and decision making. But usually only some aspects of a process are susceptible to probabilistic prediction. Using examples from science and mathematical statistics, this chapter shows how testing protocols can focus on the predictable aspects while respecting the unpredictability or unobservability of other aspects.

The role of information that is not forecast depends on whether we consider goals for Skeptic or goals for Forecaster. As we learned in Parts I and II, a typical result in game-theoretic probability says that Skeptic can guarantee a disjunction: either Forecaster and Reality agree in some respect or the capital Skeptic risks is multiplied by a large factor. According to our mathematics, Skeptic can guarantee this disjunction regardless of what other events, predictable or not, transpire – regardless of anything else Skeptic or the other players already know or witness in the course of play. In contrast, what Forecaster can accomplish is an empirical or scientific question. Given all his information, including information that may unexpectedly become available in the course of play, can Forecaster keep Skeptic from multiplying the capital he risks by a large factor? If so, we say that Forecaster is *reliable*. If he manages to be consistently reliable by following a strategy given by a theory or a model, we may claim that the theory or model is *valid*. But here *reliable* and *valid* have only an empirical meaning; they are not terms defined within a mathematical theory.

Game-Theoretic Foundations for Probability and Finance, First Edition. Glenn Shafer and Vladimir Vovk.
© 2019 John Wiley & Sons, Inc. Published 2019 by John Wiley & Sons, Inc.

Section 10.1 discusses how additional information that arises in the course of play can be included as signals in a testing protocol. This can be important in applications, because the strategy for Forecaster given by a scientific theory or other model often uses such information.

Section 10.2 further discusses the notion of validity for forecasting strategies in testing protocols. The thesis that this notion of validity is appropriate for scientific theories and models is the game-theoretic version of Cournot's principle.

Section 10.3 uses a classical model for the inheritance of Daltonism – a form of color-blindness – to study how a statistician or other analyst can reframe a probabilistic theory as a testing protocol and use it to test and predict. This is a toy example; it ignores practical difficulties as well as advances in our ability to identify genes and our understanding of gene expression. But it is complex enough to cast light on some general issues, including the multiplicity of possible tests and the statistician's relation to the protocol. The statistician does not always stand inside the protocol she uses, seeing everything the players see. Perhaps she does not see all the probabilities or other betting offers announced by Forecaster. Perhaps she does not directly see the outcomes. She may therefore be unable to implement tests that the forecasts are supposed to withstand. But she may still be able to observe statistics that establish that a given test has succeeded to a given degree, and this may be enough to permit testing, estimation, and prediction.

Section 10.4 uses an elaboration of Protocol 1.1, the protocol with which we began this book, to justify the most venerable of statistical methods, least squares. Here again, the statistician does not stand entirely inside the protocol.

Section 10.5 looks at parametric statistical models in terms of protocols with signals, with particular attention to Cox's proportional hazards model for survival analysis.

Section 10.6 considers quantum mechanics. Probability's role in this branch of physics has been endlessly debated, but some aspects of its role, especially von Neumann's formalism for measurement and the related formalism for quantum computing, are uncontroversial and are readily described by testing protocols.

Section 10.7 considers the reliability of competing forecasters. Forecasters with different information can be expected to give different forecasts for the same outcomes. Can two forecasters both be reliable – i.e. both pass statistical tests – if they persist in differing? As we will see, they cannot both succeed if they are tested by the same Skeptic. If they differ too much, Skeptic will be able to discredit at least one of them. This is *Jeffreys's law*.

10.1   SIGNALS IN OPEN PROTOCOLS

The protocols we study in this chapter, like nearly all the protocols in this book, are perfect-information protocols; the players see each other's moves as they are made. But they are also open to additional information, which may be available to all the players or only to some of them. In this section, we discuss the advantages of making additional information that is seen by all the players explicit in the protocol. We emphasize the case of probability forecasting, where a strategy for Skeptic can be interpreted as a difference of opinion with Forecaster.

A natural way to make additional information explicit in a testing protocol is to have Reality announce the new information available on each round at the beginning of that round. We write $x_1, x_2, \ldots$ for these announcements and call them *signals*. We assume that such signals are seen by all the players, so that the protocol remains a perfect-information protocol. Making the signals explicit does not change what Skeptic can accomplish, but it may allow us to interpret events that Skeptic can force as statements about the relation between the signals and the outcomes.

Consider, for example, Protocol 1.8, the elementary testing protocol for probability forecasting. Adding signals to this protocol produces the following perfect-information protocol.

Protocol 10.1

PARAMETER: Nonempty set $\mathcal{X}$

> $\mathcal{K}_0 := 1$.
> FOR $n = 1, 2, \ldots$:
>> Reality announces $x_n \in \mathcal{X}$.
>> Forecaster announces $p_n \in [0, 1]$.
>> Skeptic announces $M_n \in \mathbb{R}$.
>> Reality announces $y_n \in \{0, 1\}$.
>> $\mathcal{K}_n := \mathcal{K}_{n-1} + M_n(y_n - p_n)$.

In Chapter 2, we defined upper probabilities for events in Protocol 1.8. These events concern $p_1, y_1, p_2, y_2, \ldots$, but when Forecaster plays a strategy that makes each p_n a function of Reality's previous moves $x_1, y_1, \ldots, x_n$, they become upper probabilities for $x_1, y_1, x_2, y_2, \ldots$. If Forecaster's strategy represents a scientific theory, these upper probabilities can be used to test the theory and make predictions, even though they do not determine a probability measure for $x_1, y_1, x_2, y_2, \ldots$.

The ease with which information not itself forecast can be added to a testing protocol contrasts with the awkwardness with which measure-theoretic probability handles such information. We can fit a sequence $x_1, y_1, x_2, y_2, \ldots$ of signals and outcomes into a probability measure only if the signals are constants fixed at the outset or themselves have probabilities.

The explicit representation of different players' different roles in a testing protocol can also help analysts understand their own role. As analyst, are you taking the role of Forecaster – i.e. trying to build a theory or make good probability forecasts? (We will learn more about this role in Section 10.7 and Chapter 12.) Are you taking the role of Skeptic – i.e. trying to test a probabilistic theory or an actual forecaster in real time? Or are you standing outside the protocol, seeing only some of Reality's moves and using hypotheses about what Forecaster and Skeptic can accomplish to estimate, test, or predict? The answers to these questions differ from one statistical study to another.

In most of the applications we consider in this chapter (our study of least squares in Section 10.4 is an exception), Forecaster gives probability forecasts. This means that he announces a probability measure for each outcome. The following protocol is an abstract testing protocol of this type, stated in a form somewhat different than we have used so far.

Protocol 10.2
PARAMETERS: Nonempty set $\mathcal{X}$ and measurable space $\mathcal{Y}$
    $\mathcal{K}_0 := 1$.
    FOR $n = 1, 2, \ldots$:
        Reality announces $x_n \in \mathcal{X}$.
        Forecaster announces $P_n \in \mathcal{P}(\mathcal{Y})$.
        Skeptic announces $f_n \in [0, \infty]^{\mathcal{Y}}$ such that $P_n(f_n) = 1$.
        Reality announces $y_n \in \mathcal{Y}$.
        $\mathcal{K}_n := \mathcal{K}_{n-1} f_n(y_n)$.

Here we require f_n to be nonnegative, thus making explicit the assumption that Skeptic does not risk more than his initial unit capital. With this requirement, the rules $P_n(f_n) = 1$ and $\mathcal{K}_n := \mathcal{K}_{n-1} f_n(y_n)$ allow the same capital processes for Skeptic as the rules $P_n(f_n) = \mathcal{K}_{n-1}$ and $\mathcal{K}_n := f_n(y_n)$ used in some of our previous protocols. The condition $P_n(f_n) = 1$ implicitly requires that f_n be measurable, but the slackening of the protocol would allow any $f_n \in [0, \infty]^{\mathcal{Y}}$ for which the upper integral is 1 or less (see Exercise 7.1).

To fix ideas, suppose Forecaster always chooses a probability measure P_n that has a density, say p_n, with respect to a fixed underlying probability measure on $\mathcal{Y}$. If we set $q_n = f_n p_n$, then the condition $P_n(f_n) = 1$ in Protocol 10.2, which can also be written as $\int f_n dP_n = 1$, implies that q_n is also a density. We have

$$f_n(y_n) = \frac{q_n(y_n)}{p_n(y_n)}$$

(assuming $p_n(y_n) \neq 0$) and

$$\mathcal{K}_n = \frac{\prod_{i=1}^n q_i(y_i)}{\prod_{i=1}^n p_i(y_i)}, \tag{10.1}$$

generalizing the result given for Protocol 1.8 in Exercise 1.5. This suggests that Skeptic's choice of f_n might be inspired by disagreement with Forecaster's p_n. If he thinks the best forecast is q_n rather than p_n, then he sets $f_n := q_n/p_n$ and benefits to the extent that his q_n gives the outcome y_n greater probability than Forecaster's p_n does.

The identity (10.1) holds on the particular path $x_1 y_1 x_2 y_2 \ldots$ taken by Reality whether or not Forecaster and Skeptic follow strategies selected in advance. But if they both follow strategies that make their moves functions of Reality's previous moves, say $p_n(x_1, y_1, \ldots, x_n)$ and $f_n(x_1, y_1, \ldots, x_n)$, then (10.1) becomes

$$\mathcal{K}_n = \frac{\prod_{i=1}^{n} q_i(x_1, y_1, \ldots, x_i)(y_i)}{\prod_{i=1}^{n} p_i(x_1, y_1, \ldots, x_i)(y_i)}. \tag{10.2}$$

Using the word *probability* in an extended sense, we can restate (10.2) by saying that the nonnegative martingale $\mathcal{K}_n$ is the ratio of Skeptic's probability for the outcomes to Forecaster's probability for the outcomes. In Section 10.5, we consider the generalization of (10.2) to parametric statistical models.

The simplicity of (10.2) is appealing, but its generality should not be exaggerated. It does not always describe how we test a hypothesis represented by a strategy for Forecaster, first because Forecaster may announce less than a probability measure for each outcome, and second, as we will see in Sections 10.3–10.5, because the picture can become more complicated when the statistician does not stand inside the protocol in Skeptic's shoes.

10.2 COURNOT'S PRINCIPLE

As explained in the introduction to this chapter, a strategy for Forecaster is *valid* if it consistently keeps Skeptic from multiplying his capital by a large factor. This concept of validity is relative to Skeptic's knowledge and information.

We call the thesis that the validity of a forecasting strategy consists of its ability to withstand betting strategies *Cournot's principle*. In the mid-twentieth century, this name was sometimes given to the principle that a probabilistic theory makes predictions only by giving certain events high probability and can be tested only by checking such predictions (see the references in Section 10.9). Our game-theoretic version of the principle generalizes the older version by acknowledging that probabilities and other betting offers can be tested by a series of bets that avoid risking bankruptcy, whether or not the bets boil down to a bet on a single event and even whether or not the betting follows a prespecified strategy.

Discrediting a probabilistic theory via Cournot's principle often involves showing that a frequency fails to approximate the value predicted by the theory. This does not quite bring us, however, in alignment with the notion that probabilities can be directly interpreted as frequencies. In general – in Protocol 10.1, for example – it is only an average probability that is supposed to be approximated by a frequency. More importantly, discrepancy between frequency and average probability appears only as a symptom. When the discrepancy occurs, the theory's fundamental demerit is its failure to stand up to a betting strategy that Skeptic identified in advance and implemented.

It is also true, as we saw in Chapter 5, that a probabilistic theory can be discredited not only by the failure of observations to match probabilistic predictions on average but also by their failure to approximate these predictions in the way the theory predicts – oscillating nearer and nearer at a particular rate. But this may be of limited practical importance; see the further discussion in Section 12.2.

Cournot's principle does not say that a theory is discredited merely because an event to which it gave small probability happens. The happening discredits the theory only if the statistician testing the theory singled the event out in advance as a test. This permits betting on the event and so justifying rejection game-theoretically. It is possible to bet on several events (or more generally, to divide one's capital among several strategies), and so the simplicity with which we can describe an event of small probability or a successful betting strategy can be relevant. If there are only a few equally simple events or strategies, then the happening of the event or the success of the strategy may arouse suspicion even if the event or strategy was not identified in advance. But as statisticians often point out, such arguments are tricky; it is easy to underestimate how many events or strategies would seem simple after the fact.

Moreover, Cournot's principle leaves a theorist free to decide which aspects of a probabilistic theory to assert or test for empirical validity. The game-theoretic form of the principle says only that a probabilistic theory makes predictions by asserting that certain betting strategies will not succeed in multiplying the capital risked by a large or infinite factor. The theorist may assert this about a class of betting strategies without asserting it about others. This flexibility comes into play in Part IV, where we sometimes make predictions about what happens during very short periods of time without making predictions about the longer run.

10.3 DALTONISM

In this section, we use the genetic theory concerning *Daltonism*, a form of red-green color-blindness, to study different ways a statistician can use a testing protocol.

According to the accepted theory, Daltonism is caused by a faulty gene on the X-chromosome. A man suffers from Daltonism if he has the faulty gene on his single X-chromosome. But the gene is recessive in women; a woman suffers from Daltonism only if she has the gene on both her X-chromosomes. These facts, together with the random sorting of chromosomes in meiosis, give rise to a probability model for the inheritance of Daltonism.

Let us say that a person's status with respect to Daltonism is N (normal) if the person does not have the faulty gene, C (carrier) if the person is a woman with the faulty gene on only one X-chromosome, and A (affected) if the person suffers from Daltonism (is a man with the faulty gene or a woman with the faulty gene on both X-chromosomes).

The theory tells us that it is an even bet which of her two X-chromosomes a mother will pass to her child. So a son's status depends on his mother's status as follows:

If mother is N, son is N.

If mother is C, son is N or A, with probabilities 1/2 each.

If mother is A, son is A.

A daughter's status depends on the status of both parents:

If father is N and mother is N, daughter is N.

If father is N and mother is C, daughter is N or C, with probabilities $1/2$ each.

If father is N and mother is A, daughter is C.

If father is A and mother is N, daughter is C.

If father is A and mother is C, daughter is C or A, with probabilities $1/2$ each.

If father is A and mother is A, daughter is A.

Suppose, as was once the case, that we do not have the technology to study the chromosomes and the genes directly; we can only observe whether an individual suffers from Daltonism. Thus we cannot observe directly whether a woman is a carrier. How, in this case, can we test the theory?

Statistician in the Role of Skeptic

Although we cannot be sure that a given woman is not a carrier, we can use the theory to identify some who are: those whose mothers were A and fathers were N. A statistician can test the theory by studying children of such carriers, along with children of women who are As.

Let us write p_{FMS} for the probability that a child will be affected by Daltonism when the father's status is F, the mother's status is M, and the child's sex is S. The theory gives the following values for p_{FMS} when the mother is C or A:

$$p_{NCB} = 1/2, p_{NCG} = 0,$$
$$p_{NAB} = 1, \quad p_{NAG} = 0,$$
$$p_{ACB} = 1/2, p_{ACG} = 1/2,$$
$$p_{AAB} = 1, \quad p_{AAG} = 1. \tag{10.3}$$

A statistician can use the following specialization of Protocol 10.1 to study a sequence of births to women who are C or A.

Protocol 10.3
   $\mathcal{K}_0 = 1.$
   FOR $n = 1, 2, \ldots$:
       Reality announces $(F_n, M_n, S_n) \in \{N, A\} \times \{C, A\} \times \{B, G\}$.
       Forecaster announces $p_n := p_{F_n M_n S_n} \in [0, 1]$.
       Skeptic announces $M_n \in \mathbb{R}$.
       Reality announces $y_n \in \{0, 1\}$.
       $\mathcal{K}_n := \mathcal{K}_{n-1} + M_n(y_n - p_n).$

Reality's move specifies the status of the father (F_n), the status of the mother (M_n), and the sex of the child (S_n). The outcome y_n is 1 if the child is A and 0 if the child is not A. Forecaster, who represents the theory, is required to make the move $p_{F_n M_n S_n}$ given by (10.3).

The statistician participates in Reality's moves (F_n, M_n, S_n) by finding births to include in the study and observing the parents' Daltonism status and the child's sex. She can also play the role of Skeptic, testing the theory by betting at the odds it gives. She *discredits the theory at level α* if she always keeps her capital nonnegative and eventually multiplies her initial capital, which we have set equal to 1 for simplicity, by the factor $1/\alpha$. In principle, she can continue betting as long as she likes, so long as she keeps her capital nonnegative. She can decide how to bet (and when to stop) as she goes along, or she can follow a strategy, such as one of the strategies that we studied in Part I.

The following proposition points to some strategies Skeptic might use. We state it for Protocol 10.1; it also holds for that protocol's specializations, including Protocol 10.3. Statement 1 derives from Hoeffding's inequality, while Statement 2 derives from the central limit theorem.

Proposition 10.4   *Suppose $0 < \alpha < 1$. Then these two statements hold for Protocol 10.1.*

1. For every $N \in \mathbb{N}$,

$$\overline{\mathbb{P}}\left(|\bar{y}_N - \bar{p}_N| \geq \sqrt{\frac{1}{2} \ln \frac{2}{\alpha}} \, N^{-1/2} \right) \leq \alpha. \tag{10.4}$$

2. For $B > 0$, let τ_B be the stopping time

$$\tau_B := \min\left\{ N | \sum_{n=1}^{N} p_n(1 - p_n) \geq B \right\}.$$

Then

$$\lim_{B \to \infty} \overline{\mathbb{P}}(|\bar{y}_{\tau_B} - \bar{p}_{\tau_B}| \geq z_{\alpha/2} \sqrt{B}/\tau_B) = \alpha, \tag{10.5}$$

where z_δ is the upper δ-quantile of the standard Gaussian distribution.

Proof   Statement 1 is a form of Hoeffding's inequality (Corollary 3.8). Statement 2 can be proved similarly to, and essentially follows from, Proposition 2.10.   □

For every $N \in \mathbb{N}$ and every $\alpha > 0$, according to Statement 1 of Proposition 10.4, the statistician qua Skeptic has a strategy based on Hoeffding's inequality that multiplies her capital by $1/\alpha$ or more and hence discredits the theory at level α if

$$|\bar{y}_N - \bar{p}_N| \geq \sqrt{\frac{1}{2} \ln \frac{2}{\alpha}} \, N^{-1/2}.$$

We may assume that the statistician includes in the test only rounds where $p_n = 1/2$, because when the theory makes the prediction $p_n = 0$ or $p_n = 1$, she can bet against the prediction without risk, making more than enough money to discredit the theory immediately if the prediction fails.

Under the assumption that p_n is always $1/2$, $\tau_B = \lceil 4B \rceil$. So if we write N for τ_B, then Statement 2 of Proposition 10.4 says that when N is sufficiently large, the statistician has a strategy based on the central limit theorem that multiplies her capital by approximately α or more if

$$|\bar{y}_N - \bar{p}_N| \geq \frac{1}{2} z_{\alpha/2} N^{-1/2}.$$

Although the difference is not great (see Figure 3.2 and Exercise 10.1),

$$\frac{1}{2} z_{\alpha/2} < \sqrt{\frac{1}{2} \ln \frac{2}{\alpha}},$$

and so the test based on the central limit theorem is generally more powerful than the test based on Hoeffding's inequality.

On the other hand, the statistician qua Skeptic in Protocol 10.3 can test the theory using Hoeffding's supermartingale without selecting N and α in advance. Instead, she can continue collecting additional births and continue betting, stopping only when she has enough evidence to discredit the theory or finally decides to abandon the effort. The strategy that leads to Hoeffding's supermartingale, (3.9), reduces to

$$M_n := \mathcal{K}_{n-1} (e^{\kappa(1-p_n)} - e^{-\kappa p_n}) e^{-\kappa^2/8} \tag{10.6}$$

in Protocol 10.3. One way for the statistician to use this strategy is to choose a small positive number ϵ, put half her initial unit capital on (10.6) with $\kappa := 4\epsilon$ and the other half on (10.6) with $\kappa := -4\epsilon$. This does not risk bankruptcy, and by Hoeffding's inequality her capital will be at least $e^{2n\epsilon^2}/2$ at the end of any round n such that $|\bar{y}_n - \bar{p}_n| \geq \epsilon$. If this is large even though ϵ is small, she can stop and declare the theory to be discredited at level $2e^{-2n\epsilon^2}$.

Statistician Standing Outside a Protocol

Instead of using Protocol 10.3 to test the theory of Daltonism, the statistician could study a larger population, which includes women whose status with respect to Daltonism is unknown. This requires only a slight change in Protocol 10.3; we enlarge the choices for the mother's status M_n from $\{C, A\}$ to $\{N, C, A\}$:

Protocol 10.5
  $\mathcal{K}_0 = 1$.
  FOR $n = 1, 2, \ldots$:
      Reality announces $(F_n, M_n, S_n) \in \{N, A\} \times \{N, C, A\} \times \{B, G\}$.
      Forecaster announces $p_{F_n M_n S_n} \in [0, 1]$.
      Skeptic announces $M_n \in \mathbb{R}$.
      Reality announces $y_n \in \{0, 1\}$.
      $\mathcal{K}_n := \mathcal{K}_{n-1} + M_n(y_n - p_{F_n M_n S_n})$.

This involves adding four additional probabilities to (10.3):

$$p_{\text{NNB}} = p_{\text{NNG}} = p_{\text{ANB}} = p_{\text{ANG}} = 0.$$

The practical task of the statistician may include identifying and perhaps even recruiting the people who appear on successive rounds of Protocol 10.5, but she no longer stands fully inside the protocol, always seeing Reality's and Forecaster's moves. We may assume that she sees the sex of each child, that she sees the outcomes $y_1, y_2, \ldots$, and that she also sees whether or not each parent is an A. But she does not always know whether a non-A mother is a C or an N. So in general she does not see Forecaster's move $p_{F_n M_n S_n}$ and hence cannot implement a strategy for Skeptic that requires knowing this move.

Given what she does observe, the statistician can still implement some betting strategies, which will entail some predictions and tests. For example, on rounds where neither parent suffers from Daltonism and the child is a boy, she knows that Forecaster's move is either 0 or $1/2$. She can bet against the boy suffering from Daltonism on these rounds (choose M_n positive), and although she cannot see exactly how her capital changes (whether it changes by $M_n(y_n - 1/2)$ or by $M_n y_n$), she has strategies that she knows do not risk bankruptcy and for which she can calculate a lower bound showing that her capital will be large if there are many such rounds and the frequency with which the boys suffer from Daltonism exceeds $1/2$ by too much. See Exercise 10.2 for details.

A simpler way to think about how the statistician can use her partial knowledge of the information in the protocol is to imagine that she can instruct Skeptic to follow a strategy that she herself may not be able to implement. In this case, she can predict an event E on the basis of the theory provided that she knows that Skeptic has a strategy that multiplies his capital by a large factor, say $1/\alpha$, if E happens, and she can then use E to test the theory provided only that she can observe whether it happens (or at least possibly observe that it has happened). For example (again see Exercise 10.2), Skeptic has a strategy that will multiply the capital he risks by 20 if there are at least 100 boys whose parents do not suffer from Daltonism and the frequency of Daltonism in the first 100 such boys is more than 63%. If the statistician singles this event out in advance as a test (we can make this requirement vivid by insisting that she direct her imaginary Skeptic to play the strategy), then she is entitled to say that she has discredited the theory at level 0.05.

In order to list some of the other predictions that the statistician can make in Protocol 10.5, we consider the sets

$$\mathcal{F} := \{n \in \mathbb{N} | F_n = A\} \qquad \mathcal{M} := \{n \in \mathbb{N} | M_n = A\}$$

$$\mathcal{B} := \{n \in \mathbb{N} | S_n = B\} \qquad \mathcal{A} := \{n \in \mathbb{N} | y_n = 1\}$$

of natural numbers and write

$$\overline{\mathbf{P}}(E_1 | E_2) := \limsup_{n \to \infty} \frac{|\{k \le n | k \in E_1 \cap E_2\}|}{|\{k \le n | k \in E_2\}|}$$

when E_1 is a subset of $\mathbb{N}$ and E_2 is an infinite subset of $\mathbb{N}$. The statistician observes the sets $\mathcal{F}$, $\mathcal{M}$, $\mathcal{B}$, and $\mathcal{A}$ inasmuch as she sees, on each round n, whether n is in them.

Corollary 10.6 *In Protocol 10.5, Skeptic can force*

$$\mathcal{F}^c \cap \mathcal{B}^c \subseteq \mathcal{A}^c, \qquad \mathcal{M} \cap \mathcal{B} \subseteq \mathcal{A}, \qquad \mathcal{F} \cap \mathcal{M} \subseteq \mathcal{A},$$

and

$$|\mathcal{F}^c \cap \mathcal{M}^c \cap \mathcal{B}| = \infty \implies \overline{\mathbf{P}}(\mathcal{A} | \mathcal{F}^c \cap \mathcal{M}^c \cap \mathcal{B}) \le 1/2,$$

$$|\mathcal{F} \cap \mathcal{M}^c \cap \mathcal{B}^c| = \infty \implies \overline{\mathbf{P}}(\mathcal{A} | \mathcal{F} \cap \mathcal{M}^c \cap \mathcal{B}^c) \le 1/2,$$

$$|\mathcal{F} \cap \mathcal{M}^c \cap \mathcal{B}| = \infty \implies \overline{\mathbf{P}}(\mathcal{A} | \mathcal{F} \cap \mathcal{M}^c \cap \mathcal{B}) \le 1/2.$$

Proof The corollary follows from Exercise 1.3. $\qquad\qquad\qquad\qquad\qquad\qquad$ □

The inequality $\overline{\mathbf{P}}(\mathcal{A} | \mathcal{F}^c \cap \mathcal{M}^c \cap \mathcal{B}) \le 1/2$ formalizes the prediction that the long-run frequency of Daltonism among sons of parents that do not suffer from it will not exceed $1/2$, and the other inequalities formalize similar predictions.

10.4 LEAST SQUARES

Consider the following testing protocol, in which $\langle \cdot, \cdot \rangle$ stands for the dot product.

Protocol 10.7
PARAMETER: $w \in \mathbb{R}^K$
   FOR $n = 1, 2, \ldots$:
      Reality announces $x_n \in \mathbb{R}^K$.
      Skeptic announces $M_n \in \mathbb{R}$.
      Reality announces $\epsilon_n \in [-1, 1]$ and sets $y_n := \langle w, x_n \rangle + \epsilon_n$.
      $\mathcal{K}_n := \mathcal{K}_{n-1} + M_n \epsilon_n$.

We will show that under reasonable conditions on the signals, Skeptic has a strategy that forces the accuracy of the least squares estimate of w.

We call y_n the *outcome* of the nth round and ϵ_n the *noise* in y_n. The protocol treats ϵ_n just as Protocol 1.3 treats its outcome; it is bounded (the bound being set to 1 without loss of generality), and Skeptic can buy it in any amount, positive or negative. The main result of this section, Theorem 10.8, can be generalized to the case of unbounded ϵ_n if we add to the protocol a parameter $\alpha > 2$ and an announcement by Forecaster of upper prices for $|\epsilon_n|^\alpha$ analogous to Protocol 4.1's upper prices for $(y_n - m_n)^2$ (see Exercise 10.6).

The statistician stands outside the protocol. She does not see w; otherwise she would not need to estimate it. She also does not see the ϵ_n. She sees only the signals x_n and the outcomes y_n. But she can use least squares to estimate w and specify a strategy for Skeptic that will multiply his capital by a large factor unless w is within bounds that she can calculate from the x_n and y_n. For the sake of brevity, we prove only an asymptotic result, leaving to the reader the task of deriving corresponding finite-horizon results with explicit bounds for given confidence levels. (In accordance with standard statistical terminology, we call $1 - \alpha$ the *confidence level*, where $1/\alpha$ is the factor by which the capital is multiplied if the bounds fail.)

We treat the vectors w and x_n in Protocol 10.7 as column vectors. For $n \in \mathbb{N}$, let X_n be the $n \times K$ matrix whose ith row is x_i' (x_i considered as a row vector), and let Y_n be the n-dimensional column vector whose ith element is y_i. The *least squares estimate* of w is the element w_n of $\mathbb{R}^K$ defined by

$$w_n := (X_n' X_n)^{-1} X_n' Y_n$$

if the matrix $X_n' X_n$ is invertible. Its inverse $V_n := (X_n' X_n)^{-1}$ and the estimate w_n can be updated easily on each round:

$$w_n = w_{n-1} + \frac{y_n - x_n' w_{n-1}}{1 + x_n' V_{n-1} x_n} V_{n-1} x_n, \tag{10.7}$$

$$V_n = V_{n-1} - \frac{V_{n-1} x_n x_n' V_{n-1}}{1 + x_n' V_{n-1} x_n}. \tag{10.8}$$

See Exercises 10.4 and 10.5. When $X_n' X_n$ is not invertible, we set $w_n := 0$. We write $\lambda_n^{\max}$ and $\lambda_n^{\min}$ for $X_n' X_n$'s largest and smallest eigenvalues, respectively.

Theorem 10.8 *Skeptic can force the event*

$$(\lambda_n^{\min} \to \infty \ \& \ \ln \lambda_n^{\max}/\lambda_n^{\min} \to 0)$$

$$\Rightarrow ||w_n - w|| = O(\sqrt{\ln \lambda_n^{\max}/\lambda_n^{\min}}) = o(1) \tag{10.9}$$

in Protocol 10.7, where $||\cdot||$ is the Euclidean norm.

Proof The antecedent of (10.9) implies that $X_n'X_n$ is invertible from some n on. Let E_n be the n-dimensional vector whose ith element is ϵ_i. Because

$$
\begin{aligned}
||w_n - w||^2 &= ||(X_n'X_n)^{-1}X_n'(X_nw + E_n) - w||^2 \\
&= ||(X_n'X_n)^{-1}X_n'E_n||^2 \\
&\le ||(X_n'X_n)^{-1/2}||^2||(X_n'X_n)^{-1/2}X_n'E_n||^2 \\
&= (\lambda_n^{\min})^{-1}E_n'X_n(X_n'X_n)^{-1}X_n'E_n,
\end{aligned}
$$

where the norm $||A||$ of a $K \times K$ matrix A is defined as $\sup_{x \in \mathbb{R}^K : ||x|| \le 1} ||Ax||$ (i.e. as the square root of the largest eigenvalue of $A'A$), it suffices to prove that Skeptic can force (10.9) with the consequent replaced by

$$Q_n = O(\ln \lambda_n^{\max}), \tag{10.10}$$

where

$$Q_n := E_n'X_n(X_n'X_n)^{-1}X_n'E_n.$$

This is the squared Euclidean norm of the projection of E_n onto the columns of X_n.

Consider a particular path of Protocol 10.7 where $X_n'X_n$ is invertible from some n on, say for $n \ge N$. By (10.8),

$$x_k'V_k = x_k'V_{k-1} - \frac{x_k'V_{k-1}x_kx_k'V_{k-1}}{1 + x_k'V_{k-1}x_k} = \frac{x_k'V_{k-1}}{1 + x_k'V_{k-1}x_k}$$

for $k > N$. So we can rewrite the expression for Q_k for $k > N$ as

$$
\begin{aligned}
Q_k := E_k'X_k(X_k'X_k)^{-1}X_k'E_k &= \left(\sum_{i=1}^{k} \epsilon_i x_i'\right) V_k \left(\sum_{i=1}^{k} \epsilon_i x_i\right) \\
&= \left(\sum_{i=1}^{k-1} \epsilon_i x_i'\right) V_k \left(\sum_{i=1}^{k-1} \epsilon_i x_i\right) + 2\epsilon_k x_k'V_k \sum_{i=1}^{k-1} \epsilon_i x_i + \epsilon_k^2 x_k'V_k x_k \\
&= Q_{k-1} - \frac{\left(x_k'V_{k-1}\sum_{i=1}^{k-1} \epsilon_i x_i\right)^2}{1 + x_k'V_{k-1}x_k} + 2\epsilon_k \frac{x_k'V_{k-1}\sum_{i=1}^{k-1} \epsilon_i x_i}{1 + x_k'V_{k-1}x_k} + \epsilon_k^2 x_k'V_k x_k.
\end{aligned}
$$

Summing over $k = N + 1, \ldots, n$ for $n > N$, we obtain

$$Q_n - Q_N = - \sum_{k=N+1}^{n} \frac{\left(x_k' V_{k-1} \sum_{i=1}^{k-1} \epsilon_i x_i \right)^2}{1 + x_k' V_{k-1} x_k}$$

$$+ 2 \sum_{k=N+1}^{n} \epsilon_k \frac{x_k' V_{k-1} \sum_{i=1}^{k-1} \epsilon_i x_i}{1 + x_k' V_{k-1} x_k} + \sum_{k=N+1}^{n} \epsilon_k^2 x_k' V_k x_k. \tag{10.11}$$

We need to show that Skeptic can force the right-hand side of (10.11) to grow asymptotically as the right-hand side of (10.10) when the antecedent of (10.9) is satisfied. We will analyze each of the three sums $\sum_{k=N+1}^{n}$ in (10.11) separately. The key observation is that the first sum is always nonnegative, and we will see that the second sum grows more slowly than the first and that the third sum grows as the right-hand side of (10.10).

Setting

$$u_k := \begin{cases} \frac{x_k' V_{k-1} \sum_{i=1}^{k-1} \epsilon_i x_i}{1 + x_k' V_{k-1} x_k} & \text{if } k > N \\ 0 & \text{otherwise} \end{cases}$$

and applying (10.12) in Lemma 10.9, we can see that Skeptic can force the second sum to grow asymptotically more slowly than the first one, which is positive. By Lemma 10.10, the third sum is $O(\ln \lambda_n^{\max})$ whenever the antecedent of (10.9) holds. □

The following lemma, which we just used, is the probabilistic core of the argument.

Lemma 10.9   *In Protocol 10.7, let u_n be a predictable process (i.e. u_n depends only on x_n and the players' moves on the previous rounds). Skeptic can force*

$$\sum_{i=1}^{n} u_i \epsilon_i = o \left(\sum_{i=1}^{n} u_i^2 \right) + O(1). \tag{10.12}$$

Proof We will show separately that Skeptic can force

$$\sum_{i=1}^{\infty} u_i^2 < \infty \implies \sum_{i=1}^{n} u_i \epsilon_i = O(1) \tag{10.13}$$

and that he can force

$$\sum_{i=1}^{\infty} u_i^2 = \infty \implies \lim_{n\to\infty} \frac{\sum_{i=1}^{n} u_i \epsilon_i}{\sum_{i=1}^{n} u_i^2} = 0. \tag{10.14}$$

Equation (10.13) follows immediately from Lemma 4.9. Notice that Lemma 4.9 is also applicable in our current protocol: since Protocol 4.4 is one-sided in that $V_n \geq 0$, we can make v_n arbitrarily large.

Let us now check (10.14). By Kronecker's lemma

$$0 < b_n \uparrow \infty \ \& \ \sum_{n=1}^{\infty} \frac{y_n}{b_n} \text{ converges in } \mathbb{R} \implies \lim_{n\to\infty} \frac{1}{b_n} \sum_{i=1}^{n} y_i = 0$$

(whose special case (4.17) we have already used), we can replace the consequent of (10.14) by the convergence of

$$\sum_{n=1}^{\infty} \frac{u_n \epsilon_n}{\sum_{i=1}^{n} u_i^2}$$

in $\mathbb{R}$. By Lemma 4.9, it suffices to check

$$\sum_{n=1}^{\infty} \frac{u_n^2}{\left(\sum_{i=1}^{n} u_i^2\right)^2} < \infty$$

assuming the antecedent of (10.14). Denoting $b_n := \sum_{i=1}^{n} u_i^2$, we obtain an increasing sequence of nonnegative numbers such that $b_n \to \infty$ and our goal is to prove

$$\sum_{n=2}^{\infty} \frac{b_n - b_{n-1}}{b_n^2} < \infty. \tag{10.15}$$

Assuming, without loss of generality, that $b_n \geq 1$ for all n, we can see that the left-hand side of (10.15) does not exceed $\int_1^{\infty} x^{-2} \, dx = 1$. □

The following lemma was also used in the proof of Theorem 10.8.

Lemma 10.10   *Let $w_1, w_2, \ldots \in \mathbb{R}^K$, let $A_n := \sum_{i=1}^{n} w_i w_i'$, and let λ_n^* be the maximum eigenvalue of A_n. Assume that A_n is invertible for $n = N$ (and so for all $n \geq N$). If $\lim_{n \to \infty} \lambda_n^* = \infty$, then*

$$\sum_{k=N}^{n} w_k' A_k^{-1} w_k = O(\ln \lambda_n^*).$$

Proof This lemma is part of Lemma 2 in [236].   □

10.5 PARAMETRIC STATISTICS WITH SIGNALS

Just as probability theory often begins with the study of a single probability measure P, statistical theory often begins with the study of an indexed class of probability measures $(P_\theta)_{\theta \in \Theta}$. In an article published in 1922 that is still influential [144], R.A. Fisher asserted that the task of theoretical statistics is to estimate θ from data $y_1, \ldots, y_n$ hypothesized to be a random sample from one of these probability measures. Fisher introduced the name *parameter* for θ, which is typically an unknown real number or vector of real numbers.

In many of the statistical models subsequently developed by Fisher and his successors, the probabilities for y_n depend not on θ alone but also on a signal x_n and perhaps even on previous signals and outcomes $x_1, y_n, \ldots, x_{n-1}, y_{n-1}$. In general, such models can be represented as strategies for Forecaster in specializations of the following protocol, which elaborates Protocol 10.2 by adding an announcement of θ by Reality.

Protocol 10.11
PARAMETERS: Nonempty sets Θ and $\mathcal{X}$, measurable space $\mathcal{Y}$
    Reality announces $\theta \in \Theta$.
    $\mathcal{K}_0 := 1$.
    FOR $n = 1, 2, \ldots$:
        Reality announces $x_n \in \mathcal{X}$.
        Forecaster announces $P_n \in \mathcal{P}(\mathcal{Y})$.
        Skeptic announces $f_n \in [0, \infty]^{\mathcal{Y}}$ such that $P_n(f_n) = 1$.
        Reality announces $y_n \in \mathcal{Y}$.
        $\mathcal{K}_n := \mathcal{K}_{n-1} f_n(y_n)$.

The statistician does not quite stand inside the protocol; she observes the x_n and y_n as they are announced, but she does not observe Reality's announcement of θ.

Suppose Forecaster follows a strategy, known to the statistician, that specifies each of his moves as a function of Reality's previous moves. To fix ideas, we again assume, as in our discussion of Protocol 10.2, that the strategy specifies a density $p_n(\theta, x_1, y_1, \ldots, x_n)$ for P_n with respect to a fixed underlying probability measure. If Skeptic also follows a strategy that is a function of Reality's previous moves, say $f_n(\theta, x_1, y_1, \ldots, x_n)$, and $q_n := f_n p_n$, then (10.1) becomes

$$\mathcal{K}_n = \frac{\prod_{i=1}^n q_i(\theta, x_1, y_1, \dots, x_i)(y_i)}{\prod_{i=1}^n p_i(\theta, x_1, y_1, \dots, x_i)(y_i)}. \tag{10.16}$$

We call the denominator on the right-hand side of (10.16) θ's *likelihood* under Fore-caster's strategy. It is more conventional to call it the *partial likelihood*, reserving *likelihood* for the case considered by Fisher, where θ alone determines a probability distribution for $y_1, y_2, \dots$, but from our game-theoretic point of view there is nothing partial about the protocol or Forecaster's strategy. Similarly, we call the numerator θ's likelihood under the alternative forecasting strategy favored by Skeptic. Thus the capital $\mathcal{K}_n$ is a *likelihood ratio*.

There is an extensive literature in mathematical statistics that uses (10.16) to test hypotheses about θ and to estimate θ in models that can be represented as strategies for specializations of Protocol 10.11 [432]. This literature is formulated, more or less punctiliously, in measure-theoretic terms, but it may be possible to reformulate much of it in terms of testing protocols.

Survival Analysis

In survival analysis, statisticians study successive failures (deaths or failures of some other type) in a group of individuals whose composition may vary unpredictably. To illustrate how Protocol 10.2 applies to survival analysis, consider a study of longevity that monitors K variables, such as blood pressure, weight, and sex, on a group of people that changes as new individuals are recruited and others, because they die or for whatever other reason, are no longer available for monitoring. Let B_n be the set of individuals in the study just before the nth death, and let b_n be the individual who dies. (We ignore the possibility that two individuals die simultaneously.) For each $b \in B_n$, including b_n, let $z_n^b \in \mathbb{R}^K$ be the vector of values for the K variables just before b_n's death.

Supposing that Reality announces a vector $\theta \in \mathbb{R}^K$ that can help Forecaster use the K variables to predict which individual of the group will be the next to die, we have the following protocol:

Protocol 10.12
PARAMETER: $K \in \mathbb{N}$
    Reality announces $\theta \in \mathbb{R}^K$.
    $\mathcal{K}_0 := 1$.
    FOR $n = 1, 2, \dots$:
        Reality announces a finite nonempty set B_n and $z_n^b \in \mathbb{R}^K$ for every $b \in B_n$.
        Forecaster announces $P_n \in \mathcal{P}(B_n)$.
        Skeptic announces $f_n \in [0, \infty]^{B_n}$ such that $P_n(f_n) = 1$.
        Reality announces $b_n \in B_n$.
        $\mathcal{K}_n := \mathcal{K}_{n-1} f_n(b_n)$.

Here $P_n(f_n) = \sum_{b \in B_n} f_n(b) p_n(b)$, where p_n is the density for the probability measure P_n on the finite set B_n.

The proportional hazards model for this type of study, proposed by David Cox in 1972 [86], can be represented in Protocol 10.12 by the following strategy for Forecaster:

$$p_n(b) := \frac{e^{\langle z_n^b, \theta \rangle}}{\sum_{j \in B_n} e^{\langle z_n^j, \theta \rangle}}, \quad b \in B_n. \tag{10.17}$$

This model has been studied extensively in the measure-theoretic framework, and it would be interesting to develop the corresponding game-theoretic picture (see Exercise 10.7).

To illustrate the possibilities, consider the simplest case, where $K = 1$ (a single variable z is monitored), the group does not change except for deaths ($B_n = B_{n-1} \setminus \{b_{n-1}\}$), the value of the variable z does not change ($z_n^b = z^b$), and we know in advance that the study will stop after N deaths (N necessarily being no greater than $|B_1|$). Suppose Forecaster plays the strategy (10.17) with $\theta = 0$, which corresponds to the hypothesis that members of the group die at random, the variable z making no difference. Then we can provide a game-theoretic derivation of a test of $\theta = 0$ given by Cox and David Oakes [88, section 7.4]. For any sequence $j_1, \ldots, j_N$ of different elements of B_1, set

$$T(j_1, \ldots, j_N) := \sum_{n=1}^{N} \left(z^{j_n} - \frac{1}{|R_n|} \sum_{j \in R_n} z^j \right), \tag{10.18}$$

where $R_n := B_1 \setminus \{j_1, \ldots, j_{n-1}\}$. If $j_1, \ldots, j_N$ are the individuals who die, in order of death, so that $(b_1, \ldots, b_N) = (j_1, \ldots, j_N)$, then $T(j_1, \ldots, j_N)$ adds up how much the value of z for each individual dying exceeds the average value of z for the other individuals in the group at that moment. We expect T to be large if $\theta > 0$ but closer to zero if $\theta = 0$. Under our simplifying assumptions, the protocol is essentially a special case of Protocol 9.2. So, by Ville's theorem (Theorem 9.3), we obtain probabilities for T that are both game-theoretic and measure-theoretic:

$$\mathbb{P}\left(T(b_1, \ldots, b_N) \geq u \right) = \frac{\left| \{ (j_1, \ldots, j_N) | T(j_1, \ldots, j_N) \geq u \} \right|}{\left| \{ (j_1, \ldots, j_N) \} \right|}$$

for any $u \in \mathbb{R}$, where $(j_1, \ldots, j_N)$ ranges over all sequences of length N of distinct elements of B_1. We can choose a level α (such as 1%), define u_α as the smallest u for which $\overline{\mathbb{P}}(T(b_1, \ldots, b_N) > u) \leq \alpha$, and reject $\theta = 0$ at level α if $T(b_1, \ldots, b_N) > u_\alpha$.

We can also apply the law of the iterated logarithm. If we set $K = 1$ in Protocol 10.12, constrain Reality to continue play indefinitely and always choose $|z_n^b| \leq 1$, and constrain Forecaster to follow Cox's strategy with $\theta = 0$, then Corollary 5.4 (see (5.19)) implies that Skeptic can force

$$\limsup_{N \to \infty} \frac{T_N}{\sqrt{2N \ln \ln N}} \leq 1,$$

where (cf. (10.18))

$$T_N := \sum_{n=1}^{N} \left(z_n^{b_n} - \frac{1}{|B_n|} \sum_{b \in B_n} z^b \right).$$

This game-theoretic result is not so easily obtained by measure-theoretic arguments, because there is no probability measure in view for the number of individuals in B_n and their values for the variable z.

10.6 QUANTUM MECHANICS

The probabilities that quantum theory provides for the outcomes of measurements on an isolated physical system depend not only on the quantum state of the system but also on decisions about what to measure and when. This is sometimes thought to distinguish quantum mechanics from other fields where probability is used, because the theory does not appear to provide probabilities for these decisions. This feature of quantum mechanics fits easily, however, into our game-theoretic picture. A physicist's decision about what to measure, like a physician's decision about what individuals to monitor in a study of longevity (see Protocol 10.12), can be treated as an unprobabilized signal in a testing protocol. In this section, we briefly elaborate this point, first for John von Neumann's formalism for quantum measurement and then for quantum computing.

A Protocol for von Neumann's Formalism

Consider an isolated physical system known to be in a particular state at time t_0. According to von Neumann's account, first published in 1932 [290, chapter III], quantum mechanics tells how the state evolves so long as the system is not disturbed by measurement. It also gives probabilities for the outcome of a measurement of any particular observable (such as position, momentum, energy, or angular momentum) at time $t > t_0$ and tells how the state changes as a result of this outcome.

Following tradition, let us call the individual who decides what to measure and when Observer. Without recapitulating any details of von Neumann's formalism, we can write as follows a protocol for testing the theory by making successive measurements on a single isolated system.

Protocol 10.13

PARAMETER: t_0 (initial time)

$\quad \mathcal{K}_0 := 1.$

$\quad$ FOR $n = 1, 2, \ldots$:

$\qquad$ Observer announces an observable A_n and a time $t_n > t_{n-1}$.

$\qquad$ Quantum Mechanics announces $P_n \in \mathcal{P}(\mathbb{R})$.

$\qquad$ Skeptic announces $f_n \in [0, \infty]^{\mathbb{R}}$ such that $P_n(f_n) = 1$.

$\qquad$ At time t_n, Reality announces the measurement $y_n \in \mathbb{R}$.

$\qquad \mathcal{K}_n := \mathcal{K}_{n-1} f_n(y_n).$

Here Observer is part of Reality (she provides the signal) and Forecaster has been named Quantum Mechanics. For information on von Neumann's formalism, which describes how Quantum Mechanics computes the probability measure P_n, see [200, chapter 1] or [349, section 8.4].

Our limit theorems can be applied to Protocol 10.13 in many different ways. As an example, here is a corollary of Proposition 1.2.

Corollary 10.14 *Suppose Observer repeatedly measures observables bounded in absolute value by some constant C. Skeptic can force the event*

$$\lim_{N \to \infty} \frac{1}{N} \sum_{n=1}^{N} \left(y_n - \int y P_n(dy) \right) = 0.$$

Quantum Computing

A quantum computation takes as input a sequence of bits, say $a_n \in \{0, 1\}^K$ where $K \in \mathbb{N}$, and produces as output another such sequence, say y_n. Its advantage over conventional computation is its speed; a disadvantage is that the result y_n is not deterministic. If the goal is to compute a function of a_n, then the quantum computation may produce a wrong result, even if it is engineered to produce the correct result with a very high probability. But if the computation is easily repeated, it can be repeated until the correct result becomes evident.

Making quantum computation practical remains a daunting task, but the mathematical formalism is relatively simple. A quantum computer for strings of K bits consists of K two-state quantum systems, called *qubits*. The computer's *state space* $\mathcal{H}$ is the finite-dimensional complex Hilbert space consisting of functions $\phi : \{0, 1\}^K \to \mathbb{C}$ with the inner product

$$\langle \phi, \psi \rangle := \sum_{x \in \{0,1\}^K} \phi(x)^* \psi(x),$$

where superscript "*" stands for the complex conjugate. Elements of $\mathcal{H}$ are called *states*. For each $y \in \{0, 1\}^K$, we write $|y\rangle$ for the state defined by $|y\rangle(z) := \mathbf{1}_{z=y}$ for all $z \in \{0, 1\}^K$. Given a state ϕ, ϕ's *measure* is the measure on $\{0, 1\}^K$ that assigns each $y \in \{0, 1\}^K$ the value $|\phi(y)|^2$. A linear operator $U : \mathcal{H} \to \mathcal{H}$ is *unitary* if it is bijective and preserves the inner product.

A quantum computer must be able to implement a set $\mathcal{U}$ of allowed unitary operators. Usually these are the compositions of elementary unitary operators that are called *logical gates* and involve at most two qubits. The programmer puts the computer in the initial state $|a\rangle \in \mathcal{H}$, where $a \in \{0, 1\}^K$, and then runs a computer program U, which is a composition of elementary operators. The initial quantum state $|a\rangle$ has norm 1, and so the final quantum state $U|a\rangle$ (and all intermediate quantum states) will also have norm 1. So the measure corresponding to $U|a\rangle$ will be a probability measure.

To test whether the computer is performing correctly, or to make inferences from repeated instances of the same computation, we can use the following protocol:

Protocol 10.15
PARAMETERS: K (the number of qubits) and the allowed unitary operators $\mathcal{U}$
$\quad \mathcal{K}_0 := 0$.
$\quad$ FOR $n = 1, 2, \ldots$:
$\qquad$ Programmer announces $a_n \in \{0, 1\}^K$ and $U_n \in \mathcal{U}$.
$\qquad$ Quantum Mechanics announces $U_n|a_n\rangle$'s probability measure as P_n.
$\qquad$ Skeptic announces $f_n : \{0, 1\}^K \to [0, \infty]$ such that $P_n(f_n) = 1$.
$\qquad$ Computer announces $y_n \in \{0, 1\}^K$.
$\qquad \mathcal{K}_n := \mathcal{K}_{n-1}f_n(y_n)$.

Any attempt to understand such tests in terms of measure-theoretic probability faces, of course, the difficulty that the theory does not provide probabilities for the actions of the programmer.

10.7 JEFFREYS'S LAW

Jeffreys's law is the thesis that two competing forecasters cannot both succeed in the long run if they give persistently different probability forecasts based on the same information. In this section, we establish several results that substantiate this thesis.

We begin with the following protocol, in which Forecaster I is tested by Skeptic I and Forecaster II is tested by Skeptic II.

Protocol 10.16
PARAMETER: Measurable space $\mathcal{Y}$
$\quad \mathcal{K}_0^{\mathrm{I}} := 1$ and $\mathcal{K}_0^{\mathrm{II}} := 1$.
$\quad$ FOR $n = 1, 2, \ldots$:
$\qquad$ Forecaster I announces $P_n^{\mathrm{I}} \in \mathcal{P}(\mathcal{Y})$.
$\qquad$ Forecaster II announces $P_n^{\mathrm{II}} \in \mathcal{P}(\mathcal{Y})$.

Skeptic II announces $f_n^{II} \in [0, \infty]^{\mathcal{Y}}$ such that $P_n^{II}(f_n^{II}) = 1$.
Skeptic I announces $f_n^I \in [0, \infty]^{\mathcal{Y}}$ such that $P_n^I(f_n^I) = 1$.
Reality announces $y_n \in \mathcal{Y}$.
$\mathcal{K}_n^I := \mathcal{K}_{n-1}^I f_n^I(y_n)$ and $\mathcal{K}_n^{II} := \mathcal{K}_{n-1}^{II} f_n^{II}(y_n)$.

We will show that the two Skeptics have a joint strategy that will make at least one of them rich, in senses that we will make precise, on any play of this protocol in which the Forecasters persist in making their moves too different. This result obviously does not depend on the order in which the Forecasters move or on the order in which the Skeptics move. We will also show that Skeptic I has a strategy that will make him rich on any play of the protocol in which Skeptic II becomes rich, the two Forecasters eventually make their moves sufficiently similar, and Forecaster II never assigns probability 0 to a subset of $\mathcal{Y}$ to which Forecaster I assigned positive probability. This result does require that Skeptic I play after Skeptic II, and it is interesting only if Forecaster II plays after Forecaster I.

We measure the difference between the two Forecasters' moves by Hellinger distance. Given two probability measures P^I and P^{II} on $\mathcal{Y}$ the *Hellinger distance* $\rho(P^I, P^{II})$ between them is given by

$$\rho^2(P^I, P^{II}) := 1 - H(P^I, P^{II}) := 1 - \int_{\mathcal{Y}} \sqrt{\beta^I(y)\beta^{II}(y)} Q(dy), \tag{10.19}$$

where Q is any probability measure on $\mathcal{Y}$ such that $P^I \ll Q$, $P^{II} \ll Q$, and β^I and β^{II} are densities with respect to Q for P^I and β^{II}, respectively. A convenient choice for Q is $(P^I + P^{II})/2$, but the value of the integral does not depend on the choice of Q, β^I, and β^{II}. The expression (10.19) is always nonnegative. The integral $H(P^I, P^{II})$ is the *Hellinger integral*.

Here is one way of making precise our assertion that the two Skeptics have a joint strategy that makes at least one of them rich if the two Forecasters make their moves too different.

Proposition 10.17   *In Protocol 10.16, the Skeptics have a joint strategy guaranteeing that, for all N,*

$$\ln \mathcal{K}_N^I + \ln \mathcal{K}_N^{II} \geq \sum_{n=1}^{N} 2\rho^2(P_n^I, P_n^{II}). \tag{10.20}$$

By our convention that $\infty - \infty = \infty$, (10.20) will be true if one of the logarithms is ∞ and the other $-\infty$.

Proof of Proposition 10.17   Let Skeptic I play the strategy

$$f_n^I := \frac{\sqrt{\beta_n^{II}/\beta_n^I}}{\int \sqrt{\beta_n^I \beta_n^{II}} \, dQ_n} \tag{10.21}$$

(the Hellinger integral in the denominator is just the normalizing constant) and Skeptic II play the strategy

$$f_n^{\mathrm{II}} := \frac{\sqrt{\beta_n^{\mathrm{I}}/\beta_n^{\mathrm{II}}}}{\int \sqrt{\beta_n^{\mathrm{I}}\beta_n^{\mathrm{II}}} \, dQ_n}. \tag{10.22}$$

Using the inequality $\ln x \le x - 1$ (for $x \ge 0$), we obtain

$$f_n^{\mathrm{I}} f_n^{\mathrm{II}} = H^{-2}(P^{\mathrm{I}}, P^{\mathrm{II}}) = \exp(-2 \ln H(P^{\mathrm{I}}, P^{\mathrm{II}}))$$

$$\ge \exp(2 - 2H(P^{\mathrm{I}}, P^{\mathrm{II}})) = \exp(2\rho^2(P^{\mathrm{I}}, P^{\mathrm{II}})),$$

which implies (10.20).

Let us now look more carefully at the case where some of the denominators or numerators in (10.21) or (10.22) are zero and so the above argument is not applicable directly. If the Hellinger integral at time n,

$$\int_y \sqrt{\beta_n^{\mathrm{I}}(y)\beta_n^{\mathrm{II}}(y)}Q(dy), \tag{10.23}$$

is zero, the probability measures P_n^{I} and P_n^{II} are mutually singular. Choose a local event E such that $P_n^{\mathrm{I}}(E) = P_n^{\mathrm{II}}(\mathcal{Y} \setminus E) = 0$. If the Skeptics choose

$$f_n^{\mathrm{I}}(y) := \begin{cases} \infty & \text{if } y \in E, \\ 1 & \text{otherwise,} \end{cases} \qquad f_n^{\mathrm{II}}(y) := \begin{cases} 1 & \text{if } y \in E, \\ \infty & \text{otherwise,} \end{cases}$$

(10.20) will be guaranteed to hold: both sides will be ∞.

Let us now suppose that the Hellinger integral (10.23) is nonzero. In (10.21) and (10.22), we interpret $0/0$ as 1 (and, of course, $t/0$ as ∞ for $t > 0$). As soon as the local event

$$\left(\beta_n^{\mathrm{I}} = 0 \ \& \ \beta_n^{\mathrm{II}} > 0\right) \ \text{or} \ \left(\beta_n^{\mathrm{I}} > 0 \ \& \ \beta_n^{\mathrm{II}} = 0\right)$$

happens for the first time (if it ever happens), the Skeptics stop playing, in the sense of selecting $f_N^{\mathrm{I}} = f_N^{\mathrm{II}} := 1$ for all $N > n$. This will make sure that (10.20) always holds (in the sense of our convention reiterated after the statement of the proposition). □

For our next result we will need the following definitions:

- Skeptic I *becomes infinitely rich* if $\lim_{n \to \infty} \mathcal{K}_n^{\mathrm{I}} = \infty$. Similarly, Skeptic II *becomes infinitely rich* if $\lim_{n \to \infty} \mathcal{K}_n^{\mathrm{II}} = \infty$.
- Forecaster II *plays cautiously* if, for all n, $P_n^{\mathrm{I}} \ll P_n^{\mathrm{II}}$.

These definitions concern the particular play of the game; they do not refer to strategies.

Proposition 10.18   *In Protocol 10.16:*

1. *The Skeptics have a joint strategy guaranteeing that at least one of them will become infinitely rich if*

$$\sum_{n=1}^{\infty} \rho^2(P_n^{\mathrm{I}}, P_n^{\mathrm{II}}) = \infty.$$

2. *Skeptic I has a strategy guaranteeing that he will become infinitely rich if*

$$\sum_{n=1}^{\infty} \rho^2(P_n^{\mathrm{I}}, P_n^{\mathrm{II}}) < \infty, \tag{10.24}$$

Skeptic II becomes infinitely rich, and Forecaster II plays cautiously.

The condition that Forecaster II play cautiously is essential to Statement 2 of the proposition (see Exercise 10.8). But it will be evident from the proof that the proposition remains true if we change the definition of becoming infinitely rich from $\lim_{n\to\infty} \mathcal{K}_n = \infty$ to $\limsup_{n\to\infty} \mathcal{K}_n = \infty$.

Recall the notions of reliability and validity introduced at the beginning of this chapter: Forecaster is *reliable* if he keeps Skeptic from multiplying his capital by a large factor, and a probabilistic theory is *valid* if Forecaster keeps Skeptic from multiplying his capital by a large factor by using it as his strategy. These are not mathematical definitions. Reliability is a practical claim we might make for a forecaster, and validity is a scientific claim we might make for a theory. But Proposition 10.17 and Statement 1 of Proposition 10.18 make it clear that two reliable forecasters forecasting the same outcomes or two valid probabilistic theories about the same outcomes must agree in the long run.

Suppose, for example, that we consider Forecaster I reliable and that we make this more precise by claiming that Reality will keep $\mathcal{K}_n^{\mathrm{I}}$ bounded. If the Skeptics invest a fraction (arbitrarily small) of their initial capital in strategies whose existence is guaranteed in Proposition 10.18, then this claim implies

$$\left(\limsup_{n\to\infty} \mathcal{K}_n^{\mathrm{II}} < \infty\right) \Longleftrightarrow \left(\sum_{n=1}^{\infty} \rho^2(P_n^{\mathrm{I}}, P_n^{\mathrm{II}}) < \infty\right),$$

assuming Forecaster II plays cautiously. In the context of the algorithmic theory of randomness, this equivalence is called a criterion of randomness [157, 389].

If two competing theories give different probabilistic predictions in the same experimental conditions, say $P^{\mathrm{I}} \neq P^{\mathrm{II}}$, and scientists can repeat these conditions as many times as they want, say N, then (10.20) tells us that $\ln \mathcal{K}_N^I + \ln \mathcal{K}_N^{II} \geq Nc$, where $c := 2\rho^2(P^I, P^{II}) > 0$. Thus the product $\mathcal{K}_N^I \mathcal{K}_N^{II}$ tends to ∞ exponentially fast as $N \to \infty$, and the scientists should be able to reject at least one of the two theories at any level they want to use.

We can relax the condition, in Statement 2 of Proposition 10.18, that Forecaster II plays cautiously. This makes the protocol and the statement more complicated, but it clarifies the argument. Let us say that (E^I, E^{II}), where E^I and E^{II} are local events, is an *exceptional pair* for $P^I, P^{II} \in \mathcal{P}(\mathcal{Y})$ if $P^I(E^I) = 0$, $P^{II}(E^{II}) = 0$, and

$$P^I(E) = 0 \iff P^{II}(E) = 0$$

for all $E \subseteq (E^I \cup E^{II})^c$.

Protocol 10.19

> $\mathcal{K}_0^I := 1$ and $\mathcal{K}_0^{II} := 1$.
> FOR $n = 1, 2, \dots$:
> > Forecaster I announces $P_n^I \in \mathcal{P}(\mathcal{Y})$.
> > Forecaster II announces $P_n^{II} \in \mathcal{P}(\mathcal{Y})$.
> > Statistician announces an exceptional pair (E_n^I, E_n^{II}) for P_n^I, P_n^{II}.
> > Skeptic II announces $f_n^{II} : \mathcal{Y} \to [0, \infty]$ such that $P_n^{II}(f_n^{II}) = 1$.
> > Skeptic I announces $f_n^I : \mathcal{Y} \to [0, \infty]$ such that $P_n^I(f_n^I) = 1$.
> > Reality announces $y_n \in \mathcal{Y}$.
> > $\mathcal{K}_n^I := \mathcal{K}_{n-1}^I f_n^I(y_n)$ and $\mathcal{K}_n^{II} := \mathcal{K}_{n-1}^{II} f_n^{II}(y_n)$.

The identity of the player who announces an exceptional pair does not matter as long as it is not one of the Skeptics. One way to choose (E^I, E^{II}) is to choose β^I and β^{II} first and then set $E^I := \{\beta^I = 0\}$ and $E^{II} := \{\beta^{II} = 0\}$.

Without the condition that Forecaster II plays cautiously, we replace (10.24) with

$$\sum_{n=1}^{\infty} \rho^2(P_n^I, P_n^{II}) < \infty \quad \text{and} \quad \forall n : y_n \notin E_n^I \cup E_n^{II}. \tag{10.25}$$

Theorem 10.20 *In Protocol 10.19:*

1. *The Skeptics have a joint strategy guaranteeing that at least one of them will become infinitely rich if (10.25) is violated.*

2. *Skeptic I has a strategy guaranteeing that he will become infinitely rich if (10.25) holds and Skeptic II becomes infinitely rich.*

Proof of Proposition 10.18 and Theorem 10.20

Statement 1 of Proposition 10.18 and Theorem 10.20 immediately follows from Proposition 10.17. (In Theorem 10.20, redefine the Forecasters' moves f_n^I and f_n^{II} by setting them to ∞ on E^I and E^{II}, respectively.) So we need only prove their Statement 2.

First we establish in the following auxiliary protocol several simple analogs of standard measure-theoretic results.

Protocol 10.21
$\mathcal{K}_0 := 1.$
FOR $n = 1, 2, \ldots$:
    Forecaster announces $P_n \in \mathcal{P}(\mathcal{Y})$.
    Reality announces measurable $\xi_n : \mathcal{Y} \to \mathbb{R}$.
    Skeptic announces $f_n : \mathcal{Y} \to [0, \infty]$ such that $P_n(f_n) = \mathcal{K}_{n-1}$.
    Reality announces $y_n \in \mathcal{Y}$.
    $\mathcal{K}_n := f_n(y_n)$.

Lemma 10.22   *In Protocol 10.21, Skeptic can force*

$$\left.\begin{array}{l} \forall n : P_n(\xi_n) = 0 \\ \sum_{n=1}^{\infty} P_n(\xi_n^2) < \infty \end{array}\right\} \Longrightarrow \sup_N \sum_{n=1}^{N} \xi_n(y_n) < \infty. \qquad (10.26)$$

The condition $\forall n : P_n(\xi_n) = 0$ in (10.26) is analogous to the condition that the ξ_n be measure-theoretic martingale differences. In Lemma 10.23, we relax this to their being submartingale differences.

Proof It is easy to see that for each $C > 0$ there is a strategy, say $\mathcal{S}_C$, for Skeptic leading to capital

$$\mathcal{K}_N^{\mathcal{S}_C} = \begin{cases} 1 + \dfrac{1}{C}\left(\left(\displaystyle\sum_{n=1}^{N} \xi_n(y_n)\right)^2 - \sum_{n=1}^{N} P_n(\xi_n^2)\right) & \text{if } \displaystyle\sum_{n=1}^{N} P_n(\xi_n^2) \le C, \\ \mathcal{K}_{N-1}^{\mathcal{S}_C} & \text{otherwise,} \end{cases}$$

for $N = 0, 1, \ldots$. It remains to apply Statement 4 of Lemma 7.4 to mix all $\mathcal{K}^{\mathcal{S}_C}, C \in \mathbb{N}$ (with arbitrary positive weights), remembering that weak forcing is equivalent to forcing. $\qquad\square$

Lemma 10.23   *In Protocol 10.21, Skeptic can force*

$$\left.\begin{array}{l} \forall n : P_n(\xi_n) \ge 0 \\ \sum_{n=1}^{\infty} (P_n(\xi_n) + P_n(\xi_n^2)) < \infty \end{array}\right\} \Longrightarrow \sup_N \sum_{n=1}^{N} \xi_n(y_n) < \infty.$$

Proof It suffices to apply Lemma 10.22 to $\tilde{\xi}_n := \xi_n - P_n(\xi_n)$ (notice that $P_n(\tilde{\xi}_n^2) \leq P_n(\xi_n^2)$). $\qquad\square$

The following lemma is a version of the Borel–Cantelli–Lévy lemma (another version was called the first Borel–Cantelli lemma in Exercise 6.20). We may assume that the events E_n are encoded by $\xi_n := \mathbf{1}_{E_n}$ in Protocol 10.21.

Lemma 10.24 *In Protocol 10.21, Skeptic can force*

$$\left(\sum_{n=1}^{\infty} P_n(E_n) < \infty \right) \implies (y_n \in E_n \text{ for finitely many } n),$$

where $E_n := \{y \in \mathcal{Y} | \xi_n \neq 0\}$.

Proof This is a special case of Lemma 10.23. $\qquad\square$

Our application of the Borel–Cantelli–Lévy lemma will be made possible by the following lemma.

Lemma 10.25 *There exists a constant C such that, for all $P^{\mathrm{I}}, P^{\mathrm{II}} \in \mathcal{P}(\mathcal{Y})$,*

$$P^{\mathrm{I}}\{\beta^{\mathrm{I}} > e\beta^{\mathrm{II}}\} \leq C\rho^2(P^{\mathrm{I}}, P^{\mathrm{II}}),$$

where β^{I} and β^{II} are the densities of P^{I} and P^{II}, respectively.

Proof Let E denote the event $\{\beta^{\mathrm{I}} > e\beta^{\mathrm{II}}\}$. Note that

$$
\begin{aligned}
\rho^2(P^{\mathrm{I}}, P^{\mathrm{II}}) &= 1 - \int_{\mathcal{Y}} \sqrt{\beta^{\mathrm{I}}(y)\beta^{\mathrm{II}}(y)} Q(dy) \\
&= \int_{\mathcal{Y}} \left(\frac{1}{2}\beta^{\mathrm{I}}(y) + \frac{1}{2}\beta^{\mathrm{II}}(y) - \sqrt{\beta^{\mathrm{I}}(y)\beta^{\mathrm{II}}(y)} \right) Q(dy) \\
&\geq \int_{E} \left(\frac{1}{2}\beta^{\mathrm{I}}(y) + \frac{1}{2}\beta^{\mathrm{II}}(y) - \sqrt{\beta^{\mathrm{I}}(y)\beta^{\mathrm{II}}(y)} \right) Q(dy) \\
&\geq \int_{E} \left(\frac{1}{2}\beta^{\mathrm{I}}(y) + \frac{1}{2e}\beta^{\mathrm{I}}(y) - \sqrt{\beta^{\mathrm{I}}(y)\beta^{\mathrm{I}}(y)/e} \right) Q(dy) \\
&= \left(\frac{1}{2} + \frac{1}{2e} - e^{-1/2} \right) P^{\mathrm{I}}(E).
\end{aligned}
$$

The first inequality follows from the fact that the geometric mean never exceeds the arithmetic mean, and the second inequality uses the fact that for each $\beta > 0$ the function

$$\frac{1}{2}\beta + \frac{1}{2}x - \sqrt{\beta x}$$

is decreasing in $x \in [0, \beta/e]$; this can be checked by differentiation. So we can set

$$C := \left(\frac{1}{2} + \frac{1}{2e} - e^{-1/2} \right)^{-1} > 0. \qquad \qquad \square$$

We will also need the following elementary inequality (Exercise 10.9).

Lemma 10.26   *There exist $A, B > 0$ such that, for all $x > 0$,*

$$x(1 \wedge \ln x) + x(1 \wedge \ln x)^2 \le A(x - 1) + B(1 - \sqrt{x}). \qquad (10.27)$$

Now we can prove Statement 2 of Theorem 10.20. First assume that the functions β_n^{I} and β_n^{II} are always positive and that $E_n^{\mathrm{I}} = E_n^{\mathrm{II}} = \emptyset$ for all n. Our goal is to prove that Skeptic I can force

$$\sum_{n=1}^{\infty} \rho^2(P_n^{\mathrm{I}}, P_n^{\mathrm{II}}) < \infty \implies \liminf_{n \to \infty} \mathcal{K}_n^{\mathrm{II}} < \infty.$$

Substituting $x := \beta_n^{\mathrm{I}} / \beta_n^{\mathrm{II}}$ in (10.27), multiplying by β_n^{II}, integrating over Q_n, and summing over $n \in \mathbb{N}$, we obtain

$$\sum_{n=1}^{\infty} \int \beta_n^{\mathrm{I}} \left(1 \wedge \ln \frac{\beta_n^{\mathrm{I}}}{\beta_n^{\mathrm{II}}} \right) + \beta_n^{\mathrm{I}} \left(1 \wedge \ln \frac{\beta_n^{\mathrm{I}}}{\beta_n^{\mathrm{II}}} \right)^2 dQ_n \le B \sum_{n=1}^{\infty} \rho^2(P_n^{\mathrm{I}}, P_n^{\mathrm{II}}).$$

Let us combine this inequality with:

- Lemma 10.23 applied to Skeptic I and $\xi_n := 1 \wedge \ln \frac{\beta_n^{\mathrm{I}}}{\beta_n^{\mathrm{II}}}$. It is applicable because the inequality $x(1 \wedge \ln x) \ge x - 1$, valid for all $x > 0$, implies

$$P_n^{\mathrm{I}}(\xi_n) = \int \frac{\beta_n^{\mathrm{I}}}{\beta_n^{\mathrm{II}}} \left(1 \wedge \ln \frac{\beta_n^{\mathrm{I}}}{\beta_n^{\mathrm{II}}} \right) \beta_n^{\mathrm{II}} dQ_n \ge \int \left(\frac{\beta_n^{\mathrm{I}}}{\beta_n^{\mathrm{II}}} - 1 \right) \beta_n^{\mathrm{II}} dQ_n = 0$$

   (and ξ_n is P_n^{I}-integrable since $\xi_n \le 1$).
- Lemma 10.24 applied to Skeptic I and $E_n := \{ \beta_n^{\mathrm{I}} > e\beta_n^{\mathrm{II}} \}$. It will be applicable by Lemma 10.25.

We can now see that it suffices to prove that Skeptic I can force

$$\left.
\begin{array}{l}
\displaystyle \sup_N \sum_{n=1}^{N} \left(1 \wedge \ln \frac{\beta_n^{\mathrm{I}}(y_n)}{\beta_n^{\mathrm{II}}(y_n)} \right) < \infty \\[2ex]
\beta_n^{\mathrm{I}}(y_n) \le e\beta_n^{\mathrm{II}}(y_n) \text{ from some } n \text{ on}
\end{array}
\right\} \implies \liminf_{n \to \infty} \mathcal{K}_n^{\mathrm{II}} < \infty. \qquad (10.28)$$

Forcing (10.28) can be achieved by forcing

$$\sup_N \sum_{n=1}^N \ln \frac{\beta_n^{\mathrm{I}}(y_n)}{\beta_n^{\mathrm{II}}(y_n)} < \infty \implies \liminf_{n \to \infty} \mathcal{K}_n^{\mathrm{II}} < \infty,$$

and the latter can be done with the strategy

$$f_n^{\mathrm{I}} := \frac{\beta_n^{\mathrm{II}}}{\beta_n^{\mathrm{I}}} f_n^{\mathrm{II}}.$$

It remains to eliminate the assumptions $\beta_n^{\mathrm{I}} > 0$, $\beta_n^{\mathrm{II}} > 0$, and $E_n^{\mathrm{I}} = E_n^{\mathrm{II}} = \emptyset$. Our argument so far remains valid if the outcome space $\mathcal{Y} = \mathcal{Y}_n$ is allowed to depend on n (it can be chosen by Skeptic I's opponents at any time prior to his move). In particular, we can set $\mathcal{Y}_n := (E_n^{\mathrm{I}} \cup E_n^{\mathrm{II}})^c$ and assume, without loss of generality, that $\beta_n^{\mathrm{I}} > 0$ and $\beta_n^{\mathrm{II}} > 0$ on $\mathcal{Y}_n$. This proves Statement 2 of Theorem 10.20.

To deduce Statement 2 of Proposition 10.18 from Statement 2 of Theorem 10.20, notice that, for all n, $P_n^{\mathrm{I}}(E_n^{\mathrm{II}}) = 0$ (as $P_n^{\mathrm{I}} \ll P_n^{\mathrm{II}}$). Therefore, $P_n^{\mathrm{I}}(E_n^{\mathrm{I}} \cup E_n^{\mathrm{II}}) = 0$, and mixing any of the strategies for Skeptic I whose existence is asserted in Statement 2 of Theorem 10.20 with the strategy

$$f_n^{\mathrm{I}}(y) := \begin{cases} \infty & \text{if } y \in E_n^{\mathrm{I}} \cup E_n^{\mathrm{II}}, \\ 1 & \text{otherwise,} \end{cases}$$

we obtain a strategy for Skeptic I satisfying the condition of Statement 2 of Proposition 10.18.

10.8 EXERCISES

Exercise 10.1 Compare tests of the Daltonism theory based on (10.4) and (10.5) for $N = 100$ and for $\alpha \in \{0.05, 0.01\}$. Assume that $p_n = 1/2$ on all rounds, so that $\tau_B = 4B$. □

Exercise 10.2

1. Let $\mathcal{G}$ be the set of rounds in Protocol 10.5 in which neither parent suffers from Daltonism and the child is a boy:

$$\mathcal{G} := \{n \in \mathbb{N} | F_n \neq A, M_n \neq A, \text{ and } S_n = B\}.$$

   Let E be the event that $|\mathcal{G}| \geq 100$ and 63% or more of the boys born on the first 100 rounds in $\mathcal{G}$ will suffer from Daltonism. Show that if Skeptic plays (10.6) with $\kappa = 0.5$ on rounds in $\mathcal{G}$ (not betting on other rounds), and if E happens, then Skeptic will have multiplied his capital by more than 20 at the end of the first 100 rounds in $\mathcal{G}$.

2. Show that if Skeptic plays in the same way, except that he pretends that $p_n = 1/2$ on all the rounds in G (this may make the capital resulting from (10.6) less than his actual capital), then the conclusion still holds. □

Exercise 10.3 Deduce Corollary 10.6 using measure-theoretic probability. *Hint.* There is no need to fit the whole protocol into one measure-theoretic probability space. Consider separately all 12 possibilities for the father's and mother's status with respect to Daltonism (N, C, or A) and for the sex of the child. Apply the results of Section 1.3 to each of the 12 resulting probability spaces. □

Exercise 10.4 Verify Eq. (10.8). *Hint.* Multiply the right-hand side by $X'_{n-1}X_{n-1} + x_n x'_n$ (which is equal to $X'_n X_n$) on the left and simplify. This equation is a special case of the Sherman–Morrison formula. □

Exercise 10.5 Deduce (10.7) from (10.8). □

Exercise 10.6 (research project). Generalize Theorem 10.8 as suggested after Protocol 10.7. *Hint.* Adapt the argument in [236]. □

Exercise 10.7 (research project). Develop game-theoretic analogs of some of the measure-theoretic results that have been developed for Cox's proportional hazards model. For example (assuming $K = 1$):

1. Let $U(\theta)$ (the *efficient score*) be the derivative of the log-likelihood $L(\theta)$ and $-T(\theta)$ be the second derivative of $L(\theta)$ (for very intuitive explicit expressions, see [86, p. 191]). In the measure-theoretic framework, it has been shown that the random variable $X := U(\theta_0)/\sqrt{T(\theta_0)}$ is close to standard Gaussian for large N, and this provides another way of testing $\theta = 0$ (cf. [86, section 2]). If Forecaster plays Cox's strategy with $\theta = \theta_0$, are there natural conditions on the path under which the upper probability of the event $X \geq C$ is close to $\int_C^\infty \mathcal{N}_{0,1}$?

2. It is often assumed that $Y := (\hat{\theta} - \theta_0)\sqrt{T(\hat{\theta})}$, where $\hat{\theta} := \arg\max_\theta L(\theta)$ is the maximum likelihood estimate, is approximately standard Gaussian under $\theta = \theta_0$ (cf. [87, equation (11)]); this assumption can be used for finding confidence limits for θ. Under what conditions does this hold in the protocol? □

Exercise 10.8 Show that Statement 2 of Proposition 10.18 becomes false if the condition that Forecaster II play cautiously is dropped. □

Exercise 10.9 Prove Lemma 10.26. □

10.9   CONTEXT

Cournot's Principle

The principle that probability theory becomes relevant to phenomena only by making predictions with high probability was first stated by Antoine Augustin Cournot in 1843 [83, section 43]. See [343, 350] for the subsequent history of this principle.

The game-theoretic version of Cournot's principle says a strategy for probabilistic forecasting is a valid description of phenomena if it is impossible in practice to multiply substantially the capital one risks using the forecasts as betting odds. In [349], we called this the *fundamental interpretative hypothesis* of game-theoretic probability. It can be understood as a way of making more precise von Mises's "principle of the impossibility of a gambling system." (This phrase appears on p. 26 of the 1957 English edition of von Mises's *Probability, Statistics, and Truth*. It translates *Prinzip vom ausgeschlossenen Spielsystem*, which we find on p. 26 of the book's first edition, published in 1928 [282].) Von Mises said that it should be impossible to beat the odds by selecting rounds on which to bet; Ville generalized this by saying that it should be impossible to multiply the capital you risk infinitely no matter how you vary your bet [387].

Cournot may also have been the first to argue that probability can represent the knowledge of an idealized observer who knows more than we do [83, section 45]. This is echoed by our picture of a forecaster who knows more than a statistician standing outside the testing protocol.

Probability in Genetics

The discovery of the genetic causes of Daltonism, hemophilia, and many other inherited conditions is recounted by Peter Harper [180]. Daltonism is named after John Dalton, who argued for its hereditary nature in work published in 1798. It is briefly discussed in Feller's classical textbook [135, section V.6]. See [313] for a history of the limits of genetic prediction and [186] for a recent review of evidence for the imperfect predictability of gene expression.

Least Squares

The method of least squares was championed by Carl Friedrich Gauss and Adrien-Marie Legendre in the early nineteenth century [307, 357, 366]. In mathematical statistics its properties are usually studied under the assumption that the signals x_n are fixed vectors [8]. In this case it is sufficient for the strong consistency of least squares estimates that $\lambda_{\min} \to \infty$ [234, 235].

Theorem 10.8 is a pathwise version of a theorem by Tze Leung Lai and Ching Zong Wei [236, Theorem 1]; our proof follows theirs, with simplifications because we assume bounded noise. The Lai–Wei condition

$$\lambda_n^{\min} \to \infty \quad \text{and} \quad \ln \lambda_n^{\max} / \lambda_n^{\min} \to 0$$

in the antecedent of (10.9) is essentially optimal, as they show in [236, section 2, Example 1]; in their example, the noise variables ϵ_n are independent and identically distributed; $\epsilon_n := \pm 1$ with equal probabilities.

Survival Analysis

Cox introduced his proportional hazards model in 1972 [86]. See also [87] and [88, chapter 7]. The history of martingales in survival analysis is discussed in [1].

Quantum Mechanics

For more information on von Neumann's formalism, see [200, chapter 1] and von Neumann's original exposition [290, chapter III]. A more general version of von Neumann's axioms is given in [292, section 2.2], and a more general version of our protocol for quantum computation, Protocol 10.15, is described in [292, section 4.6]. For more general notions of measurement, see, e.g. [292, section 2.2.3] and [219, section 11].

Recent discussions of QBism [185, 277] and Bohmian mechanics [52] are also relevant.

Jeffreys's Law

The name *Jeffreys's law* was coined by A. Philip Dawid in honor of Harold Jeffreys [97, p. 281]. Proposition 10.18 generalizes Dawid's version of the law, which he used to support his theory of empirical probability [98, Theorem 7.1].

The counterpart of Proposition 10.18 in the algorithmic theory of randomness was obtained in [389] and generalized to a wide set of distances between probability measures (including (10.19) as a special case) in [157, Theorem 3]. This work was inspired by the work by Kabanov et al. [204] on the absolute continuity and singularity of probability measures (see also [358, Theorem VII.6.4]).

The proof of Proposition 10.17 mainly follows [406], which in turn follows [389] and [157]. There are two important differences from the results of [389] and [157], which are stated in terms of the algorithmic theory of randomness. First, our result is much more precise than the $O(1)$ accuracy of the algorithmic theory of randomness. Second, we pay careful attention to the exceptional cases where $\beta_n^{\mathrm{I}} = 0$ or $\beta_n^{\mathrm{II}} = 0$; this corresponds to eliminating the assumption of local absolute continuity in measure-theoretic probability (accomplished by Pukelsheim [315]).

The proof of Statement 2 of Proposition 10.18 and Theorem 10.20 adapts a proof given by Shiryaev [358, Theorem VII.6.4]. The proof of analogous statements in [389] and [157] is completely different and uses a non-asymptotic version of those statements similar to Proposition 10.17.

Hellinger Distance

The name *Hellinger distance* is sometimes given to $2\rho^2(P^{\mathrm{I}}, P^{\mathrm{II}})$, as in [389], or to another similar expression. In statistics, the Hellinger integral $H(P^{\mathrm{I}}, P^{\mathrm{II}})$ is sometimes called the Bhattacharyya coefficient, its logarithm then being called the Bhattacharyya distance. The Bhattacharyya distance and the squared Hellinger distance are close when either is small.

11

Calibrating Lookbacks and p-Values

As we learned in Chapters 1 and 10, game-theoretic probability provides a framework for testing. Skeptic tests Forecaster by trying to multiply the capital he risks by a large factor. In general, this testing is dynamic. Skeptic's degree of success – the factor by which he has multiplied his capital – varies over time. He might multiply it by a huge factor, apparently definitively discrediting Forecaster, but then lose most or all of his gains, casting this verdict into question. How can we insure against such a complete reversal? How can we make sure that when we look back and see the maximum capital that Skeptic has attained, we can claim at least some of that capital as evidence against Forecaster?

There is a simple way to retain at least a portion of Skeptic's gains. We can decide in advance on several target levels of capital at which we might stop betting, allocate our capital among these targets, and then imitate Skeptic with each of the resulting accounts until its target is reached. We call this a *lookback trading strategy*. As we will see in this chapter, this turns out to be the best we can do. We will apply the idea both to financial markets and to statistical testing.

To formalize the picture, we consider a game in which two players test Forecaster. We call them Skeptic and Rival Skeptic. Skeptic plays first, and Rival Skeptic can therefore imitate him with different accounts until different targets are reached. We develop this picture abstractly in Sections 11.1 and 11.2. Section 11.1 introduces the notion of a *lookback calibrator* using a protocol in which Forecaster has complete freedom. Section 11.2 broadens the picture to protocols in which Forecaster has less

Game-Theoretic Foundations for Probability and Finance, First Edition. Glenn Shafer and Vladimir Vovk.
© 2019 John Wiley & Sons, Inc. Published 2019 by John Wiley & Sons, Inc.

freedom and proves the propositions stated in Section 11.1 in this broader context. Section 11.3 considers compromises between the goal of matching Skeptic's current capital and the goal of matching his maximum previous capital.

In Section 11.4, we consider how a participant in an actual financial market can use lookback trading strategies to insure against the loss of capital. This leads to alternatives to existing lookback options.

In Section 11.5, we discuss how lookback calibrators can be applied to statistical testing, even when the testing is static rather than dynamic. Classical statistical testing implicitly involves different targets because a test statistic can be significant to different degrees – i.e. can produce different p-values. The Neyman–Pearson theory of testing avoids equivocation about the target by insisting on a single level of significance, chosen in advance. But we can instead spread our capital over bets on different levels, and this leads to methods of calibrating p-values to make them comparable with Neyman–Pearson significance levels or Bayes factors.

11.1 LOOKBACK CALIBRATORS

Because it allows us to define the notion of a lookback calibrator in a very simple way, we begin with a protocol in which the outcome space $\mathcal{Y}$ is fixed as a parameter but Forecaster can vary the upper expectation on $\mathcal{Y}$ as he wishes from round to round.

A Convenient Protocol

Consider the following protocol, in which $\mathcal{K}$ and $\mathcal{L}$ represent Skeptic's and Rival Skeptic's capital, respectively. As usual, we suppose that all players have perfect information. Thus Rival Skeptic sees Skeptic's moves and can imitate them.

Protocol 11.1
PARAMETER: Set $\mathcal{Y}$ with at least two elements
   $\mathcal{K}_0 := 1$ and $\mathcal{L}_0 := 1$.
   FOR $n = 1, 2, \ldots$:
      Forecaster announces an upper expectation $\overline{\mathbf{E}}_n$ on $\mathcal{Y}$.
      Skeptic announces $f_n \in [0, \infty]^{\mathcal{Y}}$ such that $\overline{\mathbf{E}}_n(f_n) \leq \mathcal{K}_{n-1}$.
      Rival Skeptic announces $g_n \in [0, \infty]^{\mathcal{Y}}$ such that $\overline{\mathbf{E}}_n(g_n) \leq \mathcal{L}_{n-1}$.
      Reality announces $y_n \in \mathcal{Y}$.
      $\mathcal{K}_n := f_n(y_n)$ and $\mathcal{L}_n := g_n(y_n)$.

Because Skeptic and Rival Skeptic are required to choose their moves from $[0, \infty]^{\mathcal{Y}}$, their capital necessarily remains nonnegative. To keep the statement of the protocol as simple as possible, we require Forecaster to announce an upper expectation on $\mathcal{Y}$, but all that is really needed is a $[0, \infty]$-upper expectation on $\mathcal{Y}$ (see the comments at the end of Section 6.4). In this chapter we do not use the continuity axiom.

Defining and Characterizing Lookback Calibrators

For $n = 0, 1, \ldots$, set

$$\mathcal{K}_n^* := \max_{0 \le i \le n} \mathcal{K}_i.$$

We are interested in how much of this maximum past capital Rival Skeptic can count on retaining. We call an increasing function $H : [1, \infty] \to [0, \infty]$ a *lookback calibrator* if $H(\infty) = \infty$ and Rival Skeptic has a strategy in Protocol 11.1 that guarantees

$$\mathcal{L}_n \ge H(\mathcal{K}_n^*) \tag{11.1}$$

for $n = 0, 1, \ldots$. The condition that H be increasing is imposed by the idea that H measures an amount that Rival Skeptic retains permanently; he should still retain it even after Skeptic attains a yet higher maximum. The condition $H(\infty) = \infty$ is justified by Rival Skeptic being able to match infinite gains by Skeptic without giving up anything else: see Exercise 11.1.

The following proposition characterizes lookback calibrators in a simple way.

Proposition 11.2 *An increasing function $H : [1, \infty] \to [0, \infty]$ is a lookback calibrator if and only if $H(\infty) = \infty$ and*

$$\int_1^\infty \frac{H(v)}{v^2} \, dv \le 1. \tag{11.2}$$

It follows from (11.2) that $H(v) \le v$ for all v and that $H(v)/v \to 0$ as $v \to \infty$ (Exercise 11.2).

Because a lookback calibrator necessarily maps $[1, \infty)$ to $[0, \infty)$ and ∞ to ∞, it suffices, when defining a lookback calibrator or discussing its particular properties, to consider how it maps $[1, \infty)$ to $[0, \infty)$. In (11.3)–(11.6), for example, we define lookback calibrators by specifying their values on $[1, \infty)$.

We establish Proposition 11.2 in Section 11.2 by proving more general results concerning what Rival Skeptic can accomplish in protocols obtained from Protocol 11.1 by constraining Forecaster.

A lookback calibrator H is *maximal* if there is no other lookback calibrator G such that $G(v) \ge H(v)$ for all $v \in [1, \infty)$. The following corollary of Proposition 11.2 characterizes maximal lookback calibrators.

Corollary 11.3 *A lookback calibrator is maximal if and only if it is right-continuous and (11.2) holds with equality. For every lookback calibrator G, there is a maximal lookback calibrator H such that $H \ge G$.*

Proof An increasing function $H : [1, \infty) \to [0, \infty)$ satisfying (11.2) can be increased to an increasing function that satisfies (11.2) with equality (just multiply it by the appropriate constant), and then it can be increased to an increasing right-continuous function without further changing the value of the integral in (11.2) (remember that an increasing function has only countably many jumps). On the other hand, an increasing right-continuous function $H : [1, \infty) \to [0, \infty)$ satisfying (11.2) with equality cannot be increased further without increasing the value of this integral.   □

Some Maximal Lookback Calibrators

Selecting a lookback calibrator involves balancing the desire to retain as much as possible of relatively modest values of $\mathcal{K}_n^*$ and the ambition to retain a substantial fraction of much larger values of $\mathcal{K}_n^*$. We want $H(v)/v$ to tend to zero slowly as $v \to \infty$.

Table 11.1 gives us some insight into the possibilities by showing values of five maximal lookback calibrators. The first two are given by

$$\mathrm{RT}_1(v) := \begin{cases} \sqrt{v} & \text{if } v \geq 4 \\ 0 & \text{otherwise} \end{cases} \tag{11.3}$$

and

$$\mathrm{RT}_2(v) := \begin{cases} 2\sqrt{v} & \text{if } v \geq 16 \\ 0 & \text{otherwise} \end{cases} \tag{11.4}$$

Table 11.1   Values of some maximal lookback calibrators, rounded down to the next integer.

v	$\mathrm{RT}_1(v)$	$\mathrm{RT}_2(v)$	$G_\kappa(v)$		$H_\kappa(v)$
			$\kappa = 0.22$	$\kappa = 0.05$	$\kappa = 1$
10	3	0	1	0	3
20	4	8	2	0	4
50	7	14	4	2	6
100	10	20	7	3	9
200	14	28	13	7	14
500	22	44	28	18	25
10^3	31	63	48	35	41
10^4	100	200	290	315	235
10^5	316	632	1 747	2 811	1 508
10^6	1 000	2 000	10 259	25 059	10 478
10^7	3 162	6 324	63 448	223 341	76 984

The value of $G_\kappa(100)$ is maximized when $\kappa = 1/\ln 100 \approx 0.22$, and $G_{0.22}$ is roughly similar to RT_1 for v between 50 and 1000. The function H_1 is roughly similar to RT_1 for v between 10 and 1000.

for $v \in [1, \infty)$. These two lookback calibrators are so easily computed that they can be used as mental rules of thumb (hence the symbol RT). The other four are chosen from two parametric families of maximal lookback calibrators:

$$G_\kappa(v) := \kappa v^{1-\kappa} \quad \text{for all } v \in [1, \infty), \tag{11.5}$$

where $\kappa \in (0, 1)$, and

$$H_\kappa(v) := \begin{cases} \kappa(1 + \kappa)^\kappa v \ln^{-1-\kappa} v & \text{if } v \geq e^{1+\kappa}, \\ 0 & \text{otherwise,} \end{cases} \tag{11.6}$$

where $\kappa \in (0, \infty)$. The smaller the κ, the slower $G_\kappa(v)/v$ and $H_\kappa(v)/v$ tend to zero as $v \to \infty$.

The reader may verify that RT_1, RT_2, G_κ, and H_κ are maximal lookback calibrators by checking that they are increasing and satisfy (11.2) with equality (see Exercise 11.3). Notice also that these particular lookback calibrators all tend to ∞ with v.

The concept of maximality for lookback calibrators has limited usefulness. The definition says that we cannot improve the bound $\mathcal{L}_n \geq H(\mathcal{K}_n^*)$ to a better lower bound on $\mathcal{L}_n$ of the same form – i.e. a better lower bound that is also a function of $\mathcal{K}_n^*$ alone. But this does not exclude improving it to a better lower bound of a different form. In fact, the strategy for Rival Skeptic we construct in Section 11.2 to prove Proposition 11.2 establishes a better lower bound, one that is a function not just of $\mathcal{K}_n^*$ but also of $\mathcal{K}_n$ (see the discussion at the end of Section 11.2 and Exercises 11.5 and 11.8). In the case of RT_1, for example, the improvement is from

$$\mathcal{L}_n \geq \sqrt{\mathcal{K}_n^*}$$

to

$$\mathcal{L}_n \geq \sqrt{\mathcal{K}_n^*} + \frac{\mathcal{K}_n}{\sqrt{\mathcal{K}_n^*}}, \tag{11.7}$$

assuming $\mathcal{K}_n^* \geq 4$. This is no improvement at all when $\mathcal{K}_n = 0$, but it is a doubling of the lower bound when $\mathcal{K}_n = \mathcal{K}_n^*$. As we will see in Section 11.3, Rival Skeptic cannot improve on (11.7) as a function of $\mathcal{K}_n^*$ and $\mathcal{K}_n$. He cannot guarantee $\mathcal{L}_n \geq H(\mathcal{K}_n^*, \mathcal{K}_n)$ for any function $H(v, w)$ that is different from $\sqrt{v} + w/\sqrt{v}$ and satisfies $H(v, w) \geq \sqrt{v} + w/\sqrt{v}$ for all $v \geq 4$ and all w (see Exercise 11.14).

Using Lookback Strategies as Insurance

Rival Skeptic might well be disappointed by the numbers in Table 11.1. The fraction of Skeptic's maximum capital that he can be sure to retain, $H(v)/v$, decreases rapidly as v grows. If Rival Skeptic suspects that Skeptic's capital will continue to grow indefinitely, he may want to take greater advantage of this possibility by using some of his capital simply to imitate Skeptic. If he chooses $c \in [0, 1)$ and devotes c of his initial unit capital to imitating Skeptic and $1 - c$ to a strategy that guarantees $H(\mathcal{K}_n^*)$, then

$$\mathcal{L}_n \geq c\mathcal{K}_n + (1 - c)H(\mathcal{K}_n^*) \tag{11.8}$$

for all $n \in \{0, 1, \dots\}$. We can think of $1 - c$ as a premium Rival Skeptic pays at the outset to insure against Skeptic losing all or nearly all of the capital or evidence he accumulates.

Our proof of Proposition 11.2 in the next section also establishes the following generalization.

Proposition 11.4   *If $c \in [0, 1)$ and $H : [1, \infty] \to [0, \infty]$ is increasing and satisfies $H(\infty) = \infty$, then Rival Skeptic has a strategy guaranteeing (11.8) in Protocol 11.1 if and only if H satisfies (11.2).*

This reduces to Proposition 11.2 when $c = 0$.

Proposition 11.4 implies that if Rival Skeptic plays so that he is guaranteed to have at least $H(\mathcal{K}_n^*)$ for all n, where $H : [1, \infty] \to [0, \infty]$ is increasing and satisfies $H(\infty) = \infty$, then the largest fraction of Skeptic's current capital $\mathcal{K}_n$ that he can also be sure of having for all n is

$$1 - \int_1^\infty \frac{H(v)}{v^2} \, dv.$$

In this sense, the compromise (11.8) is the best Rival Skeptic can do.

On the other hand, the strategies we construct for Rival Skeptic in the next section allow him to improve on the right-hand side of (11.8) in a less ambitious way. Although he cannot increase the term $c\mathcal{K}_n$ in (11.8) to $a\mathcal{K}_n$ for a value of a greater than c, he can increase it to a function of $\mathcal{K}_n$ that is strictly greater than $c\mathcal{K}_n$ when $\mathcal{K}_n$ is nonzero. In the case of the maximal lookback calibrator RT_1, for example, the improvement (11.7) that we noted earlier improves

$$\mathcal{L}_n \geq c\mathcal{K}_n + (1 - c)\sqrt{\mathcal{K}_n^*}$$

to

$$\mathcal{L}_n \geq c\mathcal{K}_n + (1 - c)\left(\sqrt{\mathcal{K}_n^*} + \frac{\mathcal{K}_n}{\sqrt{\mathcal{K}_n^*}} \right)$$

$$= \left(c + \frac{1 - c}{\sqrt{\mathcal{K}_n^*}} \right) \mathcal{K}_n + (1 - c)\sqrt{\mathcal{K}_n^*},$$

assuming $\mathcal{K}_n^* \geq 4$.

11.2 LOOKBACK PROTOCOLS

Recall our usual notion of a situation: given a nonempty set $\mathcal{Y}$, we call any finite sequence of elements of $\mathcal{Y}$ a *situation*, and we write $\mathbb{S}$ for the set of all situations. With this notation, we define the following protocol.

Protocol 11.5
PARAMETERS: Nonempty set $\mathcal{Y}$,
        family $(\mathcal{E}_s)_{s \in \mathbb{S}}$ of nonempty sets of upper expectations on $\mathcal{Y}$
$\mathcal{K}_0 := 1$ and $\mathcal{L}_0 := 1$.
FOR $n = 1, 2, \ldots$:
    Forecaster announces $\overline{\mathbf{E}}_n \in \mathcal{E}_{y_1 \ldots y_{n-1}}$.
    Skeptic announces $f_n \in [0, \infty]^{\mathcal{Y}}$ such that $\overline{\mathbf{E}}_n(f_n) \leq \mathcal{K}_{n-1}$.
    Rival Skeptic announces $g_n \in [0, \infty]^{\mathcal{Y}}$ such that $\overline{\mathbf{E}}_n(g_n) \leq \mathcal{L}_{n-1}$.
    Reality announces $y_n \in \mathcal{Y}$.
    $\mathcal{K}_n := f_n(y_n)$ and $\mathcal{L}_n := g_n(y_n)$.

We call Protocol 11.5 and any specialization of it a *lookback protocol*. Thus Protocol 11.1 is a lookback protocol; it is the specialization in which $\mathcal{Y}$ has at least two elements and $\mathcal{E}_s$, for each $s \in \mathbb{S}$, is the set of all upper expectations on $\mathcal{Y}$. Nearly all the testing protocols used so far in this book become lookback protocols once they are put in slackened form and Rival Skeptic is added. In particular, putting Protocol 7.1 in slackened form and adding Rival Skeptic produces a protocol simpler than but essentially equivalent to Protocol 11.5, as the argument of Section 7.5 shows.

We will prove two theorems about lookback protocols. Theorem 11.6 says that in a lookback protocol in which Forecaster can offer Skeptic and Rival Skeptic a sufficiently rich set of betting opportunities, Rival Skeptic cannot achieve (11.8) for any $c \in [0, 1)$ if (11.2) is violated. Theorem 11.9 says that Rival Skeptic can achieve (11.1) in any lookback protocol if (11.2) is satisfied. Propositions 11.2 and 11.4 follow immediately.

Rich Lookback Protocols

We call a lookback protocol *rich* if for every $a \in (0, 1)$ and every $s \in \mathbb{S}$ there exists $\overline{\mathbf{E}}_s \in \mathcal{E}_s$ and $E_s \subseteq \mathcal{Y}$ such that

$$\overline{\mathbf{P}}_s(E_s) = a. \tag{11.9}$$

Protocol 11.1 is rich. To see this, choose two distinct elements of $\mathcal{Y}$, say y and y'. Given $a \in (0, 1)$, define $\overline{\mathbf{E}}_s$ by

$$\overline{\mathbf{E}}_s(f) := af(y) + (1 - a)f(y')$$

and set $E_s := \{y\}$ for all s. Many other testing protocols also become rich lookback protocols when Rival Skeptic is added. Suppose, for example, that we set $\mathcal{Y} := [0, 1]$ and require Forecaster to announce the upper integral with respect to Lebesgue measure on $[0, 1]$ on every round (see Exercise 6.12). The resulting lookback protocol is rich, because $\overline{\mathbf{P}}([0, a]) = a$ for all $a \in (0, 1)$. We will see additional rich lookback protocols in Sections 11.4 and 11.5.

Theorem 11.6 *Suppose $H : [1, \infty] \rightarrow [0, \infty]$ is increasing and satisfies $H(\infty) = \infty$, $c \in [0, 1)$, and Rival Skeptic has a strategy in a rich lookback protocol that guarantees*

$$\mathcal{L}_n \geq c\mathcal{K}_n + (1 - c)H(\mathcal{K}_n^*)$$

for all $n \in \{0, 1, \dots\}$. Then H satisfies (11.2).

Proof Fix $a \in (0, 1)$, and for all $s \in \mathbb{S}$, choose $\overline{\mathbf{E}}_s$ and E_s that satisfy (11.9). Let $\overline{\mathbb{E}}$ be the global upper expectation in the protocol obtained by requiring Forecaster to announce $\overline{\mathbf{E}}_s$ in each situation s, removing him as a player, and allowing Skeptic to announce his own nonnegative initial capital $\mathcal{K}_0$ before the first round instead of setting $\mathcal{K}_0 := 1$.

Define global events $A_1, A_2, \dots$ and global variables $X_1, X_2, \dots$ and $Y_1, Y_2, \dots$ by

$$A_n := \{y_n \in E_{y_1 \dots y_{n-1}}\},$$

$$X_n := \max\{k | k \in \{0, 1, \dots, n\} \text{ and } A_1, \dots, A_k \text{ all happen}\},$$

and

$$Y_n := ca^{-n} \prod_{k=1}^{n} \mathbf{1}_{A_k} + (1 - c)H(a^{-X_n}). \tag{11.10}$$

If Skeptic plays $a^{-1}\mathcal{K}_{n-1}\mathbf{1}_{E_{y_1 \dots y_{n-1}}}$ on the nth round, then

$$\mathcal{K}_n = a^{-n} \prod_{k=1}^{n} \mathbf{1}_{A_k} \quad \text{and} \quad \mathcal{K}_n^* = a^{-X_n}.$$

So by hypothesis, Rival Skeptic has a strategy that guarantees $\mathcal{L}_n \geq Y_n$ for all n. Skeptic can play the same strategy, thus guaranteeing $\mathcal{K}_n \geq Y_n$ for all n. This implies that

$$\overline{\mathbb{E}}(Y_n) \leq 1 \tag{11.11}$$

for all n.

We can rewrite (11.10) as

$$Y_n = ca^{-n} \prod_{k=1}^{n} \mathbf{1}_{A_k} + (1-c)\left(H(1) + \sum_{k=1}^{n}(H(a^{-k}) - H(a^{-k+1}))\prod_{i=1}^{k}\mathbf{1}_{A_i}\right) \quad (11.12)$$

and then deduce that

$$\overline{\mathbb{E}}(Y_n) = c + (1-c)\left(H(1) + \sum_{k=1}^{n}(H(a^{-k}) - H(a^{-k+1}))a^k\right). \quad (11.13)$$

To deduce (11.13) from (11.12), recall that the global expected value of a global variable can be calculated iteratively:

$$\overline{\mathbb{E}}(Y_n) = \overline{\mathbb{E}}_0(\overline{\mathbb{E}}_1(\dots \overline{\mathbb{E}}_{n-1}(Y_n)\dots)). \quad (11.14)$$

(See Exercises 7.9 and 7.10.) At the end of round $n-1$, $\mathbf{1}_{A_n}$ is the only global variable on the right-hand side of (11.12) that is unknown, and

$$\overline{\mathbb{E}}_{n-1}(\mathbf{1}_{A_n}) = \overline{\mathbf{P}}_{y_1\dots y_{n-1}}(E_{y_1\dots y_{n-1}}) = a$$

by Lemma 7.6. So by Axiom E2 and Statement 1 of Proposition 6.4, we obtain $\overline{\mathbb{E}}_{n-1}(Y_n)$ by replacing $\mathbf{1}_{A_n}$ with a in the right-hand side of (11.12). Similarly, we obtain $\overline{\mathbb{E}}_{n-2}(\overline{\mathbb{E}}_{n-1}(Y_n))$ by replacing $\mathbf{1}_{A_{n-1}}$ with a. By implementing all n steps in (11.14), we reduce (11.12) to (11.13).

Combining (11.13) with (11.11), we obtain

$$H(1) + \sum_{k=1}^{n}(H(a^{-k}) - H(a^{-k+1}))a^k \leq 1,$$

which can be rewritten as

$$\sum_{k=1}^{n} H(a^{-k+1})(a^{k-1} - a^k) + H(a^{-n})a^n \leq 1.$$

Ignoring the last addend on the left-hand side, letting $n \to \infty$, and letting $a \to 1$, we can see that $\int_0^1 H(1/w)\,dw \leq 1$, and therefore,

$$\int_1^\infty \frac{H(v)}{v^2}\,dv = \int_0^1 H\left(\frac{1}{w}\right)\,dw \leq 1.$$

$\square$

Lookback Trading Strategies

We now prove two lemmas. The first is purely measure-theoretic. The second describes lookback trading strategies.

Lemma 11.7 *A function $H : [1, \infty) \to [0, \infty)$ is increasing and right-continuous if and only if there exists a measure Q on $[1, \infty)$ such that*

$$H(v) = \int_{[1,v]} uQ(du) \tag{11.15}$$

for all $v \in [1, \infty)$. In this case,

$$\int_1^\infty \frac{H(v)}{v^2} \, dv = Q([1, \infty)).$$

Proof If H is increasing and right-continuous, we can define a measure R on $[1, \infty)$ by setting $R([1, v]) := H(v)$ for all $v \in [1, \infty)$. Let Q be the measure on $[1, \infty)$ defined by $Q(du) := (1/u)R(du)$. Then (11.15) holds for all $v \in [1, \infty)$.

On the other hand, if there exists a measure Q on $[1, \infty)$ such that (11.15) holds, then H is evidently increasing and right-continuous, and

$$\int_1^\infty \frac{H(v)}{v^2} \, dv = \int_1^\infty \int_{[1,v]} \frac{u}{v^2} Q(du) dv$$
$$= \int_{[1,\infty)} \int_u^\infty \frac{u}{v^2} \, dv Q(du) = \int_{[1,\infty)} Q(du). \qquad \square$$

Now consider a probability measure P on $[1, \infty]$, and define $H^P : [1, \infty) \to [0, \infty)$ by

$$H^P(v) := \int_{[1,v]} uP(du). \tag{11.16}$$

Consider an arbitrary lookback protocol, which may or may not be rich. The players' moves are $\overline{E}_n, f_n, g_n,$ and y_n, and Skeptic and Rival Skeptic begin with unit capital and have the capital $\mathcal{K}_n$ and $\mathcal{L}_n$, respectively, at the end of the nth round. Set

$$k(u) := \inf\{n \in \{0, 1, \dots\} | \mathcal{K}_n \geq u\}$$

(with $\inf \emptyset := \infty$) for all $u \in [1, \infty)$ and

$$g_n^P := \int_{[1,\mathcal{K}_{n-1}^*]} \mathcal{K}_{k(u)} P(du) + P((\mathcal{K}_{n-1}^*, \infty]) f_n \tag{11.17}$$

for all $n \in \mathbb{N}$.

As the following lemma tells us, the sequence $g_1^P, g_2^P, \ldots$ qualifies as a strategy for Rival Skeptic. We call it a *lookback trading strategy*.

Lemma 11.8 *The sequence $g_1^P, g_2^P, \ldots$ constitutes a strategy for Rival Skeptic. When Rival Skeptic follows this strategy,*

$$\mathcal{L}_n = \int_{[1, \mathcal{K}_n^*]} \mathcal{K}_{k(u)} P(du) + P((\mathcal{K}_n^*, \infty]) \mathcal{K}_n, \tag{11.18}$$

and therefore (because $\mathcal{K}_{k(u)} \geq u$)

$$\mathcal{L}_n \geq H^P(\mathcal{K}_n^*) + P((\mathcal{K}_n^*, \infty]) \mathcal{K}_n \tag{11.19}$$

for $n = 0, 1, \ldots$.

Proof If $\mathcal{K}_{n-1}^* \geq u$, then $k(u) < n$, and so Rival Skeptic knows the values of $k(u)$ and $\mathcal{K}_{k(u)}$ and can calculate the right-hand side of (11.17) when it is his turn to move on the nth round. We have

$$\overline{\mathbf{E}}_1(g_1^P) = \overline{\mathbf{E}}_1 \left(\int_{[1,1]} \mathcal{K}_{k(u)} P(du) + P((1, \infty]) f_1 \right)$$

$$= \int_{[1,1]} \mathcal{K}_0 P(du) + P((1, \infty]) \overline{\mathbf{E}}_1(f_1)$$

$$= P(\{1\}) + P((1, \infty]) \overline{\mathbf{E}}_1(f_1) \leq 1 \tag{11.20}$$

and

$$\overline{\mathbf{E}}_n(g_n^P) = \overline{\mathbf{E}}_n \left(\int_{[1, \mathcal{K}_{n-1}^*]} \mathcal{K}_{k(u)} P(du) + P((\mathcal{K}_{n-1}^*, \infty]) f_n \right)$$

$$= \int_{[1, \mathcal{K}_{n-1}^*]} \mathcal{K}_{k(u)} P(du) + P((\mathcal{K}_{n-1}^*, \infty]) \overline{\mathbf{E}}_n(f_n)$$

$$\leq \int_{[1, \mathcal{K}_{n-1}^*]} \mathcal{K}_{k(u)} P(du) + P((\mathcal{K}_{n-1}^*, \infty]) \mathcal{K}_{n-1}$$

$$= \int_{[1, \mathcal{K}_{n-2}^*]} \mathcal{K}_{k(u)} P(du) + P((\mathcal{K}_{n-2}^*, \infty]) \mathcal{K}_{n-1} = g_{n-1}^P(y_{n-1}) \tag{11.21}$$

for $n \geq 2$. The penultimate equality in (11.21) holds because $\mathcal{K}_{k(u)} = \mathcal{K}_{n-1}$ for any $u \in (\mathcal{K}_{n-2}^*, \mathcal{K}_{n-1}^*]$. By (11.20) and (11.21), $g_1^P, g_2^P, \ldots$ constitute a strategy for Rival

Skeptic. If Rival Skeptic follows this strategy, then by (11.21),

$$\mathcal{L}_n = g_n^P(y_n) = \int_{[1,\mathcal{K}_{n-1}^*]} \mathcal{K}_{k(u)}P(du) + P((\mathcal{K}_{n-1}^*, \infty])\mathcal{K}_n$$

$$= \int_{[1,\mathcal{K}_n^*]} \mathcal{K}_{k(u)}P(du) + P((\mathcal{K}_n^*, \infty])\mathcal{K}_n.$$

□

Theorem 11.9   *Suppose $H : [1, \infty] \to [0, \infty]$ is increasing and satisfies $H(\infty) = \infty$ and (11.2). Then in every lookback protocol and for every $c \in [0, 1]$, Rival Skeptic has a strategy that guarantees (11.8) for $n = 0, 1, \ldots$.*

Proof We can assume without loss of generality that H is right-continuous and satisfies (11.2) with equality, because we can increase H to an increasing right-continuous function that satisfies these conditions. Then $(1 - c)H$ is also increasing and right-continuous. By Lemma 11.7, there is a measure Q on $[1, \infty)$ such that

$$(1 - c)H(v) = \int_{[1,v]} uQ(du)$$

for all $v \in [1, \infty)$ and $Q([1, \infty)) = 1 - c$. Extend Q to a probability measure P on $[1, \infty]$ by setting $P(\{\infty\}) := c$. Then $(1 - c)H$ is identical on $[1, \infty)$ to the function H^P defined by (11.16), and $P((\mathcal{K}_n^*, \infty]) \geq c$. So (11.19) implies (11.8).   □

Discussion

As promised, Propositions 11.2 and 11.4 follow from Theorems 11.6 and 11.9. From Theorems 11.6 and 11.9, we see that Proposition 11.4 is true as stated for Protocol 11.1 and also true for any other rich lookback protocol. As we have already noted, Proposition 11.4 implies Proposition 11.2.

We have not changed the definition of *lookback calibrator* stated in Section 11.1: a lookback calibrator is an increasing function $H : [1, \infty] \to [0, \infty]$ such that $H(\infty) = \infty$ and Rival Skeptic can guarantee $\mathcal{L}_n \geq H(\mathcal{K}_n^*)$ for all n in Protocol 11.1. But we have seen that Rival Skeptic can guarantee $\mathcal{L}_n \geq H(\mathcal{K}_n^*)$ for all n in Protocol 11.1 if and only if he can do so in any other rich lookback protocol.

The proofs of Lemma 11.8 and Theorem 11.9 also make it clear that the right-hand sides of (11.1) and (11.8) understate what Rival Skeptic can achieve, even when the lookback calibrator H in these expressions is maximal. There are three obvious ways he might do better:

1. If Skeptic unnecessarily loses capital by making a move f_n such that $\overline{\mathbf{E}}_n(f_n) < \mathcal{K}_{n-1}$, Rival Skeptic can refuse to imitate this, instead adding a constant to the move recommended by the lookback trading strategy, so that $\overline{\mathbf{E}}_n(g_n) = \mathcal{L}_{n-1}$.

2. When he follows the lookback trading strategy, Rival Skeptic's capital $\mathcal{L}_n$ is given by (11.18), not by the possibly smaller right-hand side of (11.19).

3. Even when H is maximal, so that $P(\{\infty\}) = 0$, the term $P((\mathcal{K}_n^*, \infty])\mathcal{K}_n$ in the right-hand side of (11.19) will usually be nonzero.

The benefits of the first two points depend on the details of play, but we can calculate $P((\mathcal{K}_n^*, \infty])\mathcal{K}_n$ and thereby obtain better bounds. We gave the result in the case of the maximal lookback calibrator RT_1 in (11.7) (see also Exercise 11.8).

11.3 LOOKBACK COMPROMISES

We have formulated conditions on H and c under which Rival Skeptic can guarantee $\mathcal{L}_n \geq c\mathcal{K}_n + (1 - c)H(\mathcal{K}_n^*)$. It is natural to consider more general functions of $\mathcal{K}_n$ and $\mathcal{K}_n^*$. Under what conditions on a bivariate function H can Rival Skeptic guarantee $\mathcal{L}_n \geq H(\mathcal{K}_n^*, \mathcal{K}_n)$?

We call a nonnegative function $H(v, w)$ with domain $v \in [1, \infty]$ and $w \in [0, v]$ a *lookback compromise* if $H(\infty, w) = \infty$ for all $w \in [0, \infty]$ and Rival Skeptic has a strategy in Protocol 11.1 that guarantees $\mathcal{L}_n \geq H(\mathcal{K}_n^*, \mathcal{K}_n)$. A lookback compromise H is *maximal* if there is no other lookback compromise G such that $G \geq H$. The next theorem shows that (11.19) already gives us all lookback compromises.

Theorem 11.10 *A nonnegative function $H(v, w)$ with domain $v \in [1, \infty], w \in [0, v]$ and satisfying $H(\infty, w) = \infty$ for all $w \in [0, \infty]$ is a maximal lookback compromise if and only if there exists a probability measure P on $[0, \infty]$ such that*

$$H(v, w) = H^P(v) + P((v, \infty])w$$

for all $v \in [0, \infty)$ and $w \in [0, v]$, in the notation of (11.19). For every lookback compromise G, there is a maximal lookback compromise H such that $G \leq H$.

Given a bivariate function $H(v, w)$, we write $H^=$ for the univariate function given by $H^=(v) := H(v, v)$. Given a univariate function f, we write f_r for its right derivative, if this exists (it always will under our conditions). With this notation, we have another characterization of lookback compromises.

Theorem 11.11 *A nonnegative function $H(v, w)$ with domain $v \in [1, \infty], w \in [0, v]$ and satisfying $H(\infty, w) = \infty$ for all $w \in [0, \infty]$ is a maximal lookback compromise if and only if*

- *$H^=$ is increasing, concave, and satisfies $H^=(1) = 1$ and $H_r^=(1) \leq 1$ and*
- *for each $v \in [1, \infty)$, the function $H(v, w)$ is linear in $w \in [0, v]$, with slope equal to $H_r^=(v)$.*

For proofs of Theorems 11.10 and 11.11, see [100, sections 4 and 5]. Their "if" parts, however, are either already known to us or easy to show if we ignore the statement about maximality. The "if" part of Theorem 11.10 without maximality is claimed in Lemma 11.8. Let us replace, without loss of generality, the two entries of "$\leq$" in Protocol 11.1 by "$=$." The "if" part of Theorem 11.11 without maximality is witnessed by any strategy for Rival Skeptic that guarantees, for each n,

$$\mathcal{L}_n - \mathcal{L}_{n-1} \geq H_r^=(\mathcal{K}_{n-1}^*)(\mathcal{K}_n - \mathcal{K}_{n-1}). \tag{11.22}$$

11.4 LOOKBACKS IN FINANCIAL MARKETS

In some financial and commodity markets, an investor can buy lookback options, which insure against particular commodities, stocks, or other financial instruments losing their value during particular periods. To fix ideas, suppose y_n is the price of a particular stock at the end of the nth trading day, and set

$$y_n^* = \max_{0 \leq i \leq n} y_i.$$

One example of a lookback option on the stock is a *floating lookback call* maturing at time n; at time n, it pays the investor the difference $y_n^* - y_n$. By buying the stock and the option at time 0 and holding it to time n, the investor can count on having y_n^*, the largest amount the stock was worth over the period of his investment. But his initial investment, the cost of the stock and the cost of the floating lookback call, must be deducted from this amount, and lookback options are very expensive.

The lookback trading strategies defined in Section 11.2 provide an alternative to buying lookback options. They guarantee less, as $H(y_n^*)$ may be only a small fraction of y_n^*, but aside from their transaction cost, which is limited to the cost of periodically selling parts of the investment in the stock, they are free. They also do not depend on any statistical assumptions.

Insuring Against Loss of Capital

The case where the investor holds only a single stock can be described by this protocol:

Protocol 11.12
    $y_0 := 1$ and $\mathcal{L}_0 := 1$.
    FOR $n = 1, 2, \dots$:
        Rival Skeptic announces $M_n \in [0, \infty)$ such that $M_n y_{n-1} \leq \mathcal{L}_{n-1}$.
        Reality announces $y_n \in [0, \infty)$.
        $\mathcal{L}_n := \mathcal{L}_{n-1} + M_n(y_n - y_{n-1})$.

Here we assume for simplicity that the stock's initial price, y_0, is 1. Reality is the market, which determines the prices. Rival Skeptic is the investor; his move M_n is the number of shares he holds on nth round. By holding these shares, he is investing $M_n y_{n-1}$ in the stock while keeping the remainder of his capital, $\mathcal{L}_{n-1} - M_n y_{n-1}$, in cash. Forecaster and Skeptic are absent, because they are playing specified strategies. Forecaster has nothing to add to Reality's announcement on the preceding round, and Skeptic is merely holding the stock ($\mathcal{K}_n = y_n$ for all n). As the following proposition confirms, this means that the protocol's slackening is a lookback protocol.

Proposition 11.13   *The slackening of Protocol 11.12 is the specialization of Protocol 11.5 in which (i) $\mathcal{Y} := [0, \infty)$, (ii) for every situation $s = y_1 \ldots y_{n-1}$, $\mathcal{E}_s$ consists of the single upper expectation $\overline{\mathbf{E}}_s$ defined by*

$$\overline{\mathbf{E}}_s(f) := \inf\{\alpha | \exists M \in [0, \infty) \,\forall y \in \mathcal{Y} : f(y) \le \alpha + M(y - y_{n-1})\}, \qquad (11.23)$$

and (iii) Skeptic is constrained to always make the move $f_n(y) = y$, so that his capital $\mathcal{K}_n$ is always equal to y_n.

Proof Exercise 11.11 asks the reader to verify that (11.23) defines an upper expectation. Notice that $\overline{\mathbf{E}}_s(\alpha + M(y - y_{n-1})) = \alpha$. Assuming Rival Skeptic plays efficiently in this specialization, his move on the nth round will be of the form $\alpha + M(y - y_{n-1})$, say $\alpha + M_n(y - y_{n-1})$, and it will satisfy $\overline{\mathbf{E}}_s(\alpha + M_n(y - y_{n-1})) = \mathcal{L}_{n-1}$ or $\alpha = \mathcal{L}_{n-1}$. So $\mathcal{L}_n = \mathcal{L}_{n-1} + M_n(y - y_{n-1})$, as in Protocol 11.12. □

The lookback protocol described by Proposition 11.13 is rich, because (11.9) is satisfied by

$$E_s := \left\{ \frac{y_n}{y_{n-1}} \ge \frac{1}{a} \right\},$$

the event that the price of the stock is multiplied by $1/a$. So by Theorems 11.6 and 11.9, the lookback trading strategy for a particular $c \in [0, 1)$ and a particular lookback calibrator H allows Rival Skeptic to guarantee

$$\mathcal{L}_n \ge cy_n + (1 - c)H(y_n^*) \qquad (11.24)$$

for all n in Protocol 11.12, and this is, in the sense we have described, the best Skeptic can do.

Proposition 11.13 generalizes immediately to the case where Skeptic remains in the protocol and trades in J stocks. In this case, $\mathcal{Y} := [0, \infty)^J$, an element y of $\mathcal{Y}$ being the vector of prices of the J stocks, and Skeptic and Rival Skeptic decide how much to invest in each stock on each round. We can still use (11.23) to define the upper expectation Forecaster is required to announce, provided that $M(y - y_{n-1})$ is interpreted as the inner product between the vector M representing the number of shares of each stock the player holds and the vector $y - y_{n-1}$.

Lookback trading strategies can be applied to any security, mutual fund, or commodity traded in a market that is sufficiently liquid that an investor can find a buyer at the quoted price on any round. They can also be applied to any strategy for trading in such a market or to any other investor whose behavior the investor can imitate, provided that the strategy or investor being imitated only goes long in the securities or commodities. Any short position is liable to unlimited loss, at least in theory, and is therefore incompatible with the assumption that Skeptic and Rival Skeptic are risking only their initial capital.

Superhedging Lookback Options

Financial theorists and practitioners say that a function of the prices of certain securities is *superhedged* when a strategy for trading in securities is used to guarantee a payoff at least equal to that function. We can use this language to describe Rival Skeptic's use of a lookback trading strategy to guarantee a payoff at least equal to a lookback compromise $H(y_n^*, y_n)$: he superhedges this payoff, and the cost of the superhedging is at most 1.

More generally, an investor can use a lookback trading strategy to superhedge $G(y_n^*, y_n)$, where G is not a lookback compromise. In this case, the superhedging may have a cost exceeding 1. (In the worst case, the cost will be infinite, so that the superhedging cannot be implemented.) With this in mind, we will call the cost of superhedging $G(y_n^*, y_n)$ the upper price of G.

Consider then an arbitrary function $G(v, w)$ defined for $v \geq 1$ and $w \in [0, v]$ and a contract made at the beginning of Protocol 11.12 that pays $G(y_n^*, y_n)$ at any time n of Rival Skeptic's choosing. Such a contract is called a *perpetual American lookback* with payoff G. Its *upper price* is the infimum of the amounts that allow Rival Skeptic to superhedge the payoff:

$$\overline{\mathbb{E}}(G) := \inf\{\mathcal{T}_0 | \mathcal{T} \in \mathbf{T}, \forall n \, \forall y_1 y_2 \ldots : \mathcal{T}_n(y_1 y_2 \ldots) \geq G(y_n^*, y_n)\},$$

$\mathbf{T}$ being the set of supermartingales in Protocol 11.12 (i.e. capital processes in the slackening of Protocol 11.12 modified by allowing Skeptic to choose $\mathcal{L}_0$). This equation resembles (7.9), our definition of global expected value, but $\overline{\mathbb{E}}(G)$ is not an upper expected value, because G is a process rather than a single variable.

Corollary 11.14 *If the payoff G of a perpetual American lookback is of the form $G(v, w) = H(v) + cw$ for a nonnegative Borel function H and $c \geq 0$, then*

$$\overline{\mathbb{E}}(G) = c + \int_1^\infty H(v) v^{-2} \, dv. \tag{11.25}$$

This corollary follows from the guarantee 11.24. It is not useful for the American version of the floating lookback call: if $G(v, w) = v - w$, then $\overline{\mathbb{E}}(G) = \infty$.

We can also imagine a *perpetual European lookback*, with payoff at time ∞ equal to $G(y_\infty^*, y_\infty)$, where $y_\infty^* := \lim_{n \to \infty} y_n^*$ and

$$y_\infty := \begin{cases} \lim_{n \to \infty} y_n & \text{if the limit exists,} \\ \infty & \text{if not.} \end{cases}$$

The *upper price* of such an option is just the global upper expected value $\overline{\mathbb{E}}(G)$ as defined by (7.19).

Corollary 11.15 *If the payoff G of a perpetual European lookback has the form $G(v, w) = H(v) + cw$ for a nonnegative Borel function H and $c \geq 0$, its upper price $\overline{\mathbb{E}}(G)$ is also given by (11.25).*

The proof is left as Exercise 11.12.

11.5 CALIBRATING *P*-VALUES

We have been using the size of Skeptic's maximum past capital, $\mathcal{K}_n^*$, as our measure of his maximum past evidence against Forecaster so far. We can equivalently measure this maximum past evidence by how small $1/\mathcal{K}_n^*$ is (see the discussion of Markov's inequality in Section 6.2).

Let us write p_n^* for $1/\mathcal{K}_n^*$. We will call p_n^* Skeptic's *p-value* at the end of the nth round. We can restate Proposition 11.2 in terms of Rival Skeptic's ability to guarantee

$$\mathcal{L}_n \geq h(p_n^*) \tag{11.26}$$

for various functions $h : [0, 1] \to [0, \infty]$.

To this end, call a function $h : [0, 1] \to [0, \infty]$ a *p-value calibrator* if it is decreasing, $h(0) = \infty$, and Rival Skeptic has a strategy in Protocol 11.1 that guarantees (11.26) for $n = 0, 1, \dots$. This is equivalent to there being a lookback calibrator H such that

$$h(p) = H(1/p) \tag{11.27}$$

for all $p \in [0, 1]$. Our restatement of Proposition 11.2 is that a decreasing function $h : [0, 1] \to [0, \infty]$ is a *p*-value calibrator if and only if $h(0) = \infty$ and

$$\int_0^1 h(u) \, du \leq 1 \tag{11.28}$$

(see Exercise 11.13).

Table 11.2   Some values of the p-value calibrators rt_1 and rt_2 and of the Vovk–Sellke maximum p-ratio (VS).

p	$1/p$	$\mathrm{rt}_1(p)$	$\mathrm{rt}_2(p)$	VS(p)
0.05	20	4.5	8.9	2.5
0.01	100	10	20	8.0
0.005	200	14	28	14
0.001	1 000	32	63	53
0.0005	2 000	45	89	97
0.0001	10 000	100	200	400

All values are to two significant figures.

Let us write rt_1 and rt_2 for the p-value calibrators

$$\mathrm{rt}_1(p) := \begin{cases} p^{-1/2} & \text{if } p \le 0.25, \\ 0 & \text{otherwise,} \end{cases} \qquad \mathrm{rt}_2(p) := \begin{cases} 2p^{-1/2} & \text{if } p \le 0.0625, \\ 0 & \text{otherwise,} \end{cases}$$

obtained from the lookback calibrators RT_1 and RT_2 defined in (11.3) and (11.4) by means of the relation (11.27). Table 11.2 gives some values for these p-value calibrators. It also gives the lowest upper bound for the p-value calibrators obtained from the lookback calibrators G_κ defined in (11.5):

$$\mathrm{VS}(p) := \sup_\kappa G_\kappa(1/p) = \sup_\kappa \kappa p^{\kappa-1} = \begin{cases} -1/(ep \ln p) & \text{if } p \ge 1/e, \\ 1 & \text{otherwise.} \end{cases}$$

This bound is sometimes referred to as the Vovk–Sellke (VS) maximum p-ratio [216]; it is not a p-value calibrator, but it shows the best result that can be achieved by the p-value calibrators in the family (11.5).

The relevance of p-value calibrators to hypothesis testing in mathematical statistics can be explained using the following one-round specialization of Protocol 10.2.

Protocol 11.16
PARAMETERS: Probability measure P on measurable space $\mathcal{Y}$
  $\mathcal{K}_0 := 1$.
  Skeptic announces $f \in [0, \infty]^{\mathcal{Y}}$ such that $P(f) \le \mathcal{K}_0$.
  Reality announces $y \in \mathcal{Y}$.
  $\mathcal{K}_1 := f(y)$.

Here are two ways of testing whether the outcome y in Protocol 11.16 is consistent with the probability measure P. Both are widely used in statistical practice, but only the first qualifies as an instance of game-theoretic testing.

1. *Testing with a fixed significance level.* Skeptic tests the hypothesis that y is consistent with P by choosing a small number p (the *significance level* of the test) and a measurable subset E of $\mathcal{Y}$ such that $P(E) = p$ and betting on E by setting $f := \mathbf{1}_E/p$. If E happens (i.e. if $y \in E$), then Skeptic claims that this discredits the hypothesis and that p is a measure of the discredit. This is an instance of game-theoretic testing, because Skeptic has multiplied his initial capital of 1 by the large number $\mathcal{K}_1 = 1/p$.

2. *Testing with a p-value obtained from a test statistic.* Skeptic tests P by choosing a measurable function $T : \mathcal{Y} \to \mathbb{R}$ (called the *test statistic*) and then, after observing y, calculating the quantity

$$\pi(y) := P(T \geq T(y)). \tag{11.29}$$

If $T(y)$ is so large that $\pi(y)$ is very small, then Skeptic claims that the happening of the event $\{T \geq T(y)\}$ discredits the hypothesis and that its probability $\pi(y)$ (called the *p-value*) is a measure of the discredit. This is not an instance of game-theoretic testing, because Skeptic did not single out the event $\{T \geq T(y)\}$ in advance and bet on it. He singled out the test statistic T in advance but not the cutoff value $T(y)$.

As the following proposition shows, we obtain a measure of evidence $h(\pi(y))$ with a game-theoretic interpretation when we select in advance a *p*-value calibrator h and then, after observing y, apply it to the *p*-value $\pi(y)$.

Proposition 11.17 *Suppose P is a probability measure on $\mathcal{Y}$, $T : \mathcal{Y} \to \mathbb{R}$ is a measurable function on $\mathcal{Y}$, and h is a p-value calibrator. Then Skeptic can achieve*

$$\mathcal{K}_1 \geq h(\pi(y)) \tag{11.30}$$

in Protocol 11.16 with parameters $\mathcal{Y}$ and P.

Proof It suffices to show that $P(f) \leq 1$, where $f : \mathcal{Y} \to [0, \infty)$ is given by $f(y) := h(\pi(y))$. But when y is a random variable on $\mathcal{Y}$ with probability distribution P, the random variable π on $[0, 1]$ defined by $\pi := \pi(y)$ satisfies $P(\pi \leq p) \leq p$ for all $p \in [0, 1]$. So

$$P(f) = P(h(\pi)) = \int_0^\infty P(h(\pi) \geq t) \, dt$$

$$\leq \int_0^\infty Q(h(q) \geq t) \, dt = \int_0^1 h(p) \, dp \leq 1,$$

where Q is the uniform probability measure on $[0, 1]$ and q is a random variable with this distribution; the last inequality coincides with (11.28). $\square$

It can be shown (Exercise 11.15) that Skeptic can achieve (11.30) only when h is a p-value calibrator.

11.6 EXERCISES

Exercise 11.1   Suppose $f, g \in [0, \infty]^{\mathcal{Y}}$, $\overline{\mathbf{E}}$ is an upper expectation on $\mathcal{Y}$, and $\overline{\mathbf{E}}(f) < \infty$. Show that $\overline{\mathbf{E}}(g + \infty 1_{f=\infty}) = \overline{\mathbf{E}}(g)$.   □

Exercise 11.2   Verify that, for an increasing $H : [1, \infty) \to [0, \infty]$, (11.2) implies $H(v) \leq v$ for all v and $H(v)/v \to 0$ as $v \to \infty$.   □

Exercise 11.3   Verify that $RT_1, RT_2, 'G_\kappa$, and H_κ are increasing and right-continuous and satisfy (11.2) with equality and hence are maximal lookback calibrators.   □

Exercise 11.4   Show that $RT^\dagger(v)$ is a maximal lookback calibrator, where

$$RT^\dagger(v) := \sqrt{v} - 1 \quad \text{for all } v \in [1, \infty).$$ □

Exercise 11.5

1. Show that if P is a probability measure on $[1, \infty)$ and the function $H : [1, \infty) \to [0, \infty)$ is given by

$$H(v) := \int_{[1,v]} uP(du)$$

   (cf. Lemma 11.7) for all $v \in [1, \infty)$, then P is given by the Lebesgue–Stieltjes integral

$$P([1, v]) = \int_{[1,v]} \frac{H(du)}{u}, \quad v \in [1, \infty). \tag{11.31}$$

   Hint. See the proof of Lemma 11.7.

2. Find the probability measures that (11.31) associates with the lookback calibrators RT_1, RT_2, G_κ, H_κ, and $RT^\dagger$ (the last one defined in Exercise 11.4) and calculate their densities with respect to the Lebesgue measure.   □

Exercise 11.6   Consider the following protocol:

PARAMETERS: Nonempty sets $\mathcal{Y}_1, \mathcal{Y}_2, \ldots$
  $\mathcal{K}_0 := 1$ and $\mathcal{L}_0 := 1$.
  FOR $n = 1, 2, \ldots$:
    Forecaster announces an upper expectation $\overline{\mathbf{E}}_n$ on $\mathcal{Y}_n$.
    Skeptic announces $f_n \in [0, \infty)^{\mathcal{Y}}$ such that $\overline{\mathbf{E}}_n(f_n) \leq \mathcal{K}_{n-1}$.
    Rival Skeptic announces $g_n \in [0, \infty)^{\mathcal{Y}}$ such that $\overline{\mathbf{E}}_n(g_n) \leq \mathcal{L}_{n-1}$.
    Reality announces $y_n \in \mathcal{Y}_n$.
    $\mathcal{K}_n := f_n(y_n)$ and $\mathcal{L}_n := g_n(y_n)$.

This protocol does not qualify as a lookback protocol according to the definition given in Section 11.2, because its outcome space varies from round to round. Show, however, that Rival Skeptic has a strategy that achieves (11.1) in this protocol if and only if he has a strategy that achieves it in the lookback protocol in which $\mathcal{Y} := \bigcup_{n=1}^{\infty} \mathcal{Y}_n$ and Forecaster is constrained to choose for $\overline{\mathbf{E}}_n$ an upper expectation on $\mathcal{Y}$ such that $\overline{\mathbf{E}}(f)$ depends only on the values f takes in $\mathcal{Y}_n$. □

Exercise 11.7 Can you give an example of a lookback protocol that is not rich and of increasing functions $H : [1, \infty] \to [0, \infty]$ with $H(\infty) = \infty$ such that (11.2) does not hold yet Rival Skeptic can guarantee (11.1)? □

Exercise 11.8 The inequality (11.7) improved on the inequality $\mathcal{L}_n \geq \mathrm{RT}_1(\mathcal{K}_n^*)$ by retaining the additional term in (11.19). What are the analogous improvements for the other maximal lookback calibrators defined in Section 11.1 and Exercise 11.4? □

Exercise 11.9

1. Discuss the implications of Theorems 11.6 and 11.9 for supermartingales in a lookback protocol.
2. Consider the protocol obtained from a lookback protocol by allowing Skeptic to announce his own initial capital $\mathcal{K}_0$ before the first round. Show that the global upper expectation $\overline{\mathbb{E}}$ in this protocol satisfies

$$\overline{\mathbb{E}}(\sup_{n=0,1,\dots} H(\mathcal{T}_n)) \leq \int_1^{\infty} \frac{H(v)}{v^2}\, dv \tag{11.32}$$

for any lookback calibrator H and for any supermartingale $\mathcal{T}$ with $\mathcal{T}_0 = 1$. Show that if the lookback protocol is rich, then there is a supermartingale $\mathcal{T}$ with $\mathcal{T}_0 = 1$ such that (11.32) holds with equality. □

Exercise 11.10

1. Show that Rival Skeptic can force (11.22).
2. Explain exactly how forcing (11.22) for all n guarantees $\mathcal{L}_n \geq H(\mathcal{K}_n^*, \mathcal{K}_n)$ for all n. □

Exercise 11.11 Show that (11.23) defines an upper expectation and that the infimum is attained. □

Exercise 11.12 Prove Corollary 11.15. *Hint.* The main difficulty is how to justify the convention $y_\infty := \infty$ when the limit $\lim_{n\to\infty} y_n$ does not exist. Apply Doob's convergence theorem (Section 4.2) to show that Rival Skeptic can become infinitely rich at time ∞ (starting from an arbitrarily small initial capital) on the paths where $\lim_{n\to\infty} y_n$ does not exist. □

Exercise 11.13 Verify that if $H : [1, \infty) \to [0, \infty)$ is an integrable function and $h : (0, 1] \to (0, \infty]$ is defined by (11.27), then

$$\int_1^\infty \frac{H(v)}{v^2} \, dv = \int_0^1 h(u) \, du.$$

□

Exercise 11.14 According to (11.7), the function

$$H(v, w) := \left(\sqrt{v} + \frac{w}{\sqrt{v}} \right) \mathbf{1}_{v \geq 4}$$

qualifies as a lookback compromise. Confirm in two ways that it is maximal: using Theorem 11.10 and using Theorem 11.11.   □

Exercise 11.15 Prove that Skeptic cannot achieve (11.30) unless h is a p-value calibrator.   □

Exercise 11.16 Show any function π satisfying (11.29) for some test statistic T is a p-test in the sense that it satisfies, for each $\epsilon > 0$,

$$P(\pi \leq \epsilon) \leq \epsilon.$$

Give an example showing that not every p-test can be obtained in this way.   □

11.7 CONTEXT

The first article on lookback calibration, [101], appeared in 2011. This chapter draws on the results of that article and the more general results in [100]. The concept of a lookback trading strategy is related to previous work by a number of authors, including work on algorithmic trading by El-Yaniv et al. [124] and work on complexity and randomness by Chernov et al. [70]. We learned about the benefits of the second term in (11.19) from discussions with Alexander Shen and Nikolai Vereshchagin.

Superhedging Lookback Options

David Hobson studied the pricing of lookback options when ordinary ("plain vanilla") call and put options are priced by the market [187]. The market is too incomplete even in this case to determine exact prices for lookbacks without probabilistic assumptions, but the availability of the calls and puts makes probability-free superhedging for lookbacks more efficient, and Hobson's work has generated an extensive literature on the topic. Wouter Koolen and Vovk have explored how lookback calibration might be used by an investor who wants to buy low and sell high [228].

Harmonic Functions

The problem of finding lookback compromises discussed in Section 11.3 is closely related to the problem of characterizing functions $H : [0, \infty) \times \mathbb{R} \to \mathbb{R}$ such that $H(B_t^*, B_t)$ is a local martingale, where B is Brownian motion (or any continuous local martingale with initial value 0 and infinite quadratic variation over $[0, \infty)$). Jan Obłój and Marc Yor refer to such functions H as (B^*, B)-*harmonic*, in analogy with the standard harmonic functions $h : \mathbb{R}^d \to \mathbb{R}$, for which $h(B_t)$ is a local martingale (by Itô's formula) [296]. The processes $H(B_t^*, B_t)$ obtained for

$$H(v, w) := f(v) - f'(v)(v - w), \tag{11.33}$$

where f is a smooth function, are local martingales known as *Azéma–Yor martingales* [16]. Therefore, (11.33) is an example of a (B^*, B)-harmonic function. Notice that the functions H satisfying the conditions of Theorem 11.11 are exactly of this form, with $H^= = f$ and f' understood as right derivative. Obłój [295] has shown that all (B^*, B)-harmonic functions are of the form (11.33) for an absolutely continuous f.

Calibrating p-Values

The calculation of p-values goes back to the eighteenth century. John Arbuthnot's 1710 argument concerning male and female births in London, mentioned in Section 1.8, is sometimes cited as an example, even though the argument cites only the probability of an extreme value actually observed. In 1735 [32], Daniel Bernoulli calculated a p-value corresponding to a non-extreme value of a test statistic.

The definition (11.29) works for measurable T taking values in a linearly ordered set. For an example of what can go wrong if the codomain of T is not the real line and for careful analysis of the equivalence and non-equivalence of various definitions of p-values, see [173].

The idea of calibrating p-values can be traced back at least to 1987, when James Berger and Mohan Delampady argued that calibration cannot rescue non-Bayesian significance testing from Bayesian criticisms [30]. The p-value calibrators corresponding to the lookback calibrators G_κ given by (11.5) were first introduced in 1993 [394]; they were rediscovered in [338]. The p-value calibrators corresponding to the lookback calibrators H_κ given by (11.6) were first introduced in [348], where the relationship between p-values and Bayes factors in static and dynamic hypothesis testing is further explored.

12

Defensive Forecasting

In Part I, we emphasized strategies Skeptic can use to test the agreement between Forecaster and Reality. In this chapter, we consider how Forecaster can turn knowledge of such strategies to his own advantage. As we show, he can defeat any of the strategies for Skeptic that we studied in Part I. He can do this in a very strong sense: he can keep Skeptic's capital from growing at all. When he does this, we say that he is practicing *defensive forecasting*.

Defensive forecasting enables Forecaster to produce good forecasts when the tests the forecasts need to pass can be represented by a single strategy for Skeptic. This is often the case. In particular, there is often a single strategy for Skeptic that forces a wide range of types of calibration, so that Forecaster can produce reasonably calibrated forecasts by playing against it. If the forecasts take the form of probability measures calibrated with respect to a particular loss function, then they can be used to make decisions that perform well on average with respect to that loss function.

The main message of this chapter is that good probability forecasting and decision making – probability forecasting and decision making that achieve the $\sqrt{N}$ rate of success we expect when outcomes have known or partly known probabilities – is possible even when probabilities are not known and even if we deny their existence, provided that successive forecasts or decisions are made with knowledge of earlier outcomes. The computational complexity of algorithms suggested by our proofs may keep them from competing with less optimal but easily implemented algorithms when there are many observations. Moreover, our theorems do not claim optimality for

Game-Theoretic Foundations for Probability and Finance, First Edition. Glenn Shafer and Vladimir Vovk.
© 2019 John Wiley & Sons, Inc. Published 2019 by John Wiley & Sons, Inc.

the constants by which $\sqrt{N}$ is multiplied in their error terms. But these constants are sometimes reasonably close both to optimality and to what can be achieved in practice, and the theorems clarify the very possibility of good performance without the kind of knowledge represented by a statistical model.

While we do not assume a statistical model for outcomes, we make topological assumptions not usually used in statistical modeling. We assume that testing strategies are continuous in the forecasts, and we assume that the decision rules with which our decision strategies compete are continuous. These are reasonable assumptions. The tests and decision rules used in practice in statistical work are in fact continuous, and intuitionistic mathematics, as formulated by L.E.J. Brouwer, holds that continuity is implicit in the use of real-valued functions as idealized computable objects [53, 270].

Section 12.1 introduces defensive forecasting. We show that if the outcome space is compact and metrizable, then Forecaster can defeat a strategy for Skeptic that is continuous, or at least lower semicontinuous, in Forecaster's moves. As we explain, lower semicontinuity is a minimal idealization of the assumption that the strategy can be implemented.

Section 12.2 shows how defensive forecasting can produce calibrated forecasts. We state two theorems, one concerning asymptotic calibration, and one giving bounds on miscalibration for each finite number of rounds that can all be achieved by a single forecasting strategy. We prove the two theorems, using results from the theory of reproducing kernel Hilbert spaces (RKHSs), in Section 12.3.

Section 12.4 explains how calibrated forecasts can be used to make good decisions with respect to a specified loss function. By minimizing expected loss with respect to probability forecasts whose calibration is tailored to the loss function, we can often achieve low average loss in a sequence of decision problems. We again state both an asymptotic result and a finite-horizon result. We show that under reasonable conditions on the loss function, decisions derived from calibrated probability forecasts can perform asymptotically as well as any continuous decision rule, and we give bounds on the extent to which any particular continuous decision rule in a specified class can do better for each finite number of rounds. We prove these results in Section 12.5.

Section 12.6 discusses the extent to which the strategies whose existence we prove in Sections 12.3 and 12.5 can be identified and implemented as practical forecasting and decision-making algorithms. This depends in part on identifying kernels whose RKHSs satisfy the conditions of our theorems. We discuss kernels that can be used in our problem and describe the forecasting algorithm that can be extracted from Section 12.3's proofs when the outcome space is finite.

Section 12.7 takes a closer look at continuity for testing strategies. We show that Forecaster can defeat even a discontinuous strategy if he is allowed to announce a probability measure from which his forecast will be selected, even if the uncertainty introduced by the probability measure is arbitrarily slight. We develop this idea in a completely game-theoretic way.

This chapter's theory makes contact with various ways of thinking about universal estimation and prediction that have been developed in mathematical statistics, machine learning, and the theory of algorithmic complexity. Some of these connections are sketched in Section 12.9.

12.1 DEFEATING STRATEGIES FOR SKEPTIC

At first glance, it seems impossible to make good probability forecasts without some knowledge of the process that produces outcomes. To make this point in a precise way, consider this simple perfect-information forecasting protocol:

Protocol 12.1
> FOR $n = 1, 2, \ldots$:
> > Forecaster announces $p_n \in [0, 1]$.
> > Reality announces $y_n \in \{0, 1\}$.

If Forecaster knows nothing about what Reality will do, then how can he possibly give p_n that will necessarily accord with Reality's y_n well enough to pass statistical tests? Reality can make forecasts different from $1/2$ look systematically flawed by making each outcome fall on the opposite side of $1/2$; she sets $y_n := 1$ when $p_n < 1/2$ and $y_n := 0$ when $p_n > 1/2$. She can make forecasts equal to $1/2$ look equally bad in the aggregate by not varying her response to them, say by always responding to $p_n = 1/2$ with $y_n := 0$.

Forecaster's situation is not as hopeless as this argument makes it appear, however, because the statistical tests we usually want the p_n to pass do not take advantage of maximally contrary behavior by Reality. As we saw in Part I, these tests can be represented by strategies for Skeptic, or equivalently by nonnegative supermartingales, that are continuous as functions of the forecasts. To take advantage of Reality's switching between 1 and 0 as Forecaster makes p_n less or greater than $1/2$, Skeptic would need to switch between betting on 1 and betting on 0 with equal abruptness – i.e. discontinuously.

To illustrate how Forecaster can defeat any single given strategy for Skeptic that is continuous in Forecaster's moves, consider Protocol 12.2.

Protocol 12.2
PARAMETER: Nonempty set $\mathcal{X}$
> Skeptic announces $\mathcal{K}_0 \in \mathbb{R}$.
> FOR $n = 1, 2, \ldots$:
> > Reality announces $x_n \in \mathcal{X}$.
> > Skeptic announces a continuous function $f_n : [0, 1] \to \mathbb{R}$.
> > Forecaster announces $p_n \in [0, 1]$.
> > Reality announces $y_n \in \{0, 1\}$.
> > $\mathcal{K}_n := \mathcal{K}_{n-1} + f_n(p_n)(y_n - p_n)$.

This testing protocol differs from Protocol 1.8, which we used in Chapter 1 for testing binary probability forecasts, in two ways:

- Reality announces a signal at the beginning of each round.
- Before Forecaster announces his move, Skeptic announces how he will move as a function of Forecaster's move. This makes Skeptic's move after Forecaster superfluous, and we have removed it from the protocol.

For the purposes of this chapter, we could assume that Skeptic announces a complete strategy at the outset rather than only revealing on each round what he is going to do as a function of Forecaster's move. Our results do not require this stronger assumption, however, and it would only serve to make our notation more cumbersome, as Skeptic's move $f_n(p_n)$ in the formula for $\mathcal{K}_n$ would become an expression involving all of his opponents' previous moves, $x_1, p_1, y_1, \ldots, x_{n-1}, p_{n-1}, y_{n-1}, x_n, p_n$.

The proof of the following proposition describes the method of *defensive forecasting* for Protocol 12.2.

Proposition 12.3 (**Takemura**) *Forecaster can play in Protocol 12.2 in such a way that Skeptic's capital never increases, no matter how Skeptic and Reality play.*

Proof It suffices for Forecaster to use this simple strategy:

- if the function f_n takes the value 0, choose p_n so that $f_n(p_n) = 0$;
- if f_n is always positive, take $p_n := 1$;
- if f_n is always negative, take $p_n := 0$.   □

As this proof shows, defensive forecasting is exceedingly simple once we describe the strategy for Skeptic it defends against. We can put this simplicity on stark display by applying the method to some of the strategies for Skeptic that we emphasized in Part I. For example, if we apply it to strategies for Skeptic that force laws of large numbers in Protocol 12.2, such as the Kumon–Takemura strategy or the strategies that produce Kolmogorov's martingale and Doléans's supermartingale, we find that it often produces extreme forecasts: $p_n = 0$ if $\overline{y}_{n-1} < \overline{p}_{n-1}$ and $p_n = 1$ if the inequality goes the other way (see Exercises 12.1 and 12.2). The fact that Forecaster can make his forecasts satisfy the law of large numbers in such an uninteresting way merely underscores, however, the fact that the law of large numbers, writ large, is only one of many properties that good probability forecasts should satisfy. It is hardly enough that the overall average probability should approximate the overall frequency of 1s. We also want a good match between average probability and frequency for various subsets of the rounds, such as those where x_n has a particular value or falls in a particular subset of $\mathcal{X}$. We might even be so exiguous as to demand more subtle properties associated with the law of the iterated logarithm.

Defensive forecasting becomes interesting when it is used to defend against a strategy for Skeptic that simultaneously forces many properties of the forecasts. Here is a very rough-hewn example of how this might work. Suppose

- Forecaster has identified 10 properties that he wants his forecasts to satisfy,
- for each property, he has identified a strategy that tests the property while risking only one unit of capital, and
- he will be satisfied if none of the tests multiplies the capital it risks by more than 10.

Then Forecaster can achieve his goal by defending against the average of the 10 strategies. This average is the sum of 10 strategies, each of which begins with capital 0.1 and keeps its capital nonnegative. Because Forecaster keeps the total capital of these 10 strategies from exceeding its initial unit value, no individual strategy can increase its initial capital of 0.1 more than 10-fold. In fact, as we will see, Forecaster can usually force multiple properties much more efficiently than this.

For a general theory of defensive forecasting, we need to generalize Proposition 12.3 beyond Protocol 12.2. This can be done in a variety of ways. We might, for example, use Protocol 4.1 with a signal x_n added, thus obtaining a picture in which Forecaster uses past outcomes and other data to give a predictive mean and variance for each outcome; see the references in Section 12.9. But in this chapter we consider only protocols in which Forecaster gives an entire probability distribution for the outcome on each round. Throughout the chapter, we will assume that Reality chooses outcomes from a metrizable topological space $\mathcal{Y}$ (as usual, we omit "topological" in this context) and that Forecaster chooses a probability measure from $\mathcal{P}(\mathcal{Y})$, the set of all probability measures on the Borel σ-algebra on $\mathcal{Y}$ (see the section on terminology and notation at the end of the book).

Here is our basic forecasting protocol.

Protocol 12.4
PARAMETERS: Nonempty set $\mathcal{X}$ and metrizable space $\mathcal{Y}$
  FOR $n = 1, 2, \ldots$:
    Reality announces $x_n \in \mathcal{X}$.
    Forecaster announces $P_n \in \mathcal{P}(\mathcal{Y})$.
    Reality announces $y_n \in \mathcal{Y}$.

When we discuss testing Forecaster's P_n in Protocol 12.4, we will treat $\mathcal{P}(\mathcal{Y})$ and $\mathcal{Y} \times \mathcal{P}(\mathcal{Y})$ as topological spaces, endowing $\mathcal{P}(\mathcal{Y})$ with the topology of weak convergence and $\mathcal{Y} \times \mathcal{P}(\mathcal{Y})$ with the product topology. Most of our results will assume that $\mathcal{Y}$ is compact and metrizable. In this case, $\mathcal{P}(\mathcal{Y})$ will also be compact and metrizable [39, Theorem 6 in Appendix III], and $\mathcal{Y} \times \mathcal{P}(\mathcal{Y})$ will be as well [128, theorems 3.2.4 and 4.2.2].

The following generalization of Protocol 12.2 will serve as our framework for testing Forecaster's P_n.

Protocol 12.5

PARAMETERS: Nonempty set $\mathcal{X}$ and compact metrizable space $\mathcal{Y}$

Skeptic announces $\mathcal{K}_0 \in \mathbb{R}$.

FOR $n = 1, 2, \ldots$:

Reality announces $x_n \in \mathcal{X}$.

Skeptic announces lower semicontinuous $f_n : \mathcal{Y} \times P(\mathcal{Y}) \to \mathbb{R}$
such that $\int_{\mathcal{Y}} f_n(y, P)P(dy) \leq \mathcal{K}_{n-1}$ for all $P \in P(\mathcal{Y})$.

Forecaster announces $P_n \in P(\mathcal{Y})$.

Reality announces $y_n \in \mathcal{Y}$.

$\mathcal{K}_n := f_n(y_n, P_n)$.

Here we require Skeptic's f_n in Protocol 12.5 to be lower semicontinuous in the forecast and outcome rather than imposing the stronger condition of continuity that we used in Protocol 12.2 and Proposition 12.3. Lower semicontinuity is sufficient for our results, and it is the essential aspect of continuity when Skeptic's goal is to discredit Forecaster by achieving capital that exceeds some level c. Suppose $f_n(y_n, P_n) > c$, and suppose Skeptic wants to confirm this even though he can measure y_n and P_n with only indefinitely great precision, not infinite precision. Lower semicontinuity says that there exists a neighborhood O of (y_n, P_n) such that, for all (y, P) in that neighborhood, $f_n(y, P) > c$. So once Skeptic's measurement of (y_n, P_n) is sufficiently precise for him to know that $(y_n, P_n) \in O$, he will also know that $f_n(y_n, P_n) > c$, and this is all we can ask. We made a similar appeal to the concept of upper semicontinuity in Chapter 9 (Protocol 9.5).

Since every lower semicontinuous function on a compact set is bounded below [128, Problem 3.12.23(g)], the integral $\int_{\mathcal{Y}} f_n(y, P)P(dy)$ exists in Protocol 12.5.

Proposition 12.6 (defensive forecasting) *Forecaster can play in Protocol 12.5 in such a way that Skeptic's capital never increases, no matter how Skeptic and Reality play.*

Proof For all $P, Q \in P(\mathcal{Y})$ set

$$\phi(Q, P) := \int_{\mathcal{Y}} f_n(y, P)Q(dy).$$

By hypothesis,

$$\sup_{P \in P(\mathcal{Y})} \phi(P, P) \leq \mathcal{K}_{n-1}.$$

Because $\phi(Q, P)$ is linear in Q and lower semicontinuous in P (lower semicontinuity follows from Lemma 12.7), Ky Fan's minimax theorem (see, e.g. [6, Theorem 11.4]) implies the existence of P^* such that

$$\phi(Q, P^*) \leq \sup_{P \in P(\mathcal{Y})} \phi(P, P)$$

for all $Q \in \mathcal{P}(\mathcal{Y})$ and in particular for all Q that give probability 1 to some $y \in \mathcal{Y}$. So $f_n(y, P^*) \leq \mathcal{K}_{n-1}$ for all $y \in \mathcal{Y}$. So it suffices to set $P_n := P^*$. □

This proof used the following topological lemma.

Lemma 12.7 *Suppose $f : X \times Y \to \mathbb{R}$ is a lower semicontinuous function defined on the product of two compact metrizable spaces, X and Y. If Q is a probability measure on Y, the function $x \in X \mapsto \int_Y f(x, y)Q(dy)$ is also lower semicontinuous.*

Proof Without loss of generality we assume that f is nonnegative: every lower semicontinuous function on a compact set is bounded below [128, Problem 3.12.23(g)]. According to Hahn's theorem [128, Problem 1.7.15(c)], there exists an increasing sequence of (nonnegative) continuous functions f_n such that $f_n(x, y) \to f(x, y)$ as $n \to \infty$ for all $(x, y) \in X \times Y$. Fix metrics on X and Y that induce the given topologies. Since each f_n is uniformly continuous [128, theorem 4.3.32], the functions $\int_Y f_n(x, y)Q(dy)$ are continuous, and by the monotone convergence theorem [121, theorem 4.3.2], they converge to $\int_Y f(x, y)Q(dy)$. Therefore, again by Hahn's theorem, $\int_Y f(x, y)Q(dy)$ is lower semicontinuous. □

As we will confirm in Section 12.7, results in the spirit of Proposition 12.6 can be obtained without assuming that f_n is lower semicontinuous in P if Forecaster is allowed to randomize his move slightly. But in the meantime we will use Proposition 12.6 as stated here, applying it to strategies for Skeptic under which f_n is always continuous and therefore lower semicontinuous.

12.2 CALIBRATED FORECASTS

Proposition 12.6 gives conditions under which Forecaster can use defensive forecasting to guarantee certain points of agreement between the forecasts P_n and the outcomes y_n. What points of agreement are most important, and to what extent can Forecaster guarantee many of them at once?

Many of the points of agreement that we might seek between the P_n and the y_n can be considered instances, in a broad sense, of *calibration*, a concept that we introduced in Section 1.5. In this section we elaborate on the meaning of calibration and discuss senses in which defensive forecasting makes good calibration possible.

Consider a function $h : \mathcal{X} \times \mathcal{P}(\mathcal{Y}) \times \mathcal{Y} \to \mathbb{R}$. We say that Forecaster is *h-calibrated* if the two averages

$$\frac{1}{N} \sum_{n=1}^{N} h(x_n, P_n, y_n) \quad \text{and} \quad \frac{1}{N} \sum_{n=1}^{N} \int_{\mathcal{Y}} h(x_n, P_n, y)P_n(dy) \quad (12.1)$$

are approximately equal for large N. One way to make this precise is to require that

$$\lim_{N \to \infty} \frac{1}{N} \sum_{n=1}^{N} \left(h(x_n, P_n, y_n) - \int_{\mathcal{Y}} h(x_n, P_n, y) P_n(dy) \right) = 0, \qquad (12.2)$$

in which case we will say that Forecaster is *asymptotically h-calibrated*. We will also be interested in a stronger version of calibration: Forecaster is *h-calibrated with* $\sqrt{N}$ *accuracy* if

$$\left| \sum_{n=1}^{N} \left(h(x_n, P_n, y_n) - \int_{\mathcal{Y}} h(x_n, P_n, y) P_n(dy) \right) \right| = O(\sqrt{N}). \qquad (12.3)$$

In theorems that we will state shortly, we give conditions under which Forecaster can guarantee these types of calibration for wide classes of functions. Theorem 12.8 is concerned with asymptotic calibration, Theorem 12.9 with calibration with $\sqrt{N}$ accuracy. We prove the two theorems in Section 12.3.

The Meaning of Calibration

Why do we want the two averages in (12.1) to be approximately equal, and why do we call this calibration?

These questions are most easily answered with an example of binary probability forecasting. As usual in this case, we set $\mathcal{Y} = \{0, 1\}$ and represent a probability measure P on $\mathcal{Y}$ by its probability p for $y = 1$. To fix ideas, suppose $\mathcal{X}$ consists of images of human faces and $y = 1$ means female. Suppose we classify the images in $\mathcal{X}$ into two groups, those in which the hair appears long and those in which it appears short. Represent this classification as a mapping

$$\text{hair} : \mathcal{X} \to \{\text{long}, \text{short}\},$$

and consider the function $h : \mathcal{X} \times P(\mathcal{Y}) \times \mathcal{Y} \to \mathbb{R}$ given by

$$h(x, P, y) := \begin{cases} 1 & \text{if hair}(x) = \text{long}, \ p \in [0.7, 0.8], \text{ and } y = 1, \\ 0 & \text{otherwise.} \end{cases} \qquad (12.4)$$

Set

$$C_N := \{n \in \{1, \ldots, N\} \mid \text{hair}(x_n) = \text{long and } p_n \in [0.7, 0.8]\},$$

and consider the two averages

$$\frac{1}{|C_N|} \sum_{n=1}^{N} h(x_n, P_n, y_n) = \frac{1}{|C_N|} \sum_{n \in C_N} \mathbf{1}_{y_n=1} \qquad (12.5)$$

and

$$\frac{1}{|C_N|} \sum_{n=1}^{N} \int_{\mathcal{Y}} h(x_n, P_n, y) P_n(dy) = \frac{1}{|C_N|} \sum_{n \in C_N} p_n. \qquad (12.6)$$

The two averages in (12.1) will be approximately equal if and only if (12.5) and (12.6) are approximately equal when $|C_N|$ is non-negligible as a fraction of N. (When $|C_N|$ is negligible, the two averages in (12.1) are approximately equal because they are both approximately zero.) But (12.5) is the frequency with which the face is female on the rounds in C_N, while (12.6) is the average probability Forecaster gives to female on these rounds; being an average of $|C_N|$ numbers between 0.7 and 0.8, (12.6) is also between 0.7 and 0.8. If Forecaster is doing a good job, the frequency should approximate Forecaster's average probability. As we learned from Corollary 1.9 and Proposition 1.13, Skeptic can force this approximation. If the frequency does not approximate the average probability, say if the face turns out to be female in 90% of the cases where the hair in the image appears long and Forecaster gives a probability for female between 70% and 80%, then it is natural to say that Forecaster is uncalibrated for such cases and should recalibrate by changing his forecasts for them from the 70% to 80% range to around 90%.

Asymptotic Calibration

The following theorem gives conditions under which asymptotic calibration is possible.

Theorem 12.8 *Suppose $\mathcal{X}$ and $\mathcal{Y}$ are locally compact metrizable spaces, with $\mathcal{P}(\mathcal{Y})$ equipped with the topology of weak convergence. Then Forecaster has a strategy in Protocol 12.4 that guarantees that, for every continuous function $h : \mathcal{X} \times \mathcal{P}(\mathcal{Y}) \times \mathcal{Y} \to \mathbb{R}$,*

$$(\{x_1, x_2, \ldots\} \text{ and } \{y_1, y_2, \ldots\} \text{ are precompact})$$

$$\implies \lim_{N \to \infty} \frac{1}{N} \sum_{n=1}^{N} \left(h(x_n, P_n, y_n) - \int_{\mathcal{Y}} h(x_n, P_n, y) P_n(dy) \right) = 0. \qquad (12.7)$$

Here we call a set in a topological space *precompact* if its closure is compact. In Euclidean spaces, precompactness means boundedness. The requirement that $\{x_1, x_2, \ldots\}$ and $\{y_1, y_2, \ldots\}$ be precompact means that the information on which forecasts can be calibrated cannot be expanded indefinitely. When $\mathcal{X}$ and $\mathcal{Y}$ are Euclidean spaces, the requirement means that Forecaster knows in advance that x_n and y_n are bounded, even though he may not know bounds for them.

Although Theorem 12.8 asserts only that calibration will be achieved for continuous functions, this implies that it will also be achieved for functions that can be approximated arbitrarily closely by continuous functions. This includes the discontinuous function h given by (12.4). (The value of $h(x, p, 1)$ for hair(x) = long drops from 1 to 0 when p leaves the closed interval $[0.7, 0.8]$, but the idea involved is equally well represented by a function that makes this change sharply but continuously.)

Theorem 12.8, being asymptotic, leaves us asking how well Forecaster can be calibrated in a given finite number of rounds. This question is addressed by Theorem 12.9.

Calibration with $\sqrt{N}$ Accuracy

To bound the difference between the two averages in (12.1) for a whole class of functions, we must control the magnitude of the values taken by these functions. This leads us to consider RKHSs.

A Hilbert space $\mathcal{H}$ consisting of real-valued functions (with the standard operations of addition and scalar multiplication) on a nonempty set Z is called a *reproducing kernel Hilbert space (RKHS)* if for every $z \in Z$ the evaluation functional $h \in \mathcal{H} \mapsto h(z) \in \mathbb{R}$ is bounded. By the Riesz representation theorem, this implies that for every $z \in Z$ there exists an element $\mathsf{k}_z \in \mathcal{H}$ (z's *representer*) such that

$$h(z) = \langle \mathsf{k}_z, h \rangle_{\mathcal{H}}, \quad \forall h \in \mathcal{H}.$$

Set

$$\|\mathcal{H}\|_\infty := \sup\{h(z) | z \in Z, h \in \mathcal{H}, \text{and } \|h\|_{\mathcal{H}} \leq 1\}. \tag{12.8}$$

The Riesz representation theorem also implies that

$$\|\mathsf{k}_z\|_{\mathcal{H}} = \sup_{\|h\|_{\mathcal{H}} \leq 1} \langle \mathsf{k}_z, h \rangle_{\mathcal{H}} = \sup_{\|h\|_{\mathcal{H}} \leq 1} h(z). \tag{12.9}$$

So we can rewrite (12.8) as

$$\|\mathcal{H}\|_\infty = \sup_{z \in Z} \|\mathsf{k}_z\|_{\mathcal{H}}. \tag{12.10}$$

We say that an RKHS $\mathcal{H}$ is *bounded* if $\|\mathcal{H}\|_\infty$ is finite. This is true if and only if all the functions in $\mathcal{H}$ are bounded [363, Lemma 4.23]. If Z is a topological space and the mapping $z \in Z \mapsto \mathsf{k}_z$ is continuous, then we say that $\mathcal{H}$ is *continuous*.

Suppose $\mathcal{X}$ is a nonempty set and $\mathcal{Y}$ is a compact metrizable space. Then we call an RKHS $\mathcal{H}$ on $\mathcal{X} \times \mathcal{P}(\mathcal{Y}) \times \mathcal{Y}$ *fixed-signal continuous* if for each $x \in \mathcal{X}$, the mapping $(P, y) \in \mathcal{P}(\mathcal{Y}) \times \mathcal{Y} \mapsto \mathsf{k}_{x, P, y} \in \mathcal{H}$ is continuous. (As already mentioned in Section 12.1, $\mathcal{P}(\mathcal{Y}) \times \mathcal{Y}$ is a compact metrizable space when $\mathcal{Y}$ is.)

Theorem 12.9 *Suppose the outcome space $\mathcal{Y}$ in Protocol 12.4 is a compact metrizable space, and suppose $\mathcal{H}$ is a bounded and fixed-signal continuous RKHS on $\mathcal{X} \times P(\mathcal{Y}) \times \mathcal{Y}$. Then Forecaster has a strategy that guarantees*

$$\left| \sum_{n=1}^{N} \left(h(x_n, P_n, y_n) - \int_{\mathcal{Y}} h(x_n, P_n, y) P_n(dy) \right) \right| \leq 2\|\mathcal{H}\|_\infty \|h\|_{\mathcal{H}} \sqrt{N} \qquad (12.11)$$

for all $N \in \mathbb{N}$ and $h \in \mathcal{H}$.

The integral in (12.11) exists because $h(x, P, y)$ is continuous in $y \in \mathcal{Y}$ and $\mathcal{Y}$ is compact.

The asymptotics of Theorem 12.9 may seem surprisingly strong. By the central limit theorem, we can expect a total error of order $O(\sqrt{N})$ when we use expected values to forecast N bounded random quantities, but this is true only when we consider a single number or stopping time N. When we want to bound the total error simultaneously for all N, the law of the iterated logarithm suggests that we can hope only for a bound of the form $O(\sqrt{N \ln \ln N})$. Here we obtain bounds for all N, of order $O(\sqrt{N})$, and under the hypotheses of the theorem Forecaster can make them all hold for certain, not merely with high probability. Forecaster accomplishes this by tailoring his P_n to a particular type of calibration and to the particular RKHS.

Is *h*-Calibration Enough?

Under the regularity conditions of Proposition 12.6, our general result on defensive forecasting, Forecaster can guarantee any property of his forecasts that Skeptic can test with a strategy that uses only the information

$$x_1, P_1, y_1, \ldots, x_{n-1}, P_{n-1}, y_{n-1}, x_n, P_n \qquad (12.12)$$

to choose its bet on y_n, provided only that the outcome space $\mathcal{Y}$ is compact and metrizable and the bet on y_n is lower semicontinuous in (y_n, P_n). Only some of these properties can be expressed in terms of h-calibration. So how content should we be with Theorems 12.8 and 12.9? Will the calibrated forecasts $P_1, P_2, \ldots$ that exist according to these theorems always be good enough for our purposes?

One important limitation of h-calibration is that the difference between $h(x_n, P_n, y_n)$ and $\int_{\mathcal{Y}} h(x_n, P_n, y) P_n(dy)$ tests the agreement between y_n and P_n only conditional on the current signal x_n. It does not take the other xs and ys in (12.12) into account in the same way that it takes x_n into account. Consider, for example, a business that is trying to authenticate transactions. If y_n tells us whether or not the nth transaction is fraudulent, and fraudulent transactions come in spurts, then the business may want its probability p_n for y_n = fraud to be calibrated not only on the information x_n for the nth transaction but also on whether other recent similar transactions are fraudulent.

The practical remedy is to include in the signal x_n relevant information about the earlier xs and ys. The business trying to detect fraud could include measures of the frequency of recent fraudulent transactions in x_n. This maneuver will always have its limits, however, because our theorems cannot accommodate the inclusion of more and more information in x_n as n increases; this would violate the condition that $\{x_1, x_2, \ldots\}$ be precompact in Theorem 12.8 and inflate the constants $\|\mathcal{H}\|_\infty$ and $\|h\|_{\mathcal{H}}$ in Theorem 12.9.

A more theoretical challenge for h-calibration is offered by the law of the iterated logarithm, which goes beyond mandating the convergence in (12.2) or even the convergence rate in (12.3) by demanding that the two averages in (12.1) fluctuate above and below each other in a particular way. We might express some aspects of this demand in terms of h-calibration if we include information about such fluctuation so far in x_n, but this would fall short of providing a full test of the law.

We cannot claim, then, that all properties we might ever want probability forecasts to satisfy can be expressed in terms of h-calibration. But as we will see in Section 12.4, probabilities that are h-calibrated with respect to a particular loss function can do as well as any decision rule that uses the same signal x_n.

12.3 PROVING THE CALIBRATION THEOREMS

We will use Billy J. Pettis's definition of the integral $\int_Z f \, dP$ when $f : Z \to \mathcal{H}$ is a function taking values in a Hilbert space $\mathcal{H}$ and P is a finite measure on Z. According to this definition, $\int_Z f \, dP$ is the element I of $\mathcal{H}$ such that

$$\langle I, g \rangle_{\mathcal{H}} = \int_Z \langle f(z), g \rangle_{\mathcal{H}} P(dz)$$

for all $g \in \mathcal{H}$. The existence and uniqueness of this element is assured if Z is a compact topological space, P is defined on its Borel σ-algebra, and f is continuous [323, Definition 3.26, Theorems 3.27 and 3.20].

Proof of Theorem 12.9

We start from a corollary of Proposition 12.6 that can be thought of as a version of Kolmogorov's 1929 weak law of large numbers [221]; its proof will use a version of Kolmogorov's martingale (Section 3.1).

Lemma 12.10 *Suppose $\mathcal{X}$ is a nonempty set and $\mathcal{Y}$ is a compact metrizable space. Let $\Phi_n : \mathcal{X} \times \mathcal{P}(\mathcal{Y}) \times \mathcal{Y} \to \mathcal{H}$, $n \in \mathbb{N}$, be functions taking values in a Hilbert space*

$\mathcal{H}$ such that, for all n and x, $\Phi_n(x, P, y)$ is a continuous function of $(P, y) \in P(\mathcal{Y}) \times \mathcal{Y}$. Then Forecaster has a strategy in Protocol 12.4 that guarantees

$$\left\| \sum_{n=1}^{N} \Psi_n(x_n, P_n, y_n) \right\|_{\mathcal{H}}^2 \le \sum_{n=1}^{N} \|\Psi_n(x_n, P_n, y_n)\|_{\mathcal{H}}^2 \tag{12.13}$$

for all N, where

$$\Psi_n(x, P, y) := \Phi_n(x, P, y) - \int_y \Phi_n(x, P, y) P(dy).$$

Proof According to Proposition 12.6, it suffices to check that

$$\mathcal{M}_N := \left\| \sum_{n=1}^{N} \Psi_n(x_n, P_n, y_n) \right\|_{\mathcal{H}}^2 - \sum_{n=1}^{N} \|\Psi_n(x_n, P_n, y_n)\|_{\mathcal{H}}^2$$

is the capital process of a strategy for Skeptic in Protocol 12.5. But

$$\mathcal{M}_N - \mathcal{M}_{N-1} = \left\| \sum_{n=1}^{N-1} \Psi_n(x_n, P_n, y_n) + \Psi_N(x_N, P_N, y_N) \right\|_{\mathcal{H}}^2$$

$$- \left\| \sum_{n=1}^{N-1} \Psi_n(x_n, P_n, y_n) \right\|_{\mathcal{H}}^2 - \|\Psi_N(x_N, P_N, y_N)\|_{\mathcal{H}}^2$$

$$= \left\langle 2 \sum_{n=1}^{N-1} \Psi_n(x_n, P_n, y_n), \Psi_N(x_N, P_N, y_N) \right\rangle_{\mathcal{H}}$$

$$= \langle A, \Psi_N(x_N, P_N, y_N) \rangle_{\mathcal{H}},$$

where $A := 2 \sum_{n=1}^{N-1} \Psi_n(x_n, P_n, y_n)$. The element A of $\mathcal{H}$ is known at the beginning of the Nth round. So it suffices to show that the function

$$f_N : (y, P) \mapsto \mathcal{K}_{N-1} + \langle A, \Psi_N(x_N, P, y) \rangle_{\mathcal{H}}$$

is a valid move for Skeptic on the Nth round. First, it is lower semicontinuous: by Lemma 12.11, $\int_y \Phi_N(x, P, y) P(dy)$ is continuous in P; whence Ψ_N and hence also $\langle A, \Psi_N(x_N, P, y) \rangle_{\mathcal{H}}$ are continuous in (P, y). Second, it satisfies $\int_y f_N(y, P) P(dy) \le \mathcal{K}_{N-1}$ for all $P \in P(\mathcal{Y})$, because

$$\int_y \langle A, \Psi_N(x_N, P, y) \rangle_{\mathcal{H}} P(dy) = \left\langle A, \int_y \Psi_N(x_N, P, y) P(dy) \right\rangle_{\mathcal{H}} = \langle A, 0 \rangle_{\mathcal{H}} = 0$$

by the definition of the Pettis integral. □

The proof of Lemma 12.10 used the following lemma.

Lemma 12.11   *Suppose $\mathcal{Y}$ is a compact metrizable space and $\Phi : P(\mathcal{Y}) \times \mathcal{Y} \to \mathcal{H}$ is a continuous mapping into a Hilbert space $\mathcal{H}$. The mapping $P \in P(\mathcal{Y}) \mapsto \int_{\mathcal{Y}} \Phi(P, y) P(dy)$ is also continuous.*

Proof   We want to show that $\int_{\mathcal{Y}} \Phi(P_n, y) P_n(dy) \to \int_{\mathcal{Y}} \Phi(P, y) P(dy)$ as $P_n \to P$. We have

$$\left\| \int_{\mathcal{Y}} \Phi(P_n, y) P_n(dy) - \int_{\mathcal{Y}} \Phi(P, y) P(dy) \right\|_{\mathcal{H}}$$

$$\leq \left\| \int_{\mathcal{Y}} \Phi(P_n, y) P_n(dy) - \int_{\mathcal{Y}} \Phi(P, y) P_n(dy) \right\|_{\mathcal{H}}$$

$$+ \left\| \int_{\mathcal{Y}} \Phi(P, y) P_n(dy) - \int_{\mathcal{Y}} \Phi(P, y) P(dy) \right\|_{\mathcal{H}} .$$

The first addend on the right-hand side is bounded above by

$$\int_{\mathcal{Y}} \| \Phi(P_n, y) - \Phi(P, y) \|_{\mathcal{H}} P_n(dy)$$

[323, theorem 3.29], and this tends to zero since Φ is uniformly continuous for any metric inducing the topology [128, theorem 4.3.32]. The second addend tends to zero by the continuity of the mapping $Q \in P(\mathcal{Y}) \mapsto \int_{\mathcal{Y}} f(y) Q(dy)$ for a continuous f [48, III.3.4, Corollary].   □

The following variation on Lemma 12.11 will be needed later.

Lemma 12.12   *Suppose $\mathcal{X}$ and $\mathcal{Y}$ are compact metrizable spaces and $\Phi : \mathcal{X} \times P(\mathcal{Y}) \times \mathcal{Y} \to \mathcal{H}$ is a continuous mapping into a Hilbert space $\mathcal{H}$. The mapping $(x, P) \in \mathcal{X} \times P(\mathcal{Y}) \mapsto \int_{\mathcal{Y}} \Phi(x, P, y) P(dy)$ is also continuous.*

Proof   Let $x_n \to x$ and $P_n \to P$ as $n \to \infty$. To prove

$$\int_{\mathcal{Y}} \Phi(x_n, P_n, y) P_n(dy) \to \int_{\mathcal{Y}} \Phi(x, P, y) P(dy)$$

we can use an argument similar to that in Lemma 12.11 applied to

$$\left\| \int_{\mathcal{Y}} \Phi(x_n, P_n, y) P_n(dy) - \int_{\mathcal{Y}} \Phi(x, P, y) P(dy) \right\|_{\mathcal{H}}$$

$$\leq \left\| \int_{\mathcal{Y}} \Phi(x_n, P_n, y) P_n(dy) - \int_{\mathcal{Y}} \Phi(x, P, y) P_n(dy) \right\|_{\mathcal{H}}$$

$$+ \left\| \int_{\mathcal{Y}} \Phi(x, P, y) P_n(dy) - \int_{\mathcal{Y}} \Phi(x, P, y) P(dy) \right\|_{\mathcal{H}} .$$ □

Theorem 12.9 will follow from the following lemma, which we derive from Lemma 12.10 by putting $k_{x,P,y}$, the representer of $(x, P, y) \in \mathcal{X} \times P(\mathcal{Y}) \times \mathcal{Y}$, in the role of $\Phi_n(x, P, y)$. Set

$$k_{x,P} := \int_{\mathcal{Y}} k_{x,P,y} P(dy); \qquad (12.14)$$

the function $P \mapsto k_{x,P}$ will be continuous in P by the fixed-signal continuity of $\mathcal{H}$ and Lemma 12.11.

Lemma 12.13 *Suppose $\mathcal{Y}$ is a compact metrizable space and $\mathcal{H}$ is a fixed-signal continuous RKHS on $\mathcal{X} \times P(\mathcal{Y}) \times \mathcal{Y}$. Then Forecaster has a strategy in Protocol 12.4 that guarantees*

$$\left| \sum_{n=1}^{N} \left(h(x_n, P_n, y_n) - \int_{\mathcal{Y}} h(x_n, P_n, y) P_n(dy) \right) \right|$$

$$\leq \|h\|_{\mathcal{H}} \sqrt{\sum_{n=1}^{N} \|k_{x_n, P_n, y_n} - k_{x_n, P_n}\|_{\mathcal{H}}^2}$$

for all N and all $h \in \mathcal{H}$.

Proof Using Lemma 12.10 (with all Ψ_n equal, $\Psi_n(x, P, y) := k_{x,P,y} - k_{x,P}$), we obtain

$$\left| \sum_{n=1}^{N} \left(h(x_n, P_n, y_n) - \int_{\mathcal{Y}} h(x_n, P_n, y) P_n(dy) \right) \right|$$

$$= \left| \sum_{n=1}^{N} \left(\langle h, k_{x_n, P_n, y_n} \rangle_{\mathcal{H}} - \int_{\mathcal{Y}} \langle h, k_{x_n, P_n, y} \rangle_{\mathcal{H}} P_n(dy) \right) \right|$$

$$= \left| \left\langle h, \sum_{n=1}^{N} (k_{x_n, P_n, y_n} - k_{x_n, P_n}) \right\rangle_{\mathcal{H}} \right| \leq \|h\|_{\mathcal{H}} \left\| \sum_{n=1}^{N} (k_{x_n, P_n, y_n} - k_{x_n, P_n}) \right\|_{\mathcal{H}}$$

$$\leq \|h\|_{\mathcal{H}} \sqrt{\sum_{n=1}^{N} \|k_{x_n, P_n, y_n} - k_{x_n, P_n}\|_{\mathcal{H}}^2}.$$

$\square$

To deduce Theorem 12.9 from Lemma 12.13, notice that $\|k_{x,P,y}\|_{\mathcal{H}} \leq \|\mathcal{H}\|_{\infty}$ (by (12.10)), $\|k_{x,P}\|_{\mathcal{H}} \leq \int_{\mathcal{Y}} \|k_{x,P,y}\|_{\mathcal{H}} P(dy) \leq \|\mathcal{H}\|_{\infty}$ (by [323, theorem 3.29]), and therefore,

$$\sum_{n=1}^{N} \|k_{x_n, P_n, y_n} - k_{x_n, P_n}\|_{\mathcal{H}}^2 \leq 4 \|\mathcal{H}\|_{\infty}^2 N.$$

This completes the proof.

Proof of Theorem 12.8

Let us say that an RKHS $\mathcal{H}$ on a topological space Z is *universal* if it is continuous and for every compact subset A of Z every continuous function on A can be arbitrarily well approximated in the $C(A)$ metric by functions in $\mathcal{H}$: for any $\epsilon > 0$ and any continuous function $f : A \to \mathbb{R}$, there is $h \in \mathcal{H}$ such that $\sup_{z \in A} |f(z) - h(z)| \le \epsilon$. (When Z is compact, this agrees with [363, Definition 4.52].)

Lemma 12.14   *For any compact metrizable space Z, there is a universal bounded RKHS on Z.*

A sketch of a proof of this lemma is given in [363, Exercise 4.13]. The lemma allows us to derive the conclusion of Theorem 12.8 easily from Theorem 12.9 under the additional assumption that $\mathcal{X}$ and $\mathcal{Y}$ are compact. As we mentioned in Section 12.1, $P(\mathcal{Y})$ and, therefore, $\mathcal{X} \times P(\mathcal{Y}) \times \mathcal{Y}$ are compact and metrizable when $\mathcal{X}$ and $\mathcal{Y}$ are. So by the lemma there is a universal and bounded RKHS $\mathcal{H}$ on $\mathcal{X} \times P(\mathcal{Y}) \times \mathcal{Y}$. Let h be a continuous real-valued function on $\mathcal{X} \times P(\mathcal{Y}) \times \mathcal{Y}$. The antecedent of (12.7) being true by the assumption that the spaces are compact, our task is to prove that the consequent holds for h. By the universality of $\mathcal{H}$, we can choose $g \in \mathcal{H}$ at a distance at most ϵ from h in the $C(\mathcal{X} \times P(\mathcal{Y}) \times \mathcal{Y})$ metric, and then by Theorem 12.9 we have

$$\limsup_{N \to \infty} \left| \frac{1}{N} \sum_{n=1}^{N} \left(h(x_n, P_n, y_n) - \int_{\mathcal{Y}} h(x_n, P_n, y) P_n(dy) \right) \right|$$

$$\le \limsup_{N \to \infty} \left| \frac{1}{N} \sum_{n=1}^{N} \left(g(x_n, P_n, y_n) - \int_{\mathcal{Y}} g(x_n, P_n, y) P_n(dy) \right) \right| + 2\epsilon = 2\epsilon.$$

Since this can be done for any $\epsilon > 0$, the proof for the case of compact $\mathcal{X}$ and $\mathcal{Y}$ is complete.

The rest of the proof is based on the following *game of removal* $G(Z)$, an abstract version of the doubling trick [64].

Protocol 12.15
PARAMETER: Topological space Z
> FOR $n = 1, 2, \ldots$:
>> Remover announces compact $K_n \subseteq Z$.
>> Evader announces $z_n \notin K_n$.
> WINNER: Evader if the set $\{z_1, z_2, \ldots\}$ is precompact; Remover otherwise.

Intuitively, the goal of Evader is to avoid being pushed to infinity. Without loss of generality we will assume that Remover always announces an increasing sequence of compact sets: $K_1 \subseteq K_2 \subseteq \cdots$.

Lemma 12.16   **(Gruenhage)** *Remover has a winning strategy in $G(Z)$ if Z is a locally compact and paracompact space.*

Proof We will follow the proof of Theorem 4.1 from [172] (the easy direction). If Z is locally compact and σ-compact, there exists an increasing sequence $K_1 \subseteq K_2 \subseteq \cdots$ of compact sets covering Z, and each K_n can be extended to compact K_n^* so that $\text{Int } K_n^* \supseteq K_n$ [128, theorem 3.3.2]. Remover will win $G(Z)$ choosing $K_1^*, K_2^*, \ldots$ as his moves: indeed, $\text{Int } K_n^*$ will form an open cover of $\overline{\{z_1, z_2, \ldots\}}$, and, if Evader won, it would have a finite subcover, thus violating the rules of the game.

If Z is the sum of locally compact σ-compact spaces Z_s, $s \in S$, Remover plays, for each $s \in S$, the strategy described in the previous paragraph on the subsequence of Evader's moves belonging to Z_s. If Evader chooses $z_n \in Z_s$ for infinitely many Z_s, those Z_s will form an open cover of $\overline{\{z_1, z_2, \ldots\}}$ without a finite subcover. If z_n are chosen from only finitely many Z_s, there will be infinitely many z_n chosen from some Z_s, and the result of the previous paragraph can be applied. It remains to remember that each locally compact paracompact space can be represented as the sum of locally compact σ-compact subsets [128, theorem 5.1.27].   □

Now we can complete our proof of the theorem. Forecaster's strategy ensuring (12.7) will be constructed from his strategies $S(A, B)$ ensuring the consequent of (12.7) under the condition $\forall n : (x_n, y_n) \in A \times B$ for given compact sets $A \subseteq \mathcal{X}$ and $B \subseteq \mathcal{Y}$ and from Remover's winning strategy in $G(\mathcal{X} \times \mathcal{Y})$. By Stone's theorem [128, theorem 5.1.3], all metrizable spaces are paracompact, and the product of two locally compact spaces is locally compact [128, theorem 3.3.13]; therefore, Lemma 12.16 is applicable to $G(\mathcal{X} \times \mathcal{Y})$. Without loss of generality we assume that Remover's moves are always of the form $A \times B$ for $A \subseteq \mathcal{X}$ and $B \subseteq \mathcal{Y}$. Forecaster will be playing two games in parallel: the probability forecasting game and the auxiliary game of removal $G(\mathcal{X} \times \mathcal{Y})$ (in the role of Evader). We will use the fact that the restriction to $A \times P(B) \times B$ of any continuous function $g : \mathcal{X} \times P(\mathcal{Y}) \times \mathcal{Y}$ is continuous (Exercise 12.4), where $P(B)$ is identified with a subset of $P(\mathcal{Y})$ in the natural way (the probability measures assigning measure zero to $\mathcal{Y} \setminus B$).

Forecaster asks Remover to make his first move $A_1 \times B_1$ in the game of removal. He then plays the probability forecasting game using the strategy $S(A_1, B_1)$ until Reality chooses $(x_n, y_n) \notin A_1 \times B_1$ (forever if Reality never chooses such (x_n, y_n)). As soon as such (x_n, y_n) is chosen, Forecaster, in his Evader hat, announces (x_n, y_n) and notes Remover's move (A_2, B_2). He then plays the probability forecasting game using the strategy $S(A_2, B_2)$ until Reality chooses $(x_n, y_n) \notin A_2 \times B_2$, etc.

Let us check that this strategy for Forecaster will always ensure (12.7). If Reality chooses (x_n, y_n) outside Forecaster's current $A_k \times B_k$ finitely often, the consequent of (12.7) will be satisfied. If Reality chooses (x_n, y_n) outside Forecaster's current $A_k \times B_k$ infinitely often, the set $\{(x_n, y_n) | n \in \mathbb{N}\}$ will not be precompact, and so the antecedent of (12.7) will be violated.

12.4 USING CALIBRATED FORECASTS FOR DECISION MAKING

As we explained in the introduction to this chapter, calibrated probability forecasts can be used to make decisions whose degree of success is evaluated by a loss function. The decision maker selects a decision by minimizing expected loss with respect to probability forecasts that are calibrated with respect to the loss function. In this section we discuss the extent to which this can succeed.

As in Section 12.2, we state both an asymptotic and a finite-horizon result. Theorem 12.18 gives conditions on move spaces and the loss function under which a decision maker has a strategy that performs just as well asymptotically as any decision rule that is continuous in the signal. Theorem 12.21 gives bounds for each finite number of rounds such that the decision maker has a strategy whose performance will be within all the bounds from the performance of any decision rule in a particular class; as in the case of calibration, the bounds are of order $\sqrt{N}$. In both cases, we explain how the conditions on the move spaces and the loss function can be relaxed when both the decision maker and the decision rules with which she competes are allowed to randomize. We postpone proofs to Section 12.5.

Here is our basic protocol for decision making.

Protocol 12.17
PARAMETERS: Nonempty topological spaces $\mathcal{X}$, Γ, and $\mathcal{Y}$
   FOR $n = 1, 2, \ldots$:
      Reality announces $x_n \in \mathcal{X}$.
      Decision Maker announces $\gamma_n \in \Gamma$.
      Reality announces $y_n \in \mathcal{Y}$.

We call Γ the *decision space*, and we call Decision Maker's moves $\gamma_1, \gamma_2, \ldots$ her *decisions*. The success of each decision is measured by a *loss function* $\lambda : \mathcal{X} \times \Gamma \times \mathcal{Y} \to \mathbb{R}$. In most applications, $\lambda(x, \gamma, y)$ does not depend on x. We are mainly interested in the case where $\mathcal{X}$, Γ, and $\mathcal{Y}$ are locally compact metrizable spaces, the prime examples being Euclidean spaces and their open and closed subsets.

We call any mapping from $\mathcal{X}$ to Γ a *decision rule*. We suppose that Decision Maker competes with a wide class of decision rules. Her goal is to achieve

$$\sum_{n=1}^{N} \lambda(x_n, \gamma_n, y_n) \lesssim \sum_{n=1}^{N} \lambda(x_n, D(x_n), y_n)$$

for $N = 1, 2, \ldots$ and as many decision rules $D : \mathcal{X} \to \Gamma$ as possible. This is the usual goal in the literature on online learning [64].

Asymptotically Optimal Decisions

Let us call a loss function λ *large at infinity* if for all $x \in \mathcal{X}$ and $y \in \mathcal{Y}$,

$$\lim_{\substack{x' \to x, \, y' \to y \\ \gamma \to \infty}} \lambda(x', \gamma, y') = \infty.$$

This means that for every $M \in \mathbb{R}$, there is a neighborhood O_x of x, a neighborhood O_y of y, and a compact subset C of Γ such that $\lambda(x', \gamma, y') \geq M$ when $x' \in O_x$, $y' \in O_y$, and $\gamma \notin C$. Faraway γ perform poorly. The loss functions usually used in statistics and machine learning satisfy this condition, and of course all loss functions satisfy it when Γ is compact.

Theorem 12.18 *Suppose $\mathcal{X}$ and $\mathcal{Y}$ are locally compact metrizable spaces, Γ is a convex subset of a Fréchet space, and the loss function $\lambda(x, \gamma, y)$ is continuous, large at infinity, and convex in $\gamma \in \Gamma$. Then Decision Maker has a strategy in Protocol 12.17 that guarantees that*

($\{x_1, x_2, \dots\}$ and $\{y_1, y_2, \dots\}$ are precompact)

$$\Longrightarrow \limsup_{N \to \infty} \left(\frac{1}{N} \sum_{n=1}^{N} \lambda(x_n, \gamma_n, y_n) - \frac{1}{N} \sum_{n=1}^{N} \lambda(x_n, D(x_n), y_n) \right) \leq 0 \qquad (12.15)$$

for every continuous decision rule $D : \mathcal{X} \to \Gamma$.

Here is a simple example that satisfies the hypotheses of Theorem 12.18: $\mathcal{X} = \mathbb{R}^K$, $\Gamma = \mathcal{Y} = \mathbb{R}^L$, and $\lambda(x, \gamma, y) := \|y - \gamma\|$.

For examples showing that Theorem 12.18 can fail if we drop the assumption that D is continuous and that the consequent of (12.15) can fail if its antecedent does not hold, see Exercise 12.5. The theorem also fails if the assumption that λ is convex in γ is dropped (see [208, Theorem 9]).

The usefulness of Theorem 12.18 can also be limited by a paucity of continuous decision rules. For example, if $\Gamma = \{0, 1\}$ and $\mathcal{X}$ is connected, there are no non-constant continuous decision rules. However, this paucity of decision rules is more apparent than real (see [208, 391]).

We can obtain a more broadly applicable result by allowing the decision rules and Decision Maker's strategy to be randomized. Let us call a function $D : \mathcal{X} \to \mathcal{P}(\Gamma)$ a *randomized decision rule*; we are usually interested in continuous randomized decision rules. To allow a randomized strategy for Decision Maker, we substitute $\mathcal{P}(\Gamma)$ for Γ in Protocol 12.17.

Protocol 12.19

PARAMETERS: Nonempty topological spaces $\mathcal{X}$, Γ, and $\mathcal{Y}$
 FOR $n = 1, 2, \ldots$:
  Reality announces $x_n \in \mathcal{X}$.
  Decision Maker announces $\gamma_n \in \mathcal{P}(\Gamma)$.
  Reality announces $y_n \in \mathcal{Y}$.

Corollary 12.20 *Suppose $\mathcal{X}$ and $\mathcal{Y}$ are locally compact metrizable spaces, Γ is a metrizable space, and λ is a loss function that is continuous and large at infinity. Then Decision Maker has a strategy in Protocol 12.19 that guarantees*

$$(\{x_1, x_2, \ldots\} \text{ and } \{y_1, y_2, \ldots\} \text{ are precompact})$$

$$\implies \left(\limsup_{N \to \infty} \left(\frac{1}{N} \sum_{n=1}^{N} \lambda(x_n, g_n, y_n) - \frac{1}{N} \sum_{n=1}^{N} \lambda(x_n, d_n, y_n) \right) \le 0 \text{ a.s.} \right) \quad (12.16)$$

for every continuous randomized decision rule D, where $g_1, g_2, \ldots, d_1, d_2, \ldots$ are independent random variables with g_n distributed as γ_n and d_n distributed as $D(x_n)$, $n \in \mathbb{N}$.

The "a.s." in (12.16) means "almost surely" in the measure-theoretic sense; (12.16) has probability 1 when these random variables are independent and distributed as specified. The statement of the theorem assumes in particular that Decision Maker's randomization is independent of D's randomization.

For any metrizable space $\mathcal{X}$, any discrete (e.g. finite) subset $\{x_1, x_2, \ldots\}$ of $\mathcal{X}$, and any sequence $\gamma_n \in \mathcal{P}(\Gamma)$ of probability measures on Γ, there is a continuous randomized decision rule D such that $D(x_n) = \gamma_n$ for all n. (For example, set $D(x) := \sum_n \phi_n(x)\gamma_n$, where $\phi_n : \mathcal{X} \to [0, 1]$, $n \in \mathbb{N}$, are continuous functions with disjoint supports such that $\phi_n(x_n) = 1$ for all n.) So there is no shortage of randomized decision rules.

In principle, we can replace the measure-theoretic notions (independent random variables and probability 1) with their game-theoretic counterparts in Corollary 12.20. But this would require a further elaboration of Protocol 12.19; we need a player who tests the probabilities. Instead of pausing to develop this elaboration in detail, we refer the reader to Section 12.7, where a similar elaboration is developed to explain game-theoretically the randomization we use to deal with discontinuous strategies for defensive forecasting's Skeptic.

Decisions with $\sqrt{N}$ Regret

In order to state a finite-horizon counterpart to Theorem 12.18, we introduce the following notation:

- Given a bounded RKHS $\mathcal{G}$ on $\mathcal{X} \times \mathcal{Y}$, we define $\|g\|_{\mathcal{G}}$ for all $g : \mathcal{X} \times \mathcal{Y} \to \mathbb{R}$ by giving it its usual meaning when $g \in \mathcal{G}$ and setting $\|g\|_{\mathcal{G}} := \infty$ when $g \notin \mathcal{G}$.
- Given a decision rule $D : \mathcal{X} \to \Gamma$ and a loss function $\lambda : \mathcal{X} \times \Gamma \times \mathcal{Y} \to \mathbb{R}$, we define $\lambda_D : \mathcal{X} \times \mathcal{Y} \to \mathbb{R}$ by

$$\lambda_D(x, y) := \lambda(x, D(x), y).$$

- We use the usual uniform norm for functions on $\mathcal{X} \times \Gamma \times \mathcal{Y}$:

$$\|\lambda\|_{\infty} := \sup_{x \in \mathcal{X}, \gamma \in \Gamma, y \in \mathcal{Y}} |\lambda(x, \gamma, y)|.$$

This is finite if λ is continuous and $\mathcal{X}$, Γ, and $\mathcal{Y}$ are compact. But see Exercise 12.11.

In consonance with our definition of fixed-signal continuity right before Theorem 12.9, we call an RKHS $\mathcal{G}$ on $\mathcal{X} \times \mathcal{Y}$ *fixed-signal continuous* if $\mathcal{Y}$ is a topological space and the mapping $y \in \mathcal{Y} \mapsto k_{x,y} \in \mathcal{G}$ is continuous for each $x \in \mathcal{X}$.

Theorem 12.21 *Suppose $\mathcal{X}$ and $\mathcal{Y}$ are compact metrizable spaces, Γ is a convex compact subset of a topological vector space, and $\lambda(x, \gamma, y)$ is continuous in (x, γ, y) and convex in γ. Suppose $\mathcal{G}$ is a bounded and fixed-signal continuous RKHS on $\mathcal{X} \times \mathcal{Y}$. Then Decision Maker has a strategy in Protocol 12.17 that guarantees*

$$\sum_{n=1}^{N} \lambda(x_n, \gamma_n, y_n) \le \sum_{n=1}^{N} \lambda(x_n, D(x_n), y_n)$$

$$+ 2\sqrt{\|\lambda\|_{\infty}^2 + \|\mathcal{G}\|_{\infty}^2}(\|\lambda_D\|_{\mathcal{G}} + 1)\sqrt{N} + 1 \qquad (12.17)$$

for all decision rules D and all $N \in \mathbb{N}$.

As we will see in Section 12.5, an application of Hoeffding's inequality gives the following corollary.

Corollary 12.22 *Suppose $\mathcal{X}$, Γ, and $\mathcal{Y}$ are compact metrizable spaces, λ is continuous, and $\mathcal{G}$ is a bounded and fixed-signal continuous RKHS on $\mathcal{X} \times \mathcal{Y}$. Suppose $N \in \mathbb{N}$ and $\delta \in (0, 1)$. Then Decision Maker has a strategy in Protocol 12.19 that guarantees*

$$\sum_{n=1}^{N} \lambda(x_n, g_n, y_n) \le \sum_{n=1}^{N} \lambda(x_n, d_n, y_n)$$

$$+ 2\sqrt{\|\lambda\|_{\infty}^2 + \|\mathcal{G}\|_{\infty}^2}(\|\lambda_D\|_{\mathcal{G}} + 1)\sqrt{N} + \|\lambda\|_{\infty}\sqrt{8 \ln \frac{1}{\delta}}\sqrt{N} + 1$$

with probability at least $1 - \delta$ for every continuous randomized decision rule $D : \mathcal{X} \to \mathcal{P}(\Gamma)$, where g_n and d_n are independent random variables distributed as γ_n and $D(x_n)$, respectively.

Theorem 12.21 and Corollary 12.22 are nonvacuous only when $\lambda_D \in \mathcal{G}$; otherwise the right-hand sides of the inequalities are infinite. One case where $\lambda_D \in \mathcal{G}$ is guaranteed is where $\mathcal{G}$ is a Sobolev space H^m (to be defined in Section 12.6, m being a measure of required smoothness) and D and λ are very smooth. For example, it suffices to assume that $\mathcal{X}$ is a compact set in a Euclidean space $\mathbb{R}^K$, Γ is a Euclidean space, $\mathcal{Y}$ is finite, D is the restriction to $\mathcal{X}$ of an infinitely differentiable function defined on a neighborhood of $\mathcal{X}$ (in $\mathbb{R}^K$), and $\lambda = \lambda(x, P, y)$ is the restriction to $\mathcal{X} \times \Gamma \times \mathcal{Y}$ of a smooth (in x and P) function defined on a neighborhood of $\mathcal{X} \times \Gamma \times \mathcal{Y}$.

12.5   PROVING THE DECISION THEOREMS

Our proofs of Theorems 12.18 and 12.21 will use the calibration theorems to show that Forecaster has a strategy for choosing the $P_1, P_2, \ldots$ that guarantees for all N both

$$\sum_{n=1}^{N} \lambda(x_n, D(x_n), y_n) \approx \sum_{n=1}^{N} \int_{\mathcal{Y}} \lambda(x_n, D(x_n), y) P_n(dy) \qquad (12.18)$$

for all the decision rules D under consideration and

$$\sum_{n=1}^{N} \lambda(x_n, G(x_n, P_n), y_n) \approx \sum_{n=1}^{N} \int_{\mathcal{Y}} \lambda(x_n, G(x_n, P_n), y) P_n(dy), \qquad (12.19)$$

where $G(x, P)$ is the element of Γ that minimizes the expected value of $\lambda(x, P, y)$ with respect to P (assuming it exists). Decision Maker's strategy will be to take $G(x_n, P_n)$ as her move γ_n; the left-hand side of (12.19) will be approximately less than or equal to the left-hand side of (12.18) because the right hand side of (12.19) is less than or equal to the right hand side of (12.18) by the definition of the function G.

In our informal discussion, we will call (12.18) *calibration for the decision rules*, and we will call (12.19) *calibration for Decision Maker*.

One difficulty encountered by the argument just sketched is that there may not be an element $\gamma = G(x, P)$ of Γ that exactly minimizes the expected value of $\lambda(x, \gamma, \cdot)$ with respect to P, and even if there is, the function $G(x, P)$ thus defined may be discontinuous, perhaps making $\lambda(x, G(x, P), y)$ a discontinuous function of (x, P, y) and blocking the application of the calibration theorems. But as we will show, under compactness and convexity conditions that are natural but stronger than the assumptions made in the theorems we are undertaking to prove, there do exist functions $G(x, P)$ that are continuous and approximately minimize the expected loss with respect to P. This will be enough for us to turn our informal argument into a proof of a preliminary weak form of Theorem 12.18.

Existence of a Continuous Approximate Choice Function

Let us write $\lambda(x, \gamma, P)$ for the expected value of the loss $\lambda(x, \gamma, y)$ with respect to a probability measure P on $\mathcal{Y}$:

$$\lambda(x, \gamma, P) := \int_{\mathcal{Y}} \lambda(x, \gamma, y) P(dy). \tag{12.20}$$

Let us say that $G : \mathcal{X} \times P(\mathcal{Y}) \to \Gamma$ is a (precise) *choice function* if it satisfies

$$\lambda(x, G(x, P), P) = \inf_{\gamma \in \Gamma} \lambda(x, \gamma, P), \quad \forall x \in \mathcal{X}, P \in P(\mathcal{Y}). \tag{12.21}$$

As we already noted, there may not be a continuous function that satisfies (12.21); cf. Exercises 12.6 and 12.7.

The following lemma establishes the existence of approximate choice functions under natural compactness and convexity conditions.

Lemma 12.23 *Suppose X is a paracompact topological space, Y is a nonempty compact convex subset of a topological vector space, and $f : X \times Y \to \mathbb{R}$ is a continuous function convex in $y \in Y$ for each $x \in X$. Then for any $\epsilon > 0$ there exists a continuous function $g : X \to Y$ such that*

$$\forall x \in X : \quad f(x, g(x)) \leq \inf_{y \in Y} f(x, y) + \epsilon. \tag{12.22}$$

Proof Each $(x, y) \in X \times Y$ has a neighborhood $A_{x,y} \times B_{x,y}$ such that $A_{x,y}$ and $B_{x,y}$ are open sets in X and Y, respectively, and

$$\sup_{A_{x,y} \times B_{x,y}} f - \inf_{A_{x,y} \times B_{x,y}} f < \frac{\epsilon}{2}.$$

For each $x \in X$ choose a finite subcover of the cover $\{A_{x,y} \times B_{x,y} | y \in Y\}$ of $\{x\} \times Y$ and let A_x be the intersection of all $A_{x,y}$ in this subcover. The sets A_x constitute an open cover of X such that

$$(x_1 \in A_x, x_2 \in A_x) \implies |f(x_1, y) - f(x_2, y)| < \frac{\epsilon}{2} \tag{12.23}$$

for all $x, x_1, x_2 \in X$ and $y \in Y$. Since X is paracompact, there exists [128, Theorem 5.1.9] a locally finite partition $\{\phi_i | i \in I\}$ of unity subordinated to the open cover of X formed by all A_x, $x \in X$. For each $i \in I$ choose $x_i \in X$ such that $\phi_i(x_i) > 0$ (without loss of generality we can assume that such x_i exists for each $i \in I$) and choose $y_i \in \arg\min_y f(x_i, y)$. Now we can set

$$g(x) := \sum_{i \in I} \phi_i(x) y_i.$$

Inequality (12.22) follows, by (12.23) and the convexity of $f(x, y)$ in y, from

$$\forall y \in Y : \quad f(x, g(x)) = f\left(x, \sum_i \phi_i(x)y_i\right) \leq \sum_i \phi_i(x)f(x, y_i)$$

$$\leq \sum_i \phi_i(x)f(x_i, y_i) + \frac{\epsilon}{2} \leq \sum_i \phi_i(x)f(x_i, y) + \frac{\epsilon}{2}$$

$$\leq \sum_i \phi_i(x)f(x, y) + \epsilon = f(x, y) + \epsilon,$$

where i ranges over the finite number of $i \in I$ for which $\phi_i(x)$ is nonzero.   □

A Weak Form of Theorem 12.18

To further develop the informal argument at the beginning of this section, we prove the following weak form of Theorem 12.18. (The antecedent of (12.15) disappears because of the proposition's compactness assumptions.)

Proposition 12.24 *Suppose $\mathcal{X}$ and $\mathcal{Y}$ are compact metrizable spaces, Γ is a nonempty compact convex subset of a topological vector space, and the loss function $\lambda(x, \gamma, y)$ is continuous in (x, γ, y), large at infinity, and convex in $\gamma \in \Gamma$. Then for every $\epsilon > 0$, Decision Maker has a strategy in Protocol 12.17 that guarantees that*

$$\limsup_{N \to \infty} \left(\frac{1}{N}\sum_{n=1}^{N} \lambda(x_n, \gamma_n, y_n) - \frac{1}{N}\sum_{n=1}^{N} \lambda(x_n, D(x_n), y_n)\right) \leq \epsilon$$

for every continuous decision rule $D : \mathcal{X} \to \Gamma$.

Proof The hypotheses of the proposition imply that $\lambda(x, \gamma, P)$ is convex in γ, and by Lemma 12.12 it is continuous in $(x, \gamma, P) \in \mathcal{X} \times \Gamma \times P(\mathcal{Y})$. Taking $\mathcal{X} \times P(\mathcal{Y})$ as X and Γ as Y, we see from Lemma 12.23 that for each $\epsilon > 0$ there exists a continuous function G satisfying

$$\lambda(x, G(x, P), P) \leq \inf_{\gamma \in \Gamma} \lambda(x, \gamma, P) + \epsilon, \quad \forall x \in \mathcal{X}, P \in P(\mathcal{Y}). \qquad (12.24)$$

This function G is our approximate choice function.

If Decision Maker calculates P_n satisfying (12.7) in Theorem 12.8, and sets $\gamma_n := G(x_n, P_n)$, then, for every continuous $D : \mathcal{X} \to \Gamma$,

$$\sum_{n=1}^{N} \lambda(x_n, \gamma_n, y_n) = \sum_{n=1}^{N} \lambda(x_n, G(x_n, P_n), y_n)$$

$$= \sum_{n=1}^{N} \lambda(x_n, G(x_n, P_n), P_n)$$

$$+ \sum_{n=1}^{N} (\lambda(x_n, G(x_n, P_n), y_n) - \lambda(x_n, G(x_n, P_n), P_n))$$

$$= \sum_{n=1}^{N} \lambda(x_n, G(x_n, P_n), P_n) + o(N) \le \sum_{n=1}^{N} \lambda(x_n, D(x_n), P_n) + \epsilon N + o(N)$$

$$= \sum_{n=1}^{N} \lambda(x_n, D(x_n), y_n)$$

$$- \sum_{n=1}^{N} (\lambda(x_n, D(x_n), y_n) - \lambda(x_n, D(x_n), P_n)) + \epsilon N + o(N)$$

$$= \sum_{n=1}^{N} \lambda(x_n, D(x_n), y_n) + \epsilon N + o(N). \tag{12.25}$$

The third and last equalities in the chain (12.25) used Theorem 12.8. The third inequality used calibration for Decision Maker while the last equality used calibration for the decision rules. The inequality used (12.24). $\qquad \square$

Proof of Theorem 12.21

To establish the bounds stated in Theorem 12.21, we need to develop bounds for the two types of calibration and then combine them. The combination amounts to constructing an RKHS from two separate RKHSs. The following corollary of Lemma 12.10 allows this combination of calibration for Decision Maker and calibration for the decision rules.

Corollary 12.25 *Suppose a_0 and a_1 are positive constants, $\mathcal{H}_0$ and $\mathcal{H}_1$ are Hilbert spaces, $\mathcal{X}$ is a nonempty set, and $\mathcal{Y}$ is a compact metrizable space. Suppose further, for $n \in \mathbb{N}$ and $j = 0, 1$, that $\Phi_{n,j} : \mathcal{X} \times P(\mathcal{Y}) \times \mathcal{Y} \to \mathcal{H}_j$ is continuous in $(P, y) \in P(\mathcal{Y}) \times \mathcal{Y}$ for fixed $x \in \mathcal{X}$. Then Forecaster has a strategy in Protocol 12.4 that guarantees*

$$\left\| \sum_{n=1}^{N} \Psi_{n,j}(x_n, P_n, y_n) \right\|_{\mathcal{H}_j}^2$$

$$\le \frac{1}{a_j} \sum_{n=1}^{N} (a_0 \|\Psi_{n,0}(x_n, P_n, y_n)\|_{\mathcal{H}_0}^2 + a_1 \|\Psi_{n,1}(x_n, P_n, y_n)\|_{\mathcal{H}_1}^2)$$

for all N and for both $j = 0$ and $j = 1$, where

$$\Psi_{n,j}(x, P, y) := \Phi_{n,j}(x, P, y) - \int_{\mathcal{Y}} \Phi_{n,j}(x, P, y) P(dy).$$

Proof Let $\mathcal{H}$ be the Cartesian product $\mathcal{H}_0 \times \mathcal{H}_1$ equipped with the inner product

$$\langle h, h' \rangle_{\mathcal{H}} = \langle (h_0, h_1), (h_0', h_1') \rangle_{\mathcal{H}} := \sum_{j=0}^{1} a_j \langle h_j, h_j' \rangle_{\mathcal{H}_j}. \tag{12.26}$$

Define $\Phi : \mathcal{X} \times P(\mathcal{Y}) \times \mathcal{Y} \to \mathcal{H}$ by

$$\Phi_n(x, P, y) := (\Phi_{n,0}(x, P, y), \Phi_{n,1}(x, P, y)). \tag{12.27}$$

It is clear that $\Phi_n(x, P, y)$ is continuous in (P, y) for all n. Applying the strategy of Lemma 12.10 to it and using (12.13), we obtain

$$a_j \left\| \sum_{n=1}^{N} \Psi_{n,j}(x_n, P_n, y_n) \right\|_{\mathcal{H}_j}^2 \leq \left\| \left(\sum_{n=1}^{N} \Psi_{n,0}(x_n, P_n, y_n), \sum_{n=1}^{N} \Psi_{n,1}(x_n, P_n, y_n) \right) \right\|_{\mathcal{H}}^2$$

$$= \left\| \sum_{n=1}^{N} \Psi_n(x_n, P_n, y_n) \right\|_{\mathcal{H}}^2 \leq \sum_{n=1}^{N} \| \Psi_n(x_n, P_n, y_n) \|_{\mathcal{H}}^2$$

$$= \sum_{n=1}^{N} \sum_{j=0}^{1} a_j \| \Psi_{n,j}(x_n, P_n, y_n) \|_{\mathcal{H}_j}^2. \qquad \square$$

Suppose $\mathcal{X}$, Γ, and $\mathcal{Y}$ are compact metrizable spaces and $\mathcal{G}$ is a fixed-signal continuous RKHS on $\mathcal{X} \times \mathcal{Y}$. Let $G_n : \mathcal{X} \times P(\mathcal{Y}) \to \Gamma$ be a sequence of continuous functions (approximate choice functions) satisfying

$$\lambda(x, G_n(x, P), P) < \inf_{\gamma \in \Gamma} \lambda(x, \gamma, P) + 2^{-n}, \quad \forall x \in \mathcal{X}, P \in P(\mathcal{Y}). \tag{12.28}$$

They exist by (12.24). We will use Corollary 12.25 with $\mathcal{H}_0 = \mathbb{R}$, $\mathcal{H}_1 = \mathcal{G}$, $a_0 = a_1 = 1$, and

$$\Psi_{n,0}(x, P, y) := \lambda(x, G_n(x, P), y) - \lambda(x, G_n(x, P), P), \tag{12.29}$$

$$\Psi_{n,1}(x, P, y) := \mathsf{k}_{x,y} - \mathsf{k}_{x,P}, \tag{12.30}$$

where $\mathsf{k}_{x,y}$ is the representer of (x, y) in $\mathcal{G}$ and $\mathsf{k}_{x,P} := \int_{\mathcal{Y}} \mathsf{k}_{x,y} P(dy)$. Notice that $\mathsf{k}_{x,P}$ is continuous in P and, therefore, $\Psi_{n,1}(x, P, y)$ is continuous in (P, y) by Lemma 12.11. We have

$$\| \Psi_{n,0}(x, P, y) \|_{\mathbb{R}} = | \Psi_{n,0}(x, P, y) | \leq 2 \| \lambda \|_{\infty},$$

$$\| \Psi_{n,1}(x, P, y) \|_{\mathcal{G}} \leq 2 \| \mathcal{G} \|_{\infty}, \tag{12.31}$$

where the second inequality follows from (12.10) and $\|k_{x,P}\|_{\mathcal{G}} \le \|\mathcal{G}\|_\infty$ (see Exercise 12.10). This instantiation of Corollary 12.25 yields the two following lemmas, the first stipulating calibration for Decision Maker and the second stipulating calibration for the decision rules.

Lemma 12.26 *The strategy for Forecaster in Corollary 12.25, based on (12.29) and (12.30), guarantees*

$$\left| \sum_{n=1}^{N} (\lambda(x_n, G_n(x_n, P_n), y_n) - \lambda(x_n, G_n(x_n, P_n), P_n)) \right| \le 2\sqrt{\|\lambda\|_\infty^2 + \|\mathcal{G}\|_\infty^2} \sqrt{N}.$$

Proof This follows from

$$\left| \sum_{n=1}^{N} (\lambda(x_n, G_n(x_n, P_n), y_n) - \lambda(x_n, G_n(x_n, P_n), P_n)) \right|^2 \le 4 \sum_{n=1}^{N} (\|\lambda\|_\infty^2 + \|\mathcal{G}\|_\infty^2). \qquad \square$$

Lemma 12.27 *The strategy for Forecaster in Corollary 12.25, based on (12.29) and (12.30), guarantees*

$$\left| \sum_{n=1}^{N} (\lambda(x_n, D(x_n), y_n) - \lambda(x_n, D(x_n), P_n)) \right| \le 2\sqrt{\|\lambda\|_\infty^2 + \|\mathcal{G}\|_\infty^2} \|\lambda_D\|_{\mathcal{G}} \sqrt{N}.$$

Proof We reason as in the proof of Lemma 12.13:

$$\left| \sum_{n=1}^{N} (\lambda(x_n, D(x_n), y_n) - \lambda(x_n, D(x_n), P_n)) \right|$$

$$= \left| \sum_{n=1}^{N} (\lambda_D(x_n, y_n) - \lambda_D(x_n, P_n)) \right|$$

$$= \left| \sum_{n=1}^{N} \langle \lambda_D, k_{x_n, y_n} - k_{x_n, P_n} \rangle_{\mathcal{G}} \right| \le \|\lambda_D\|_{\mathcal{G}} \left\| \sum_{n=1}^{N} (k_{x_n, y_n} - k_{x_n, P_n}) \right\|_{\mathcal{G}}$$

$$\le 2\|\lambda_D\|_{\mathcal{G}} \sqrt{\sum_{n=1}^{N} (\|\lambda\|_\infty^2 + \|\mathcal{G}\|_\infty^2)} = 2\sqrt{\|\lambda\|_\infty^2 + \|\mathcal{G}\|_\infty^2} \|\lambda_D\|_{\mathcal{G}} \sqrt{N}. \qquad \square$$

We can now complete the proof of Theorem 12.21 by setting $\gamma_n := G_n(x_n, P_n)$, where the P_n are produced by the strategy for Forecaster in Corollary 12.25 based on (12.29) and (12.30). Following (12.25) and using the two preceding lemmas, we obtain

$$\sum_{n=1}^{N} \lambda(x_n, \gamma_n, y_n) = \sum_{n=1}^{N} \lambda(x_n, G_n(x_n, P_n), y_n)$$

$$= \sum_{n=1}^{N} \lambda(x_n, G_n(x_n, P_n), P_n)$$

$$+ \sum_{n=1}^{N} (\lambda(x_n, G_n(x_n, P_n), y_n) - \lambda(x_n, G_n(x_n, P_n), P_n))$$

$$\leq \sum_{n=1}^{N} \lambda(x_n, G_n(x_n, P_n), P_n) + 2\sqrt{\|\lambda\|_\infty^2 + \|\mathcal{G}\|_\infty^2} \sqrt{N}$$

$$\leq \sum_{n=1}^{N} \lambda(x_n, D(x_n), P_n) + 2\sqrt{\|\lambda\|_\infty^2 + \|\mathcal{G}\|_\infty^2} \sqrt{N} + 1$$

$$= \sum_{n=1}^{N} \lambda(x_n, D(x_n), y_n) + 2\sqrt{\|\lambda\|_\infty^2 + \|\mathcal{G}\|_\infty^2} \sqrt{N} + 1$$

$$- \sum_{n=1}^{N} (\lambda(x_n, D(x_n), y_n) - \lambda(x_n, D(x_n), P_n))$$

$$\leq \sum_{n=1}^{N} \lambda(x_n, D(x_n), y_n) + 2\sqrt{\|\lambda\|_\infty^2 + \|\mathcal{G}\|_\infty^2} (\|\lambda_D\|_\mathcal{G} + 1)\sqrt{N} + 1,$$

the second inequality using (12.28).

Proof of Corollary 12.22

The difference $\lambda(x_n, g_n, y_n) - \lambda(x_n, d_n, y_n)$ never exceeds $2\|\lambda\|_\infty$ in absolute value. So by Hoeffding's inequality ([64, Corollary A.1] or, in the game-theoretic form, Corollary 3.8),

$$\mathbb{P}\left\{ \sum_{n=1}^{N}(\lambda(x_n, g_n, y_n) - \lambda(x_n, d_n, y_n)) - \sum_{n=1}^{N}(\lambda(x_n, \gamma_n, y_n) - \lambda(x_n, D(x_n), y_n)) > t \right\}$$

$$\leq \exp\left(-\frac{t^2}{8\|\lambda\|_\infty^2 N}\right)$$

when $t > 0$. Choosing t satisfying

$$\exp\left(-\frac{t^2}{8\|\lambda\|_\infty^2 N}\right) = \delta,$$

i.e.

$$t := \|\lambda\|_\infty \sqrt{8 \, \ln\frac{1}{\delta}}\sqrt{N},$$

we obtain the statement of the corollary.

Proof of Theorem 12.18

To complete the proof of Theorem 12.18, we use the following variation on Lemma 12.27.

Lemma 12.28 *The strategy for Forecaster in Corollary 12.25, based on (12.29) and (12.30), guarantees*

$$\left|\sum_{n=1}^{N}\left(f(x_n, y_n) - \int_{\mathcal{Y}} f(x_n, y) P_n(dy)\right)\right| \leq 2\sqrt{\|\lambda\|_\infty^2 + \|\mathcal{G}\|_\infty^2}\,\|f\|_{\mathcal{G}}\sqrt{N}$$

for any $f \in \mathcal{G}$.

Proof Following the proof of Lemma 12.27:

$$\left|\sum_{n=1}^{N}\left(f(x_n, y_n) - \int_{\mathcal{Y}} f(x_n, y) P_n(dy)\right)\right|$$

$$= \left|\sum_{n=1}^{N}\langle f, \mathsf{k}_{x_n,y_n} - \mathsf{k}_{x_n,P_n}\rangle_{\mathcal{G}}\right| \leq \|f\|_{\mathcal{G}}\left\|\sum_{n=1}^{N}(\mathsf{k}_{x_n,y_n} - \mathsf{k}_{x_n,P_n})\right\|_{\mathcal{G}}$$

$$\leq 2\|f\|_{\mathcal{G}}\sqrt{\sum_{n=1}^{N}(\|\lambda\|_\infty^2 + \|\mathcal{G}\|_\infty^2)} = 2\sqrt{\|\lambda\|_\infty^2 + \|\mathcal{G}\|_\infty^2}\,\|f\|_{\mathcal{G}}\sqrt{N}.$$

$\square$

As in the proof of Theorem 12.8, we first assume that the spaces $\mathcal{X}$, Γ, and $\mathcal{Y}$ in the statement of Theorem 12.18 are compact. In this case, as we now show, Theorem 12.21 guarantees the consequent of (12.15) for all continuous decision rules when it is applied to a universal and bounded RKHS on $\mathcal{X} \times \mathcal{Y}$.

Indeed, let $\mathcal{G}$ be such an RKHS, and fix a continuous decision rule $D : \mathcal{X} \to \Gamma$. For any $\epsilon > 0$, we can find a function $f \in \mathcal{G}$ that is ϵ-close in $C(\mathcal{X} \times \mathcal{Y})$ to

$\lambda_D = \lambda(x, D(x), y)$. Reasoning as in (12.25) and in the similar chain in the proof of Theorem 12.21 (with the same definition of approximate choice functions), we obtain

$$\sum_{n=1}^{N} \lambda(x_n, \gamma_n, y_n) = \sum_{n=1}^{N} \lambda(x_n, G_n(x_n, P_n), y_n)$$

$$= \sum_{n=1}^{N} \lambda(x_n, G_n(x_n, P_n), P_n)$$

$$+ \sum_{n=1}^{N} (\lambda(x_n, G_n(x_n, P_n), y_n) - \lambda(x_n, G_n(x_n, P_n), P_n))$$

$$\leq \sum_{n=1}^{N} \lambda(x_n, G_n(x_n, P_n), P_n) + 2\sqrt{\|\lambda\|_\infty^2 + \|G\|_\infty^2} \sqrt{N}$$

$$\leq \sum_{n=1}^{N} \lambda(x_n, D(x_n), P_n) + 2\sqrt{\|\lambda\|_\infty^2 + \|G\|_\infty^2} \sqrt{N+1}$$

$$= \sum_{n=1}^{N} \lambda(x_n, D(x_n), y_n) + 2\sqrt{\|\lambda\|_\infty^2 + \|G\|_\infty^2} \sqrt{N+1}$$

$$- \sum_{n=1}^{N} (\lambda(x_n, D(x_n), y_n) - \lambda(x_n, D(x_n), P_n))$$

$$\leq \sum_{n=1}^{N} \lambda(x_n, D(x_n), y_n) + 2\sqrt{\|\lambda\|_\infty^2 + \|G\|_\infty^2} \sqrt{N+1}$$

$$- \sum_{n=1}^{N} \left(f(x_n, y_n) - \int_{\mathcal{Y}} f(x_n, y) P_n(y) \right) + 2\epsilon N$$

$$\leq \sum_{n=1}^{N} \lambda(x_n, D(x_n), y_n) + 2\sqrt{\|\lambda\|_\infty^2 + \|G\|_\infty^2} (\|f\|_G + 1) \sqrt{N+1} + 2\epsilon N.$$

We can see that

$$\limsup_{N\to\infty} \left(\frac{1}{N} \sum_{n=1}^{N} \lambda(x_n, \gamma_n, y_n) - \frac{1}{N} \sum_{n=1}^{N} \lambda(x_n, D(x_n), y_n) \right) \leq 2\epsilon;$$

since this is true for any $\epsilon > 0$, the consequent of (12.15) holds.

Now we must remove the assumption that $\mathcal{X}$, Γ, and $\mathcal{Y}$ are compact. We will need the following lemmas.

Lemma 12.29 *Let λ be a loss function that is large at infinity. For each pair of compact sets $A \subseteq \mathcal{X}$ and $B \subseteq \mathcal{Y}$ and each $M \in \mathbb{R}$, there exists a compact set $C \subseteq \Gamma$ such that*

$$\forall x \in A, \gamma \notin C, y \in B : \quad \lambda(x, \gamma, y) > M. \tag{12.32}$$

Proof For each pair of points $x \in A$ and $y \in B$, fix the neighborhoods $O_x \ni x$ and $O_y \ni y$ and a compact set $C(x, y) \subseteq \Gamma$ such that $\lambda(x', \gamma, y') > M$ when $x' \in O_x$, $y' \in O_y$, and $\gamma \notin C$. Since the sets O_x form an open cover of A and the sets O_y form an open cover of B, we can find finite subcovers $\{O_{x_1}, \ldots, O_{x_m}\}$ and $\{O_{y_1}, \ldots, O_{y_n}\}$. Clearly

$$C := \bigcup_{\substack{i=1,\ldots,m \\ j=1,\ldots,n}} C(O_{x_i}, O_{y_j})$$

satisfies (12.32). $\qquad\square$

Lemma 12.30 *Under the conditions of Theorem 12.18, for each pair of compact sets $A \subseteq \mathcal{X}$ and $B \subseteq \mathcal{Y}$, there exists a compact set $C = C(A, B) \subseteq \Gamma$ such that for each continuous decision rule $D : \mathcal{X} \to \Gamma$, there exists a continuous decision rule $D' : \mathcal{X} \to C$ that dominates D in the sense that*

$$\forall x \in A, y \in B : \quad \lambda(x, D'(x), y) \le \lambda(x, D(x), y). \tag{12.33}$$

Proof Assume, without loss of generality, that A and B are nonempty. Fix any $\gamma_0 \in \Gamma$. Let

$$M_1 := \sup_{(x,y)\in A\times B} \lambda(x, \gamma_0, y),$$

let $C_1 \subseteq \Gamma$ be a compact set such that

$$\forall x \in A, \gamma \notin C_1, y \in B : \quad \lambda(x, \gamma, y) > M_1 + 1,$$

let

$$M_2 := \sup_{(x,\gamma,y)\in A\times C_1\times B} \lambda(x, \gamma, y),$$

and let $C_2 \subseteq \Gamma$ be a compact set such that

$$\forall x \in A, \gamma \notin C_2, y \in B : \quad \lambda(x, \gamma, y) > M_2 + 1.$$

(We have been using repeatedly Lemma 12.29.) It is obvious that $M_1 \le M_2$ and $\gamma_0 \in C_1 \subseteq C_2$.

Let us now check that C_1 lies inside the interior of C_2. Indeed, for any fixed $(x, y) \in A \times B$ and $\gamma \in C_1$, we have $\lambda(x, \gamma, y) \leq M_2$; since $\lambda(x, \gamma', y) > M_2 + 1$ for all $\gamma' \notin C_2$, some neighborhood of γ will lie completely in C_2.

Let $D : \mathcal{X} \to \Gamma$ be a continuous decision rule. We will show that (12.33) holds for some continuous decision rule D' taking values in the compact set C_2. Namely, we define

$$D'(x) := f(D(x))D(x) + (1 - f(D(x)))\gamma_0,$$

where $f : \Gamma \to [0, 1]$ is a continuous function such that $f = 1$ on C_1 and $f = 0$ on C_2^c (such an f exists by the Tietze–Uryson theorem [128, theorem 2.1.8]). Assuming C_2 convex (which can be done by [323, Theorem 3.20(c)]), we can see that D' indeed takes values in C_2.

It remains to check (12.33):

$$\lambda(x, D'(x), y) = \lambda(x, f(D(x))D(x) + (1 - f(D(x)))\gamma_0, y)$$

$$\leq f(D(x))\lambda(x, D(x), y) + (1 - f(D(x)))\lambda(x, \gamma_0, y)$$

$$= \lambda(x, D(x), y) + (1 - f(D(x)))(\lambda(x, \gamma_0, y) - \lambda(x, D(x), y))$$

$$\leq \lambda(x, D(x), y),$$

where the first inequality follows from the convexity of $\lambda(x, \gamma, y)$ in γ and the second inequality follows from

$$f(D(x)) < 1 \implies D(x) \notin C_1 \implies \lambda(x, \gamma_0, y) \leq M_1 < M_1 + 1 \leq \lambda(x, D(x), y). \quad \square$$

For each pair of compact sets $A \subseteq \mathcal{X}$ and $B \subseteq \mathcal{Y}$, fix a compact set $C(A, B) \subseteq \Gamma$ as in Lemma 12.30. In analogy with the proof of Theorem 12.8, Decision Maker's strategy ensuring (12.15) is constructed from Remover's winning strategy in $G(\mathcal{X} \times \mathcal{Y})$ (see Protocol 12.15) and from Decision Maker's strategies $S(A, B)$ outputting predictions $\gamma_n \in C(A, B)$ and ensuring the consequent of (12.15) for $D : A \to C(A, B)$ under the assumption that $(x_n, y_n) \in A \times B$. Remover's moves are assumed to be of the form $A \times B$ for compact $A \subseteq \mathcal{X}$ and $B \subseteq \mathcal{Y}$. Decision Maker is simultaneously playing the game of removal $G(\mathcal{X} \times \mathcal{Y})$ as Evader.

Decision Maker asks Remover to make his first move $A_1 \times B_1$ in the game of removal. Decision Maker then plays the prediction game using the strategy $S(A_1, B_1)$ until Reality chooses $(x_n, y_n) \notin A_1 \times B_1$ (forever if Reality never chooses such (x_n, y_n)). As soon as such (x_n, y_n) is chosen, Decision Maker announces (x_n, y_n) in the game of removal and notes Remover's response (A_2, B_2). She then continues playing the prediction game using the strategy $S(A_2, B_2)$ until Reality chooses $(x_n, y_n) \notin A_2 \times B_2$, etc.

Let us check that this strategy for Decision Maker will always ensure (12.15). If Reality chooses (x_n, y_n) outside Decision Maker's current $A_k \times B_k$ finitely often, the consequent of (12.15) will be satisfied for all continuous $D : \mathcal{X} \to C(A_K, B_K)$ $((A_K, B_K)$ being Remover's last move) and so, by Lemma 12.30, for all continuous

$D : \mathcal{X} \to \Gamma$. If Reality chooses (x_n, y_n) outside Decision Maker's current $A_k \times B_k$ infinitely often, the set of (x_n, y_n), $n \in \mathbb{N}$, will not be precompact, and so the antecedent of (12.15) will be violated.

Proof of Corollary 12.20

Set

$$\lambda(x, \gamma, y) := \int_\Gamma \lambda(x, g, y)\gamma(dg), \tag{12.34}$$

where γ is a probability measure on Γ. Take this as the loss function for a new instance of Protocol 12.17 with the decision space $\mathcal{P}(\Gamma)$. When γ ranges over $\mathcal{P}(C)$ (identified with the subset of $\mathcal{P}(\Gamma)$ consisting of the measures concentrated on C) for compact C, the loss function (12.34) is continuous by Lemma 12.12 (namely, its special case where Φ does not depend on P). We need the following analogue of Lemma 12.30.

Lemma 12.31 *Under the conditions of Corollary 12.20, for each pair of compact sets $A \subseteq \mathcal{X}$ and $B \subseteq \mathcal{Y}$, there exists a compact set $C = C(A, B) \subseteq \Gamma$ such that for each continuous randomized decision rule $D : \mathcal{X} \to \mathcal{P}(\Gamma)$, there exists a continuous randomized decision rule $D' : \mathcal{X} \to \mathcal{P}(C)$ such that (12.33) holds (D' dominates D "on average").*

Proof Define γ_0, C_1, and C_2 as in the proof of Lemma 12.30. Fix a continuous function $f_1 : \Gamma \to [0, 1]$ such that $f_1 = 1$ on C_1 and $f_1 = 0$ on C_2^c (it exists by the Tietze–Uryson theorem [128, theorem 2.1.8]). Set $f_2 := 1 - f_1$. Let $D : \mathcal{X} \to \mathcal{P}(\Gamma)$ be a continuous randomized decision rule. For each $x \in \mathcal{X}$, split $D(x)$ into two measures on Γ absolutely continuous with respect to $D(x)$: $D_1(x)$ with Radon–Nikodym density f_1 and $D_2(x)$ with Radon–Nikodym density f_2; set

$$D'(x) := D_1(x) + |D_2(x)|\delta_{\gamma_0}$$

(letting $|P| := P(\Gamma)$ for $P \in \mathcal{P}(\Gamma)$). It is clear that D' is continuous (in the topology of weak convergence, as usual), takes values in $\mathcal{P}(C_2)$, and

$$
\begin{aligned}
\lambda(x, D'(x), y) &= \int_\Gamma \lambda(x, \gamma, y)f_1(\gamma)D(x)(d\gamma) + \lambda(x, \gamma_0, y)\int_\Gamma f_2(\gamma)D(x)(d\gamma) \\
&\le \int_\Gamma \lambda(x, \gamma, y)f_1(\gamma)D(x)(d\gamma) + \int_\Gamma M_1 f_2(\gamma)D(x)(d\gamma) \\
&\le \int_\Gamma \lambda(x, \gamma, y)f_1(\gamma)D(x)(d\gamma) + \int_\Gamma \lambda(x, \gamma, y)f_2(\gamma)D(x)(d\gamma) \\
&= \lambda(x, D(x), y)
\end{aligned}
$$

for all $(x, y) \in A \times B$. $\qquad\square$

Fix one of the mappings $(A, B) \mapsto C(A, B)$ whose existence is asserted by the lemma. We will prove that Decision Maker's strategy of the previous subsection with $P(C(A, B))$ in place of $C(A, B)$ satisfies the requirement of Corollary 12.20 in the new instance of Protocol 12.17. Let $D : \mathcal{X} \to P(\Gamma)$ be a continuous randomized decision rule, i.e. a continuous decision rule in the new instance. Let (A_K, B_K) be Remover's last move (if Remover makes infinitely many moves, the antecedent of (12.16) is false, and there is nothing to prove), and let $D' : \mathcal{X} \to P(C(A_K, B_K))$ be a continuous randomized decision rule satisfying (12.33) with $A := A_K$ and $B := B_K$. From some n on the randomized strategy produces $\gamma_n \in P(\Gamma)$ concentrated on $C(A_K, B_K)$, and they will satisfy

$$\limsup_{N \to \infty} \left(\frac{1}{N} \sum_{n=1}^{N} \lambda(x_n, \gamma_n, y_n) - \frac{1}{N} \sum_{n=1}^{N} \lambda(x_n, D(x_n), y_n) \right)$$

$$\leq \limsup_{N \to \infty} \left(\frac{1}{N} \sum_{n=1}^{N} \lambda(x_n, \gamma_n, y_n) - \frac{1}{N} \sum_{n=1}^{N} \lambda(x_n, D'(x_n), y_n) \right) \leq 0. \quad (12.35)$$

The loss function is bounded in absolute value on the compact set $A_K \times (C(A_K, B_K) \cup D(A_K)) \times B_K$ by a constant c. The measure-theoretic law of the iterated logarithm for martingales (see [370] or the stronger game-theoretic version in Chapter 5) implies that

$$\limsup_{N \to \infty} \frac{\left| \sum_{n=1}^{N} (\lambda(x_n, g_n, y_n) - \lambda(x_n, \gamma_n, y_n)) \right|}{\sqrt{2c^2 N \ln \ln N}} \leq 1,$$

$$\limsup_{N \to \infty} \frac{\left| \sum_{n=1}^{N} (\lambda(x_n, d_n, y_n) - \lambda(x_n, D(x_n), y_n)) \right|}{\sqrt{2c^2 N \ln \ln N}} \leq 1$$

with probability 1. Combining the last two inequalities with (12.35) gives

$$\limsup_{N \to \infty} \left(\frac{1}{N} \sum_{n=1}^{N} \lambda(x_n, g_n, y_n) - \frac{1}{N} \sum_{n=1}^{N} \lambda(x_n, d_n, y_n) \right) \leq 0 \text{ a.s.,}$$

which is the consequent of (12.16).

12.6 FROM THEORY TO ALGORITHM

Our four theorems concerning calibration and decision making, the asymptotic Theorems 12.8 and 12.18 and the finite-horizon Theorems 12.9 and 12.21, assert the existence of strategies that achieve certain goals without saying explicitly how to find such strategies and implement them as practical algorithms. But some guidance can be extracted from the proofs.

The greatest obstacle to deriving practical algorithms from our theory is the mathematical programming problem posed by Proposition 12.6, the fundamental result on defensive forecasting. All four theorems rely on Forecaster finding the minimax strategy whose existence it asserts. This is feasible when the outcome space $\mathcal{Y}$ is finite; as we saw in the proof of Proposition 12.3, it comes down to finding a root of a single equation in one real variable when $\mathcal{Y}$ is binary. But it can be infeasible when $\mathcal{Y}$ is not finite.

The strategies whose existence is asserted by Theorems 12.9 and 12.21 are tailored to particular RKHSs. To think about finding these strategies and implementing them as concrete algorithms, we must turn to the RKHSs' reproducing kernels. In general, kernels provide an equivalent and more concrete language for RKHSs. The notion of a kernel can be defined abstractly without reference to an RKHS, any kernel can be used to construct an RKHS, all RKHSs are obtained in this way, and the kernel for each RKHS is unique.

In this section we review the relationship between RKHSs and their reproducing kernels, list a few important kernels, discuss the role of Sobolev spaces, and extract a forecasting algorithm from the proof of Theorem 12.9 and a decision-making algorithm from the proof of Theorem 12.21 for the case where $\mathcal{Y}$ is finite.

Reproducing Kernels

In this context, a *kernel* on a nonempty set Z is a function $\mathsf{k} : Z^2 \to \mathbb{R}$ that satisfies these two conditions:

- It is symmetric: $\mathsf{k}(z, z') = \mathsf{k}(z', z)$ for all $(z, z') \in Z^2$.
- It is nonnegative definite:

$$\sum_{i=1}^{m} \sum_{j=1}^{m} t_i t_j \mathsf{k}(z_i, z_j) \geq 0$$

for all $m \in \mathbb{N}$, $(t_1, \ldots, t_m) \in \mathbb{R}^m$, and $(z_1, \ldots, z_m) \in Z^m$.

If $\mathcal{H}$ is an RKHS on Z, then the function $\mathsf{k} : Z^2 \to \mathbb{R}$ defined by

$$\mathsf{k}(z, z') := \langle \mathsf{k}_z, \mathsf{k}_{z'} \rangle_{\mathcal{H}} = \mathsf{k}_z(z') = \mathsf{k}_{z'}(z)$$

is $\mathcal{H}$'s *reproducing kernel*. The RKHS with k as reproducing kernel is unique and can be constructed from k. It includes all functions of the form $z' \in Z \mapsto \mathsf{k}(z, z')$ for $z \in Z$ and all finite linear combinations of them. These linear combinations form a vector space, the kernel can be used to define an inner product on this vector space, and the RKHS is its completion with respect to the resulting norm. For details, see [363, chapter 4].

We can define $\|\mathcal{H}\|_\infty$ directly in terms of $\mathcal{H}$'s reproducing kernel k:

$$\|\mathcal{H}\|_\infty = \sup_{z \in Z} \sqrt{\mathsf{k}(z,z)} = \sup_{z,z' \in Z} \sqrt{|\mathsf{k}(z,z')|}.$$

This relation implies that an RKHS is bounded according to the definition we gave in Section 12.2 if and only if its reproducing kernel is bounded [363, Lemma 4.23].

When Z is a topological space, an RKHS $\mathcal{H}$ on Z is continuous if and only if its reproducing kernel is continuous [363, Lemma 4.29] (see also Lemma 12.32). The property of fixed-signal continuity can be characterized similarly: an RKHS on $\mathcal{X} \times \mathcal{P}(\mathcal{Y}) \times \mathcal{Y}$ is fixed-signal continuous if and only if its reproducing kernel $\mathsf{k}((x, P, y), (x', P', y'))$ is continuous in $(P, y, P', y') \in (\mathcal{P}(\mathcal{Y}) \times \mathcal{Y})^2$ (see Exercise 12.14 for further details).

A kernel is *universal* if its RKHS is universal. Each of the following functions is a universal kernel when restricted to Z^2, Z being a compact set in a Euclidean space $\mathbb{R}^K$ [363, Corollary 4.58]. (Here $\|\cdot\|$ is the Euclidean norm and $\langle\cdot,\cdot\rangle$ the Euclidean inner product.)

- The *Gaussian kernel*

$$\mathsf{k}(z,z') := \exp\left(-\frac{\|z - z'\|^2}{\sigma^2}\right),$$

σ being an arbitrary positive constant. The corresponding RKHS is defined explicitly in [363, section 4.4].

- The *exponential kernel*

$$\mathsf{k}(z,z') := \exp(\langle z, z'\rangle).$$

- The *binomial kernel*

$$\mathsf{k}(z,z') := (1 - \langle z, z'\rangle)^{-\alpha}, \tag{12.36}$$

α being an arbitrary positive constant. In this case we assume that the compact subset Z of $\mathbb{R}^K$ is a subset of the ball $\{z \in \mathbb{R}^K \mid \|z\| < 1\}$; this assumption can always be made true by scaling Z. Cf. Exercise 12.13.

For a discussion of how kernels can be combined to construct additional kernels, and in particular how kernels on Cartesian products can be constructed, see [363, section 4.1].

The Gaussian, exponential, and binomial kernels can be used for our problem when $\mathcal{X}$ is a compact subset of a Euclidean space and $\mathcal{Y}$ is a finite set, because in this case $\mathcal{P}(\mathcal{Y})$ can be represented as a simplex in a Euclidean space. This simplex being compact, the product $\mathcal{X} \times \mathcal{P}(\mathcal{Y}) \times \mathcal{Y}$ can be regarded as the disjoint union of $|\mathcal{Y}|$ copies of $\mathcal{X} \times \mathcal{P}(\mathcal{Y})$, and this is a compact set in a Euclidean space of dimension $\dim \mathcal{X} + |\mathcal{Y}| - 1$. RKHSs with Gaussian, exponential, and binomial reproducing kernels are complicated to describe, but this is not a problem for the algorithms, which use the kernels directly.

Sobolev Spaces

Additional examples of universal bounded RKHSs are provided by Sobolev spaces. These spaces are easily described, but their reproducing kernels are known only in some special cases. Examples of Sobolev reproducing kernels on several important subsets of $\mathbb{R}^K$ are given in [31, section 7.4]; a general expression for Sobolev reproducing kernels on $\mathbb{R}^K$ itself is given by [294]. The unavailability of other Sobolev reproducing kernels limits the usefulness of Theorem 12.21 (see the comments following Corollary 12.22).

In the theory of Sobolev spaces, "domain" is used for a nonempty open set in Euclidean space, often required to satisfy further regularity conditions [5]. Given a domain Z and an integer $m \in \{0, 1, \ldots\}$, the Sobolev spaces $H^m(Z)$ can be defined as follows. For a smooth real-valued function u on Z, set

$$\|u\|_m := \sqrt{\sum_{0 \leq |\alpha| \leq m} \int_Z (D^\alpha u)^2}, \tag{12.37}$$

where $\int_Z$ is the integral with respect to Lebesgue measure on Z, α runs over the multi-indices $\alpha = (\alpha_1, \ldots, \alpha_K) \in \{0, 1, \ldots\}^K$, and

$$|\alpha| := \alpha_1 + \cdots + \alpha_K \quad \text{and} \quad D^\alpha u := \frac{\partial^{|\alpha|} u}{\partial_{t_1}^{\alpha_1} \cdots \partial_{t_K}^{\alpha_K}},$$

$(t_1, \ldots, t_K)$ being a typical point of $\mathbb{R}^K$. The Sobolev space $H^m(Z)$ is the completion of the set of smooth functions on Z with respect to the norm (12.37). If $m > K/2$ and Z satisfies appropriate regularity properties, $H^m(Z)$ is a universal and bounded RKHS:

- When $m > K/2$ and Z is regular, the Sobolev embedding theorem asserts that $H^m(Z)$ can be identified with a bounded RKHS of continuous functions on Z's closure $\overline{Z}$ [5, Theorem 4.12].
- If continuous, $H^m(Z)$ is universal; even the functions in $C^\infty(\mathbb{R}^K)$, all of which belong to all Sobolev spaces on Z, are dense in $C(\overline{Z})$ [5, theorem 2.29].
- When $m > K/2$ and Z satisfies a slightly stronger regularity condition, the Sobolev embedding theorem further asserts that $H^m(Z)$ embeds into a Hölder function class [5, Theorem 4.12]. In combination with the following lemma and (12.9), this immediately implies that $H^m(Z)$ is a continuous RKHS.

Lemma 12.32   *The reproducing kernel* k *of an RKHS* $\mathcal{H}$ *on a topological space* Z *is continuous if and only if* $k(\cdot, z)$ *is continuous for each* $z \in Z$ *and the function* $z \in Z \mapsto \|k_z\|$ *is continuous.*

The continuity of $k(\cdot, z)$ for each $z \in Z$ is known as the *separate continuity* of the kernel.

Proof A kernel k is continuous if and only if it is separately continuous and $z \in Z \mapsto$ $k(z, z)$ is continuous [363, Lemma 4.29]. But $k(z, z) = \langle k_z, k_z \rangle_{\mathcal{H}} = \|k_z\|_{\mathcal{H}}^2$.   □

Extracting a Forecasting Algorithm from the Proof of Theorem 12.9

The proof of Theorem 12.9 puts the RKHS $\mathcal{H}$ on $\mathcal{X} \times P(\mathcal{Y}) \times \mathcal{Y}$ to work using its representers $k_{x,P,y}$, from which it defines $k_{x,P}$ by (12.14) and then (in the proof of Lemma 12.13) $\Psi_n : \mathcal{X} \times P(\mathcal{Y}) \times \mathcal{Y} \to \mathbb{R}$ by $\Psi_n(x, P, y) := k_{x,P,y} - k_{x,P}$.

To make defensive forecasting's minimax strategy practical, let us assume that $\mathcal{Y}$ is finite. In this case, the integral in (12.14) reduces to a sum:

$$\Psi_n(x, P, y) = k_{x,P,y} - \sum_{y' \in \mathcal{Y}} k_{x,P,y'} P\{y'\}. \tag{12.38}$$

Proposition 12.6 requires that Forecaster ensure $\mathcal{K}_N \leq \mathcal{K}_{N-1}$ for all N. In the context of Lemma 12.10, this becomes $\mathcal{M}_N - \mathcal{M}_{N-1} \leq 0$, or

$$\left\langle \sum_{n=1}^{N-1} \Psi_n(x_n, P_n, y_n), \Psi_N(x_N, P_N, y_N) \right\rangle_{\mathcal{H}} \leq 0. \tag{12.39}$$

Given all the previous moves, Forecaster must choose P_N so that (12.39) holds for all y_N.

Substituting (12.38) in (12.39) and expanding the inner product, we obtain an expression involving only inner products with representers as arguments – i.e. values of $\mathcal{H}$'s reproducing kernel k. This gives the following recipe for Forecaster: given the previous moves, choose P_N such that for all $y_N \in \mathcal{Y}$,

$$\sum_{n=1}^{N-1} k((x_n, P_n, y_n), (x_N, P_N, y_N))$$

$$- \sum_{n=1}^{N-1} \sum_{y \in \mathcal{Y}} k((x_n, P_n, y_n), (x_N, P_N, y)) P_N\{y\}$$

$$- \sum_{n=1}^{N-1} \sum_{y \in \mathcal{Y}} k((x_n, P_n, y), (x_N, P_N, y_N)) P_n\{y\}$$

$$+ \sum_{n=1}^{N-1} \sum_{y,y' \in \mathcal{Y}} k((x_n, P_n, y), (x_N, P_N, y')) P_n\{y\} P_N\{y'\} \leq 0.$$

This may be combined with any of the simple kernels listed above.

Extracting a Decision-Making Algorithm from the Proof of Theorem 12.21

In the proof of Theorem 12.21 we used the direct sum (12.26) (with $a_0 = a_1 = 1$) of the RKHSs $\mathbb{R}$ and $\mathcal{H}$, and the mappings Ψ in (12.39) were defined by

$$\Psi_n(x, P, y) = (\Psi_{n,0}(x, P, y), \Psi_{n,1}(x, P, y))$$

and (12.29)–(12.30). Plugging this into (12.39) gives the condition

$$\sum_{n=1}^{N-1} \lambda(x_n, G_n(x_n, P_n), y_n) \lambda(x_N, G_N(x_N, P_N), y_N)$$

$$- \sum_{n=1}^{N-1} \sum_{y \in \mathcal{Y}} \lambda(x_n, G_n(x_n, P_n), y_n) \lambda(x_N, G_N(x_N, P_N), y) P_N\{y\}$$

$$- \sum_{n=1}^{N-1} \sum_{y \in \mathcal{Y}} \lambda(x_n, G_n(x_n, P_n), y) \lambda(x_N, G_N(x_N, P_N), y_N) P_n\{y\}$$

$$+ \sum_{n=1}^{N-1} \sum_{y,y' \in \mathcal{Y}} \lambda(x_n, G_n(x_n, P_n), y) \lambda(x_N, G_N(x_N, P_N), y') P_n\{y\} P_N\{y'\}$$

$$+ \sum_{n=1}^{N-1} \mathsf{k}((x_n, y_n), (x_N, y_N)) - \sum_{n=1}^{N-1} \sum_{y \in \mathcal{Y}} \mathsf{k}((x_n, y_n), (x_N, y)) P_N\{y\}$$

$$- \sum_{n=1}^{N-1} \sum_{y \in \mathcal{Y}} \mathsf{k}((x_n, y), (x_N, y_N)) P_n\{y\}$$

$$+ \sum_{n=1}^{N-1} \sum_{y,y' \in \mathcal{Y}} \mathsf{k}((x_n, y), (x_N, y')) P_n\{y\} P_N\{y'\} \le 0, \tag{12.40}$$

where k is the reproducing kernel of the RKHS $\mathcal{G}$ on $\mathcal{X} \times \mathcal{Y}$ mentioned in Theorem 12.21. The forecast P_N to be used on the Nth round is found from the condition that (12.40) hold for all $y_N \in \mathcal{Y}$. Decision Maker should then make the decision $G_N(x_N, P_N)$ recommended by the choice function G_N given the signal x_N.

12.7 DISCONTINUOUS STRATEGIES FOR SKEPTIC

In the introduction to this chapter, we argued that probabilistic forecasts are adequately tested by strategies for Skeptic that are continuous in those forecasts. Defensive forecasting can nevertheless be extended to handle strategies for Skeptic that are discontinuous in the forecasts. This requires only a dose – a tiny dose will do – of randomization by Forecaster. Instead of announcing his forecast before

Reality announces the outcome, Forecaster announces a probability measure for the forecast. Even if the probability measure is very concentrated, this randomization will smooth the picture that Reality sees, allowing a minimax strategy for Forecaster to succeed.

We will describe Forecaster's randomization game-theoretically. The forecast to which Skeptic's strategy is applied is chosen by another player in the game, whom we call Random Number Generator. To assure Random Number Generator's faithfulness to Forecaster's probability measure, we need a player who bets against this faithfulness; we assign this task to Forecaster himself. Rather than demanding that Forecaster keep Skeptic's capital from growing at all, we now ask him only to limit Skeptic's capital to the extent that Random Number Generator similarly limits Forecaster's own capital. By Cournot's principle, Random Number Generator should not allow Forecaster to multiply his capital by a large factor.

In the following protocol, $\mathcal{K}_0, \mathcal{K}_1, \dots$ represents as usual Skeptic's capital process, while $\mathcal{F}_0, \mathcal{F}_1, \dots$ represents Forecaster's capital process (from betting on Random Number Generator's choice of the P_n). The outcome space $\mathcal{Y}$ may be finite; in this case any function $S_n : \mathcal{Y} \times \mathcal{P}(\mathcal{Y}) \to \mathbb{R}$ is continuous in $y \in \mathcal{Y}$. We call a probability measure Q on a finite set *positive* if it assigns positive probability to every element of the set.

Protocol 12.33
PARAMETERS: Nonempty set $\mathcal{X}$ and compact metrizable space $\mathcal{Y}$
> $\mathcal{K}_0 := 1$.
> $\mathcal{F}_0 := 1$.
> FOR $n = 1, 2, \dots$:
>> Reality announces $x_n \in \mathcal{X}$.
>> Skeptic announces $f_n : \mathcal{Y} \times \mathcal{P}(\mathcal{Y}) \to [0, \infty)$ continuous in $y \in \mathcal{Y}$ and
>>> satisfying $\int_{\mathcal{Y}} f_n(y, P) P(dy) \leq \mathcal{K}_{n-1}$ for all $P \in \mathcal{P}(\mathcal{Y})$.
>> Forecaster announces a finite $D_n \subseteq \mathcal{P}(\mathcal{Y})$ and a positive $Q_n \in \mathcal{P}(D_n)$.
>> Reality announces $y_n \in \mathcal{Y}$.
>> Forecaster announces $g_n : D_n \to [0, \infty)$ such that
>>> $\int_{\mathcal{P}(\mathcal{Y})} g_n \, dQ_n := \sum_{P \in D_n} g_n(P) Q_n \{P\} \leq \mathcal{F}_{n-1}$.
>> Random Number Generator announces $P_n \in D_n$.
>> $\mathcal{K}_n := f_n(y_n, P_n)$.
>> $\mathcal{F}_n := g_n(P_n)$.

Adopting the terminology we introduced for Skeptic in Section 1.2, we say that Forecaster *can force* an event defined by his opponents' moves in this protocol if he has a strategy that guarantees that his capital will tend to infinity when the event does not happen, and we say that he *can weakly force* such an event if he has a strategy that guarantees that his capital is unbounded when the event does not happen. By the argument given in Section 1.2, Forecaster can force any event that he can weakly force.

The following proposition says that Forecaster can weakly force and hence force Skeptic's capital $\mathcal{K}_n$ to stay bounded in Protocol 12.33. It is therefore a randomized counterpart of Proposition 12.6.

Proposition 12.34 *For any $\epsilon > 0$ and any sequence $\mathcal{A}_1, \mathcal{A}_2, \ldots$ of open covers of $\mathcal{P}(\mathcal{Y})$, Forecaster has a strategy in Protocol 12.33 that guarantees*

- $\mathcal{K}_n \leq (1 + \epsilon)\mathcal{F}_n$ *for all n,*
- D_n *lies completely in one element of $\mathcal{A}_n$, and*
- $|D_n| \leq |\mathcal{Y}|$.

The first statement in Proposition 12.34, $\mathcal{K}_n \leq (1 + \epsilon)\mathcal{F}_n$ for all n, says that Forecaster can weakly force $\mathcal{K}_n$ to stay bounded. The other two statements say that he can do so with probability distributions Q_n that are very concentrated, putting all their probability within a region that is small in terms of $\mathcal{P}(\mathcal{Y})$'s topology (second statement for a fine cover $\mathcal{A}_n$) and, when $\mathcal{Y}$ is finite, on a number of probability measures that is no greater than the number of elements in $\mathcal{Y}$ (third statement).

Proposition 12.34 asserts that Forecaster has a strategy that guarantees the three statements regardless of how the other players move. Random Number Generator can, however, defeat our interpretation of the proposition (that Forecaster will defeat Skeptic by keeping his capital $\mathcal{K}_n$ bounded) by making $\mathcal{F}_n$ unbounded. So the interpretation depends on supposing that Random Number Generator will validate Forecaster's predictions of his behavior.

We require Reality to move before Forecaster chooses his bet g_n to test Random Number Generator, because, as we will see in the proposition's proof, the choice of g_n required by his winning strategy depends on knowing Reality's y_n. This gives the application of Cournot's principle to Random Number Generator a particular twist; it requires that Random Number Generator be oblivious to the moves by Reality – that he not conspire with Reality – and that Reality, for her part, not be aware of any strategy or habit of Random Number Generator's that she might exploit. For further discussion, see [416, 398].

When $\mathcal{Y}$ is finite, and more generally when it is a compact metric space, we can make an even sharper statement about how little randomization Forecaster needs to use. Write dist for the Euclidean distance on the simplex $\mathcal{P}(\mathcal{Y})$ when $\mathcal{Y}$ is finite and more generally for the Prokhorov metric on $\mathcal{P}(\mathcal{Y})$ when $\mathcal{Y}$ is a compact metric space [39, Appendix III, Theorem 6], and define the diameter of a finite subset D of $\mathcal{P}(\mathcal{Y})$ in the natural way:

$$\text{diam } D := \max_{p,q \in D} \text{dist}(p, q).$$

Then the following corollary follows immediately from Proposition 12.34.

Corollary 12.35 *Suppose $\mathcal{Y}$ is finite (or a compact metric space). For any $\epsilon > 0$ and any sequence $\epsilon_1, \epsilon_2, \ldots$ of positive real numbers, Forecaster has a strategy in Protocol 12.33 that guarantees*

- *$\mathcal{K}_n \leq (1 + \epsilon)\mathcal{F}_n$ for all n,*
- *diam $D_n \leq \epsilon_n$, and*
- *$|D_n| \leq |\mathcal{Y}|$.*

The condition $\mathcal{K}_n \leq (1 + \epsilon)\mathcal{F}_n$ says that Forecaster can guarantee $\mathcal{F}_n \geq \mathcal{K}_n$ to any approximation required. Every dollar gained by Skeptic can be attributed to the poor performance of Random Number Generator. The condition diam $D_n \leq \epsilon_n$ brings out how little randomization is required.

Proof of Proposition 12.34 We will repeatedly use the fact that $\mathcal{P}(\mathcal{Y})$ is paracompact [128, Theorem 5.1.1].

Fix a round n of the game. Let $\delta > 0$ be a small constant (how small will be determined later). For each $P \in \mathcal{P}(\mathcal{Y})$ set

$$A_P := \left\{ T \in \mathcal{P}(\mathcal{Y}) \mid \int_{\mathcal{Y}} f_n(y, P)T(dy) < \mathcal{K}_{n-1} + \delta \right\}. \tag{12.41}$$

Notice that $P \in A_P$ and that A_P is an open set. Let $\mathcal{B}$ be any open star refinement of $\mathcal{A}_n$ (it exists by [128, Theorem 5.1.12]), let $\mathcal{C}$ be any locally finite open refinement of $\mathcal{B}$ (it exists by the definition of paracompactness), and let B_P be the intersection of A_P with an arbitrary element of $\mathcal{C}$ containing P. Notice that the B_P form an open cover of $\mathcal{P}(\mathcal{Y})$. If $\mathcal{Y}$ is finite, replace $\{B_P\}_{P \in \mathcal{P}(\mathcal{Y})}$ by its open shrinking of order $|\mathcal{Y}| - 1$ (it exists by the Dowker theorem [128, Theorem 7.2.4], since $\mathcal{Y}$ is normal [128, Theorem 3.1.9] and $\dim \mathcal{P}(\mathcal{Y}) = |\mathcal{Y}| - 1$ [128, Theorem 7.3.19]); we will use the same notation $\{B_P\}_{P \in \mathcal{P}(\mathcal{Y})}$ for the shrinking. Let $\{g_s\}_{s \in S}$ be a locally finite partition of unity subordinated to the open cover $\{B_P\}_{P \in \mathcal{P}(\mathcal{Y})}$ [128, Theorem 5.1.9]. For each $s \in S$ choose a $P_s \in \mathcal{P}(\mathcal{Y})$ such that $\{P | g_s(P) > 0\} \subseteq B_{P_s}$. For $y \in \mathcal{Y}$ and $P \in \mathcal{P}(\mathcal{Y})$, set

$$f^*(y, P) := \sum_{s \in S} f_n(y, P_s)g_s(P).$$

This sum is well defined because only a finite number of addends are nonzero.

It is clear that $f^*(y, P)$ is continuous in P; let us check that it almost satisfies $\int_{\mathcal{Y}} f^*(y, P)P(dy) \leq \mathcal{K}_{n-1}$ for all $P \in \mathcal{P}(\mathcal{Y})$. We have

$$\int_{\mathcal{Y}} f^*(y, P)P(dy) = \int_{\mathcal{Y}} \sum_{s \in S_P} f_n(y, P_s)g_s(P)P(dy)$$

$$= \sum_{s \in S_P} \int_{\mathcal{Y}} f_n(y, P_s)P(dy)g_s(P) \leq \mathcal{K}_{n-1} + \sum_{s \in S_P} \delta g_s(P) = \mathcal{K}_{n-1} + \delta, \tag{12.42}$$

where S_P is the finite set of all s for which $g_s(P) > 0$; the inequality in (12.42) uses the fact that $P \in B_{P_s} \subseteq A_{P_s}$ and the definition (12.41). Therefore, $\int_y f^*(y, P)P(dy) \le \mathcal{K}_{n-1} + \delta$ for all P. Applying Ky Fan's theorem (see the proof of Proposition 12.6), we see that there exists $P^* \in \mathcal{P}(\mathcal{Y})$ satisfying $f^*(y, P^*) \le \mathcal{K}_{n-1} + \delta$, for all $y \in \mathcal{Y}$.

Forecaster chooses a Q_n concentrated on the P_s with positive $g_s(P^*)$ and assigning weight $g_s(P^*)$ to each of these P_s. This will ensure that Q_n is concentrated on a finite subset D_n of an element of the cover $\mathcal{A}_n$ and that $|D_n| \le |\mathcal{Y}|$.

Let δ be $\epsilon 2^{-n}$ or less. This will ensure

$$\int f_n(y, P)Q_n(dP) \le \mathcal{K}_{n-1} + \epsilon 2^{-n} \tag{12.43}$$

for all $y \in \mathcal{Y}$. Let Forecaster's strategy further tell him to use as his second move the function g_n given by

$$g_n(P) := \mathcal{F}_{n-1} + \frac{1}{1 + \epsilon}(f_n(y_n, P) - \mathcal{K}_{n-1} - \epsilon 2^{-n}) \tag{12.44}$$

for $P \in D_n$ and defined arbitrarily for $P \notin D_n$. The condition $\int g_n dQ_n \le \mathcal{F}_{n-1}$ is then guaranteed by (12.43).

It remains to check $\mathcal{K}_n \le (1 + \epsilon)\mathcal{F}_n$ and that $\mathcal{F}_n$ is never negative. This can be done by a formal calculation (see the proof of Theorem 3 in [416]), but we prefer the following intuitive picture. We would like Forecaster to use $g_n(P) := \mathcal{F}_{n-1} + f_n(y_n, P) - \mathcal{K}_{n-1} - \epsilon 2^{-n}$ (for $P \in D_n$) as his second move; this would always keep his capital $\mathcal{F}_n$ above $\mathcal{K}_n - \epsilon$. To make sure that $\mathcal{F}_n$ is never negative, Forecaster would have to start with initial capital $\mathcal{F}_0 = 1 + \epsilon$, leading to $\mathcal{F}_n \ge \mathcal{K}_n$ for all n, but our protocol requires $\mathcal{F}_0 = 1$. So Forecaster's strategy must be scaled down to the initial capital 1, leading to (12.44); $\mathcal{F}_n \ge \mathcal{K}_n$ becomes $(1 + \epsilon)\mathcal{F}_n \ge \mathcal{K}_n$. (Recall that scaling a strategy to a smaller initial capital means multiplying all its subsequent moves by the same factor as the initial capital, thus assuring that the capital on succeeding rounds is also multiplied by this factor.)   □

The results of this section open the way to generalizing this chapter's results on calibration and decision making to discontinuous test functions h and discontinuous decision rules D. These are topics for future research.

12.8 EXERCISES

Exercise 12.1 In Chapter 1, we noted Kumon and Takemura's simple strategy for weakly forcing $\bar{x}_n \to 0$ in Protocol 1.3 (see (1.18) and Exercise 1.11).

1. Translate the Kumon–Takemura strategy for Skeptic to Protocol 12.2 and find the defensive-forecasting strategy for Forecaster that keeps it from increasing Skeptic's capital. *Hint.* Adapting the strategy to Protocol 1.8, we obtain

$$M_n := \frac{1}{2}(\bar{y}_{n-1} - \bar{p}_{n-1})\mathcal{K}_{n-1},$$

and so the strategy for Skeptic in Protocol 12.2 is

$$f_n(p_n) := \frac{1}{2}(\bar{y}_{n-1} - \bar{p}_{n-1})\mathcal{K}_{n-1}.$$

Notably, this does not depend on p_n. Skeptic's resulting gain, $f_n(p_n)(y_n - p_n)$, does depend on p_n.

2. Show that when Forecaster defends against the Kumon–Takemura strategy,

$$\left| \sum_{i=1}^{n} (y_i - p_i) \right| \leq 1$$

for all n. It follows that $(\bar{y}_n - \bar{p}_n) \to 0$ as $n \to \infty$ at the rate $1/n$ rather than the slower rate $1/\sqrt{n}$ expected when the y_n are considered random with respect to the p_n. □

Exercise 12.2 Spell out the defensive-forecasting strategies against Kolmogorov's and Doléan's supermartingales in Protocol 12.2. □

Exercise 12.3 The strategy of Lemma 12.13 generalizes the strategy implemented by the K29 algorithm of [423]. Develop an analogous generalization of the algorithm of large numbers in [401, 396]. □

Exercise 12.4 In the proof of Theorem 12.8 we used the fact that the restriction to $A \times P(B) \times B$ of any continuous function $g : \mathcal{X} \times P(\mathcal{Y}) \times \mathcal{Y}$ is continuous, where A and B are compact. Prove this fact. □

Exercise 12.5 Theorem 12.18 assumes that the decision rule D is continuous, and it also relies on a compactness assumption for x_n and y_n. Consider the extent to which these assumptions can be relaxed, taking account of the following examples.

1. Let $\mathcal{X} := \Gamma := \mathcal{Y} := [-1, 1]$ and $\lambda(x, \gamma, y) := |y - \gamma|$ and suppose that Reality follows the strategy

$$x_n := \sum_{i=1}^{n-1} 3^{-i} \operatorname{sign} \gamma_i \quad \text{and} \quad y_n := -\operatorname{sign} \gamma_n, \quad \text{for } n = 1, 2, \dots,$$

where γ_n are Forecaster's moves and $\operatorname{sign} : \Gamma \to \{-1, 1\}$ is the function given by

$$\operatorname{sign} \gamma := \begin{cases} 1 & \text{if } \gamma \geq 0, \\ -1 & \text{otherwise.} \end{cases}$$

Consider the discontinuous decision rule $D : \mathcal{X} \to \Gamma$ given by

$$D(x) := \begin{cases} -1 & \text{if } x < \sum_{i=1}^{\infty} 3^{-i} \operatorname{sign} \gamma_i, \\ 1 & \text{otherwise.} \end{cases}$$

Show that (12.15) fails.

2. Suppose that $\mathcal{X} = \mathbb{R}$, $\Gamma = \mathcal{Y} = [-1, 1]$, and $\lambda(x, \gamma, y) = |y - \gamma|$. Show that the consequent of (12.15) will fail for some continuous decision rule D when $x_n = n$ for all n. □

The following exercise introduces three classical loss functions (see, e.g. [64, chapter 3] and [412]).

Exercise 12.6 Consider the instantiation of Protocol 12.17 in which there are no signals, $\mathcal{Y} := \{0, 1\}$, and $\Gamma := [0, 1]$. (We can square the absence of $\mathcal{X}$ with Protocol 12.17 by saying that $\mathcal{X}$ has a single element and is omitted from the notation because the choice of $x \in \mathcal{X}$ makes no difference.) Show that there is a unique choice function and find an explicit formula for it for the following loss functions:

• the *square loss function* $\lambda(\gamma, y) := (y - \gamma)^2$;

• the *log loss function*

$$\lambda(\gamma, y) := \begin{cases} -\ln \gamma & \text{if } y = 1, \\ -\ln(1 - \gamma) & \text{if } y = 0; \end{cases}$$

• the *spherical loss function*

$$\lambda(\gamma, y) := \begin{cases} 1 - \dfrac{\gamma}{\sqrt{\gamma^2 + (1-\gamma)^2}} & \text{if } y = 1, \\ 1 - \dfrac{1-\gamma}{\sqrt{\gamma^2 + (1-\gamma)^2}} & \text{if } y = 0. \end{cases}$$

Hint. Represent $\mathcal{P}(\mathcal{Y})$ as the interval $[0, 1]$. *Answer.* These are *proper* loss functions, in the sense that $G(p) := p$ is a choice function (in fact, the only choice function). □

Exercise 12.7 Consider the instantiation of Protocol 12.17 with no signals, $\mathcal{Y} := \{0, 1\}$, $\Gamma := [0, 1]$, and the loss function $\lambda(\gamma, y) := |y - \gamma|$ (called the *absolute loss function*). Find all choice functions. Show that there is no continuous choice function. □

Exercise 12.8 Suppose J is a finite set,

$$\mathcal{Y} := [0, \infty)^J, \quad \Gamma := \left\{ \gamma \in [0, \infty)^J \mid \sum_{j \in J} \gamma_j = 1 \right\},$$

$$\text{and} \quad \lambda(\gamma, y) := -\ln \sum_{j \in J} \gamma_j y_j; \quad (12.45)$$

as in Exercise 12.7, $\mathcal{X}$ is omitted: there is no signal. Show that there is a choice function G. □

If each outcome y in Exercise 12.8 is interpreted as the ratio of the closing to opening price of $|J|$ stocks and the decision (or *portfolio*) γ is the proportions of Decision Maker's capital invested in different stocks at the beginning of the round, then $-\lambda(\gamma, y)$ is the logarithmic increase in Decision Maker's capital. This is Cover's game of sequential investment [64, chapter 10]. We will study continuous-time versions in Part IV. The output of the choice function is known as the *growth-optimal portfolio* or *log-optimal portfolio* (for the given probability model) and will be discussed further in Chapter 16. The following exercise states its key property.

Exercise 12.9   ([50], second section 3). In the context (12.45) of Exercise 12.8, let $P \in \mathcal{P}(\mathcal{Y})$, $\delta := G(P)$ be the corresponding log-optimal portfolio, and let $\gamma \in \Gamma$ be any other portfolio. Prove that δ is a *numeraire portfolio*, in the sense of

$$\int_{y} \frac{\sum_{j \in J} \gamma_j y_j}{\sum_{j \in J} \delta_j y_j} P(dy) \leq 1.$$

Hint. Since δ is a growth-optimal portfolio,

$$\int_{y} \ln \sum_{j \in J} \gamma'_j y_j P(dy) \leq \int_{y} \ln \sum_{j \in J} \delta_j y_j P(dy)$$

for any portfolio $\gamma' \in \Gamma$. Set $\gamma'_j := (1 - \epsilon)\delta_j + \epsilon\gamma_j$ for $\epsilon \in (0, 1)$ and let $\epsilon \to 0$.   □

Exercise 12.10   In the proof of (12.31) we used the inequality $\|k_{x,P}\|_G \leq \|G\|_\infty$. Check this inequality. *Hint.* Apply Pettis's definition repeatedly to $\|k_{x,P}\|^2 = \langle k_{x,P}, k_{x,P} \rangle$.   □

Exercise 12.11   Set

$$c_\lambda := \sup_{x \in \mathcal{X}, \gamma \in \Gamma, y \in \mathcal{Y}} \lambda(x, \gamma, y) - \inf_{x \in \mathcal{X}, \gamma \in \Gamma, y \in \mathcal{Y}} \lambda(x, \gamma, y)$$

for a loss function $\lambda : \mathcal{X} \times \mathcal{P}(\mathcal{Y}) \times \mathcal{Y} \to \mathbb{R}$. Show that $c_\lambda/2 \leq \|\lambda\|_\infty$. Show that Theorem 12.21, Corollary 12.22, and the lemmas used in their proofs can be strengthened by replacing $\|\lambda\|_\infty$ by $c_\lambda/2$.   □

Exercise 12.12   Show that for any mapping $\Phi : Z \to \mathcal{H}$ from a set Z to a Hilbert space $\mathcal{H}$, the function $k : Z^2 \to \mathbb{R}$ defined by

$$k(z, z') := \langle \Phi(z), \Phi(z') \rangle \tag{12.46}$$

is a kernel. Such a mapping Φ is often called a *feature mapping*.   □

Exercise 12.13   Consider the special case of the binomial kernel (12.36) for $\alpha := 1$, $K := 1$, and $Z \subseteq (-1, 1)$ (this is called Vovk's infinite polynomial in [334, Example 4.24]). Show that this kernel is generated, in the sense of (12.46) in Exercise 12.12, by the feature mapping $\Phi : Z \to \ell_2$ defined by

$$\Phi(z) := (1, z, z^2, \ldots),$$

where $z \in Z$. (Remember that the inner product in ℓ_2 is defined by $\langle z, z' \rangle := \sum_n z_n z'_n$.) Using the fact that every continuous function on Z can be approximated by a polynomial (the Weierstrass approximation theorem), deduce the universality of the corresponding RKHS. □

Exercise 12.14 Let $\mathcal{X}$ and $\mathcal{Y}$ be topological spaces. Show that an RKHS on $\mathcal{X} \times \mathcal{Y}$ is fixed-signal continuous (in the sense that, for each $x \in \mathcal{X}$, the function mapping $y \in \mathcal{Y}$ to the representer $k_{x,y}$ is continuous) if and only if the following three equivalent conditions hold:

- for any $x, x' \in \mathcal{X}$, the reproducing kernel $k((x,y),(x',y'))$ is continuous in $(y,y') \in \mathcal{Y}^2$;
- for any $x \in \mathcal{X}$, $k((x,y),(x,y'))$ is continuous in (y,y');
- for any $(x,y) \in \mathcal{X} \times \mathcal{Y}$, $k((x,y),(x,y'))$ is continuous in $y' \in \mathcal{Y}$, and for any $x \in \mathcal{X}$, $k((x,y),(x,y))$ is continuous in $y \in \mathcal{Y}$.

Notice that the statement will remain true if we replace $\mathcal{Y}$ by $\mathcal{P}(\mathcal{Y}) \times \mathcal{Y}$. *Hint.* Generalize [363, the proof of Lemma 4.29]. □

Exercise 12.15 Show that Proposition 12.34 continues to hold when the condition that Skeptic's move f_n in Protocol 12.33 be continuous in $y \in \mathcal{Y}$ is relaxed to the condition that it be upper semicontinuous in $y \in \mathcal{Y}$. *Hint.* The key point in the proof of Proposition 12.34 where the continuity of f_n in y is used is the claim that the set (12.41) is open. This claim will still be true when f_n is only required to be upper semicontinuous in y. □

12.9 CONTEXT

Forecast, Prediction, or Decision?

In the preceding chapters, as in [349], we call a player who announces probabilities Forecaster rather than Predictor. For the sake of consistency, we have continued to do so in this chapter. In some related literature, however, especially the literature on online learning, our Forecaster and our Decision Maker are both called Predictor.

In the twentieth century, *forecast* became a marker of scientific expertise. In the late nineteenth century in the United States, *weather prophet* was often used to refer to both government forecasters and the entrepreneurs who published seasonal forecasts and almanacs, but the government weather service eventually succeeded in making its more scientific predictions widely known as *forecasts* [306]. In the first decades of the twentieth century, the entrepreneurs and economists who confidently predicted business cycles also used *forecast* as they claimed scientific status for their methods [155]. We still say *weather forecast* more often than *weather prediction*, and *forecast* retains an edge over *prediction* among financial analysts.

Even in the eighteenth century, weather prophets sometimes gave odds for their forecasts [288]. Systematic use of numbers to express degrees of certainty for weather forecasts has been traced back to W. Ernest Cooke, who began to attach the numbers 1 through 5 to his forecasts in 1905 [80], and Cleve Hallenbeck, who advocated "forecasting precipitation in percentages of probability" in 1920 [175]. The probabilities given by weather forecasters are *probability forecasts*, but the terms *probability prediction* and *probabilistic prediction* have recently gained ground in other fields (see [168]).

In his 1970 treatise [143, sections 3.1.2 and 5.2.3], Bruno de Finetti distinguished in Italian between *predizione* and *previsione*, using the latter as his name for subjective probabilities and more generally for subjective expected values. One might translate *previsione* into English as *forecast*, but the book's translators chose to render it as *prevision*. This choice has been echoed by others sympathetic to de Finetti's views, especially in the literature on imprecise probabilities [426].

The Impossibility of Good Probability Forecasting

The argument that Reality can always defeat Forecaster, recalled in the first paragraph of Section 12.1, was stated clearly and concisely in the context of probability forecasting by A. Philip Dawid in 1985 [99]. For many game theorists, the impossibility of predicting the behavior of an adversary is so obvious that there is no point in spelling out such an argument. Dawid's formulation was nevertheless an important step, because it makes possible the observation that Reality's strategy fails if Skeptic must follow a continuous strategy for testing the forecasts or Forecaster is allowed to randomize. Dawid attributed his insight to an earlier general argument published by Hilary Putnam in 1963 [316], but Michael Scriven, writing in 1965, may have been the first to note that Reality can force Forecaster to err at every step [248, 337]. For a detailed review of Putnam's argument and its subsequent development, see [364].

Calibration

Ever since Cooke and Hallenbeck, weather forecasters have sought to produce probabilities that are matched by subsequently observed frequencies. But it was only in the 1970s that *calibration* began to be used to name this match, by psychologists who noted that most people are overconfident when asked to supplement guesses with probabilities that the guesses are correct [249]. Such overconfidence evidently calls for recalibration.

In this chapter, we have used *calibration* in a very broad way. In the case where the outcome space is binary, say $\mathcal{Y} = \{0, 1\}$, we ask not only that the frequency of 1s approximate any given $p \in [0, 1]$ over all the rounds in which the forecast approximates p but that this also be true for the subset of these rounds in which a signal x has a given value. This broad sense of *calibration*, which seems to cover most of what we want to accomplish in probability forecasting, was introduced by Dawid in 1985 [98]. By following Dawid, we are diverging, however, from current practice. Most authors continue to use *calibration* in more restricted senses, not taking into account any additional signal x.

Our work on defensive forecasting and calibration was inspired by the pioneering work of Dean Foster and Rakesh Vohra, who showed that randomized forecasting can achieve calibration in the narrower sense, where no signal is taken into account. Their work, initially met with great skepticism, finally appeared in 1998 [149]. In a conference paper that appeared in 2004 and an article that appeared in 2008, Sham Kakade and Foster established the special case of Theorem 12.8 in which there are no signals, $\mathcal{Y}$ is finite, and the functions h are of the form $h(x, P, y) := w(P)\mathbf{1}_{y=b}$ for $b \in \mathcal{Y}$ and a Lipschitz function $w : \mathcal{P}(\mathcal{Y}) \to \mathbb{R}$ (for a natural metric on $\mathcal{P}(\mathcal{Y})$) [207, Theorem 2.1]. Their proof resembled ours, but they did not use defensive forecasting or isolate its relevance. They pointed out that the randomization needed when h is not continuous can be arbitrarily slight [207, section 2.2].

Additional references to game-theoretic work related to defensive forecasting are provided in our first article on randomized forecasting, published in 2005 [416]. Akimichi Takemura's recognition that randomization is not needed in that article's argument when Skeptic's testing strategy is continuous led to the general concept of defensive forecasting, first published in that same year [423].

Forecasts That Fall Short of Probability Measures

This chapter's protocol for defensive forecasting, Protocol 12.4, requires that Forecaster announce a probability measure. Defensive forecasting can also be applied to protocols in which Forecaster offers fewer bets. Methods similar to those of Section 12.2 can produce calibrated forecasts in *linear protocols*, a wide class that includes Protocol 1.1 (bounded forecasting) and a version of Protocol 4.1 (mean/variance forecasting) that requires outcomes to fall in a finite interval [415].

Competitive Online Prediction

Theorem 12.18 is a typical result in competitive online prediction (also called *prediction with expert advice*). In general, as in this chapter, work in this field does not assume a statistical model. Instead, one competes with a pool of decision rules, prediction strategies, or experts. In this chapter this pool is an RKHS or the set of continuous functions on $\mathcal{X}$, but it might also be a finite set of strategies or a finite panel of actual experts. In simple cases the goal is to perform almost as well as the strategy or expert that comes out best in hindsight. This goal is often achievable if the pool is finite and not too large. In other cases such success may depend on assigning each strategy in the pool a penalty that depends on its complexity.

Nicoló Cesa-Bianchi and Gábor Lugosi's book on competitive online prediction, published in 2006 [64], traces the history of the field to David Blackwell's work in the 1950s. The field blossomed in the 1990s with the appearance of exponential weights algorithms. In 1988, Alfredo DeSantis, George Markowsky, and Mark Wegman showed that Bayes mixtures of probabilistic strategies have interesting worst-case guarantees [110], and in the early 1990s the Weighted Majority Algorithm was discovered independently by Nick Littlestone and Manfred Warmuth [253] and by Vovk [392]. Cover's universal portfolio algorithm was another early contribution [85]. Vovk's Aggregating Algorithm, already published in 1990 [391], unified the picture by generalizing both Bayes mixtures and the Weighted Majority Algorithm. The Aggregating Algorithm works for a wide range of loss functions, and it has been shown to be optimal for a wide class of competitive online problems with a finite benchmark pool of experts [395].

Defensive forecasting complements exponential weights algorithms and extends the application domain of competitive online prediction (see, e.g. [72]). The results in Section 12.4 give good performance guarantees in many situations but are not optimal. On the other hand, optimal performance guarantees for finitely many experts can be attained by defensive forecasting as well as by the Aggregating Algorithm [71, 399].

Vladimir V'yugin's textbook [424] (in Russian) covers several of the topics of this chapter: calibration in Chapter 3, competitive online prediction in Chapter 5, and defensive forecasting in Chapter 7.

Levin's Neutral Probability Measure

A one-round version of defensive forecasting was introduced in the context of the algorithmic theory of randomness by Leonid Levin [241, 242] and then explained and developed by Peter Gács [158].

Levin's work was inspired by Kolmogorov's definition of randomness in terms of algorithmic complexity [225, section 4]. Kolmogorov was interested in the randomness of finite sequences, but in 1966 [269], Per Martin-Löf extended it to define an asymptotic statistical test of computable probability measures that is, in a certain sense, universal. In 1976 [241], Levin modified Martin-Löf's definition, extending it to noncomputable probability measures.

Levin's test is a function $t : Z \times P(Z) \to [0, \infty]$, where Z is a topological space. (In 1976, he considered the case $Z = \{0, 1\}^\infty$ but noted that his argument worked for any other well-behaved compact space with a countable base.) Suppose Z is compact and metrizable (this is equivalent to Levin's assumption that it is compact with a countable base [128, theorem 4.2.8]). Let us say that a function $t : Z \times P(Z) \to [0, \infty]$ is a *test of randomness* if it is lower semicontinuous and, for all $P \in P(Z)$,

$$\int_Z t(z, P)P(dz) \leq 1.$$

(Gács and, especially, Levin impose additional requirements.) The intuition behind this definition is that if we first choose a test t, and then observe z and find that $t(z, P)$ is very large, we are entitled to reject P. (The P-probability that $t(z, P) \geq C$ cannot exceed $1/C$, for any $C > 0$.)

Levin established the following fundamental result: for any test of randomness t, there exists a probability measure P such that

$$\forall z \in Z : \quad t(z, P) \leq 1. \tag{12.47}$$

(See [241, footnote 1] for a statement of the result and [158, section 5] for details of the proof.) This result follows from our general result on defensive forecasting, Proposition 12.6. Levin and Gács called P an *a priori*, or *neutral*, probability measure when (12.47) holds for P and the universal test t. In Levin's view, such a measure can be considered a universal explanation for all observations.

Universally Consistent Nonparametric Estimation

Theorem 12.18 can be compared with similar results obtained under the assumption that $(x_1, y_1), (x_2, y_2), \ldots$ are independent observations from an unknown probability measure on $\mathcal{X} \times \mathcal{Y}$. In a classical article on this topic published in 1977 [369], Charles Stone gave methods of estimating $P(Y|X)$ from such independent observations that are consistent for all probability measures P on $\mathcal{X} \times \mathcal{Y}$ and lead to decision strategies for which the consequent of (12.15) holds almost surely.

Stone's theory does not require that the loss function λ be continuous, and it does not use assumptions of compactness. The most essential aspects of the continuity and compactness hypotheses used by Theorem 12.18 more or less follow, however, from Stone's probabilistic assumptions. Under mild conditions, every measurable decision rule can be arbitrarily well approximated by a continuous one (according to Luzin's theorem [121, theorem 7.5.2] combined with the Tietze–Uryson theorem [128, theorem 2.1.8]), and every probability measure is almost concentrated on a compact set (according to Ulam's theorem [121, theorem 7.1.4]).

Reproducing Kernel Hilbert Spaces

The one-to-one correspondence between kernels and RKHSs goes back to Nachman Aronszajn [12, 13] and has been widely exploited in machine learning and statistics. Among the first to use it in machine learning was Aizerman's laboratory (including Emmanuel M. Braverman and Lev I. Rozonoer) at the Institute of Control Problems [7]. The method became popular after Vladimir Vapnik and his coauthors combined kernels with the method of generalized portrait developed at Lerner's laboratory at the same institute approximately at the same time by, first of all, Vapnik and Alexey Chervonenkis. The resulting Support Vector Machine [384] has subsequently served as a powerful impetus for the development of kernel methods in machine learning: see, e.g. [334, 355] and, especially, [363, chapter 4] (our main reference for kernels in this chapter).

In functional analysis, Banach spaces are as popular as Hilbert spaces, and the methodology of RKHSs can be partially generalized to *proper Banach functional spaces* (PBFSs), defined as Banach spaces $\mathcal{H}$ of real-valued functions on a nonempty set Z such that, for each $z \in Z$, the evaluation functional $h \in \mathcal{H} \mapsto h(x)$ is bounded. This generalization, introduced in [397], allows us to treat less regular function classes.

Perfect-Information Games

This book and its predecessor [349] promote the theory of perfect-information games as an alternative foundation for probability theory and related fields. Yet we use few results from the existing theory of such games. In Chapter 4 (Section 4.7) we used Martin's theorem. In this chapter we used another result from the existing theory, Gary Gruenhage's theorem, part of which was stated as Lemma 12.16. Gruenhage's full result [172, Theorem 4.1] is that, if Z is a locally compact topological space, Remover has a winning strategy in $G(Z)$ if and only if Z is paracompact. Gruenhage's [172] contains several other interesting topological perfect-information games. For examples of perfect-information games that are important in set theory, particularly descriptive set theory, see [214, section 21] and [202].

Part IV

Game-Theoretic Finance

We now look at how game-theoretic probability in continuous time can help us understand the emergence of Brownian motion, the foundations of the Itô calculus, and other aspects of continuous-time martingales. These ideas are widely used in physics, in other branches of science, and especially in finance theory. We emphasize finance theory, because it has inspired most of the work on continuous-time martingales in recent decades and because its game-theoretic nature is already so evident.

We develop game-theoretic finance theory at the same abstract and idealized level used in measure-theoretic finance theory. For simplicity, we assume that financial securities' price paths are continuous, even though most of the results we report can be generalized to accommodate some discontinuities. More importantly, we assume that securities do not pay dividends and that their shares can be frequently bought, sold, and shorted in arbitrary fractional amounts with no transaction costs. The implications of our results for any actual financial market must be tempered by taking into account the degree to which it departs from this extremely idealized picture.

Measure-theoretic finance theory assumes that a financial security's successive prices are described by a stochastic process in continuous time. A stochastic process being a family of random variables in a probability space, this assumption seems to involve a hugely complex probability measure. But a celebrated theorem proven by Lester Dubins and Gideon Schwarz in 1965 [120] revealed an aspect of the picture that does not use all this complexity: no matter the details of the probability measure, any continuous martingale must be Brownian motion up to a time change. If,

Game-Theoretic Foundations for Probability and Finance, First Edition. Glenn Shafer and Vladimir Vovk.
© 2019 John Wiley & Sons, Inc. Published 2019 by John Wiley & Sons, Inc.

following Paul Samuelson [325], we suppose that the price of a security should be a continuous martingale when properly discounted, then the Dubins–Schwarz theorem suggests that we should be able to say something interesting without assuming the probability measure. This is confirmed by our game-theoretic analysis. If a trader cannot become infinitely rich by exploiting the idealized trading opportunities represented by continuous price paths for securities, then each security's price path must have the properties of Brownian motion up to a time change. Purely game-theoretic assumptions are also sufficient, as we will see, to assure that price paths for different securities are almost surely related as prescribed by the Itô calculus.

In the discrete-time theory of Parts I–III, Skeptic tries, over infinitely many successive rounds, to multiply infinitely the capital he risks. Long-run strategies are still available in our continuous-time picture: we use the time interval $[0, \infty)$. But a trader may also make infinitely many trades in a small interval of time, using smaller and smaller fractions of initial capital to trade at greater and greater frequency within that interval. Such *trading in the small*, as we may call it, can make the trader infinitely rich instantly as soon as prices move in a way to which theory gives probability 0.

The game-theoretic analysis thus allows us to distinguish two kinds of probability 0 events: probability 0 events blocked by trading-in-the-small strategies and probability 0 events blocked by long-run strategies. This distinction is important in applications (where "infinitely" becomes "very"), because it permits two different ways of explaining in game-theoretic terms what it means for a market's prices to be efficient. On the one hand, we can take efficiency to mean that a trader will not become very rich instantly. On the other, we can take it to mean that a trader will not become very rich even in the long run. When we assume or test only the first kind of efficiency, the numeraire with respect to which wealth is measured (dollars, euros, some market index, etc.) hardly matters, because reasonable numeraires do not change much with respect to each other in the small. But in order to test a market's efficiency in the strong long-run sense, we must specify the numeraire. When we assume efficiency in this long-run sense for a given market and a given numeraire, we call the numeraire an *efficient numeraire* for that market. The implication is that a trader can do no better than buy and hold the numeraire.

There are only two players in our financial games: the market, which plays the role of Reality by choosing the securities' price paths, and a trader, who plays the role of Skeptic by trading in the securities at these prices. There is no Forecaster announcing probabilities or additional prices beyond those given by the market. Forecaster's absence is a key aspect of the difference between the game-theoretic and measure-theoretic pictures. The game-theoretic picture does not need probabilities or martingales beyond those derived from the game's trading opportunities.

As we showed in [349, chapters 11–15], it is possible, using nonstandard analysis, to develop a continuous-time version of the step-by-step picture developed in discrete time in Parts I–III. In this continuous yet step-by-step picture, the trader takes a position in the market's securities at the beginning of each infinitesimal trading period, and the market reveals new prices at the end of each such period. Here, however, we relinquish the step-by-step picture and suppose that each player moves only once. The trader moves first, announcing a trading strategy. The market then moves by announcing the securities' entire price paths, from time 0 to time ∞. This brings us closer to the standard mathematics of continuous-time measure-theoretic probability, even though we do not introduce any probability measure.

The first three chapters that follow, Chapters 13–15, develop the continuous-time game-theoretic framework, with an emphasis on what the trader can accomplish trading in the small. Chapter 16 uses this framework to study game-theoretic versions of some classic topics concerning the long-run behavior of stock prices, including the equity premium and capital asset pricing model (CAPM). Chapter 17 develops a game-theoretic analog to stochastic portfolio theory.

Chapter 13 studies the game-theoretic picture for a single security. Here the market's move is a continuous price path for the security, and the strategies from which the trader chooses and the capital processes they determine are described constructively. Though relatively elementary, this setting already allows us to state a game-theoretic Dubins–Schwarz theorem. The emphasis in this chapter is on trading in the small and on the local properties of Brownian motion that it can force, such as absence of derivatives. If and when those local properties fail, our trading strategies make the trader infinitely rich straight away. And these properties are indeed observed to some approximation for highly liquid securities in financial markets. We may or may not want to assert or test efficiency in the long-run sense. One interesting long-run implication is that a trader can become infinitely rich unless the quadratic variation of the security's price over the entire interval $[0, \infty)$ is finite. Finite quadratic variation means that trading eventually dwindles or stops, and it would not be unreasonable to suppose that this always happens; usually a publicly traded corporation does eventually go bankrupt or otherwise disappear.

Chapter 13's model is also rich enough that we can use it to explain why continuous time forces on us measurability assumptions that we were able to avoid in discrete time. If we do not require the trader's strategy to be measurable or restrict it in some other equally effective way, then the axiom of choice implies the existence of trading strategies that make the trader infinitely rich instantly as soon as the price path is not constant. This is the topic of Section 13.5.

Chapter 14 develops a more abstract and more powerful picture, which we use in the remaining chapters. Here the market chooses its move ω from an abstract *sample space* Ω equipped with a filtration. We consider a countable number of basic securities, with continuous price paths determined by ω, processes adapted to the filtration. This is the standard measure-theoretic setup. But instead of positing a probability measure, we interpret the notion of trading in the securities very broadly, allowing the trader to choose not only from the capital processes resulting from discrete trading strategies but also from the much broader class of capital processes obtained from these by taking limits. We call the mathematical object constructed in this way a *martingale space*. The capital processes obtained by taking limits include processes that we call *nonnegative supermartingales* and processes that we call *continuous martingales.* The nonnegative supermartingales play the role played by nonnegative supermartingales in Parts I–III, forcing properties in the sense that a given property must hold unless a particular nonnegative supermartingale becomes infinite, in this case instantly. This allows us to develop the Itô calculus.

The notion of martingale spaces in Chapter 14 assumes that the trader can move freely in and out of the market. He has, so to speak, a bank account that he can overdraw at a zero rate of interest relative to the numeraire in which prices are measured. Chapter 15 develops a theory in which changes in the trader's portfolio must be self-financing. The spaces thus constructed are called *market spaces*, and the objects analogous to nonnegative supermartingales and continuous martingales are called simply *nonnegative supergales* and *continuous gales*. If we select a positive continuous gale to serve as our numeraire, dividing all the prices at each time t by its value at time t, then we obtain a martingale space. This framework allows us to establish a game-theoretic version of Girsanov's theorem, which relates the continuous martingales and semimartingales in the martingale space obtained using one numeraire to those in the martingale space obtained using another.

In Chapter 16, we use the game-theoretic Girsanov theorem to derive a game-theoretic CAPM, a game-theoretic explanation of the equity premium, and related results. In order to apply our mathematical results on these topics to an actual market, we must specify a numeraire that we consider efficient for that market – one that we do not think a trader can outperform in the long run. Suitable numeraires have been proposed by Burton Malkiel [260] (a broad market index), by Ioannis Karatzas and Constantinos Kardaras ([210] and their forthcoming book [209]), and by Eckhard Platen and David Heath [310].

Chapter 17 considers a market in which there are only finitely many securities and develops a game-theoretic version of E. Robert Fernholz's stochastic portfolio theory [139]. In Fernholz's theory, which is stochastic in the sense that it is based on the theory of continuous-time stochastic processes, the total capitalization of a market is used as the numeraire. Fernholz assumes that the market maintains its diversity (no single stock dominates) and that trading never dies out (in particular, the total quadratic variation with respect to the numeraire is unbounded). Under these assumptions, the total capitalization turns out not to be an efficient numeraire; it can be outperformed by long-only portfolios that hold small-cap stocks in greater proportion than their capitalization. We obtain similar results in our game-theoretic framework.

13

Emergence of Randomness in Idealized Financial Markets

This chapter shows how Brownian motion emerges game-theoretically in continuous time. By trading in a security in the small – trading with higher and higher frequency – a trader can multiply the capital he risks infinitely as soon as the security's price path, modulo a time change, fails to behave locally like Brownian motion. This follows from the game-theoretic Dubins–Schwarz theorem, a central result of this chapter.

This chapter's protocol involves a single security and two players, a market and a trader. The trader moves first, announcing a trading strategy. Then the market moves, announcing a continuous price path. In Section 13.1 we explain how a strategy for the trader is constructed and how it determines his capital process. We also define the concept of instant enforcement: we say that the trader can instantly force an event if he has a strategy that multiplies the capital he risks infinitely at any time at which the event fails. In Section 13.2, we show how the trader can instantly force some of the best-known properties of Brownian motion: the existence of quadratic variation, the absence of isolated crossings, the absence of strict monotonicity, and nondifferentiability.

In Section 13.3, we define game-theoretic upper expected value for the chapter's protocol and state one version of the protocol's Dubins–Schwarz theorem: a variable that is unchanged when the price path is changed by the addition of a constant or certain deformations of the time scale has upper expected value equal to its expected value under Brownian motion. In Section 13.4 we derive some of this

Game-Theoretic Foundations for Probability and Finance, First Edition. Glenn Shafer and Vladimir Vovk.
© 2019 John Wiley & Sons, Inc. Published 2019 by John Wiley & Sons, Inc.

theorem's consequences, including some of the properties already demonstrated in Section 13.2. One of the consequences concerns quadratic variation of nonnegative price paths; it tells us that the trader can multiply the capital he risks infinitely in the long run unless the total quadratic variation of the price path from 0 to ∞ is finite.

In most of this chapter, as in the remaining chapters of this book, we use measurability: the trader's strategy and hence his capital process are required to be Borel measurable. We managed to avoid imposing conditions of measurability in the discrete-time theory of Parts I–III, but here we cannot avoid it. As we show in Section 13.5, the trader can become infinitely rich as soon as the price path ceases to be constant if the requirement of measurability is dropped from the definitions in Section 13.1. This unwelcome result, a consequence of the axiom of choice, parallels consequences of that axiom encountered by measure theory nearly a hundred years ago. The Banach–Tarski paradox is an example: we can divide the unit ball into five pieces, necessarily not measurable, and then rearrange the pieces to get two unit balls [21, 320].

13.1   CAPITAL PROCESSES AND INSTANT ENFORCEMENT

As already noted, our protocol has two players, a market (playing the role of Reality) and a trader (playing the role of Skeptic). The market chooses a continuous function $\omega : [0, \infty) \to \mathbb{R}$ (the price path of a security) after the trader chooses a trading strategy. Our first task is to define the class of trading strategies from which the trader chooses.

Let Ω be the set of all continuous real-valued functions on $[0, \infty)$, equipped with the usual *Borel σ-algebra* $\mathcal{F}$ – i.e. the smallest σ-algebra making all functions $\omega \mapsto \omega(t)$, $t \in [0, \infty)$, measurable. We call elements of $\mathcal{F}$ *events*, and we call extended real-valued functions on Ω *extended variables* if they are $\mathcal{F}$-measurable.[1] We will also be interested, however, in subsets of $\mathcal{F}$ and functions on Ω that are not necessarily $\mathcal{F}$-measurable.

A *process X* is a family of extended variables $X_t : \Omega \to \overline{\mathbb{R}}$, $t \in [0, \infty)$, such that

$$\omega|_{[0,t]} = \omega'|_{[0,t]} \implies X_t(\omega) = X_t(\omega') \tag{13.1}$$

for all $\omega, \omega' \in \Omega$ and all $t \in [0, \infty)$. The condition (13.1) says that $X_t(\omega)$ depends on ω only via $\omega|_{[0,t]}$. This agrees with Part I's definition of a process as a function of the situation: the situations are the partial paths $\omega|_{[0,t]}$.[2]

---

[1]This is standard in measure-theoretic probability, but it departs from our practice in Parts I–III, where we called any subset of Ω an event and any extended real-valued function on Ω an extended variable.

[2]In measure-theoretic probability, a family of functions satisfying (13.1) is an *adapted process* (see the discussion of the Galmarino test is Section 13.7). As in Parts I–III, we drop the adjective *adapted* because we do not consider processes that do not satisfy (13.1).

A *stopping time* is an extended variable $\tau : \Omega \to [0, \infty]$ such that

$$\omega|_{[0,\tau(\omega)]} = \omega'|_{[0,\tau(\omega)]} \implies \tau(\omega) = \tau(\omega')$$

for all $\omega, \omega' \in \Omega$. This means that $\tau(\omega)$ does not change if we change ω over $(\tau(\omega), \infty)$. An extended variable F is τ-*measurable* if

$$\omega|_{[0,\tau(\omega)]} = \omega'|_{[0,\tau(\omega)]} \implies F(\omega) = F(\omega')$$

for all $\omega, \omega' \in \Omega$. This means that $F(\omega)$ depends on ω only via $\omega|_{[0,\tau(\omega)]}$. We often simplify $\omega(\tau(\omega))$ to $\omega(\tau)$. The argument ω is sometimes omitted in other cases as well.

The class of allowed strategies for the trader is defined in two steps. A *simple trading strategy* G consists of an increasing sequence of stopping times $\tau_1 \le \tau_2 \le \cdots$ such that $\lim_{n\to\infty} \tau_n(\omega) = \infty$ for each $\omega \in \Omega$ and, for each $n \in \mathbb{N}$, a bounded τ_n-measurable variable h_n. A simple trading strategy G and *initial capital* $c \in \mathbb{R}$ determine the *simple capital process*

$$\mathcal{K}_t^{G,c}(\omega) := c + \sum_{n=2}^{\infty} h_{n-1}(\omega)(\omega(\tau_n \wedge t) - \omega(\tau_{n-1} \wedge t)), \quad t \in [0, \infty). \quad (13.2)$$

Zero terms in the sum are ignored, so that the sum is finite for each t. We call $h_n(\omega)$ the trader's *position* taken at time τ_n and $\mathcal{K}_t^{G,c}(\omega)$ his *capital at time t*.

A *nonnegative capital process* is a process X of the form

$$X_t(\omega) := \sum_{k=1}^{\infty} \mathcal{K}_t^{G_k, c_k}(\omega), \quad (13.3)$$

where the simple capital processes $\mathcal{K}_t^{G_k, c_k}(\omega)$ are nonnegative for all t and ω, and the sum $\sum_{k=1}^{\infty} c_k$, with all its terms nonnegative, is finite. Though we require (13.3) to be nonnegative, it can take the value ∞. Since $\mathcal{K}_0^{G_k, c_k}(\omega) = c_k$ does not depend on ω, $X_0(\omega)$ also does not depend on ω and is sometimes abbreviated to X_0.

We think of (13.3) as the capital process of a trader who splits his initial capital into a countable number of accounts, running on each account a simple trading strategy that guarantees it will never go into debt.

For any event E and $t \in [0, \infty)$, let $E|_t$ be the subset of E consisting of ω such that

$$\omega'|_{[0,t]} = \omega|_{[0,t]} \implies \omega' \in E. \quad (13.4)$$

Intuitively, $E|_t$ is the "event" that E happens by time t, but it may fail to be in $\mathcal{F}$ (cf. Exercise 13.1). We say that the trader *can instantly block* E and that E is *instantly blockable* if (i) $E = \cup_{t\in[0,\infty)}E|_t$ (E is *positively decidable*) and (ii) there exists a nonnegative capital process X such that $X_0 = 1$ and, for all t and ω,

$$\omega \in E|_t \implies X_t(\omega) = \infty. \quad (13.5)$$

When E is instantly blockable, we say that E^c holds *with instant enforcement*. We often abbreviate *with instant enforcement* to *w.i.e.*

13.2 EMERGENCE OF BROWNIAN RANDOMNESS

We now identify some events that the trader can instantly block or force and prove that he can do so by spelling out strategies he can use. These are events to which Brownian motion assigns probability 0 (if the trader can instantly block them) or 1 (if the trader can instantly force them). In most cases, the fact that the trader can instantly block or force these events follows from Theorem 13.9, which we state in Section 13.3. But the specific strategies described here provide insights not so easily gleaned from the proof of Theorem 13.9, which is given in [404].

Isolated Crossings

Brownian motion assigns probability 0 to the occurrence of isolated zeros. More generally, for any $b \in \mathbb{R}$, Brownian motion assigns probability 0 to there being an isolated point in the set

$$\mathcal{L}_\omega(b) := \{t \in [0, \infty) | \omega(t) = b\}.$$

Proposition 13.1 *For any $b \in \mathbb{R}$, the trader can instantly block the event that $\mathcal{L}_\omega(b)$ has an isolated point.*

Proof If $\mathcal{L}_\omega(b)$ has an isolated point, there are rational numbers $a \geq 0$ and $D \neq 0$ such that strictly after the time $\inf\{t | t \geq a, \omega(t) = b\}$ ω does not take value b before hitting the value $b + D$ (this is true even if 0 is the only isolated point of $\mathcal{L}_\omega(b)$). Let us denote this event as $E_{a,D}$. Suppose, for concreteness, that D is positive (the case of negative D is treated analogously). For each $\epsilon > 0$ (we are particularly interested in small values of ϵ), there is a nonnegative simple capital process $X^{\epsilon,a,D}$ that starts from ϵ and takes value $D + \epsilon \geq D$ when $E_{a,D}$ happens (take position 1 at the time $\inf\{t | t \geq a, \omega(t) = b\}$ and then take position 0 when the set $\{b - \epsilon, b + D\}$ is hit). For each pair of rational numbers (a, D) such that $a \geq 0$ and $D \neq 0$, fix a positive weight $w_{a,D} > 0$ such that $\sum_{a,D} w_{a,D} = 1$, the sum being over all such pairs. Set

$$X := \sum_{\epsilon,a,D} w_{a,D} X^{\epsilon,a,D}, \tag{13.6}$$

where ϵ ranges over the set $\{1/2, 1/4, \ldots\}$. It remains to notice that the nonnegative capital process X satisfies $X_0 = 1$ and $X_t = \infty$ for any $t > 0$ such that $\mathcal{L}_\omega(b)$ has an isolated point in $[0, t]$. $\qquad\square$

Nonmonotonicity

Let us say that $t \in [0, \infty)$ is a *point of increase* for $\omega \in \Omega$ if there exists $\delta > 0$ such that $\omega(t_1) \leq \omega(t) \leq \omega(t_2)$ for all $t_1 \in ((t - \delta)^+, t]$ and $t_2 \in [t, t + \delta)$. Points of decrease are defined in the same way except that $\omega(t_1) \leq \omega(t) \leq \omega(t_2)$ is replaced by $\omega(t_1) \geq \omega(t) \geq \omega(t_2)$. We say that ω is *locally constant to the right of* $t \in [0, \infty)$ if there exists $\delta > 0$ such that ω is constant over the interval $[t, t + \delta]$. The following proposition is the game-theoretic counterpart of Dvoretzky et al. [122] result for Brownian motion.

Proposition 13.2 *It is instantly enforceable that ω has no points t of increase or decrease such that ω is not locally constant to the right of t.*

Exercise 13.11 shows that the qualification about local constancy to the right of t in Proposition 13.2 is essential. We will deduce this proposition from the theorem we state in the next section, Theorem 13.9, and Dvoretzky et al.'s result. But here we prove several of its corollaries by describing a strategy the trader can use. First, the price path is nowhere monotonic (unless constant).

Corollary 13.3 *The trader can instantly block ω being monotonic in an open interval where it is not constant.*

Proof Each interval of monotonicity where ω is not constant contains a rational time point a after which ω increases (if we assume, for concreteness, that "monotonic" means "increasing") by a rational amount $D > 0$ before hitting the level $\omega(a)$ again; let us denote this event as $E_{a,D}$. As in the proof of Proposition 13.1, it is easy to see that we can profit on the event $E_{a,D}$: there is a nonnegative simple capital process $X^{\epsilon,a,D}$ that starts from ϵ and takes value $D + \epsilon \geq D$ as soon as $E_{a,D}$ happens. The nonnegative capital process (13.6), where again $w_{a,D}$ are positive weights summing to 1 and ϵ ranges over the set $\{1/2, 1/4, \dots\}$, satisfies $X_0 = 1$ and $X_t = \infty$ for any $t > 0$ such that ω is monotonic in an open interval in $[0, t]$ without being constant in that interval. $\qquad\square$

Let us say that a closed interval $[t_1, t_2] \subseteq [0, \infty)$ is an *interval of local maximum* for ω if (i) ω is constant on $[t_1, t_2]$ but not constant on any larger interval containing $[t_1, t_2]$ and (ii) there exists $\delta > 0$ such that $\omega(s) \leq \omega(t)$ for all $s \in ((t_1 - \delta)^+, t_1) \cup (t_2, t_2 + \delta)$ and all $t \in [t_1, t_2]$. In the case where $t_1 = t_2$ we can say "point" instead of "interval." A ray $[t, \infty)$, $t \in [0, \infty)$, is a *ray of local maximum* for ω if (i) ω is constant on $[t, \infty)$ but not constant on any larger ray $[s, \infty)$, $s \in (0, t)$, and (ii) there exists $\delta > 0$ such that $\omega(s) \leq \omega(t)$ for all $s \in ((t - \delta)^+, t)$. An *interval* or *ray of strict local maximum* is defined in the same way except that $\omega(s) \leq \omega(t)$ is replaced by $\omega(s) < \omega(t)$. The definitions of intervals and rays of (strict) local minimum are obtained by obvious modifications; as usual "extremum" means maximum or minimum. We say that $t \in [0, \infty)$ is a *point of constancy* for ω if there exists $\delta > 0$ such that $\omega(s) = \omega(t)$ for all $s \in ((t - \delta)^+, t + \delta)$; points $t \in [0, \infty)$ that are not points of constancy are *points of nonconstancy*. (We do not count points of constancy among points of local extremum.)

Corollary 13.4   *It is instantly enforceable that any interval of local extremum is a point, any point of local extremum is strict, the ray of local extremum (if it exists) is strict, the set of points of local extremum is countable, and any neighborhood of any point of nonconstancy contains a point of local maximum and a point of local minimum.*

Proof We will prove only the statements concerning local maxima.

If ω has an interval of local maximum $[t_1, t_2]$ with $t_1 \neq t_2$, t_2 will be a point of decrease and ω will not be locally constant to the right of t_2, and by Proposition 13.2 it is instantly enforceable that there will be no such points (alternatively, one could use the direct argument given in the proof of Corollary 13.3). We can see that no such $[t_1, t_2]$ can even be an interval of local maximum "on the right."

Now suppose that there is a point or ray of local maximum that is not strict. In this case there is a quadruple $0 < t_1 < t_2 < t_3 < t_4$ of rational numbers and another rational number $D > 0$ such that $\max_{t \in [t_1, t_2]} \omega(t) = \max_{t \in [t_3, t_4]} \omega(t) > \omega(t_4) + D$. The absence of such a quadruple is instantly enforceable: proceed as in the proof of Proposition 13.1.

The set of all points of strict local maximum is countable, as the following standard argument demonstrates: each point of strict local maximum can be enclosed in an open interval with rational end-points in which that point is a strict maximum, and all these open intervals will be different.

Finally, Corollary 13.3 immediately implies that every neighborhood of every point of nonconstancy contains a point of local maximum.   □

Here is a game-theoretic counterpart of the classical result that Brownian motion is nowhere differentiable (Paley et al. [298]).

Corollary 13.5   *It is instantly enforceable that ω does not have a nonzero derivative anywhere.*

Proof A point with a nonzero derivative is a point of increase or decrease such that ω is not locally constant to the right of it.   □

The necessity of various conditions in Corollaries 13.3–13.5 is discussed in Exercise 13.2.

Emergence of Volatility

For each $p \in (0, \infty)$, the *p-variation* of a continuous function $f : [a, b] \to \mathbb{R}$ defined over an interval $[a, b] \subset \mathbb{R}$ with $a < b$ is

$$\mathrm{var}_p(f) := \sup_{\kappa} \sum_{i=1}^{n} |f(t_i) - f(t_{i-1})|^p, \tag{13.7}$$

where n ranges over all positive integers and κ over all subdivisions $a = t_0 < t_1 < \cdots < t_n = b$ of the interval $[a, b]$. It can be shown (see Exercise 13.3) that there exists a unique number $\mathrm{vi}(f) \in [0, \infty]$, called the *variation index* of f, such that $\mathrm{var}_p(f)$ is finite when $p > \mathrm{vi}(f)$ and infinite when $p < \mathrm{vi}(f)$; notice that $\mathrm{vi}(f) \notin (0, 1)$.

The following is a game-theoretic counterpart of a property stated by Lévy for Brownian motion [246] and by Dominique Lepingle [239, Theorem 1 and Proposition 3] for continuous semimartingales.

Proposition 13.6 *It is instantly enforceable that*

$$\forall t \in [0, \infty) : \mathrm{vi}(\omega|_{[0,t]}) = 2 \text{ or } \omega \text{ is constant over } [0, t]. \tag{13.8}$$

The condition (13.8) is equivalent to $\mathrm{vi}(\omega|_{[0,t]}) \in \{0, 2\}$ for all t.

Proof of Proposition 13.6 with the "= 2" in (13.8) replaced by "≥ 2." Assume, without loss of generality, that $\omega(0) = 0$. We need to show that it is instantly blockable that, for some t, $\mathrm{vi}(\omega_{[0,t]}) < 2$ and $\mathrm{nc}(\omega_{[0,t]})$, where $\mathrm{nc}(f)$ stands for "f is not constant." Since a countable convex mixture of nonnegative capital processes is a nonnegative capital process, it suffices to show that, for each $p \in (0, 2)$, it is instantly blockable that, for some t, $\mathrm{vi}(\omega_{[0,t]}) < p$ and $\mathrm{nc}(\omega_{[0,t]})$. Fix such a p. It suffices to show that it is instantly blockable that, for some t, $\mathrm{var}_p(\omega_{[0,t]}) < \infty$ and $\mathrm{nc}(\omega_{[0,t]})$. Therefore, it suffices to show that, for each $C \in (0, \infty)$, it is instantly blockable that, for some t, $\mathrm{var}_p(\omega_{[0,t]}) < C$ and $\mathrm{nc}(\omega_{[0,t]})$. Fix such a C. Finally, it suffices to show that, for each $A > 0$, the event

$$E_{p,C,A} := \{ \omega \in \Omega | \exists t \in [0, \infty) : \mathrm{var}_p(\omega_{[0,t]}) < C \text{ and } \sup_{s \in [0,t]} |\omega(s)| > A \}$$

$$= \{ \omega \in \Omega | \exists t \in [0, \infty) : \mathrm{var}_p(\omega_{[0,t]}) < C \text{ and } |\omega(t)| > A \}$$

is instantly blockable. Fix such an A.

Choose a $\delta > 0$ such that $A/\delta \in \mathbb{N}$ and let $\Gamma := \{k\delta | k \in \mathbb{Z}\}$ be the corresponding grid. Define a sequence of stopping times τ_n inductively by $\tau_0 := 0$ and

$$\tau_n(\omega) := \inf\{t > \tau_{n-1}(\omega) | \omega(t) \in \Gamma \setminus \{\omega(\tau_{n-1})\}\}, \quad n \in \mathbb{N},$$

with $\inf \emptyset$ understood to be ∞. Set $T_A(\omega) := \inf\{t \,|\, |\omega(t)| = A\}$, again with $\inf \emptyset := \infty$, and

$$
h_n(\omega) := \begin{cases} 2\omega(\tau_n) & \text{if } \tau_n(\omega) < T_A(\omega) \text{ and } n+1 < C/\delta^p, \\ 0 & \text{otherwise.} \end{cases}
$$

The simple capital process (analogous to Kolmogorov's martingale, Section 3.1) corresponding to the simple trading strategy $G := (\tau_n, h_n)$ and initial capital $c := \delta^{2-p} C$ will satisfy

$$
\omega^2(\tau_n) - \omega^2(\tau_{n-1}) = 2\omega(\tau_{n-1})\big(\omega(\tau_n) - \omega(\tau_{n-1})\big) + \big(\omega(\tau_n) - \omega(\tau_{n-1})\big)^2
$$
$$
= \mathcal{K}^{G,c}_{\tau_n}(\omega) - \mathcal{K}^{G,c}_{\tau_{n-1}}(\omega) + \delta^2
$$

provided $\tau_n(\omega) \le T_A(\omega)$ and $n < C/\delta^p$, and so satisfy

$$
\omega^2(\tau_N) = \mathcal{K}^{G,c}_{\tau_N}(\omega) - \mathcal{K}^{G,c}_0 + N\delta^2 = \mathcal{K}^{G,c}_{\tau_N}(\omega) - \delta^{2-p}C + \delta^{2-p}N\delta^p \le \mathcal{K}^{G,c}_{\tau_N}(\omega) \quad (13.9)
$$

provided $\tau_N(\omega) \le T_A(\omega)$ and $N \le C/\delta^p$. On the event $E_{p,C,A}$ we have $T_A(\omega) < \infty$ and $N < C/\delta^p$ for the N defined by $\tau_N = T_A$. Therefore, on this event

$$
A^2 = \omega^2(T_A) \le \mathcal{K}^{G,c}_{T_A}(\omega).
$$

We can see that $\mathcal{K}^{G,c}_t(\omega)$ increases from $\delta^{2-p}C$, which can be made arbitrarily small by making δ small, to A^2 over $t \in [0, T_A]$; combining a countable number of such processes shows that the event $E_{p,C,A}$ is instantly blockable.

The only remaining gap in our argument is that $\mathcal{K}^{G,c}_t$ may become negative strictly between some $\tau_{n-1} < T_A$ and τ_n with $n < C/\delta^p$ (it will be nonnegative at all $\tau_N \in [0, T_A]$ with $N < C/\delta^p$, as can be seen from (13.9)). We can, however, bound $\mathcal{K}^{G,c}_t$ for $\tau_{n-1} < t < \tau_n$ as follows:

$$
\mathcal{K}^{G,c}_t(\omega) = \mathcal{K}^{G,c}_{\tau_{n-1}}(\omega) + 2\omega(\tau_{n-1})(\omega(t) - \omega(\tau_{n-1})) \ge 2|\omega(\tau_{n-1})|(-\delta) \ge -2A\delta,
$$

and so we can make the simple capital process nonnegative by adding a negligible amount $2A\delta$ to the initial capital. $\qquad \square$

To complete the proof of Proposition 13.6, we must show that $\mathrm{vi}(\omega|_{[0,t]}) \le 2$ w.i.e. This follows from Proposition 13.8, which gives a much more precise result.

Let $M^b_a(f)$ (respectively $D^b_a(f)$) be the number of upcrossings (respectively downcrossings) of an open interval (a, b) by a function $f : [0, t] \to \mathbb{R}$ during the time interval $[0, t]$. For each $h > 0$ set

$$
M(f, h) := \sum_{k \in \mathbb{Z}} M^{kh}_{(k-1)h}(f), \quad D(f, h) := \sum_{k \in \mathbb{Z}} D^{kh}_{(k-1)h}(f).
$$

The following lemma presents Doob's upcrossing argument in the form of an inequality.

Lemma 13.7   *Let $0 \leq a < b$ be real numbers. There exists a nonnegative simple capital process S that starts from $S_0 = a$ and satisfies, for all nonnegative $\omega \in \Omega$,*

$$S_t(\omega) \geq (b - a)M_a^b(\omega|_{[0,t]}). \tag{13.10}$$

Proof   The following standard (and familiar from Sections 4.2 and 7.2) argument will be easy to formalize. A simple trading strategy G leading, with a as initial capital, to S can be defined as follows. At first G takes position $1_{\omega(0) \leq a}$. Whenever ω hits $[0, a]$, G takes position 1. And whenever ω hits $[b, \infty)$, G takes position 0. If ω is nonnegative, S will also be nonnegative; otherwise, we should stop trading when ω hits 0.

Formally, we define $\tau_1(\omega) := \inf\{s|\omega(s) \in [0, a]\}$ and, for $n = 2, 3, \ldots,$

$$\tau_n(\omega) := \inf\{s|s > \tau_{n-1}(\omega) \text{ and } \omega(s) \in I_n\},$$

where $I_n := [b, \infty)$ for even n and $I_n := [0, a]$ for odd n; as usual, the expression $\inf \emptyset$ is interpreted as ∞. Since ω is a continuous function and $[0, a]$ and $[b, \infty)$ are closed sets, the infima in the definitions of $\tau_1, \tau_2, \ldots$ are attained, and these functions are stopping times satisfying $\tau_n \to \infty$. Therefore, $\omega(\tau_1) \leq a$, $\omega(\tau_2) \geq b$, $\omega(\tau_3) \leq a$, $\omega(\tau_4) \geq b$, and so on. The positions taken by G at the times $\tau_1, \tau_2, \ldots$ are $h_1 := 1$, $h_2 := 0$, $h_3 := 1$, $h_4 := 0$, etc.; remember that the initial capital is a. Let $t \in [0, \infty)$ and n be the largest integer such that $\tau_n \leq t$ (with $n := 0$ when $\tau_1 > t$). Now we obtain from (13.2): if n is even,

$$S_t(\omega) = \mathcal{K}_t^{G,c}(\omega)$$

$$= a + (\omega(\tau_2) - \omega(\tau_1)) + (\omega(\tau_4) - \omega(\tau_3)) + \cdots + (\omega(\tau_n) - \omega(\tau_{n-1}))$$

$$\geq a + (b - a)M_a^b(\omega|_{[0,t]}),$$

and if n is odd,

$$S_t(\omega) = \mathcal{K}_t^{G,c}(\omega)$$

$$= a + (\omega(\tau_2) - \omega(\tau_1)) + (\omega(\tau_4) - \omega(\tau_3)) + \cdots + (\omega(\tau_{n-1}) - \omega(\tau_{n-2}))$$

$$\quad + (\omega(t) - \omega(\tau_n))$$

$$\geq a + (b - a)M_a^b(\omega|_{[0,t]}) + (\omega(t) - \omega(\tau_n))$$

$$\geq a + (b - a)M_a^b(\omega|_{[0,t]}) + (0 - a) = (b - a)M_a^b(\omega|_{[0,t]});$$

in both cases, (13.10) holds. In particular, $S_t(\omega)$ is nonnegative; the same argument applied to $s \in [0, t]$ in place of t shows that $S_s(\omega)$ is nonnegative for all $s \in [0, t]$. □

Now we generalize the definition (13.7). Let $\phi : [0, \infty) \to [0, \infty)$. For a continuous function $f : [a, b] \to \mathbb{R}$, where $\emptyset \neq [a, b] \subset \mathbb{R}$, we set

$$\mathrm{var}_\phi(f) := \sup_\kappa \sum_{i=1}^n \phi(|f(t_i) - f(t_{i-1})|), \tag{13.11}$$

where κ ranges over all partitions $a = t_0 \leq t_1 \leq \cdots \leq t_n = b, n \in \mathbb{N}$, of $[a, b]$. The notation clashes with (13.7), which corresponds to the case $\phi(u) := u^p$, but it is standard and will not lead to ambiguity.

Proposition 13.8   *Suppose* $\phi : (0, \infty) \to (0, \infty)$ *satisfies*

$$\int_0^1 \frac{\phi(u)}{u^3} \, du < \infty \tag{13.12}$$

and

$$\sup_{0 < u \leq v \leq 2u} \frac{\phi(v)}{\phi(u)} < \infty. \tag{13.13}$$

Then $\forall t : \mathrm{var}_\phi(\omega|_{[0,t]}) < \infty$ *w.i.e., where* $\phi(0)$ *is set to 0.*

Informally, the condition (13.12) says that $\phi(u)$ should approach 0 somewhat faster than u^2 as $u \to 0$, and (13.13) is a weak regularity condition. To obtain the inequality $\leq$ in Proposition 13.6, set $\phi(u) := u^p$, where $p > 2$ is rational, and remember that a countable convex mixture of nonnegative capital processes is a nonnegative capital process. Another simple example of a function ϕ satisfying (13.12) is $\phi(u) := (u/\log^* u)^2$, where $\log^* u := 1 \vee |\log u|$. A better example is $\phi(u) := u^2/(\log^* u \log^* \log^* u \cdots)$, where log is binary logarithm (the product is finite if we ignore the factors equal to 1). For a proof of (13.12) for this function, see [240, Appendices B and C]; it is interesting that (13.12) fails if log is understood as natural logarithm. However, even for the last choice of ϕ, the inequality $\mathrm{var}_\phi(\omega) < \infty$ a.s. is still weaker than the inequality $\mathrm{var}_\psi(\omega) < \infty$ a.s., with ψ defined by (13.22), which we will deduce from the game-theoretic Dubins–Schwarz theorem in combination with known results about measure-theoretic Brownian motion (see Corollary 13.21).

Proof of Proposition 13.8   It suffices to prove the statement of the proposition under the assumption that $\omega(t) \in [a, b]$ for given $a, b \in \mathbb{R}$; this follows from countable convex mixtures of nonnegative capital processes being nonnegative capital processes. It will be more convenient to assume that ω is nonnegative and $\sup \omega \leq 2^L$ for a given positive integer L. Fix such an L.

Replacing the integral in the condition (13.12) by a sum over u of the form 2^{-j}, $j = 0, 1, \ldots$, we can rewrite this condition as

$$\sum_{j=0}^{\infty} 2^{2j} \phi(2^{-j}) < \infty. \tag{13.14}$$

Set $w(j) := 2^{2j} \phi(2^{-j}), j = 0, 1, \ldots$; by (13.14), $\sum_{j=0}^{\infty} w(j) < \infty$. Without loss of generality we will assume that $\sum_{j=0}^{\infty} w(j) = 1$.

Let $t \in (0, \infty)$ and $0 = t_0 \le t_1 \le \cdots \le t_n = t$ be a partition of the interval $[0, t]$; without loss of generality we replace all "$\le$" by "$<$." Fix a nonnegative $\omega \in \Omega$ satisfying $\sup \omega \le 2^L$. Split $\sum_{i=1}^{n} \phi(|\omega(t_i) - \omega(t_{i-1})|)$ into two parts:

$$\sum_{i=1}^{n} \phi(|\omega(t_i) - \omega(t_{i-1})|) = \sum_{i \in I_+} \phi(\omega(t_i) - \omega(t_{i-1})) + \sum_{i \in I_-} \phi(\omega(t_{i-1}) - \omega(t_i)),$$

where

$$I_+ := \{i \mid \omega(t_i) - \omega(t_{i-1}) > 0\},$$
$$I_- := \{i \mid \omega(t_i) - \omega(t_{i-1}) < 0\}.$$

By Lemma 13.7, for each $j = 0, 1, \ldots$ and each $k \in \{1, \ldots, 2^{L+j}\}$, there exists a nonnegative simple capital process $S^{j,k}$ that starts from $(k-1)2^{-j}$ and satisfies

$$S_t^{j,k}(\omega) \ge 2^{-j} M_{(k-1)2^{-j}}^{k2^{-j}}(\omega).$$

Summing $2^{-L-j} S^{j,k}$ over $k = 1, \ldots, 2^{L+j}$ (in other words, averaging $S^{j,k}$), we obtain a nonnegative capital process S^j such that

$$S_0^j = \sum_{k=1}^{2^{L+j}} (k-1) 2^{-L-2j} \le 2^{L-1},$$

$$S_t^j(\omega) \ge 2^{-L-2j} M(\omega, 2^{-j}).$$

For each $i \in I_+$, let $j(i)$ be the smallest nonnegative integer j satisfying

$$\exists k \in \mathbb{N} : \omega(t_{i-1}) \le (k-1)2^{-j} \le k2^{-j} \le \omega(t_i).$$

Summing $w(j)S^j$ over $j = 0, 1, \ldots$, we obtain a nonnegative capital process S such that $S_0 \le 2^{L-1}$ and

$$S_t(\omega) \ge \sum_{j=0}^{\infty} w(j) 2^{-L-2j} M(\omega, 2^{-j}) \ge \sum_{i \in I_+} w(j(i)) 2^{-L-2j(i)}$$

$$= 2^{-L} \sum_{i \in I_+} \phi(2^{-j(i)}) \ge \delta \sum_{i \in I_+} \phi(\omega(t_i) - \omega(t_{i-1})), \tag{13.15}$$

where $\delta > 0$ depends only on L and the supremum in (13.13). The second inequality in the chain (13.15) follows from the fact that to each $i \in I_+$ corresponds an upcrossing of an interval of the form $((k-1)2^{-j(i)}, k2^{-j(i)})$.

An inequality analogous to the inequality between the second and the last terms of the chain (13.15) can be proved for downcrossings instead of upcrossings, I_- instead of I_+, and $\omega(t_{i-1})$ and $\omega(t_i)$ swapped around. Using this inequality (in the third "$\geq$" below) gives

$$S_t(\omega) \geq \sum_{j=0}^{\infty} w(j)2^{-L-2j}M(\omega, 2^{-j}) \geq \sum_{j=0}^{\infty} w(j)2^{-L-2j}(D(\omega, 2^{-j}) - 2^{L+j})$$

$$\geq \delta \sum_{i \in I_-} \phi(\omega(t_{i-1}) - \omega(t_i)) - \sum_{j=0}^{\infty} w(j)2^{-j} \geq \delta \sum_{i \in I_-} \phi(\omega(t_{i-1}) - \omega(t_i)) - 1.$$

Averaging the two lower bounds for $S_t(\omega)$, we obtain

$$S_t(\omega) \geq \frac{\delta}{2} \sum_{i=1}^{n} \phi(|\omega(t_i) - \omega(t_{i-1})|) - \frac{1}{2}.$$

Taking supremum over all partitions gives

$$S_t(\omega) \geq \frac{\delta}{2} \mathrm{var}_\phi(\omega|_{[0,t]}) - \frac{1}{2}.$$

We can see that the event $\exists t : \mathrm{var}_\phi(\omega|_{[0,t]}) = \infty$ is instantly blockable. $\square$

13.3 EMERGENCE OF BROWNIAN EXPECTATION

In this section, we state a general game-theoretic Dubins–Schwarz theorem.[3] This theorem, proven in [404], concerns an extended variable F whose value $F(\omega)$ does not change when a constant is added to the price path ω or the time scale is deformed in certain natural ways. It says that when F is invariant in this way, its game-theoretic upper expected value is equal to its expected value under a Brownian motion that starts with the value 0 at time 0.

Superinvariance

We begin by defining precisely the invariance we require of the extended variable F.

Let us call a continuous increasing (not necessarily strictly increasing) function $f : [0, \infty) \to [0, \infty)$ satisfying $f(0) = 0$ a *time transformation*. Equipped with the binary operation of composition, the time transformations form a monoid, with the

---

[3] An alternative game-theoretic Dubins–Schwarz theorem says that the price path becomes Brownian motion when clock time is replaced by quadratic variation. See Section 13.7 for discussion and references.

identity $t \mapsto t$ as the unit. Given $\omega \in \Omega$ and a time transformation f, we write ω^f for the element $\omega \circ f$ of Ω.

We call a pair (c, f), where $c \in \mathbb{R}$ and f is a time transformation, a *level and time transformation*, or an *LT-transformation*. Given $\omega \in \Omega$ and an LT-transformation (c, f), we write $\omega^{c,f}$ for the element $c + \omega \circ f$ of Ω. The *trail* of $\omega \in \Omega$ is the set of all $\psi \in \Omega$ such that $\psi^{c,f} = \omega$ for some LT-transformation (c, f). An event E is *LT-superinvariant* if for any ω in E, the whole trail of ω is in E. We write $\mathcal{I}$ for the subset of $\mathcal{F}$ consisting of all LT-superinvariant events.[4] The set $\mathcal{I}$ is not closed under complementation and hence is not a σ-algebra (see Exercise 13.8).

We say that a nonnegative extended variable $F : \Omega \to [0, \infty]$ is $\mathcal{I}$-*measurable* if, for each $c \in [0, \infty)$, the set $\{\omega | F(\omega) \geq c\}$ is in $\mathcal{I}$. In other words (Exercise 13.9), F is $\mathcal{I}$-measurable if, for each $\omega \in \Omega$ and each LT-transformation (c, f),

$$F(\omega^{c,f}) \leq F(\omega). \tag{13.16}$$

Intuitively, (13.16) means that $F(\omega)$ measures some aspect of ω's variation. This is expressed by an inequality rather than an equality because if $\lim_{t \to \infty} f(t) < \infty$, the variation in ω after $\lim_{t \to \infty} f(t)$ will be missing from $\omega^{c,f}$ (see Remark 13.18 for details). But (13.16) implies that $F(\omega + c) = F(\omega)$ for all $c \in \mathbb{R}$, and we will define a group $\mathcal{G}$ of particularly interesting time transformations such that $F(\omega^{c,f}) = F(\omega)$ when F is LT-superinvariant and $f \in \mathcal{G}$ (see the last subsection of this section).

A Game-Theoretic Dubins–Schwarz Theorem

The *upper expected value* of a function $F : \Omega \to [0, \infty]$ is

$$\overline{\mathbb{E}}(F) := \inf\{X_0 | \forall \omega \in \Omega : \liminf_{t \to \infty} X_t(\omega) \geq F(\omega)\}, \tag{13.17}$$

where X ranges over the nonnegative capital processes.[5] In the terminology of finance (and neglecting the fact that the infimum in (13.17) may not be attained), $\overline{\mathbb{E}}(F)$ is the lowest price at which the trader can superhedge the European contingent claim paying F at time ∞. Remember that $\mathcal{W}$ is the Wiener measure on Ω (see the section on terminology and notation at the end of the book) and that $\mathcal{W}(F)$ stands for $\int F \, d\mathcal{W}$.

Theorem 13.9   *Each nonnegative $\mathcal{I}$-measurable extended variable F satisfies*

$$\overline{\mathbb{E}}(F) = \mathcal{W}(F). \tag{13.18}$$

---

[4] The symbol $\mathcal{I}$ is a Gothic I.
[5] This definition is analogous to (7.19), the definition of global upper expected value in discrete time, except that here we consider only nonnegative functions.

See [404] for a proof of the inequality $\leq$, the part of the equality in (13.18) that is more difficult to prove. See Exercise 13.4 for a proof of the inequality $\geq$.

We can specialize Theorem 13.9 to events using the concept of upper probability. As in Parts I and II, the *upper probability* of a subset E of Ω is

$$\overline{\mathbb{P}}(E) := \overline{\mathbb{E}}(1_E), \tag{13.19}$$

and the *lower probability* is $\underline{\mathbb{P}}(E) := 1 - \overline{\mathbb{P}}(E^c)$. We say that E is *null* if $\overline{\mathbb{P}}(E) = 0$.[6]

Corollary 13.10 *If $E \in \mathcal{I}$, then $\overline{\mathbb{P}}(E) = \mathcal{W}(E)$. In particular, $E \in \mathcal{I}$ is null if and only if $\mathcal{W}(E) = 0$.*

Notice that $\overline{\mathbb{P}}(\Omega) = 1$. This follows from Corollary 13.10, but it is also evident directly from the definition of upper probability, because a nonnegative capital process cannot strictly increase its value on a constant price path. This can be called the *coherence* of the protocol; stronger concepts of no arbitrage are discussed at the end of Chapter 16.

According to Corollary 13.10, upper probability looks like Wiener measure so long as we consider only events in $\mathcal{I}$. But as the following lemma shows, the picture is very different outside $\mathcal{I}$.

Lemma 13.11 *Suppose $P(E) = 1$, where E is an event and P is a probability measure on $(\Omega, \mathcal{F})$ that makes the process $S_t(\omega) := \omega(t)$ a martingale with respect to the filtration $(\mathcal{F}_t)$, where $\mathcal{F}_t$ is the smallest σ-algebra making all functions $\omega \mapsto \omega(s)$, $s \in [0, t]$, measurable. Then $\overline{\mathbb{P}}(E) = 1$.*

Proof It suffices to prove that (13.2) is a local martingale under P. In this case $\overline{\mathbb{P}}(E) < 1$ in conjunction with the maximal inequality for nonnegative supermartingales would contradict the assumption that $P(E) = 1$. It can be checked using the optional sampling theorem that each addend in (13.2) is a martingale, and so each partial sum in (13.2) is a martingale and (13.2) itself is a local martingale. □

Divergent and Nowhere Constant Price Paths

Because of its generality, some aspects of Theorem 13.9 may appear counterintuitive (for example, Ω is the only element of $\mathcal{I}$ that contains a constant). The picture simplifies when we consider only price paths that are divergent and nowhere constant.

---

[6] As in discrete time, a property of $\omega \in \Omega$ holds *almost surely* if the set where it fails is null. We do not use this notion directly in this chapter, but see Lemma 13.17.

A price path ω is *divergent* if there is no $c \in \mathbb{R}$ such that $\lim_{t \to \infty} \omega(t) = c$; it is *nowhere constant* if there is no interval (t_1, t_2), where $0 \le t_1 < t_2$, such that ω is constant on (t_1, t_2). We write DS for the event that the price path is divergent and nowhere constant. Applying Lemma 13.11 to Brownian motion gives

$$\overline{\mathbb{P}}(\text{DS}) = 1. \tag{13.20}$$

Intuitively, DS is the event that trading never stops completely and does not slow down asymptotically to an almost complete standstill. The condition that a price path be in DS is weaker than Dubins and Schwarz's condition that it be unbounded and nowhere constant [120].

In order to specialize Theorem 13.9 to subsets of DS, we introduce the notions of LT-invariance and restricted upper and lower probability. We write $\mathcal{G}$ for the group consisting of all unbounded and strictly increasing time transformations $f : [0, \infty) \to [0, \infty)$. An event E is *LT-invariant* if it contains the whole orbit $\{\omega^{c,f} | c \in \mathbb{R}, f \in \mathcal{G}\}$ of each of its elements $\omega \in E$. It is clear that DS is LT-invariant. The LT-invariant events form a σ-algebra: E^c is LT-invariant whenever E is (cf. Exercise 13.8).

Lemma 13.12 *An event $E \subseteq \text{DS}$ is LT-superinvariant if and only if it is LT-invariant.*

Proof If E (not necessarily $E \subseteq \text{DS}$) is LT-superinvariant, $\omega \in E$, $c \in \mathbb{R}$, and $f \in \mathcal{G}$, we have $\psi := \omega^{c,f} \in E$ as $\psi^{-c,f^{-1}} = \omega$. Therefore, LT-superinvariance always implies LT-invariance.

It is clear that, for all $\psi \in \Omega$ and LT-transformations (c, f), $\psi^{c,f} \notin \text{DS}$ unless $f \in \mathcal{G}$. Let $E \subseteq \text{DS}$ be LT-invariant, $\omega \in E$, $c \in \mathbb{R}$, f be a time transformation, and $\psi^{c,f} = \omega$. Since $\psi^{c,f} \in \text{DS}$, we have $f \in \mathcal{G}$, and so $\psi = \omega^{-c,f^{-1}} \in E$. Therefore, LT-invariance implies LT-superinvariance for subsets of DS. $\qquad \square$

Lemma 13.13 *An event $E \subseteq \text{DS}$ is LT-superinvariant if and only if $\text{DS} \setminus E$ is LT-superinvariant.*

Proof This follows immediately from Lemma 13.12. $\qquad \square$

For any subsets B and E of Ω, set

$$\overline{\mathbb{P}}(E; B) := \inf\{X_0 | \forall \omega \in B : \liminf_{t \to \infty} X_t(\omega) \ge \mathbf{1}_E(\omega)\} = \overline{\mathbb{P}}(E \cap B),$$

with X again ranging over the nonnegative capital processes. When $\overline{\mathbb{P}}(E; B) = 0$, we say that E is *B-null*. We see immediately that $\overline{\mathbb{P}}(\cdot; B)$ is countably (and hence finitely) subadditive.

Lemma 13.14   *If B and $E_1, E_2, \ldots$ are subsets of Ω, then*

$$\overline{\mathbb{P}}\left(\bigcup_{n=1}^{\infty} E_n; B \right) \leq \sum_{n=1}^{\infty} \overline{\mathbb{P}}(E_n; B).$$

In particular, a countable union of B-null sets is B-null.

Set $\underline{\mathbb{P}}(E; B) := 1 - \overline{\mathbb{P}}(E^c; B) = \mathbb{P}(E \cup B^c)$. We call the quantities $\overline{\mathbb{P}}(E; B)$ and $\underline{\mathbb{P}}(E; B)$ *E's upper and lower probabilities when restricted to B*,[7] and we think of them as *E's* upper and lower probabilities when the market is required to choose ω from B. We will use these concepts only in the case where $\overline{\mathbb{P}}(B) = 1$.

Lemma 13.15   *If $\overline{\mathbb{P}}(B) = 1$, then $\underline{\mathbb{P}}(E; B) \leq \overline{\mathbb{P}}(E; B)$ for every set $E \subseteq \Omega$.*

Proof Suppose $\underline{\mathbb{P}}(E; B) > \overline{\mathbb{P}}(E; B)$ for some E; by the definition of $\underline{\mathbb{P}}$, this would mean that $\overline{\mathbb{P}}(E; B) + \overline{\mathbb{P}}(E^c; B) < 1$. Since $\overline{\mathbb{P}}(\cdot; B)$ is finitely subadditive (Lemma 13.14), this would imply $\overline{\mathbb{P}}(\Omega; B) < 1$, which is equivalent to $\overline{\mathbb{P}}(B) < 1$ and, therefore, contradicts our assumption.   □

When $\underline{\mathbb{P}}(E; B) = \overline{\mathbb{P}}(E; B)$, we say that their common value is *E's probability when restricted to B.*

The following corollary of Theorem 13.9 is a Dubins–Schwarz result for price paths in DS. It says that every $\mathcal{I}$-measurable event has a probability when the market is required to choose a price path in DS and that this probability is equal to the probability given by the Wiener measure.

Corollary 13.16   *If $E \in \mathcal{I}$, then $\overline{\mathbb{P}}(E; \mathrm{DS}) = \underline{\mathbb{P}}(E; \mathrm{DS}) = \mathcal{W}(E)$.*

Proof Let $E \in \mathcal{I}$. Events $E \cap \mathrm{DS}$ and $E^c \cap \mathrm{DS}$ belong to $\mathcal{I}$: for the first of them, this immediately follows from $\mathrm{DS} \in \mathcal{I}$ and $\mathcal{I}$ being closed under finite intersections (cf. Exercise 13.8), and for the second, it suffices to notice that $E^c \cap \mathrm{DS} = \mathrm{DS} \setminus (E \cap \mathrm{DS}) \in \mathcal{I}$ (cf. Lemma 13.13). Applying (13.18) to these two events and making use of (13.20) and Lemma 13.15, we obtain

$$\mathcal{W}(E) = 1 - \mathcal{W}(E^c \cap \mathrm{DS}) = 1 - \overline{\mathbb{P}}(E^c; \mathrm{DS}) = \underline{\mathbb{P}}(E; \mathrm{DS}) \leq \overline{\mathbb{P}}(E; \mathrm{DS})$$

$$= \mathcal{W}(E \cap \mathrm{DS}) = \mathcal{W}(E).$$

□

---

[7]Upper and lower probabilities restricted to B are not directly related to the concept of conditional probability given B. Their analogues with respect to a probability measure P are the functions $E \mapsto P(E \cap B)$ (in the case of upper probability) and $E \mapsto P(E \cup B^c)$ (in the case of lower probability), where B is assumed to be measurable [321, section II.6, Lemma II.35(1)]. Both functions coincide with P when $P(B) = 1$ (see also [339, chapter 3]).

13.4 APPLICATIONS OF DUBINS–SCHWARZ

In this section, we demonstrate the power of Theorem 13.9 by deriving the main results of Section 13.2 and two additional corollaries (Corollaries 13.21 and 13.22) that give much more precise results. The final application (Proposition 13.24) is very different and concerns the behavior of the price path in the long run.

The following lemma tells us that "instantly enforceable" and "almost sure" are equivalent for negatively decidable events.

Lemma 13.17 *A positively decidable event E is instantly blockable if and only if it is null.*

Proof The "only if" is obvious. Let us prove the "if." Suppose E is null. For each $\epsilon > 0$, fix a nonnegative capital process X^ϵ such that $X_0^\epsilon = \epsilon$ and $\liminf_{t \to \infty} X_t^\epsilon \geq 1$ on E. The nonnegative capital process $X := \sum_{k \in \mathbb{N}} 2^{-k} X^{2^{-k}}$ will satisfy $X_0 = 1$ and $\lim_{t \to \infty} X_t = \infty$ on E. Fix a price path ω and $t \in [0, \infty)$; we are required to establish (13.5). Arguing indirectly, suppose $\omega \in E|_t$ and $X_t(\omega) < \infty$. Defining

$$\omega'(s) := \begin{cases} \omega(s) & \text{if } s \leq t, \\ \omega(t) & \text{if } s > t, \end{cases}$$

we obtain $\omega' \in E$ (by the definition (13.4)) such that

$$\liminf_{s \to \infty} X_s(\omega') = X_t(\omega') = X_t(\omega) < \infty,$$

which contradicts the definition of X. □

The proofs of some of our corollaries will use the detailed understanding of the notion of LT-superinvariance laid out in the following remark.

Remark 13.18 Let f be a time transformation. If f is the constant 0, then ω^f is the constant $\omega(0)$. Otherwise, we can change ω into ω^f in three steps: (i) replace ω by $\omega|_{[0,T)}$, where $T := \lim_{t \to \infty} f(t)$; (ii) replace $\omega|_{[0,T)}$ by $\omega|_{[0,T)} \circ g$, where $g : [0, T') \to [0, T)$ is the increasing homeomorphism whose graph is obtained from the graph of f by removing all horizontal pieces (this step continuously deforms the time interval $[0, T)$ into $[0, T')$); and (iii) insert at most countably many horizontal pieces into the resulting graph while preserving its continuity (a semi-infinite horizontal piece is also allowed when $T' < \infty$, assuming continuity can be preserved).

The *time trail* of $\omega \in \Omega$ is the set of all $\psi \in \Omega$ such that $\psi^f = \omega$ for some time transformation f. It consists of all elements of Ω that can be obtained from ω by the following steps: (i) remove any number of horizontal pieces from the graph of ω; let $[0, T)$ be the domain of the resulting function ω' (it is possible that $T < \infty$; if

$T = 0$, output any $\omega'' \in \Omega$ satisfying $\omega''(0) = \omega(0)$); (ii) assuming $T > 0$, continuously deform the time interval $[0, T)$ into $[0, T')$ for some $T' \in (0, \infty]$; let ω'' be the resulting function with the domain $[0, T')$; and (iii) if $T' = \infty$, output ω''; if $T' < \infty$ and $\lim_{t \to T'} \omega(t)$ exists in $\mathbb{R}$, extend ω'' to $[0, \infty)$ in any way making sure that the extension belongs to Ω and output the extension; otherwise output nothing. A set E is LT-superinvariant if and only if application of these last three steps, (i)–(iii), combined with adding a constant, never leads outside E.   ∎

Points of Increase

Now we can deduce Proposition 13.2 from Theorem 13.9.

Proof of Proposition 13.2   By Lemma 13.17, we can replace the "instantly enforceable" statement that we are required to prove by the corresponding "almost sure" statement; this will be done automatically in this and future propositions of this kind. Consider the set E of all $\omega \in \Omega$ that have points t of increase or decrease such that ω is not locally constant to the right of t and ω is not locally constant to the left of t (with the obvious definition of local constancy to the left of t; if $t = 0$, every ω is locally constant to the left of t). Since E is LT-superinvariant (cf. Remark 13.18), Theorem 13.9 and Dvoretzky et al.'s result show that the event E is null. And the following standard game-theoretic argument (used earlier in this chapter) shows that the event that ω is locally constant to the left but not locally constant to the right of a point of increase or decrease is null. For concreteness, we will consider the case of a point of increase. It suffices to show that for all rational numbers $b > a > 0$ and $D > 0$ the event that

$$\inf_{t \in [a,b]} \omega(t) = \omega(a) \leq \omega(a) + D \leq \sup_{t \in [a,b]} \omega(t) \tag{13.21}$$

is null (see Lemma 13.14). The simple capital process that starts from $\epsilon > 0$, takes position $h_1 := 1/D$ at time $\tau_1 = a$, and eliminates it by setting $h_2 := 0$ at time

$$\tau_2 := \min\{t \geq a | \omega(t) \in \{\omega(a) - D\epsilon, \omega(a) + D\}\}$$

is nonnegative and turns ϵ (an arbitrarily small amount) into 1 when (13.21) happens. (Notice that this argument works both when $t = 0$ and when $t > 0$.)   □

Variation Index

Recall the definitions of p-variation (see (13.7)) and variation index given earlier. Let us check that Proposition 13.6 follows from Theorem 13.9 and standard results of measure-theoretic probability. The following is its slightly stronger version; for Brownian motion it was established by Lévy [246].

Corollary 13.19 *The following is instantly enforceable. For any interval $[u, v] \subseteq$ $[0, \infty)$ such that $u < v$, either* $\mathrm{vi}(\omega|_{[u,v]}) = 2$ *or ω is constant over* $[u, v]$.

Proof Consider the set of $\omega \in \Omega$ such that, for some interval $[u, v] \subseteq [0, \infty)$, neither $\mathrm{vi}(\omega|_{[u,v]}) = 2$ nor ω is constant over $[u, v]$. This set is LT-superinvariant (cf. Remark 13.18), and so in conjunction with Theorem 13.9, Lévy's result implies that it is null. □

Corollary 13.19 says that

$$
\mathrm{var}_p(\omega|_{[u,v]}) \begin{cases} < \infty & \text{if } p > 2, \\ = \infty & \text{if } p < 2 \text{ and } \omega|_{[u,v]} \text{ is not constant} \end{cases}
$$

with instant enforcement. However, it does not say anything about the situation for $p = 2$. The following result completes the picture.

Corollary 13.20 *The following is instantly enforceable. For any interval $[u, v] \subseteq$ $[0, \infty)$ such that $u < v$, either* $\mathrm{var}_2(\omega|_{[u,v]}) = \infty$ *or ω is constant over* $[u, v]$.

Proof Lévy [246] proves for Brownian motion that $\mathrm{var}_2(\omega|_{[u,v]}) = \infty$ almost surely (for fixed $[u, v]$, which implies the statement for all $[u, v]$). Consider the set of $\omega \in \Omega$ such that, for some interval $[u, v] \subseteq [0, \infty)$, neither $\mathrm{var}_2(\omega|_{[u,v]}) = \infty$ nor ω is constant over $[u, v]$. This set is LT-superinvariant, and so in conjunction with Theorem 13.9, Lévy's result implies that it is null. □

More Precise Results

Theorem 13.9 allows us to deduce much stronger results than Corollaries 13.19 and 13.20 from known results about Brownian motion.

We specialize the notation $\log^*$ to the natural logarithm, $\ln^* u := 1 \vee |\ln u|$, $u > 0$, and let $\psi : [0, \infty) \to [0, \infty)$ be S. James Taylor's [379] function

$$
\psi(u) := \frac{u^2}{2\ln^* \ln^* u} \tag{13.22}
$$

(with $\psi(0) := 0$). Remember the definition of var_ϕ given by (13.11); we will write $\mathrm{var}_{\phi,T}(\omega)$ to mean $\mathrm{var}_\phi(\omega_{[0,T]})$.

Corollary 13.21 *It is instantly enforceable that*

$$
\forall t \in [0, \infty) : \mathrm{var}_{\psi,t}(\omega) < \infty.
$$

Suppose $\phi : [0, \infty) \to [0, \infty)$ is such that $\psi(u) = o(\phi(u))$ as $u \to 0$. It is instantly enforceable that

$$\forall t \in [0, \infty) : \omega \text{ is constant on } [0, t] \text{ or } \mathrm{var}_{\phi,t}(\omega) = \infty.$$

Corollary 13.21 refines Corollaries 13.19 and 13.20; it will be further strengthened by Corollary 13.22.

The quantity $\mathrm{var}_{\psi,T}(\omega)$ is not nearly as fundamental as

$$\mathrm{w}_T(\omega) := \lim_{\delta \to 0} \sup_{\kappa \in K_\delta[0,T]} \sum_{i=1}^{n_\kappa} \psi(|\omega(t_i) - \omega(t_{i-1})|), \tag{13.23}$$

where $\omega \in \Omega$, $T \in [0, \infty)$, and $K_\delta[0, T]$ is the set of all partitions $0 = t_0 \leq \cdots \leq t_{n_\kappa} = T$ of $[0, T]$ whose mesh is less than δ: $\max_i(t_i - t_{i-1}) < \delta$. Notice that the expression after the $\lim_{\delta \to 0}$ in (13.23) is increasing in δ; therefore, $\mathrm{w}_T(\omega) \leq \mathrm{var}_{\psi,T}(\omega)$.

The following corollary contains Corollaries 13.19–13.21 as special cases. It is similar to Corollary 13.21 but is stated in terms of the process w.

Corollary 13.22   *It is instantly enforceable that*

$$\forall T \in [0, \infty) : \omega \text{ is constant on } [0, T] \text{ or } \mathrm{w}_T(\omega) \in (0, \infty). \tag{13.24}$$

Proof First let us check that under the Wiener measure (13.24) holds for almost all ω. It is sufficient to prove that $\mathrm{w}_T = T$ for all $T \in [0, \infty)$ a.s. Furthermore, it is sufficient to consider only rational $T \in [0, \infty)$. Therefore, it is sufficient to consider a fixed rational $T \in [0, \infty)$. And for a fixed T, $\mathrm{w}_T = T$ a.s. follows from Taylor's result [379, Theorem 1].

In view of Corollary 13.10 it suffices to check that the complement of the event (13.24) is LT-superinvariant, i.e. to check (13.30), where E is the complement of (13.24). In other words, it suffices to check that $\omega^f = \omega \circ f$ satisfies (13.24) whenever ω satisfies (13.24). This follows from Lemma 13.23, which says that $\mathrm{w}_T(\omega \circ f) = \mathrm{w}_{f(T)}(\omega)$. $\qquad\square$

Lemma 13.23 *Let $T \in [0, \infty)$, $\omega \in \Omega$, and f be a time transformation. Then $\mathrm{w}_T(\omega \circ f) = \mathrm{w}_{f(T)}(\omega)$.*

Proof Fix $T \in [0, \infty)$, $\omega \in \Omega$, a time transformation f, and $c \in [0, \infty]$. Our goal is to prove

$$\lim_{\delta \to 0} \sup_{\kappa \in K_\delta[0,f(T)]} \sum_{i=1}^{n_\kappa} \psi(|\omega(t_i) - \omega(t_{i-1})|) = c$$

$$\implies \lim_{\delta \to 0} \sup_{\kappa \in K_\delta[0,T]} \sum_{i=1}^{n_\kappa} \psi(|\omega(f(t_i)) - \omega(f(t_{i-1}))|) = c \tag{13.25}$$

in the notation of (13.23). Suppose the antecedent in (13.25) holds. Notice that the two $\lim_{\delta \to 0}$ in (13.25) can be replaced by $\inf_{\delta > 0}$.

To prove that the limit on the right-hand side of (13.25) is $\leq c$, take any $\epsilon > 0$. We will assume $c < \infty$ (the case $c = \infty$ is trivial). Let $\delta > 0$ be so small that

$$\sup_{\kappa \in K_\delta[0, f(T)]} \sum_{i=1}^{n_\kappa} \psi(|\omega(t_i) - \omega(t_{i-1})|) < c + \epsilon.$$

Let $\delta' > 0$ be so small that $|t - t'| < \delta' \Longrightarrow |f(t) - f(t')| < \delta$. Since $f(\kappa) \in K_\delta[0, f(T)]$ whenever $\kappa \in K_{\delta'}[0, T]$,

$$\sup_{\kappa \in K_{\delta'}[0, T]} \sum_{i=1}^{n_\kappa} \psi(|\omega(f(t_i)) - \omega(f(t_{i-1}))|) < c + \epsilon.$$

To prove that the limit on the right-hand side of (13.25) is $\geq c$, take any $\epsilon > 0$ and $\delta' > 0$. We will assume $c < \infty$ (the case $c = \infty$ can be considered analogously). Place a finite number N of points including 0 and T onto the interval $[0, T]$ so that the distance between any pair of adjacent points is less than δ'; this set of points will be denoted κ_0. Let $\delta > 0$ be so small that $\psi(|\omega(t'') - \omega(t')|) < \epsilon/N$ whenever $|t'' - t'| < \delta$. Choose a partition $\kappa = (t_0, \ldots, t_n) \in K_\delta[0, f(T)]$ (so that $0 = t_0 \leq \cdots \leq t_n = f(T)$ and $\max_i (t_i - t_{i-1}) < \delta$) satisfying

$$\sum_{i=1}^{n} \psi(|\omega(t_i) - \omega(t_{i-1})|) > c - \epsilon.$$

Let $\kappa' = (t_0', \ldots, t_n')$ be a partition of the interval $[0, T]$ satisfying $f(\kappa') := (f(t_0'), \ldots, f(t_n')) = \kappa$. This partition will satisfy

$$\sum_{i=1}^{n} \psi(|\omega(f(t_i')) - \omega(f(t_{i-1}'))|) > c - \epsilon,$$

and the union $\kappa'' = (t_0'', \ldots, t_{N+n}'')$ (with its elements listed in the increasing order) of κ_0 and κ' will satisfy

$$\sum_{i=1}^{N+n} \psi(|\omega(f(t_i'')) - \omega(f(t_{i-1}''))|) > c - 2\epsilon.$$

Since $\kappa'' \in K_{\delta'}[0, T]$ and ϵ and δ' can be taken arbitrarily small, this completes the proof. $\qquad \square$

The value $\mathrm{w}_T(\omega)$ defined by (13.23) can be interpreted as the quadratic variation of the price path ω over the time interval $[0, T]$. Another game-theoretic definition of quadratic variation is given in Chapter 14. For the equivalence of the two definitions, see Remark 14.16.

The following proposition can be thought of as an elaboration of Corollary 13.22 for nonnegative price paths.

Proposition 13.24   *For any $c > 0$,*

$$\overline{\mathbb{P}}(\exists T \in [0, \infty] : \inf_{t \in [0,T]} \omega(t) \geq \omega(0) - 1 \text{ and } \mathrm{w}_T(\omega) \geq c)$$

$$= \sqrt{\frac{2}{\pi}} \int_0^{c^{-1/2}} e^{-x^2/2} \, dx \leq \sqrt{\frac{2}{\pi c}}. \tag{13.26}$$

In particular (letting $T := \infty$ and $c \to \infty$),

$$\overline{\mathbb{P}}(\inf_t \omega(t) \geq \omega(0) - 1 \text{ and } \mathrm{w}_\infty(\omega) = \infty) = 0.$$

When applied to a price path ω with $\omega(0) = 1$ and $\omega \geq 0$ (the last condition being satisfied for typical securities traded in real-world financial markets), this proposition shows that the trader can become infinitely rich at infinity if the security's volatility does not die out eventually. A more advanced version of this proposition is given in Proposition 17.7.

Proof of Proposition 13.24   We are only required to prove (13.26). First let us check that the event E in the parentheses on the left-hand side of (13.26) is LT-invariant. Assuming $\omega^f \in E$ we will prove that $\omega \in E$ (cf. (13.30)), where f is a time transformation. It suffices to notice that, by Lemma 13.23,

$$\exists T \in [0, \infty] : \inf_{t \in [0,T]} (\omega \circ f)(t) \geq \omega(0) - 1 \text{ and } \mathrm{w}_T(\omega \circ f) \geq c$$

is equivalent to

$$\exists T \in [0, \infty] : \inf_{t \in [0, f(T)]} \omega(t) \geq \omega(0) - 1 \text{ and } \mathrm{w}_{f(T)}(\omega) \geq c.$$

By Theorem 13.9, we can replace the $\overline{\mathbb{P}}$ on the left-hand side of (13.26) by the Wiener measure, and so the left-hand side of (13.26) becomes $\mathcal{W}\{\omega | \inf_{t \in [0,c]} \omega(t) \geq -1\}$. It remains to apply the standard measure-theoretic result given in [212, (2.6.2)]. □

13.5 GETTING RICH QUICK WITH THE AXIOM OF CHOICE

In this section only, let us remove all measurability requirements from the definitions of *process, stopping time, simple trading strategy, position, simple capital process, nonnegative capital process*, and *instant enforcement* in Section 13.1. When we drop measurability in this way, we obtain the following theorem, which says that Skeptic can become infinitely rich as soon as the price path becomes nonconstant.

Theorem 13.25 *The price path ω is constant with instant enforcement.*

In more optimistic terms, there is a nonnegative capital process X with $X_0 = 1$ that becomes infinite as soon as ω ceases to be constant: for all $t \in (0, \infty)$,

$$(\exists t_1, t_2 \in [0, t) : \omega(t_1) \neq \omega(t_2)) \implies X_t(\omega) = \infty. \tag{13.27}$$

Our proof is based on Christopher Hardin and Alan Taylor's work on hat puzzles. These authors show in [177] (see also [178, section 7.4]) that the axiom of choice provides an Occam-type strategy that can usually predict the short-term future, and this makes it easy for the trader to get rich when the price changes in a nondegenerate way.

Fix a well-order $\preccurlyeq$ of Ω, which exists by Zermelo's theorem (one of the alternative forms of the axiom of choice; see, e.g. [202, Theorem 5.1]). Let ω_a, where $\omega \in \Omega$ and $a \in [0, \infty)$, be the $\preccurlyeq$-smallest element of Ω such that $\omega_a|_{[0,a]} = \omega|_{[0,a]}$. Intuitively, using ω_a as the prediction at time a for ω is an instance of Occam's razor: out of all hypotheses compatible with the available data $\omega|_{[0,a]}$, we choose the simplest one, where simplicity is measured by the chosen well-order.

For any $\omega \in \Omega$ set

$$
\begin{aligned}
W_\omega &:= \{t \in [0, \infty) \mid \forall t' \in (t, \infty) : \omega_{t'} \neq \omega_t\} \\
&= \{t \in [0, \infty) \mid \forall t' \in (t, \infty) : \omega_{t'} \succ \omega_t\} \\
&= \{t \in [0, \infty) \mid \forall t' \in (t, \infty) : \omega_t|_{[0,t']} \neq \omega|_{[0,t']}\}.
\end{aligned} \tag{13.28}
$$

The following lemma says, intuitively, that short-term prediction of the future is usually possible.

Lemma 13.26

1. *The set W_ω is well-ordered by $\leq$. (Therefore, each of its points is isolated on the right, which implies that W_ω is countable and nowhere dense.)*

2. If $t \in [0, \infty) \setminus W_\omega$, there exists $t' > t$ such that $\omega_t|_{[t,t']} = \omega|_{[t,t']}$.

3. If $t \in W_\omega$, there exists $t' > t$ such that $\omega_s|_{[t,t']} = \omega|_{[t,t']}$ for all $s \in (t, t')$.

Statement 1 of Lemma 13.26 says that the set W_ω is small. Statement 2 says that at each time point t outside the small set W_ω, the Occam prediction system that outputs ω_t as its prediction is correct (over some nontrivial time interval). And statement 3 says that even at time points t in W_ω, the Occam prediction system becomes correct (in the same weak sense) immediately after time t.

Proof of Lemma 13.26 Let us first check that W_ω is well-ordered by $\leq$. Suppose there is an infinite strictly decreasing chain $t_1 > t_2 > \cdots$ of elements of W_ω. Then $\omega_{t_1} > \omega_{t_2} > \cdots$, which contradicts $\preceq$ being a well-order.

Each point $t \in W_\omega \setminus \{1\}$ is isolated on the right since $W_\omega \cap (t, t') = \emptyset$, where t' is the successor of t. Therefore, W_ω is nowhere dense. To check that W_ω is countable, map each $t \in W_\omega$ to a rational number in the interval (t, t'), where t' is the successor of t; this mapping is an injection.

As statement 2 is obvious (and essentially asserted in (13.28)), let us check statement 3. Suppose $t \in W_\omega$. The set of all ω_s, $s \in (t, \infty)$, has a smallest element $\omega_{t'}$, where $t' \in (t, \infty)$. It remains to notice that $\omega_s = \omega_{t'}$ for all $s \in (t, t')$. $\square$

Proof of Theorem 13.25 For each pair of rational numbers (a, b) such that $0 < a < b$, fix a positive weight $w_{a,b} > 0$ such that $\sum_{a,b} w_{a,b} = 1$, the sum being over all such pairs. For each such pair (a, b) we will define a nonnegative capital process $X^{a,b}$ such that $X_0^{a,b} = 1$ and $X_b^{a,b}(\omega) = \infty$, when $\omega|_{[a,b]} = \omega_a|_{[a,b]}$ and $\omega|_{[a,b]}$ is not constant. Let us check that the process

$$X := \sum_{a,b} w_{a,b} X^{a,b} \tag{13.29}$$

will then achieve our goal (13.27).

Let $\omega \in \Omega$ and c be the largest $t \in [0, \infty)$ such that $\omega|_{[0,t]}$ is constant. Let us assume that ω is not constant, in which case $c < \infty$ and the supremum is attained by the continuity of ω. Set $\omega_{c+} := \omega_t$ for $t \in (c, c + \epsilon)$ for a sufficiently small $\epsilon > 0$ (namely, such that $t \mapsto \omega_t$ is constant over the interval $t \in (c, c + \epsilon)$; such an ϵ exists by Lemma 13.26). Choose $d \in (c, \infty)$ such that $\omega_d = \omega_{c+}$ (and, therefore, $\omega|_{(c,d]} = \omega_{c+}|_{(c,d]}$ and $\omega_t = \omega_{c+}$ for all $t \in (c, d]$). Take rational $a, b \in (c, d)$ such that $a < b$ and $\omega|_{[a,b]}$ is not constant; since $X_b^{a,b}(\omega) = \infty$, (13.29) gives $X_b(\omega) = \infty$, and since b can be arbitrarily close to c, we obtain (13.27).

It remains to construct such a nonnegative capital process $X^{a,b}$ for fixed a and b. From now until the end of this proof, ω is a generic element of Ω. For each $n \in \mathbb{N}$, let $\mathbb{D}_n := \{k2^{-n} \mid k \in \mathbb{Z}\}$ and define a sequence of stopping times T_k^n, $k = -1, 0, 1, \ldots$, inductively by $T_{-1}^n := a$,

$$T_0^n(\omega) := \inf\{t \in [a, b] \mid \omega(t) \in \mathbb{D}_n\},$$

$$T_k^n(\omega) := \inf\{t \in [T_{k-1}^n(\omega), b] \mid \omega(t) \in \mathbb{D}_n \ \& \ \omega(t) \neq \omega(T_{k-1}^n)\}, \quad k \in \mathbb{N},$$

where we set $\inf \emptyset := b$. For each $n \in \mathbb{N}$, define a simple capital process X^n as the capital process of the simple trading strategy with the stopping times

$$\omega \in \Omega \mapsto \tau_k^n(\omega) := T_k^n(\omega) \wedge T_k^n(\omega_a), \quad k = 0, 1, \ldots,$$

the corresponding positions h_k^n that are defined as

$$h_k^n(\omega) := \begin{cases} 2^{2n}(\omega_a(\tau_{k+1}^n(\omega_a)) - \omega(\tau_k^n)) & \text{if } \omega_{\tau_k^n(\omega)} = \omega_a \text{ and } \tau_k^n(\omega) < b, \\ 0 & \text{otherwise,} \end{cases}$$

and with an initial capital of 1. Since the increments of this simple capital process never exceed 1 in absolute value (and trading stops as soon as the prediction ω_a is falsified), its initial capital of 1 ensures that it always stays nonnegative. The final value $X_b^n(\omega)$ is $\Omega(2^n)$ (in Knuth's asymptotic notation) unless $\omega|_{[a,b]} \neq \omega_a|_{[a,b]}$ or $\omega|_{[a,b]}$ is constant. If we now set

$$X^{a,b} := \sum_{n=1}^{\infty} n^{-2} X^n,$$

we obtain $X_a^{a,b} < \infty$ and $X_b^{a,b}(\omega) = \infty$ unless $\omega|_{[a,b]} \neq \omega_a|_{[a,b]}$ or $\omega|_{[a,b]}$ is constant. □

13.6 EXERCISES

Exercise 13.1 Give an example of an event $E \in \mathcal{F}$ and $t \in [0, \infty)$ such that $E|_t \notin \mathcal{F}$. Show that E can be chosen positively decidable. *Hint.* Use the existence of analytical sets in $\mathbb{R}$ that are not Borel [214, Theorem 14.2]. □

Exercise 13.2 Show that the following events have upper probability 1 (thus demonstrating the necessity of various conditions in Corollaries 13.3–13.5):

- ω is constant on $[0, \infty)$;
- for some $t \in (0, \infty)$, $[t, \infty)$ is the ray of local maximum (or minimum) for ω;
- $\omega'(t)$ exists for no $t \in [0, \infty)$.

Hint. Define suitable measure-theoretic continuous martingales and use Lemma 13.11. □

Exercise 13.3 Let $f : [a, b] \to \mathbb{R}$ be continuous.

- Prove that $\text{vi}(f)$ is well defined, i.e. there exists a unique number $c \in [0, \infty]$ such that $\text{var}_p(f)$ is finite when $p > c$ and infinite when $p < c$.
- Prove that $\text{vi}(f) \notin (0, 1)$. □

Exercise 13.4 Prove the inequality $\geq$ in (13.18). *Hint.* Apply the argument of Lemma 13.11 (suitably generalized). □

Exercise 13.5 Show that if an event E is null, there is a nonnegative capital process X such that $X_0 = 1$ and $\lim_{t \to \infty} X_t(\omega) = \infty$ for all $\omega \in E$. $\square$

Exercise 13.6 Show that $E \in \mathcal{I}$ if and only if $E \in \mathcal{F}$ and, for each $\omega \in \Omega$ and each LT-transformation (c, f), it is true that

$$\omega^{cf} \in E \implies \omega \in E. \tag{13.30}$$

Hint. Specialize (13.16) to $F := \mathbf{1}_E$. $\square$

Exercise 13.7 Define a partial order $\leq$ on Ω as follows: $\omega' \leq \omega$ if and only if there is an LT-transformation (c, f) such that $\omega' = \omega^{cf}$. (The intuition behind this definition is that some information in ω may be lost, even if the time scale is ignored: it is possible that $f(\infty) < \infty$.)

- Show that an event E is in $\mathcal{I}$ if and only if E is an upper set for this partial order.
- Show that an extended variable F is $\mathcal{I}$-measurable if and only if it is $\mathcal{F}$-measurable and increasing (with respect to this partial order). $\square$

Exercise 13.8 Show that $\mathcal{I}$ is closed under countable unions and intersections (in particular, it is a monotone class). Show that it is not closed under complementation, and so it is not a σ-algebra. $\square$

Exercise 13.9 Show that a nonnegative extended variable F is $\mathcal{I}$-measurable if and only if it satisfies (13.16). $\square$

Exercise 13.10 Prove that we can replace the $\mathcal{W}$ in the statement of Theorem 13.9 by any probability measure P on $(\Omega, \mathcal{F})$ such that the process $X_t(\omega) := \omega(t)$ is a martingale with respect to P and the filtration $(\mathcal{F}_t)$, is unbounded P-a.s., and is nowhere constant P-a.s. $\square$

Exercise 13.11 The upper probability of the following event is 1: there is a point t of increase such that ω is locally constant to the right of t. $\square$

Exercise 13.12 Three men – A, B, and C – are blindfolded and told that either a red or a green hat will be placed on each of them. After this is done, the blindfolds are removed; the men are asked to raise a hand if they see a red hat and to leave the room as soon as they are sure of the color of their own hat. All three hats happen to be red, so all three men raise a hand. Several minutes go by until C, who is more astute than the others, leaves the room. How did he deduce the color of his hat? *Hint.* See [164, pp. 138, 140]. $\square$

13.7 CONTEXT

Measurability

The unavoidability of measurability conditions in measure theory was established in the early twentieth century by a series of paradoxes, including Giuseppe Vitali's examples and Hausdorff's decomposition of the sphere and culminating in the Banach–Tarski paradox [425].

We have already mentioned that as the result of Doob's work, measure-theoretic probability also uses measurability in a substantive way to model accumulation of information as time flows. In addition to one σ-algebra $\mathcal{F}$, we have a filtration $(\mathcal{F}_t)$, where the σ-algebra $\mathcal{F}_t \subseteq \mathcal{F}$ represents the information available at time t. We use this standard setting beginning in Chapter 14, but in this chapter we prefer a more constructive and cautious approach, especially that we cannot use the standard setting in Section 13.5, where we ask whether measurability is needed at all. So throughout this chapter we use Galmarino's [162] approach to define stopping times τ and τ-measurable functions. By the Galmarino test (see, e.g. [105, theorem IV.100]), the Galmarino definitions in Section 13.1 are equivalent to the standard ones. In the standard setting, our term *τ-measurable* is replaced by $\mathcal{F}_\tau$-*measurable*, where $\mathcal{F}_\tau$ is the σ-algebra associated with the stopping time τ, as defined in Chapter 14.

The hat puzzles cited in Section 13.5 have roots going back at least to the 1950s: see Martin Gardner [164] and George Gamow and Marvin Stern [163, chapter 4], which describe the same puzzle (Exercise 13.12), although only Gardner uses the language of hats (Gamow and Stern's story is about well-bred Britons with faces smeared by soot). For a further fascinating history leading up to predicting the future, see [178].

Probabilistic Modeling of Security Prices

The variation index of price paths being 2 means in practice that the price increment scales approximately as the square root of time. In 1853, this was noted for stock-market prices by Jules Regnault [318], who called it *la loi des écarts* (the law of differences) and compared it with Newton's law of gravitation (see [20, chapter 1]). Our proof of the emergence of volatility follows [422] and [402].

As mentioned in Section 2.6, Louis Bachelier was the first to develop a mathematical model for stock-market prices. His work, which began with his dissertation in 1900, was not noticed by economists until the late 1950s. A disadvantage of his model is that the stock price can become negative with positive probability. M.F. Maury Osborne proposed geometric Brownian motion as a model in 1959 [297].

Comparing Game-Theoretic and Measure-Theoretic Results

Most of the results in Section 13.2 have measure-theoretic counterparts in Karatzas and Shreve's authoritative 1991 textbook [212]:

- Proposition 13.1 corresponds to part of their Theorem 2.9.6,
- Proposition 13.2 to their Theorem 2.9.13,
- Corollary 13.3 to their Theorem 2.9.9,

- Corollary 13.4 to part of their Theorem 2.9.12, and
- Corollary 13.5 roughly to their Theorem 2.9.18.

Versions of Proposition 13.6 and Corollary 13.19 were obtained in [402] by adapting Michel Bruneau's proof [57]. In measure-theoretic probability they were established for continuous semimartingales by Lepingle [239, Theorem 1 and Proposition 3]; as already mentioned, for Brownian motion it was done by Lévy [246]. In our proof of Proposition 13.8, we also follow Bruneau's proof [57], which is surprisingly game-theoretic in character. Bruneau's proof was extended to càdlàg processes by Christophe Stricker [372], whose argument was adapted to game-theoretic probability in [409].

The quantity $w_T(\omega)$ was introduced by S. James Taylor [379] for the purpose of establishing his beautiful measure-theoretic result, on which we modeled Corollary 13.22. See [56] for a much more explicit expression for $\mathrm{var}_{\psi,T}(\omega)$.

As we have already mentioned, Proposition 13.2, without the clause about local constancy, was established for Brownian motion by Dvoretzky et al. [122]. Dubins and Schwarz observed that their own reduction of continuous martingales to Brownian motion shows that Dvoretzky et al.'s result continues to hold for all continuous martingales that are almost surely unbounded and almost surely nowhere constant [120]. In [405], Proposition 13.2 is proven directly using an idea from Krzysztof Burdzy [58].

Dubins–Schwarz

The Dubins–Schwarz theorem is often referred to as the Dambis–Dubins–Schwarz theorem to acknowledge the contribution of Karl Dambis [92, Theorem 7]. From the point of view of measure-theoretic probability, Dambis's and Dubins and Schwarz's results are equivalent, but Dambis assumes the existence of quadratic variation in a much more explicit way, and in our mind his result appears closer to the predecessors of Theorem 13.9 in which the quadratic variation was postulated (see, e.g. [393] and [403]).

Dubins and Schwarz [120, 335, 336] make simplifying assumptions that make the monoid of time transformations a group; we make similar assumptions in Corollary 13.16. In general, the notion of trail, discussed at the beginning of Section 13.3, is usually defined for groups rather than monoids (see, e.g. [291]); in this case trails are called orbits.

The main tool in the proof of Theorem 13.9 in [404] is a move used by Lindeberg in his proof of his central limit theorem (see Sections 2.3 and 7.4): replace physical time t by intrinsic time (quadratic variation) A_t. During busy trading intrinsic time flows fast; when trading is sluggish, it flows slower.

In addition to Theorem 13.9, which is an abstract game-theoretic version of the Dubins–Schwarz theorem, there is a concrete game-theoretic version. Both are stated and proven in [404]. The concrete version expresses directly the idea that the price path becomes Brownian motion when physical time is replaced by intrinsic time. The concrete version is more awkward mathematically, but it is more intuitive and often easier to apply.

Extensions to Càdlàg Price Paths

Some of the results of this chapter, as well as other results in Part IV, have been extended to càdlàg price paths. Usually, as in the work on the volatility of càdlàg price paths [160, 254, 409] and on the Itô calculus for càdlàg price paths [254, 413], the jumps are assumed to be bounded. An important exception is the existence of quadratic variation for nonnegative càdlàg price paths without any restrictions on the size of jumps, established in [254, Theorem 3.2].

Standard and Nonstandard Analysis

In [349], we used nonstandard analysis to study market prices in continuous time. Our decision to use standard analysis instead followed the treatment of game-theoretic volatility by Takeuchi, Kumon, and Takemura in 2009 [376]. While that article's Theorem 3.1 is in the spirit of our Proposition 13.6, [377] gives more precise quantitative results. The main results of [376, 377] are extended to a multidimensional setting in [231].

Connections with Chapter 11

Corollary 11.15 implies that the upper expected value of the perpetual European lookback with payoff function $H(v, w) = F(v)$ is (11.25) with $c = 0$. This result becomes very natural in view of Theorem 13.9. Suppose, in the spirit of this chapter, that the price path of the stock, y_t, is continuous. By Theorem 13.9, $\overline{\mathbb{E}}(F(y_\infty^*))$ is equal to the expected value of $F(B_\infty^*)$ where B is a standard Brownian motion started at 1 and stopped when it hits 0. The density of B_∞^*'s distribution is v^{-2}, in agreement with Corollary 11.15. Indeed, because it is a measure-theoretic martingale, the probability that Brownian motion started at 1 hits $v \geq 1$ before hitting 0 is $1/v$ (see, e.g. [286, Theorem 2.49]); therefore, B_∞^*'s distribution function is $1 - 1/v$, and its density is v^{-2}. These connections are discussed in detail in [404, section 15.3] and [100, section 7].

14

A Game-Theoretic Itô Calculus

In this chapter we introduce and study martingale spaces, which provide an abstract framework for studying continuous-time game-theoretic martingales. Martingale spaces are sufficiently powerful to support the Itô calculus, and they will be used throughout the remaining chapters of this book.

The framework of Chapter 13 was concrete. There we considered the price path of a single security, assumed to be continuous, and our sample space Ω consisted of the security's possible price paths – i.e. all the continuous functions on $[0, \infty)$. The sample space for a martingale space, in contrast, is an abstract measurable space Ω, with a filtration as in continuous-time measure-theoretic probability, and the price paths of securities are adapted processes on Ω. The martingale space consists of the measurable space, supplemented not with a probability measure but with price paths for a countable number of securities. These price paths are the space's basic martingales. For simplicity, we assume that they are continuous, but the framework can be extended to handle some discontinuous price paths (see Section 14.11).

The power of martingale spaces derives from the bold definitions used to expand a martingale space's countable set of basic martingales. This expansion is explained in Section 14.1. We begin, as in Chapter 13, with simple capital processes obtained using simple strategies for trading in the basic martingales. But then we close the class of nonnegative simple capital processes under lim inf to obtain a class of processes that we call *nonnegative supermartingales*. We expect a nonnegative supermartingale not to instantly become infinite. So we can use the nonnegative supermartingales to

Game-Theoretic Foundations for Probability and Finance, First Edition. Glenn Shafer and Vladimir Vovk.
© 2019 John Wiley & Sons, Inc. Published 2019 by John Wiley & Sons, Inc.

single out *instantly blockable* sets, which we predict will not happen, and *instantly enforceable* sets, which we predict will happen. We then close the class of simple capital processes under lim to define, modulo instantly blockable sets, processes that we call *continuous martingales* and to predict aspects of their behavior. In Section 14.2, we show that the class of continuous martingales and trader's ability to force properties do not change if we add continuous martingales from the expanded class to the set of basic martingales.

We introduce the Itô integral in Section 14.3, covariation and quadratic variation in Section 14.4, Itô's formula in Section 14.5, and the Doléans exponential and logarithm in Section 14.6. We then return to themes considered in Chapter 13: game-theoretic expectation and probability (Section 14.7), Dubins–Schwarz (Section 14.8), and coherence (Section 14.9).

14.1 MARTINGALE SPACES

A *martingale space* is a quintuple of the form

$$(\Omega, \mathcal{F}, (\mathcal{F}_t)_{t \in [0,\infty)}, J, (S^j)_{j \in J}), \tag{14.1}$$

where $(\Omega, \mathcal{F})$ is a measurable space, $(\mathcal{F}_t)$ is a filtration, J is a countable nonempty index set, possibly finite, and S^j, for each $j \in J$, is a process such that $S^j_t(\omega)$ is a continuous real-valued function of $t \in [0, \infty)$ for each $\omega \in \Omega$. We will make the simplifying assumption that each S^j_0 is a constant (i.e. $S^j_0(\omega)$ does not depend on $\omega \in \Omega$).

The triplet $(\Omega, \mathcal{F}, (\mathcal{F}_t))$ is a measurable space with a filtration, as usually defined in continuous-time measure-theoretic probability:

- As usual, we call Ω the *sample space*.
- By a *filtration* $(\mathcal{F}_t)$, we mean a family of sub-σ-algebras $\mathcal{F}_t \subseteq \mathcal{F}$ such that $\mathcal{F}_t \subseteq \mathcal{F}_s$ whenever $t \leq s$.

This is starting point for continuous-time measure-theoretic probability, but instead of adding a probability measure P, we add only the processes $(S^j)_{j \in J}$, which we call *basic martingales*.

The notion of a martingale space is central to this and the remaining chapters, and it can be used as the foundation for any application of the Itô calculus. In the financial application we emphasize, J is a set of securities and S^j_t is the price of security j at time t. For each ω, we call the continuous function $S^j(\omega) : t \mapsto S^j_t(\omega)$ the *price path* for S^j determined by ω. As in the preceding chapter, our results can be interpreted in terms of a protocol with two players: a trader and a market. The trader chooses a strategy for trading in the securities, and then the market chooses ω. If we prefer, we may instead call the player who chooses ω Reality, as the framework allows ω to determine more than the price paths $(S^j(\omega))_{j \in J}$.

We call the elements of $\mathcal{F}$ *events*, but we will also be interested in subsets of Ω that are not in $\mathcal{F}$. A *variable* is an $\mathcal{F}$-measurable function of the type $\Omega \to \mathbb{R}$, and an *extended variable* is an $\mathcal{F}$-measurable function of the type $\Omega \to [-\infty, \infty]$. By a *process*, we mean a function $X : [0, \infty) \times \Omega \to [-\infty, \infty]$ such that for each $t \in [0, \infty)$, the function $X_t : \Omega \to [-\infty, \infty]$ defined by $X_t(\omega) := X(t, \omega)$ is $\mathcal{F}_t$-measurable. This is called an adapted process in measure-theoretic probability; as in preceding chapters, we omit the adjective "adapted." We are mostly interested in processes X with X_0 constant. We call the functions $t \mapsto X_t(\omega)$, $\omega \in \Omega$, the process X's *paths*. A *stopping time* is an extended variable τ taking values in $[0, \infty]$ such that, for any $t \in [0, \infty)$, the event $\{\tau \leq t\}$ belongs to $\mathcal{F}_t$. If τ is a stopping time, the σ-algebra $\mathcal{F}_\tau$ consists of all events E (*events preceding* τ) such that $E \cap \{\tau \leq t\} \in \mathcal{F}_t$ for each $t \in [0, \infty)$.

A *simple trading strategy* ψ in a martingale space is a triple $(c, (\tau_1, \ldots, \tau_N), h)$, where

- $c \in \mathbb{R}$ (interpreted as the initial capital) and $N \in \mathbb{N}$ (the number of trading times);
- $\tau_1 \leq \cdots \leq \tau_N$ is an increasing sequence of stopping times;
- h is a family of bounded variables $(h_{n,j})_{n \in \{1, \ldots, N\}, j \in J}$ such that each $h_{n,j}$ is $\mathcal{F}_{\tau_n}$-measurable and $h_{n,j} = 0$ for all but finitely many (n, j).

We interpret τ_n as the nth trading time and $h_{n,j}$ as the position in S^j (or *bet* on S^j) taken at time τ_n; here we allow trading only in finitely many securities. This definition, in which N is finite, might appear narrower than the definition given in Chapter 13, but the case $N = \infty$ is implicitly allowed by the additional definitions we now give.

The *simple capital process* $\mathcal{K}^\psi$ corresponding to a simple trading strategy ψ is defined by

$$\mathcal{K}_t^\psi(\omega) := c + \sum_{n=2}^{N+1} \sum_{j \in J} h_{n-1,j}(\omega)(S_{\tau_n(\omega) \wedge t}^j(\omega) - S_{\tau_{n-1}(\omega) \wedge t}^j(\omega)),$$

$$t \in [0, \infty), \omega \in \Omega, \tag{14.2}$$

where $\tau_{N+1} := \infty$ and, as usual, the zeros in the infinite sum $\sum_{j \in J}$ are ignored, so that the sum is finite. All simple capital processes have continuous paths and satisfy our general definition of a process.

Equation (14.2) implicitly assumes that the interest rate is zero; the trader can borrow without cost to go long in securities and receives no interest when he goes short in them. We will drop the assumption of zero interest rates in Section 15.1.

Let us say that a class C of nonnegative processes is lim inf-*closed* if the process

$$X_t(\omega) := \liminf_{k \to \infty} X_t^k(\omega) \tag{14.3}$$

is in C whenever each process X^k is in C. A process X is a *nonnegative supermartingale* in the martingale space (14.1) if it belongs to the smallest lim inf-closed class of nonnegative processes containing all nonnegative simple capital processes. We think of nonnegative supermartingales as nonnegative capital processes. They can lose capital, as the approximation is in the sense of lim inf.

Remark 14.1   An equivalent definition of the class C of nonnegative supermartingales can be given using transfinite induction on the countable ordinals α (see, e.g. [105, section 0.8] or [202, chapter 2]). Namely, define C^α as follows:

- C^0 is the class of all nonnegative simple capital processes;
- for $\alpha > 0$, $X \in C^\alpha$ if and only if there exists a sequence $X^1, X^2, \ldots$ of processes in $C^{<\alpha} := \cup_{\beta<\alpha} C^\beta$ such that (14.3) holds.

It is easy to check that the class of all nonnegative supermartingales is the union of the nested family C^α over all countable ordinals α. The *rank* of a nonnegative supermartingale X is defined to be the smallest α such that $X \in C^\alpha$; in this case we will also say that X is *of rank* α.   ∎

The following definition modifies the notion of instant enforcement used in Chapter 13. Now we say that a set $E \subseteq [0, \infty) \times \Omega$ (also called a property of $t \in [0, \infty)$ and $\omega \in \Omega$) is *instantly enforceable* if there exists a nonnegative supermartingale X such that $X_0 = 1$ and, for all $t \in [0, \infty)$ and $\omega \in \Omega$,

$$(t, \omega) \notin E \implies X_t(\omega) = \infty. \tag{14.4}$$

Complements of instantly enforceable sets are *instantly blockable*. We also use *with instant enforcement (w.i.e.)* to mean that a stated property is instantly enforceable.

Lemma 14.2   *A countable intersection of instantly enforceable sets in $[0, \infty) \times \Omega$ is instantly enforceable.*

Proof   Equivalently, we are required to prove that a countable union of instantly blockable sets is instantly blockable. This follows from the fact that a countable convex mixture of nonnegative supermartingales is a nonnegative supermartingale (see Exercise 14.3).   □

A sequence of processes X^k converges to a process X *uniformly on compacts with instant enforcement (ucie)* if the property

$$\lim_{k \to \infty} \sup_{s \in [0,t]} |X_s^k(\omega) - X_s(\omega)| = 0 \tag{14.5}$$

of t and ω is instantly enforceable. If continuous X^k converge ucie to X, we can consider the limit X to be continuous as well; to make this precise, we will extend the notion of a continuous process.

Let ∂ (the *cemetery state*) be any element outside the real line $\mathbb{R}$; we add it to $\mathbb{R}$ as an isolated point. Processes $X : [0, \infty) \times \Omega \to \mathbb{R} \cup \{\partial\}$ are defined analogously to $[-\infty, \infty]$-valued processes. Let us say that a process $X : [0, \infty) \times \Omega \to \mathbb{R} \cup \{\partial\}$ is a *continuous process* if:

- it takes values in $\mathbb{R}$ with instant enforcement;
- for each $\omega \in \Omega$,
  — the set of $t \in [0, \infty)$ for which $X_t(\omega) \in \mathbb{R}$ contains 0 and is connected;
  — $X_t(\omega)$ is continuous as function of t in this set.

(Even though an object that qualifies as a continuous process by this definition does not qualify as a process by our earlier definition, we will sometimes call it a process when there is no danger of confusion.) The *effective domain* of a continuous process X is defined to be

$$\mathrm{dom}\, X := \{(t, \omega) | X_t(\omega) \in \mathbb{R}\}.$$

The notion of ucie convergence is extended to continuous processes in a natural way: $X^k \to X$ ucie if the conjunction of the properties

$$(\forall k \; \forall s \in [0, t] : X_s^k(\omega) \in \mathbb{R}) \quad \text{and} \quad (\forall s \in [0, t] : X_s(\omega) \in \mathbb{R})$$

and (14.5) of t and ω is instantly enforceable.

A class C of continuous processes is lim-*closed* if it contains every continuous process X for which there exists a sequence X^k of continuous processes in C such that

- $\mathrm{dom}\, X \subseteq \mathrm{dom}\, X^k$ for each k;
- for each $(t, \omega) \in \mathrm{dom}\, X$, we have (14.5).

A continuous process is a *continuous martingale* (in the martingale space (14.1)) if it is an element of the smallest lim-closed class of continuous processes that contains all simple capital processes. The *rank* of a continuous martingale is defined as in Remark 14.1. Examples of continuous martingales (of rank 0) are the basic martingales $S^j, j \in J$.

The following lemma shows that the filtration $(\mathcal{F}_t)$ can be replaced by $(\mathcal{F}_{t+})$, where $\mathcal{F}_{t+} := \cap_{s > t} \mathcal{F}_s$, as far as the notions of nonnegative supermartingale and continuous martingale are concerned.

Lemma 14.3

- *Each nonnegative supermartingale in the martingale space*

$$(\Omega, \mathcal{F}, (\mathcal{F}_{t+})_{t\in[0,\infty)}, J, (\mathcal{S}^j)_{j\in J}) \tag{14.6}$$

is a nonnegative supermartingale in the martingale space (14.1).
- *Each continuous martingale in (14.6) is a continuous martingale in (14.1).*

In particular, Lemma 14.3 says that all nonnegative supermartingales and continuous martingales in the martingale space (14.6) are processes and continuous processes, respectively, in the martingale space (14.1) (i.e. adapted to $(\mathcal{F}_t)$). The opposite statement to Lemma 14.3 is obvious: the classes of nonnegative supermartingales and continuous martingales can only expand when we expand the filtration.

Proof of Lemma 14.3   It suffices to prove that each simple capital process in the martingale space (14.6) is a nonnegative supermartingale and continuous martingale in the martingale space (14.1); the remaining argument by transfinite induction is trivial. Let (14.2) be a simple capital process in (14.6). Replacing each $(\mathcal{F}_{t+})$-stopping time τ_n by the $(\mathcal{F}_t)$-stopping time $\tau_n + \varepsilon$ for $\varepsilon > 0$ and letting $\varepsilon \to 0$, we recover the same process $\mathcal{K}^\psi$, which is, therefore, a nonnegative supermartingale and continuous martingale in (14.6).   □

A filtration $(\mathcal{F}_t)$ is said to be *right-continuous* if $\mathcal{F}_t = \mathcal{F}_{t+}$ for all $t \in [0, \infty)$. All our results in the rest of the book depend on the underlying martingale spaces only via their families of nonnegative supermartingales and continuous martingales. Therefore, we can, and we shall, assume, with no loss of generality, that the filtration $(\mathcal{F}_t)$ in (14.1) is right-continuous: if it is not, we replace $(\mathcal{F}_t)$ by $(\mathcal{F}_{t+})$, which is always right-continuous.

The following two lemmas rely on the right-continuity of $(\mathcal{F}_t)$.

Lemma 14.4   *For any continuous process X, the function $\Sigma_X : \Omega \to [0, \infty]$ defined by*

$$\Sigma_X(\omega) := \inf\{t \in [0, \infty) | X_t(\omega) = \partial\}$$

is a stopping time.

Proof   Since, for each $t \in [0, \infty)$,

$$\{\omega | \Sigma_X(\omega) \le t\} = \bigcap_\varepsilon \{\omega | X_{t+\varepsilon}(\omega) = \partial\},$$

where ε ranges over the positive rational numbers, the event $\{\Sigma_X \le t\}$ is in $\mathcal{F}_s$ for each $s > t$. In combination with the right-continuity of the filtration this implies $\{\Sigma_X \le t\} \in \mathcal{F}_t$. □

We will refer to Σ_X as the *frontier* of dom X. The continuous process X takes real values before Σ_X, the value ∂ after Σ_X, and either value at Σ_X.

Lemma 14.5 *For any continuous process X and any open set $A \subseteq \mathbb{R}$, the function $\tau_A : \Omega \to [0, \infty]$ defined by*

$$\tau_A(\omega) := \inf\{t \in [0, \infty) | X_t(\omega) \in A\} \tag{14.7}$$

is a stopping time.

Proof Since the event

$$\{\omega | \tau_A(\omega) \le t\} = \{\omega | \exists s \in \mathbb{Q} \cap [0, t] : X_s(\omega) \in A\}$$
$$\cup \{\omega | \exists \varepsilon > 0 \, \forall s \in \mathbb{Q} \cap (t, t + \varepsilon) : X_s(\omega) \in A\}$$

is in $\mathcal{F}_s$ for each $s > t$ and the filtration is right-continuous, we have $\{\tau_A \le t\} \in \mathcal{F}_t$. □

Lemmas 14.4 and 14.5 can be combined in the following corollary; remember that ∂ is regarded as an isolated point of $\mathbb{R} \cup \{\partial\}$.

Corollary 14.6 *For any continuous process X and any open set $A \subseteq \mathbb{R} \cup \{\partial\}$, τ_A (defined by (14.7)) is a stopping time.*

The following lemma says that, essentially, each nonnegative continuous martingale is a nonnegative supermartingale. Let us say that a continuous martingale X is ∞-*dominated* by a nonnegative supermartingale Y if $X_0 = Y_0$ and $X_t(\omega) = Y_t(\omega)$ unless $Y_t(\omega) = \infty$.

Lemma 14.7 *Every continuous martingale X such that $X_t(\omega) \ge 0$ w.i.e. is ∞-dominated by a nonnegative supermartingale.*

Proof We will show that, for every continuous martingale X, the process

$$X_t^*(\omega) := \begin{cases} X_t(\omega) & \text{if } t \le \tau \text{ and } X_t(\omega) \in \mathbb{R}, \\ 0 & \text{if } t \ge \tau \text{ and } X_t(\omega) \in \mathbb{R}, \\ \infty & \text{if } X_t(\omega) = \partial, \end{cases}$$

where τ is the stopping time $\inf\{t|X_t < 0\}$ (with $\inf \emptyset := \infty$), is a nonnegative supermartingale. This implies the statement of the lemma: if X satisfies its condition and Y is a nonnegative supermartingale witnessing that $X_t(\omega) \geq 0$ w.i.e., the $\liminf_{k\to\infty}$ of the nonnegative supermartingales $X^* + Y/k$ ∞-dominates X.

If X is a simple capital process, then $X_t^* = X_{\tau\wedge t}$ is a simple capital process that is also a nonnegative supermartingale. (Remember that simple capital processes never take value ∂.)

In order to use transfinite induction, we assume that the rank of a continuous martingale X is α and that Y^* is a nonnegative supermartingale for any continuous martingale Y of a rank less than α. Let X^k be a sequence of continuous martingales of rank less than α such that $\forall k : \text{dom } X \subseteq \text{dom } X^k$ and (14.5) holds for all $(t, \omega) \in \text{dom } X$. It suffices to notice that

$$X_t^* = \liminf_{j\to\infty} \liminf_{k\to\infty} ((X^k + 1/j)_t^* + Y_t'/k),$$

where Y' is a nonnegative supermartingale satisfying $Y_0' = 1$ and $Y_t'(\omega) = \infty$ for $(t, \omega) \notin \text{dom } X$, and therefore, X^* is a nonnegative supermartingale. $\qquad\square$

A continuous process A is a *finite variation continuous process* if $A_0(\omega) = 0$ for all ω and the total variation of the function $s \in [0, t] \mapsto A_s(\omega)$ is finite for all $(t, \omega) \in \text{dom } A$. A continuous process X is a *continuous semimartingale* if there exist a continuous martingale Y and a finite variation continuous process A such that $\text{dom } X = \text{dom } Y = \text{dom } A$ and $X = Y + A$ on $\text{dom } X$. We call such a decomposition $X = Y + A$ of X the *standard decomposition*; the article *the* is justified by its uniqueness (see Proposition 14.23).

Comparison with Measure-Theoretic Probability

The motivation for our terminology is the analogy with measure-theoretic probability. We will be using the definitions of measure-theoretic nonnegative supermartingales that do not impose any integrability or continuity conditions, as in [319, Definition II.1.1].

In this subsection we will assume that the measurable space $(\Omega, \mathcal{F})$ is equipped with a probability measure P and that each $S^j, j \in J$, is a continuous local martingale with respect to $(\mathcal{F}_t)$ (we give the standard definition shortly). Now we have two notions of nonnegative supermartingales and continuous martingales: as defined earlier in the chapter, which we will prefix with "game-theoretic," and the standard ones, which we will prefix with "measure-theoretic." We start from (a version of) the standard definitions.

A process X_t is a *measure-theoretic supermartingale* if

- $P(X_t^-) < \infty$ for each $t \in [0, \infty)$, where f^- stands for $(-f) \vee 0$;
- $P(X_t|\mathcal{F}_s) \leq X_s$ P-a.s. for each pair $s, t \in [0, \infty)$ such that $s \leq t$.

We say that X is a *measure-theoretic martingale* if both X and $-X$ are measure-theoretic supermartingales. And we say that X is a *continuous local martingale* (there is no need to say "measure-theoretic" as we have only one notion of continuous local martingale) if there is a sequence of stopping times $\tau_1 \le \tau_2 \le \cdots$ such that $\lim_{n \to \infty} \tau_n = \infty$ P-a.s. and each stopped process $X_t^{\tau_n} := X_{\tau_n \wedge t}$ is a continuous measure-theoretic martingale. As mentioned earlier, we assume that each S^j is a continuous measure-theoretic local martingale. (A process is *continuous* if all, or P-almost all, of its paths are continuous.)

Each simple capital process is a continuous local martingale (Exercise 14.5). Since each nonnegative continuous local martingale is a measure-theoretic supermartingale [319, p. 123], nonnegative simple capital processes are measure-theoretic supermartingales. By Fatou's lemma, $\lim\inf_k X^k$ is a measure-theoretic supermartingale whenever X^k are nonnegative measure-theoretic supermartingales:

$$P\left(\liminf_k X_t^k \mathcal{F}_s\right) \le \liminf_k P(X_t^k | \mathcal{F}_s) \le \liminf_k X_s^k \quad \text{a.s.},$$

where $0 \le s < t$. Therefore, the game-theoretic nonnegative supermartingales are a subset of the set of all nonnegative measure-theoretic supermartingales.

To see that our definition of "instantly enforceable" agrees with the measure-theoretic picture, we will use the following continuous-time version of Ville's inequality (cf. (2.32) and (8.18)).

Proposition 14.8 *For any game-theoretic nonnegative supermartingale X such that $X_0 = 1$ and any $K \in (0, \infty)$,*

$$P\left\{ \sup_{t \in \mathbb{Q} \cap [0,\infty)} X_t \ge K \right\} \le \frac{1}{K}.$$

Proof Let $D_1 \subseteq D_2 \subseteq \cdots \subseteq \mathbb{Q}$ be an increasing sequence of finite sets of rational numbers such that $\cup_n D_n = \mathbb{Q} \cap [0, \infty)$. By Ville's inequality for discrete-time measure-theoretic supermartingales, for any $\varepsilon \in (0, 1)$,

$$P\left\{ \sup_{t \in \mathbb{Q} \cap [0,\infty)} X_t \ge K \right\} \le P\left\{ \sup_{t \in \mathbb{Q} \cap [0,\infty)} X_t > (1 - \varepsilon)K \right\}$$

$$= \lim_{n \to \infty} P\left\{ \sup_{t \in D_n} X_t > (1 - \varepsilon)K \right\} \le \frac{1}{(1 - \varepsilon)K},$$

and it remains to let $\varepsilon \to 0$. $\square$

Corollary 14.9 *If a property E of t and ω is instantly enforceable, then we have, for P-almost all ω,*

$$\forall t \in [0, \infty) : (t, \omega) \in E.$$

Proof It suffices to prove that, for any game-theoretic nonnegative supermartingale X such that $X_0 = 1$, the probability of $\sup_{t \in \mathbb{Q} \cap [0,\infty)} X_t = \infty$ is zero. This follows from Proposition 14.8. $\qquad\qquad\qquad\qquad\qquad\qquad\qquad\qquad\qquad\qquad\qquad\qquad\qquad\qquad$ $\square$

Let us now check that continuous martingales X, as defined above (but with ∂ replaced by, say, 0), are measure-theoretic continuous local martingales. Since simple capital processes are continuous local martingales and a limit of a sequence of continuous local martingales that converge in probability uniformly on compact intervals is always a continuous local martingale [73, Theorem 3.1], it suffices to apply transfinite induction on the rank of X (Exercise 14.6). Notice that we can make all sample paths of X continuous by changing it on a set of measure zero.

Comparison with the Setting of Chapter 13

The setting of Chapter 13 can be embedded in our current setting but its supply of nonnegative supermartingales is much poorer. The definitions given in Chapter 13 can be interpreted as definitions in the martingale space (14.1), where

E1 $\Omega = C[0, \infty)$ is the set of all continuous functions on $[0, \infty)$;

E2 $\mathcal{F}$ is the Borel σ-algebra;

E3 for each $t \in [0, \infty)$, $\mathcal{F}_t$ is the sub-σ-algebra of $\mathcal{F}$ generated by the functions $\omega \in \Omega \mapsto \omega(s)$, $s \in [0, t]$; notice that the filtration $(\mathcal{F}_t)$ is not right-continuous;

E4 J is a one-element set;

E5 the only basic martingale $S = S^j$ is defined by $S_t(\omega) := \omega(t) - \omega(0)$.

In this martingale space, the definitions of a stopping time in Section 13.1 and of a stopping time in this section are equivalent by the Galmarino test (see, e.g. [105, theorem IV.100]). Our definitions of τ-measurability in Section 13.1 and $\mathcal{F}_\tau$-measurability given in this section are equivalent also by the Galmarino test. Finally, the Galmarino test implies that the processes of Section 13.2 coincide with the processes of this section. The simple capital processes of Chapter 13 are clearly nonnegative supermartingales and continuous martingales in the sense of this section. The nonnegative supermartingales of Chapter 13 typically form a much smaller family as compared with the nonnegative supermartingales in this section, but it was still sufficient to establish several non-trivial properties of the basic martingale. And we did not define continuous martingales in Chapter 13.

14.2 CONSERVATISM OF CONTINUOUS MARTINGALES

In this section we derive a useful technical result showing that adding countably many continuous martingales to our market as new traded securities (in addition to the basic martingales $S^j, j \in J$) is its "conservative extension," to use a logical term: it does not increase the supply of nonnegative supermartingales and continuous martingales in an interesting way.

Suppose $X^j, j \in I$, is a family of continuous martingales indexed by a nonempty countable set I (perhaps containing all $S^j, j \in J$). Our goal in this section is to show that we can use it instead of the family $S^j, j \in J$, of basic martingales in the definition of nonnegative supermartingales and continuous martingales and this will not lead us (in any interesting way) outside the existing classes of nonnegative supermartingales and continuous martingales.

A *simple trading strategy in* $(X^j)_{j \in I}$ is a triple $\psi = (c, (\tau_1, \ldots, \tau_N), h)$, where

- $c \in \mathbb{R}$ and $N \in \mathbb{N}$;
- $\tau_1 \leq \cdots \leq \tau_N$ is an increasing sequence of stopping times;
- h is a family of bounded variables $(h_{n,j})_{n \in \{1, \ldots, N\}, j \in I}$ such that each $h_{n,j}$ is $\mathcal{F}_{\tau_n}$-measurable and $h_{n,j} = 0$ for all but finitely many (n, j).

(This is identical to the definition of a simple trading strategy except that we allow a different index set.) The *simple capital process* corresponding to a simple trading strategy ψ in $(X^j)_{j \in I}$ is the continuous process defined by

$$Y_t(\omega) := c + \sum_{n=2}^{N+1} \sum_{j \in I} h_{n-1,j}(\omega)(X^j_{\tau_n(\omega) \wedge t}(\omega) - X^j_{\tau_{n-1}(\omega) \wedge t}(\omega)) \tag{14.8}$$

(cf. (14.2)) if none of the continuous martingales X^j takes value ∂ at ω over the interval $[0, t]$; otherwise, we set $Y_t(\omega) := \partial$. A process is a *nonnegative supermartingale over* $(X^j)_{j \in I}$ if it belongs to the smallest lim inf-closed class of nonnegative processes containing all nonnegative simple capital processes over $(X^j)_{j \in I}$ with ∂ replaced by ∞. A continuous process is a *continuous martingale over* $(X^j)_{j \in I}$ if it belongs to the smallest lim-closed class of continuous processes containing all simple capital processes over $(X^j)_{j \in I}$.

Proposition 14.10 *Let $X^j, j \in I$, be a family of continuous martingales indexed by a nonempty countable set I. Each continuous martingale over $(X^j)_{j \in I}$ is a continuous martingale. Each nonnegative supermartingale over $(X^j)_{j \in I}$ is ∞-dominated by a nonnegative supermartingale.*

Proof Let us first show that each simple capital process (14.8) over $(X^j)_{j \in I}$ is a continuous martingale. Since only finitely many $h_{n,j}$ are nonzero, we can, and will, assume that I is finite. Let us proceed by transfinite induction in the largest rank of X^j, $j \in I$. If all X^j are basic martingales, the statement is trivial. The inductive step is also easy.

Now each simple capital process over $(X^j)_{j \in I}$ being a continuous martingale implies that

- each continuous martingale over $(X^j)_{j \in I}$ is a continuous martingale by the definition of continuous martingales;
- each simple capital process over $(X^j)_{j \in I}$ is ∞-dominated by a nonnegative supermartingale by Lemma 14.7; therefore, each nonnegative supermartingale over $(X^j)_{j \in I}$ is also ∞-dominated by a nonnegative supermartingale.   □

14.3 ITÔ INTEGRATION

Let H be a continuous process and X be a continuous martingale. In view of Proposition 14.10 (and the nature of this section's results, all of which are expressed in terms of nonnegative supermartingales and continuous martingales) we can assume that X is a basic martingale S^j, although most of our discussion does not depend on this assumption. (This remark is also applicable to the following sections.) To define the Itô integral of H with respect to X, we need to partition time with sufficient resolution to see arbitrarily small changes in both H and X. A *partition* is any increasing sequence T of stopping times $0 = T_0 \leq T_1 \leq T_2 \leq \cdots$. Let us say that a sequence $T^1, T^2, \ldots$ of partitions *finely covers* a continuous process Y if

1. For all $k, n \in \mathbb{N}$ and $\omega \in \Omega$,

$$\sup_t \ Y_t(\omega) - \inf_t \ Y_t(\omega) \leq 2^{-k},$$

where t ranges over the elements of $[T^k_{n-1}(\omega), T^k_n(\omega)]$ satisfying $Y_t(\omega) \in \mathbb{R}$.
2. For all $k \in \mathbb{N}$ and $\omega \in \Omega$,

$$\lim_{n \to \infty} \ T^k_n(\omega) \geq \Sigma_Y(\omega) \tag{14.9}$$

and

$$(\Sigma_Y(\omega) < \infty \ \text{and} \ Y_{\Sigma_Y}(\omega) \in \mathbb{R}) \implies \exists n \in \mathbb{N} : T^k_n(\omega) \geq \Sigma_Y(\omega), \tag{14.10}$$

where Σ_Y is the frontier of dom Y.

Condition 1 formalizes (T^k) being fine for Y and Condition 2 formalizes (T^k) covering dom Y. One sequence of partitions $T^1, T^2, \ldots$ that finely covers both H and X is

$$T^k_n(\omega) := \inf \left\{ t > T^k_{n-1}(\omega) \middle| |H_t - H_{T^k_{n-1}}| \vee |X_t - X_{T^k_{n-1}}| > 2^{-k-1} \right\} \tag{14.11}$$

for $n \in \mathbb{N}$, where the equality in (14.11) is regarded as true when $H_t = \partial$ or $X_t = \partial$. (The fact that T_n^k are stopping times follows from Corollary 14.6 and its proof.) We will refer to (14.11) as the *canonical sequence of partitions*.

Given a sequence of partitions T^k that finely covers both a continuous process H and a continuous martingale X, we set

$$(H \cdot X)_t^k := \sum_{n=1}^{\infty} H_{T_{n-1}^k \wedge t}(X_{T_n^k \wedge t} - X_{T_{n-1}^k \wedge t}) \tag{14.12}$$

for all $t \in [0, \infty)$ and $k \in \mathbb{N}$. By Proposition 14.10, (14.12) is a continuous martingale (as each $H_{T_{n-1}^k}$ is bounded).

Proposition 14.11 *For any sequence of partitions T^k that finely covers H and X, $(H \cdot X)^k$ converge ucie as $k \to \infty$. The limit will stay the same with instant enforcement if T^k is replaced by another sequence of partitions finely covering H and X.*

The limit whose existence is asserted in Proposition 14.11 for the canonical sequence of partitions (14.11) will be denoted $H \cdot X$ or $\int H \, dX$ and called the *Itô integral* of H with respect to X; its value $(H \cdot X)_t$ at time t will be also denoted as $\int_0^t H \, dX$ or $\int_0^t H_s \, dX_s$. Since the convergence is uniform over compact time intervals, $H \cdot X$ is a continuous function over $[0, t]$ with instant enforcement (which implies that $H \cdot X$ is a continuous function over the whole of $[0, \infty)$ almost surely, as defined in Section 14.7). Despite having defined the Itô integral using the canonical sequence of partitions, the second statement of Proposition 14.11 says that the definition is in fact invariant.

Proposition 14.11 will be deduced from the following lemma.

Lemma 14.12 *For any sequence X^k, $k = 1, 2, \dots$, of nonnegative continuous martingales satisfying $X_0^k \leq 1$, we have $\sup_{s \in [0,t]} X_s^k = O(k^2)$ as $k \to \infty$ with instant enforcement.*

Proof Fix such a sequence of nonnegative continuous martingales X^k. It suffices to show that $\sup_{s \in [0,t]} X_s^k \leq k^2$ from some k on w.i.e. Let $\tilde{X}^k$ be the nonnegative continuous martingale X^k stopped at the moment when it crosses level k^2: $\tilde{X}_t^k := X_{t \wedge \tau}^k$, where $\tau := \inf\{t | X_t^k > k^2\}$ (see Lemma 14.5 and Exercise 14.4). Set $\overline{X} := \sum_k k^{-2} \overline{X}^k$, where $\overline{X}^k$ is a nonnegative supermartingale ∞-dominating $\tilde{X}^k$ (see Lemma 14.7). It remains to notice that $\overline{X}_0 < \infty$ and $\overline{X}_t = \infty$ whenever $\sup_{s \in [0,t]} X_s^k > k^2$ for infinitely many k. $\qquad\square$

Proof of Proposition 14.11 The value of t will be fixed throughout the proof. It suffices to prove that the sequence of functions $(H \cdot X)_s^k$ on the interval $s \in [0, t]$ is Cauchy (in the uniform metric) with instant enforcement (i.e. in this context, unless a nonnegative supermartingale independent of t and with a finite initial value takes value ∞ at t).

Let us arrange the stopping times $T_0^k, T_1^k, T_2^k, \ldots$ and $T_0^{k-1}, T_1^{k-1}, T_2^{k-1}, \ldots$, where $k > 1$, into one strictly increasing sequence (removing duplicates if needed) a_n, $n = 0, 1, \ldots : 0 = a_0 < a_1 < a_2 < \cdots$, each a_n is equal to one of the T_n^k or one of the T_n^{k-1}, each T_n^k is among the a_n, and each T_n^{k-1} is among the a_n; if the resulting sequence is finite, $a_1, \ldots, a_N$, we set $a_n := \infty$ for $n > N$. Let us apply the strategy leading to the supermartingale (3.4) to

$$
\begin{aligned}
y_n &:= b_k(((H \cdot X)_{a_n}^k - (H \cdot X)_{a_{n-1}}^k) \\
&\quad - ((H \cdot X)_{a_n}^{k-1} - (H \cdot X)_{a_{n-1}}^{k-1})) \\
&= b_k(H_{a'_{n-1}}(X_{a_n} - X_{a_{n-1}}) - H_{a''_{n-1}}(X_{a_n} - X_{a_{n-1}})) \\
&= b_k(H_{a'_{n-1}} - H_{a''_{n-1}})(X_{a_n} - X_{a_{n-1}}),
\end{aligned}
\tag{14.13}
$$

where $b_k := k^2$ (the rationale for this choice will become clear later), $a'_{n-1} := T_{n'}^k$ with n' being the largest integer such that $T_{n'}^k \le a_{n-1}$, and $a''_{n-1} := T_{n''}^{k-1}$ with n'' being the largest integer such that $T_{n''}^{k-1} \le a_{n-1}$. (Notice that either $a'_{n-1} = a_{n-1}$ or $a''_{n-1} = a_{n-1}$.) Informally, we consider the simple capital process $\mathcal{K}^k$ with starting capital 1 corresponding to betting $\mathcal{K}_{a_{n-1}}^k$ on y_n at each time a_{n-1}, $n = 1, 2, \ldots$. Formally, the bet (on X) at time a_{n-1} is $\mathcal{K}_{a_{n-1}}^k b_k(H_{a'_{n-1}} - H_{a''_{n-1}})$.

We often do not reflect k in our notation (such as a_n and y_n), but this should not lead to ambiguities.

The condition of Proposition 3.4 is satisfied (whenever y_n is defined) as

$$
|y_n| \le b_k 2^{-k+1} 2^{-k} \le 0.5,
\tag{14.14}
$$

where the last inequality (ensuring that both (3.4) and

$$
\mathcal{T}_n' := \exp\left(\sum_{i=1}^n (-y_i) - \sum_{i=1}^n y_i^2 \right)
\tag{14.15}
$$

are supermartingales) is true for all $k \ge 1$. By Proposition 3.4, we will have

$$
\mathcal{K}_{a_N}^k \ge \prod_{n=1}^N \exp(y_n - y_n^2), \quad N = 0, 1, \ldots,
$$

whenever $a_N < \infty$. Proposition 3.4 also shows that, in addition,

$$
\mathcal{K}_s^k \ge \mathcal{K}_{a_{n-1}}^k \exp(y_{n,s} - y_{n,s}^2), \quad n \in \mathbb{N}, \quad s \in [a_{n-1}, a_n] \cap \mathbb{R},
$$

where

$$y_{n,s} := b_k(((H \cdot X)_s^k - (H \cdot X)_{a_{n-1}}^k)$$
$$- ((H \cdot X)_s^{k-1} - (H \cdot X)_{a_{n-1}}^{k-1})) \qquad (14.16)$$
$$= b_k(H_{a'_{n-1}} - H_{a''_{n-1}})(X_s - X_{a_{n-1}})$$

(cf. (14.13); notice that (14.14) remains true for $y_{n,s}$ in place of y_n). This simple capital process $\mathcal{K}^k$ is obviously nonnegative.

To cover both (14.13) and (14.16), we modify (14.16) to

$$y_{n,s} := b_k(H_{a'_{n-1}} - H_{a''_{n-1}})(X_{a_n \wedge s} - X_{a_{n-1} \wedge s}).$$

We have a nonnegative capital process $\mathcal{K}^k$ that starts from 1 and whose value at time s is at least

$$\exp\left(b_k((H \cdot X)_s^k - (H \cdot X)_s^{k-1}) - \sum_{n=1}^\infty y_{n,s}^2\right). \qquad (14.17)$$

Let us show that

$$\sup_{s \in [0,t]} \sum_{n=1}^\infty y_{n,s}^2 = o(1) \qquad (14.18)$$

as $k \to \infty$ with instant enforcement. It suffices to show that

$$\sup_{s \in [0,t]} \sum_{n=1}^\infty (k^2 2^{-k+1}(X_{a_n \wedge s} - X_{a_{n-1} \wedge s}))^2 = o(1) \qquad \text{w.i.e.}$$

Using the trading strategy leading to Kolmogorov's martingale (3.1), we obtain the simple capital process

$$\tilde{\mathcal{K}}_s^k = k^{-3} + \sum_{n=1}^\infty (k^2 2^{-k+1}(X_{a_n \wedge s} - X_{a_{n-1} \wedge s}))^2$$
$$- \left(\sum_{n=1}^\infty k^2 2^{-k+1}(X_{a_n \wedge s} - X_{a_{n-1} \wedge s})\right)^2$$
$$= k^{-3} + \sum_{n=1}^\infty k^4 2^{-2k+2}(X_{a_n \wedge s} - X_{a_{n-1} \wedge s})^2$$
$$- k^4 2^{-2k+2}(X_s - X_0)^2. \qquad (14.19)$$

Formally, this simple capital process corresponds to the initial capital $\tilde{\mathcal{K}}_0^k = k^{-3}$ and betting $-2k^2 2^{-k+1}(X_{a_{k-1}} - X_0)$ at time a_{n-1}, $n = 1, 2, \ldots$ (cf. (3.2)). Let us make this simple capital process nonnegative by stopping trading at the first moment s after which $k^4 2^{-2k+2}(X_s - X_0)^2$ exceeds k^{-3} (which will not happen before time t for sufficiently large k); notice that this will make $\tilde{\mathcal{K}}^k$ nonnegative even if the addend $\sum_{n=1}^{\infty} \cdots (\cdots)^2$ in (14.19) is ignored. Since $\tilde{\mathcal{K}}^k$ is a nonnegative simple capital process with initial value k^{-3}, applying Lemma 14.12 to $k^3 \tilde{\mathcal{K}}^k$ gives $\sup_{s \le t} \tilde{\mathcal{K}}^k_s = O(k^{-1}) = o(1)$ w.i.e. Therefore, the sum $\sum_{n=1}^{\infty} \cdots (\cdots)^2$ in (14.19) is $o(1)$ uniformly over $s \in [0, t]$ w.i.e., which completes the proof of (14.18).

In combination with (14.18), (14.17) implies

$$\mathcal{K}^k_s \ge \exp(b_k((H \cdot X)^k_s - (H \cdot X)^{k-1}_s) - 1)$$

for all $s \le t$ from some k on with instant enforcement. Applying the strategy leading to the supermartingale (3.4) to $-y_{n,s}$ in place of $y_{n,s}$ (cf. (14.15)) and averaging the resulting simple capital processes, we obtain a simple capital process $\overline{\mathcal{K}}^k$ satisfying $\overline{\mathcal{K}}^k_0 = 1$ and

$$\overline{\mathcal{K}}^k_s \ge \frac{1}{2} \exp(b_k |(H \cdot X)^k_s - (H \cdot X)^{k-1}_s| - 1)$$

for all $s \le t$ from some k on with instant enforcement.

By the definition of $\overline{\mathcal{K}}^k$ and Lemma 14.12, we obtain

$$\sup_{s \in [0,t]} \frac{1}{2} \exp(k^2 |(H \cdot X)^k_s - (H \cdot X)^{k-1}_s| - 1) = O(k^2) \qquad \text{w.i.e.}$$

The last equality implies

$$\sup_{s \in [0,t]} |(H \cdot X)^k_s - (H \cdot X)^{k-1}_s| = O\left(\frac{\log k}{k^2}\right) \qquad \text{w.i.e.}$$

Since the sum $\sum_k (\log k) k^{-2}$ is finite, we have the ucie convergence of $(H \cdot X)^k$ as $k \to \infty$.

To show that the limits coincide with instant enforcement for two sequences T_1 and T_2 of partitions finely covering H and X, the argument above should be applied to T^k_1 and T^k_2 instead of T^k and T^{k-1}. $\qquad\square$

Lemma 14.13 *The Itô integral of a continuous process with respect to a continuous martingale is a continuous martingale.*

Proof By Proposition 14.10, (14.12) are continuous martingales. By definition, their ucie limit is a continuous martingale as well. It remain to apply Proposition 14.11. $\qquad\square$

The definition of the Itô integral $\int H \, dX$ can be easily extended to the case where X is a continuous semimartingale. Namely, we set $\int H \, dX$ to $\int H \, dY + \int H \, dA$, where $Y + A$ is the standard decomposition of X (see Proposition 14.23), and $\int H \, dA$ is the Lebesgue–Stieltjes integral.

14.4 COVARIATION AND QUADRATIC VARIATION

We start from establishing the existence of the covariation between two continuous martingales, X and Y. The covariation of X and Y can be approximated by

$$[X, Y]_t^k := \sum_{n=1}^{\infty} (X_{T_n^k \wedge t} - X_{T_{n-1}^k \wedge t})(Y_{T_n^k \wedge t} - Y_{T_{n-1}^k \wedge t}), \quad k \in \mathbb{N}. \tag{14.20}$$

We show that the ucie limit of $[X, Y]^k$ as $k \to \infty$ exists for sequences of partitions finely covering X and Y, denote it $[X, Y]$ (or $[X, Y]_t(\omega)$ if we need to mention the arguments), and call it the *covariation* between X and Y.

Lemma 14.14 *The ucie limit of (14.20) exists for sequences of partitions finely covering both X and Y. Moreover, it satisfies the* integration by parts formula

$$X_t Y_t - X_0 Y_0 = \int_0^t X \, dY + \int_0^t Y \, dX + [X, Y]_t \quad w.i.e. \tag{14.21}$$

Proof The Itô integral $\int_0^t X \, dY$ was defined in the previous section as the ucie limit as $n \to \infty$ of

$$(X \cdot Y)_t^k(\omega) := \sum_{n=1}^{\infty} X_{T_{n-1}^k \wedge t}(Y_{T_n^k \wedge t} - Y_{T_{n-1}^k \wedge t}), \quad k \in \mathbb{N}.$$

Swapping X and Y we obtain the analogous expression for $\int_0^t Y \, dX$. It is easy to check that

$$X_t Y_t - X_0 Y_0 = (X \cdot Y)_t^k + (Y \cdot X)_t^k + [X, Y]_t^k.$$

Passing to the ucie limit as $n \to \infty$, we obtain the existence of $[X, Y]$ and the integration by parts formula (14.21). □

It is clear from (14.21) that $[X, Y]$ is a continuous process. Moreover, the following lemma shows that it is a finite variation continuous process.

Setting $Y := X$ leads to the definition of the *quadratic variation* $[X, X]$, which we will sometimes abbreviate to $[X]$. It is clear from the definition (14.20) that $[X]$ is an increasing and, therefore, finite variation continuous process. The following lemma shows that $[X, Y]$ is a finite variation continuous process for any continuous martingales X and Y (and, as Lemma 14.17 shows, even for any continuous semi-martingales X and Y).

Lemma 14.15   *For any continuous martingales X and Y,*

$$[X, Y] = \frac{1}{2}([X + Y] - [X] - [Y]) \quad w.i.e.,$$

and $[X, Y]$ is a finite variation continuous process.

Proof The identity $ab = \frac{1}{2}((a + b)^2 - a^2 - b^2)$ implies

$$[X, Y]^k = \frac{1}{2}([X + Y]^k - [X]^k - [Y]^k)$$

for each $k \in \mathbb{N}$, and it remains to pass to a ucie limit as $k \to \infty$.   □

Remark 14.16   In the setting of Chapter 13 (as embedded in the setting of this chapter, as discussed in Section 14.1), we have two notions of quadratic variation of the basic martingale $X_t(\omega) := \omega(t)$, namely $w_t(\omega)$ defined by (13.23) and $[X]_t(\omega)$ (defined as in this chapter or directly as in [404]). Theorem 13.9 implies that the two notions coincide for all t for almost all ω (therefore, by Lemma 13.17, with instant enforcement). Indeed, since $w_t = [X]_t = t$, $\forall t \in [0, \infty)$, holds almost surely in the case of Brownian motion, it suffices to check that the complement of the event $\forall t \in [0, \infty) : w_t = A_t$ belongs to $\mathcal{I}$. This follows from Lemma 13.23 and the analogous statement for $[X]$: if $w_t(\omega) = [X]_t(\omega)$ for all t, we also have

$$w_t(\omega \circ f) = w_{f(t)}(\omega) = [X]_{f(t)}(\omega) = [X]_t(\omega \circ f)$$

for all t.   ■

Let us now extend the notions of covariation and quadratic variation to continuous semimartingales. Our definition of $[X, Y]$ for continuous martingales carries over to the case of continuous semimartingales X and Y verbatim (using a sequence of partitions that finely covers all components of the standard decompositions of X and Y), and it is clear that Lemma 14.14 holds for any continuous semimartingales. Quadratic variation can still be defined as $[X] := [X, X]$.

As in measure-theoretic probability, the covariation between two continuous semimartingales only depends on their martingale parts.

Lemma 14.17   *If X and X' are two continuous semimartingales with standard decompositions $X = Y + A$ and $X' = Y' + A'$, then $[X, X'] = [Y, Y']$ w.i.e.*

Proof By the definition (14.20) of covariation,

$$
\begin{aligned}
[X, X']_t^k(\omega) &= \sum_{n=1}^{\infty} (X_{T_n^k \wedge t} - X_{T_{n-1}^k \wedge t})(X'_{T_n^k \wedge t} - X'_{T_{n-1}^k \wedge t}) \\
&= \sum_{n=1}^{\infty} (Y_{T_n^k \wedge t} - Y_{T_{n-1}^k \wedge t})(Y'_{T_n^k \wedge t} - Y'_{T_{n-1}^k \wedge t}) & (14.22) \\
&+ \sum_{n=1}^{\infty} (Y_{T_n^k \wedge t} - Y_{T_{n-1}^k \wedge t})(A'_{T_n^k \wedge t} - A'_{T_{n-1}^k \wedge t}) & (14.23) \\
&+ \sum_{n=1}^{\infty} (A_{T_n^k \wedge t} - A_{T_{n-1}^k \wedge t})(Y'_{T_n^k \wedge t} - Y'_{T_{n-1}^k \wedge t}) & (14.24) \\
&+ \sum_{n=1}^{\infty} (A_{T_n^k \wedge t} - A_{T_{n-1}^k \wedge t})(A'_{T_n^k \wedge t} - A'_{T_{n-1}^k \wedge t}). & (14.25)
\end{aligned}
$$

Since the first addend (14.22) in the last sum is $[Y, Y']_t^k$, we are required to show that the other three addends, (14.23)–(14.25), converge to zero as $k \to \infty$ uniformly on compact intervals. The same argument works for all three addends; e.g. (14.23) tends to zero uniformly on compact intervals because

$$
\left| \sum_{n=1}^{\infty} (Y_{T_n^k \wedge t} - Y_{T_{n-1}^k \wedge t})(A'_{T_n^k \wedge t} - A'_{T_{n-1}^k \wedge t}) \right| \le 2^{-k} O(1) \to 0 \quad (k \to \infty),
$$

where we have used the finite variation of A' and the sequence of partitions finely covering the relevant continuous processes. $\square$

14.5 ITÔ'S FORMULA

We begin by stating Itô's formula for continuous semimartingales.

Theorem 14.18 *Let $F : \mathbb{R} \to \mathbb{R}$ be a function of class C^2 and X be a continuous semimartingale. Then*

$$
F(X_t) = F(X_0) + \int_0^t F'(X_s) \, dX_s + \frac{1}{2} \int_0^t F''(X_s) \, d[X]_s \quad w.i.e.
$$

The last integral $\int_0^t F''(X_s)\, d[X]_s$ can be understood in the Lebesgue–Stieltjes sense.

Proof By Taylor's formula,

$$F(X_{T_n^k \wedge t}) - F(X_{T_{n-1}^k \wedge t}) = F'(X_{T_{n-1}^k \wedge t})(X_{T_n^k \wedge t} - X_{T_{n-1}^k \wedge t})$$

$$+ \frac{1}{2} F''(\xi_n)(X_{T_n^k \wedge t} - X_{T_{n-1}^k \wedge t})^2,$$

where $\xi_n \in [X_{T_{n-1}^k \wedge t}, X_{T_n^k \wedge t}]$ (and $[a, b]$ is understood to be $[b, a]$ when $a > b$). It remains to sum this equality over $n \in \mathbb{N}$ and to pass to the ucie limit as $k \to \infty$. $\square$

The next result is a vector version of Theorem 14.18 and is proved in a similar way. By a *vector continuous semimartingale*, we mean a finite sequence $X = (X^1, \ldots, X^d)$ of continuous semimartingales considered as a function mapping $(t, \omega) \in [0, \infty) \times \Omega$ to the vector $X_t(\omega) = (X_t^1(\omega), \ldots, X_t^d(\omega))$.

Theorem 14.19 *Let $F : \mathbb{R}^d \to \mathbb{R}$ be a function of class C^2 and $X = (X^1, \ldots, X^d)$ be a vector continuous semimartingale. Then*

$$F(X_t) = F(X_0) + \sum_{i=1}^d \int_0^t \frac{\partial F}{\partial x_i}(X_s)\, dX_s^i$$

$$+ \frac{1}{2} \sum_{i=1}^d \sum_{j=1}^d \int_0^t \frac{\partial^2 F}{\partial x_i \partial x_j}(X_s)\, d[X^i, X^j]_s \quad w.i.e. \quad (14.26)$$

The requirement that F be twice continuously differentiable can be relaxed for the components for which X^i has a special form, such as $X_t^i = t$ for all t. This, however, will not be needed in this book.

14.6 DOLÉANS EXPONENTIAL AND LOGARITHM

The following proposition introduces a game-theoretic version $\mathcal{E}(X)$ of the Doléans exponential of a continuous martingale X.

Proposition 14.20 *If X is a continuous martingale, $\mathcal{E}(X) := \exp(X - [X]/2)$ is a continuous martingale as well.*

Proof A standard trick (cf. [319, Proposition IV.3.4]) is to apply Itô's formula (14.26) to the function $F(x, y) = \exp(x - y/2)$ and vector continuous semimartingale $(X, Y) =$

$(X, [X])$. Since $[X, [X]] = 0$ (Exercise 14.9), $[[X], [X]] = 0$ (by Lemma 14.17), and

$$\frac{\partial F}{\partial y} + \frac{1}{2}\frac{\partial^2 F}{\partial x^2} = 0,$$

we have, by Itô's formula,

$$F(X_t, [X]_t) = F(X_0, 0) + \int_0^t \frac{\partial F}{\partial x}(X_s, [X]_s)\, dX_s \quad \text{w.i.e.;} \qquad (14.27)$$

therefore, $F(X_t, [X]_t)$ is a continuous martingale (by Lemma 14.13). $\qquad \square$

Since $\partial F/\partial x = F$, (14.27) can be rewritten as the stochastic differential equation

$$Y_t = Y_0 + \int_0^t Y_s\, dX_s \qquad (14.28)$$

for $Y := \exp(X - [X]/2)$; the Doléans exponential is its solution.

We are also interested in an inverse operation to taking the Doléans exponential. The *Doléans logarithm X* of a positive continuous martingale Y can be defined in two different ways: by the Itô integral

$$X_t := \ln Y_0 + \int_0^t \frac{dY_s}{Y_s} \qquad (14.29)$$

and by the more explicit formula

$$X_t := \ln Y_t + \frac{1}{2}[\ln Y]_t. \qquad (14.30)$$

The two definitions are equivalent, but we will only check that (14.30) implies (14.29) (and so (14.30) can be taken as the main definition). Applying Itô's formula to the function $F(y_1, y_2) := \ln y_1 + \frac{1}{2}y_2$ and the continuous semimartingales Y and $[\ln Y]$, we obtain the first definition (14.29) from the second definition (14.30):

$$X_t = \ln Y_t + \frac{1}{2}[\ln Y]_t = F(Y_t, [\ln Y]_t)$$

$$= F(Y_0, 0) + \int_0^t \frac{\partial F}{\partial y_1}(Y_s, [\ln Y]_s)\, dY_s + \int_0^t \frac{\partial F}{\partial y_2}(Y_s, [\ln Y]_s)\, d[\ln Y]_s$$

$$+ \frac{1}{2}\int_0^t \frac{\partial^2 F}{\partial y_1^2}(Y_s, [\ln Y]_s)\, d[Y]_s$$

$$= \ln Y_0 + \int_0^t \frac{dY_s}{Y_s} + \frac{1}{2}\int_0^t d[\ln Y]_s - \frac{1}{2}\int_0^t \frac{d[Y]_s}{Y_s^2}$$

$$= \ln Y_0 + \int_0^t \frac{dY_s}{Y_s} \quad \text{w.i.e.}$$

(The last equality follows from $[\ln Y]_t = \int_0^t d[Y]_s/Y_s^2$, which is easy to check and is generalized in (16.2).) The first definition (14.29) shows that the Doléans logarithm of a positive continuous martingale is a continuous martingale. The following proposition summarizes our discussion so far in this section adding a couple of trivial observations.

Proposition 14.21 *If Y is a positive continuous martingale, $\mathcal{L}(Y) := \ln Y + [\ln Y]/2$ is a continuous martingale. For any continuous martingale X, $\mathcal{L}(\mathcal{E}(X)) = X$ w.i.e. For any positive continuous martingale Y, $\mathcal{E}(\mathcal{L}(Y)) = Y$ w.i.e.*

Proof The second statement follows from

$$\mathcal{L}(\mathcal{E}(X)) = \mathcal{L}(\exp(X - [X]/2)) = X - [X]/2 + \frac{1}{2}[X - [X]/2] = X \quad \text{w.i.e.,}$$

where the last equality holds by Lemma 14.17. Similarly, the third statement follows from

$$\mathcal{E}(\mathcal{L}(Y)) = \exp(\mathcal{L}(Y) - [\mathcal{L}(Y)]/2)$$
$$= \exp(\ln Y + [\ln Y]/2 - [\ln Y + [\ln Y]/2]/2) = \exp(\ln Y) = Y \quad \text{w.i.e.}$$
□

Informally, (14.28) and (14.29) can be rewritten as $dY_t = Y_t \, dX_t$ and $dX_t = dY_t/Y_t$, respectively; in this form their similarity is more obvious.

14.7 GAME-THEORETIC EXPECTATION AND PROBABILITY

In this section we will define a general notion of game-theoretic upper probability in continuous time. So far the only probability-type property that we have used in this chapter was that of instant enforcement, which is closely connected with events of upper probability 0 (Exercise 14.2).

The initial value X_0 of a nonnegative supermartingale X is always a constant. Given a function $F : \Omega \to [0, \infty]$, we define its *upper expected value* as

$$\overline{\mathbb{E}}(F) := \inf\{X_0 | \forall \omega \in \Omega : \liminf_{t \to \infty} X_t(\omega) \geq F(\omega)\}, \qquad (14.31)$$

X ranging over the nonnegative supermartingales. The *upper probability* $\overline{\mathbb{P}}(E)$ of a set $E \subseteq \Omega$ is defined as $\overline{\mathbb{E}}(\mathbf{1}_E)$, where $\mathbf{1}_E$ is the indicator function of E. In other words, $\overline{\mathbb{P}}(E)$ can be defined by

$$\overline{\mathbb{P}}(E) = \inf\{X_0 | \forall \omega \in E : \liminf_{t \to \infty} X_t(\omega) \geq 1\}, \qquad (14.32)$$

X again ranging over the nonnegative supermartingales. A property of $\omega \in \Omega$ holds *almost surely* (a.s.) if its complement has upper probability 0. These are standard definitions using cash as numeraire (in the terminology of the next chapter). The projection onto Ω of the complement of a property of t and ω that holds w.i.e. always has upper probability 0 (Exercise 14.2).

We again consider the measure-theoretic setting reviewed in the penultimate subsection of Section 14.1. Let us now check that $\overline{\mathbb{E}}(F) \geq \mathbb{E}(F)$ for each nonnegative extended variable $F : \Omega \to [0, \infty]$, which will imply that $\overline{\mathbb{P}}(E) \geq \mathbb{P}(E)$ for each event $E \subseteq \Omega$. (In this sense our definition of upper probability $\overline{\mathbb{P}}$ is not too permissive, unlike the definition ignoring measurability in Section 13.5.) It suffices to prove that, for any game-theoretic nonnegative supermartingale X with $X_0 > 0$,

$$\liminf_{t \to \infty} X_t \geq F \Longrightarrow X_0 \geq \mathbb{E}(F).$$

This can be done using the Fatou lemma: assuming the antecedent and letting n range over the natural numbers $\mathbb{N}$,

$$\mathbb{E}(F) \leq \mathbb{E}(\liminf_{t \to \infty} X_t) \leq \mathbb{E}(\liminf_{n \to \infty} X_n) \leq \liminf_{n \to \infty} \mathbb{E}(X_n) \leq X_0.$$

14.8 GAME-THEORETIC DUBINS–SCHWARZ THEOREM

It is shown in Chapter 13 that, roughly, a continuous price path can be transformed into a Brownian motion by replacing physical time with quadratic variation. This time we apply this idea in a way that is closer to the measure-theoretic Dubins–Schwarz result, replacing a continuous price path by a continuous martingale and using Proposition 14.10.

We will use the notion of $\mathcal{I}$-measurability of a function $F : C[0, \infty) \to [0, \infty]$ introduced in Section 13.3. We have the following corollary of Theorem 13.9 in our current framework.

Corollary 14.22 *Let* $F : C[0, \infty) \to [0, \infty]$ *be* $\mathcal{I}$-*measurable, and let* X *be a continuous martingale. Then*

$$\overline{\mathbb{E}}(F(X)) \leq \mathcal{W}(F). \tag{14.33}$$

In (14.33) we set $F(X) := \infty$ when $X \notin C[0, \infty)$.

Proof By Proposition 14.10 we can assume, without loss of generality, that X is a basic martingale. Notice that, for every nonnegative capital process $Y_t(\omega)$, $\omega \in C[0, \infty)$, in the sense of Chapter 13 (see (13.3)), the process $Y_t(X)$ is then a nonnegative supermartingale: this is obvious when Y is a simple capital process, and the transfinite-inductive step is easy. It remains to apply the part $\leq$ of Theorem 13.9. (The simple part $\geq$ of that theorem is no longer applicable; cf. Exercise 14.12.)   □

Corollary 14.22 implies the uniqueness of the standard decomposition of continuous semimartingales (at least in the sense of a.s., which we know to be close to w.i.e.; see Lemma 13.17).

Proposition 14.23   *The decomposition of a continuous semimartingale X into the sum $X = Y + A$ of a continuous martingale Y and a finite variation continuous process A is unique (w.i.e.).*

The detailed statement of the proposition is: if $X = Y + A$ and $X = Y' + A'$ are two such decompositions, then $Y|_{[0,t]} = Y'|_{[0,t]}$ and $A|_{[0,t]} = A'|_{[0,t]}$ with instant enforcement.

Proof sketch   Let $X = Y + A = Y' + A'$ be two such decompositions; then $Y - Y' = A' - A$, and so we have a continuous martingale that is simultaneously a finite variation continuous process. Define $F := \mathbf{1}_E$, where $E \subseteq C[0, \infty)$ consists of the functions in $C[0, \infty)$ having a finite variation over some interval $[0, t]$, $t \in (0, \infty)$, while not being constant over that interval. By Corollary 14.22, there exists a nonnegative supermartingale Z with $Z_0 = 1$ that tends to ∞ when $Y - Y'$ has a finite variation over some $[0, t]$ without being constant over that interval. This shows the uniqueness of the standard decomposition in the sense of a.s. This nonnegative supermartingale Z will tend to ∞ on any $\omega \in \Omega$ such that $(Y(\omega) - Y'(\omega))|_{[0,t]} = (A'(\omega) - A(\omega))|_{[0,t]} \neq 0$. This means that already $Z_t(\omega) = \infty$ if we assume, additionally, that the martingale space is strongly coherent (as defined in Exercise 14.15). The assumption of strong coherence is, however, unnecessary (see Exercise 14.14).   □

14.9   COHERENCE

As in Chapter 13, we say that the martingale space 14.1 is *coherent* if $\overline{\mathbb{P}}(\Omega) = 1$ (equivalently, if $\underline{\mathbb{P}}(\emptyset) = 0$).

Lemma 14.24   *The martingale space 14.1 is coherent if and only if, for each set $E \subseteq \Omega$, $\underline{\mathbb{P}}(E) \leq \overline{\mathbb{P}}(E)$.*

Proof   In the "if" direction, suppose $\overline{\mathbb{P}}(\Omega) < 1$; we then have

$$\underline{\mathbb{P}}(\Omega) = 1 - \overline{\mathbb{P}}(\emptyset) = 1 > \overline{\mathbb{P}}(\Omega).$$

In the "only if" direction, suppose there is $E \subseteq \Omega$ such that $\underline{\mathbb{P}}(E) > \overline{\mathbb{P}}(E)$, i.e. $\overline{\mathbb{P}}(E) + \overline{\mathbb{P}}(E^c) < 1$. Fix such an E. There are $a_1, a_2 \in [0, 1]$ such that $a_1 + a_2 < 1$, $\overline{\mathbb{P}}(E) < a_1$, and $\overline{\mathbb{P}}(E^c) < a_2$. Therefore, there are nonnegative supermartingales X^1 and X^2 such that $X_0^1 < a_1$, $X_0^2 < a_2$, $\liminf_{t \to \infty} X_t^1 \geq \mathbf{1}_E$, and $\liminf_{t \to \infty} X_t^2 \geq \mathbf{1}_{E^c}$. Then $X := X^1 + X^2$ satisfies $X_0 < 1$ and $\liminf_{t \to \infty} X_t \geq 1$, which contradicts the coherence of the martingale space. □

A wide family of coherent martingale spaces was given in the subsection "Comparison with Measure-Theoretic Probability" of Section 14.1. For variations of Lemma 14.24, see Exercises 14.15–14.18.

14.10 EXERCISES

Exercise 14.1 Check that the notions of nonnegative supermartingale and continuous martingale will not change if the definition of a simple trading strategy is modified by replacing the stopping times $\tau_1, \ldots, \tau_N$ by nonnegative numbers. *Hint.* Each stopping time can be approximated by a stopping time taking finitely many values. □

Exercise 14.2 Check that any instantly blockable property of t and ω is *evanescent* in the sense of its projection onto Ω being a null set. □

Exercise 14.3 Prove that a countable convex mixture of nonnegative supermartingales is a nonnegative supermartingale. *Hint.* Use transfinite induction. □

Exercise 14.4 Show that $Y_t := X_{t \wedge \tau}$ is a continuous process whenever X is a continuous process and τ is a stopping time (see also Exercise 17.7). *Hint.* Prove the statement for simple capital processes and then use transfinite induction. □

Exercise 14.5 Assume that $\omega \in \Omega$ is generated stochastically, as described in the subsection "Comparison with Measure-Theoretic Probability" of Section 14.1. Show that each simple capital process is a measure-theoretic continuous local martingale. □

Exercise 14.6 Let X^k and X be continuous martingales such that $X^k \to X$ ucie. Suppose that all X^k and X (with ∂ replaced by 0) are measure-theoretic continuous local martingales under a given probability measure. Show that $X^k \to X$ uniformly on compact intervals in probability. □

Exercise 14.7 Prove that the Lebesgue sequence of dyadic partitions (14.11) indeed satisfies (14.9) and (14.10) for $Y := H$ and $Y := X$. □

Exercise 14.8 Replace the use of Doléans's supermartingale (3.4) in the proof of Proposition 14.11 by another use of Kolmogorov's martingale (3.1) (which is already used in the proof of (14.18)). □

Exercise 14.9 Show that $[X, [X]] = 0$ for any continuous martingale X. *Hint.* See the proof of Lemma 14.17. □

Exercise 14.10 Show that the definition (14.32) is robust in the following sense: the value of $\overline{\mathbb{P}}(E)$ will not change if we replace $\liminf_{t \to \infty}$ by $\sup_{t \in [0,\infty)}$, and it will not change if we replace lim inf by lim sup. *Hint.* Stop when X exceeds 1; to show that this can be done, use transfinite induction.   □

Exercise 14.11 Show that the definition (14.31) is robust in the following sense: the value of $\overline{\mathbb{E}}(F)$ will not change if we replace lim inf by lim sup. *Hint.* Use the idea in Doob's convergence theorem (Theorem 7.5).   □

Exercise 14.12 Give an example of a function F such that (14.33) holds with "<" in place of "≤." *Hint.* The range of X can contain far from all continuous paths starting at X_0.   □

The following exercise gives a game-theoretic version of Hale Trotter's result on the existence of local time [382, section 1], formalizing an intuition of Paul Lévy's.

Exercise 14.13 (local time, [300]) Let X be a continuous martingale. Show that, almost surely, there exists a continuous function $L : [0, \infty) \times \mathbb{R} \to \mathbb{R}$ such that, for all bounded Borel $k : \mathbb{R} \to [0, \infty)$ and all $t \in [0, \infty)$,

$$\int_0^t k(X_s) \, d[X]_s = \int_{-\infty}^{\infty} k(x)L(t, x) \, dx. \tag{14.34}$$

This asserts the existence of the *local time* $L : [0, \infty) \times \mathbb{R} \to \mathbb{R}$, and (14.34) can be regarded as its definition (in combination with the continuity of L). The intuition behind $L(t, x)$ is that it is the amount of intrinsic time (measured by quadratic variation) that X spends in the vicinity of x over the time interval $[0, t]$. *Hint.* Use Corollary 14.22. See [300, pp. 12–13].   □

Exercise 14.14 Prove Proposition 14.23 without any additional assumptions (such as those used in the proof sketch). *Hint.* Using Proposition 14.10 assume that Y and Y' are basic martingales. Follow the proof of the easier part of Proposition 13.6 (given right after the statement of the proposition).   □

Exercise 14.15

1. For $\omega \in \Omega$ and $t \in [0, \infty)$, define the *information about ω at time t* as

$$I_t(\omega) := \cap \{E \in \mathcal{F}_t | \omega \in E\}.$$

   Show that X_t is constant on $I_t(\omega)$ for any adapted process X.

2. The *conditional upper expected value* $\overline{\mathbb{E}}_t(F)$ of a function $F : \Omega \to [0, \infty)$ at time t is the extended variable

$$\overline{\mathbb{E}}_t(F)(\omega) := \inf\{X_t(\omega) | \forall \omega' \in I_t(\omega) : \liminf_{t \to \infty} X_t(\omega') \geq F(\omega')\} \tag{14.35}$$

   (cf. (14.31)), X ranging over the nonnegative supermartingales. The *conditional upper probability* $\overline{\mathbb{P}}_t(E)$ of a set $E \subseteq \Omega$ at time t is defined as the extended variable $\overline{\mathbb{E}}_t(\mathbf{1}_E)$. The martingale space is *strongly coherent* if, for any $t \in [0, \infty)$, $\overline{\mathbb{P}}_t(\Omega) = 1$. Prove that the martingale space of Chapter 13 (as defined at the end of Section 14.1) is strongly coherent. *Hint.* Use the measure-theoretic argument of the previous chapter (as in Lemma 13.11). □

Exercise 14.16 Suppose our martingale space (14.1) is strongly coherent. Show that, for any nonnegative bounded variable F and any $c > 0$,

$$\overline{\mathbb{E}}(F + c) = \overline{\mathbb{E}}(F) + c.$$

□

Exercise 14.17 Suppose our martingale space is strongly coherent. For a bounded variable F, set
$$\overline{\mathbb{E}}(F) := \overline{\mathbb{E}}(F + c) - c,$$

where $c \in \mathbb{R}$ is large enough for $F + c$ to be nonnegative. Show that this definition does not depend on the choice of c. *Hint.* Use the result of Exercise 14.16. □

Exercise 14.18 Suppose our martingale space is strongly coherent. For a bounded variable F, set

$$\underline{\mathbb{E}}(F) := -\overline{\mathbb{E}}(-F),$$

where $\overline{\mathbb{E}}$ is as defined in Exercise 14.17. Show that $\underline{\mathbb{E}}(F) \leq \overline{\mathbb{E}}(F)$. □

Exercise 14.19 Suppose our martingale space is strongly coherent and let $t \in [0, \infty)$. Do Exercises 14.16, 14.17, and 14.18 with $\overline{\mathbb{E}}_t$ in place of $\overline{\mathbb{E}}$ and $\underline{\mathbb{E}}_t$ in place of $\underline{\mathbb{E}}$. Remember that $\overline{\mathbb{E}}_t$ is defined by (14.35). □

Exercise 14.20 Show that all results of this chapter using the notion of upper expectation (14.32) continue to hold when the X in (14.32) is allowed to range over C^k for a finite k in place of C. What is the smallest value of such k? *Hint.* Use the same proofs. □

14.11 CONTEXT

Measure-Theoretic Probability in Continuous Time

The extension of Kolmogorov's axioms to continuous time was codified in Doob's 1953 book [116], but his terminology and many details of his definitions were modified and improved in subsequent decades. Even now there is no full consensus, with standard books and even textbooks offering slightly different definitions. For example, the definitions of supermartingales given in [319, Definition II.1.1], [212, Definition 1.3.1], and [199, Definition I.1.36] impose different continuity and integrability requirements. To avoid misunderstandings, we have made this chapter's discussion of continuous-time measure-theoretic probability self-contained, defining even the most basic notions in Section 14.1. Our definitions mostly follow those given by Revuz and Yor [319], which tend to be the least restrictive.

Foundations of Game-Theoretic Probability in Continuous Time

The basic structure (14.1) goes back to a 1993 article [393, section 4], which defines simple trading strategies, simple capital processes, a notion of coherence, and a primitive notion of game-theoretic upper expectation. Exercises 14.15–14.18 are based on [393, section 5].

On the other hand, it was only in 2016 [419, 420], after a long and slow evolution, that we suggested the bold definitions used in this chapter: define nonnegative supermartingales by closing the nonnegative simple capital processes with respect to lim inf and define continuous martingales by closing the simple capital processes with respect to lim. As recounted in Section 13.7, we studied continuous-time game-theoretic probability using nonstandard analysis in 2001 [349] and shortly thereafter [421, 422], reconsidering the use of standard analysis for this purpose only after work by Takeuchi, Kumon, and Takemura that appeared in working-paper form in 2007 [376]. The cautious definitions of nonnegative supermartingales used in Chapter 13 (and called nonnegative capital processes there) also appeared in 2007, in [405], and was further exploited in [402] and [404]. This chapter's definition was inspired by Nicolas Perkowski and David Prömel's modification of [404]'s definition of game-theoretic upper probability in work published in working-paper form in 2013 [301]. The main differences are that Perkowski and Prömel define the upper probability (14.32) using the test supermartingales in the class C^1 rather than C (in the notation of Remark 14.1) and that they consider a finite horizon (our time interval is $[0, \infty)$ instead of their $[0, T]$). In this book we only need C^k for low k (Exercise 14.20), but the full strength of our definitions might be useful for future work.

In this chapter we have used definitions as close as possible to measure-theoretic definitions, partly because these definitions will already be familiar to most of our readers. In particular, we have emphasized measurability. It is possible, however, that future research will lead to somewhat different definitions. We learned from Section 13.5 that some restrictions are needed in order to prevent the trader from instantly multiplying his capital by infinity, but there may be alternatives to measurability.

Measure-Theoretic Itô Integration

The mainstream approach was initiated by Raymond Paley et al. [298], who only considered Brownian motion as integrator and a deterministic function as integrand. A key step was made by Kiyosi Itô [197], who allowed stochastic integrands, albeit still only with Brownian motion as the integrator. An extension to square-integrable martingales as integrators is due to Hiroshi Kunita and Shinzo Watanabe [232, Definition 2.1]. Modern textbooks (such as [212] and [319]) define the Itô integral $\int H \, dX$ where the integrator is a continuous semimartingale and the integrand is a locally bounded progressively measurable process; for discontinuous X, the further assumption of H's predictability is added.

The Itô integral for a particular integrator is usually defined in three steps:

1. Define the integral for simple integrands (simply as sums), in a similar way to (14.12).
2. Show that the mapping from the simple integrands to the integrals is continuous in a natural topology or metric (e.g. this mapping can be an isometry, in which case it is referred to as *Itô's isometry*).
3. Extend the mapping from the simple integrands to a much wider class of integrands by continuity.

The Itô integral is ideally suited to financial applications, but in some other applications (particularly in engineering) Ruslan Stratonovich's [371] integral is also popular. See [201] for further information on the history of Itô and Stratonovich integration.

Game-Theoretic Itô Integration

The first game-theoretic definition of the Itô integral was given by Nicolas Perkowski and David Prömel [301]. They followed Itô's classical approach, namely steps 1–3 of the previous paragraph, using convergence in game-theoretic upper probability (which they called "Vovk's outer measure") in the third step. Unlike Itô's approach, Perkowski and Prömel's definition is pathwise: their Itô integral $\int_0^t H\,dY$ depends only on $H|_{[0,t]}$ and $Y|_{[0,t]}$, even though the H and Y in their definition are processes rather than paths.

In Section 14.3 we mainly follow [414]. Whereas Perkowski and Prömel obtained almost sure conclusions, we directly obtain instant enforcement. One worry [254, Remark 4.4] about this approach has been the possibility of dependence on the sequence of partitions, but this is dealt with by the second statement of Proposition 14.11. The definition in [414] is "purely pathwise," in that $\int_0^t H\,dY$ is defined for paths (rather than processes) H and Y. The definitions of this chapter (unlike those of Chapter 13) are not purely pathwise in this sense. This simplifies the exposition and might have advantages in applications outside finance (see the end of this section).

Other Pathwise Itô Integrals

There are several other pathwise definitions of the Itô integral, but they all require much stronger assumptions than we have made. Here are two particularly popular definitions.

- Hans Föllmer's definition [146], as developed for example by Anna Ananova and Rama Cont [9]. It postulates the existence of quadratic variation along a particular sequence of partitions, and the value of the integral it defines depends on the choice of that sequence of partitions.
- Terry Lyons's rough-path definition (see, e.g. [156, Theorem 4.4]). It begins by postulating "elementary" Itô integrals.

Neither approach justifies the existence of its postulated quantities, whereas the game-theoretic approach defines all its objects in terms of price paths in idealized financial markets. The game-theoretic approach can also be used to justify Föllmer's and Lyons's definitions in the context of idealized financial markets. Föllmer's definition is justified simply by an appeal to Lemma 14.14: the existence of quadratic variation over $[0, t]$ is instantly enforceable.

The rough-path approach complements the sample path $X : [0, \infty) \to \mathbb{R}$ by postulated values $\mathbb{X}_{s,t} := \int_s^t (X_u - X_s)\, dX_u$, $s \le t$, satisfying the very natural equality

$$\mathbb{X}_{s,t} - \mathbb{X}_{s,u} - \mathbb{X}_{u,t} = (X_u - X_s)(X_t - X_u), \qquad (14.36)$$

usually referred to as *Chen's relation*, for all $0 \le s \le u \le t < \infty$. In the case of Itô Brownian motion, we *define*

$$\mathbb{X}_{s,t} := \frac{1}{2}((X_t - X_s)^2 - (t - s)).$$

It is the complemented path $\mathbf{X} := (X, \mathbb{X})$ that is called a *rough path*. The *rough integral* is defined pathwise as

$$\int_0^t F(X_s)\, dX_s :=$$

$$\lim \sum_{n=1}^{\infty} (F(X_{t_{n-1} \wedge t})(X_{t_n \wedge t} - X_{t_{n-1} \wedge t}) + F'(X_{t_{n-1} \wedge t})\mathbb{X}_{t_{n-1} \wedge t, t_n \wedge t}), \quad (14.37)$$

where $F : \mathbb{R} \to \mathbb{R}$ is a smooth function (cf. (14.12) with $H := F(X)$) and lim is over the sequences of partitions (t_n) of $[0, t]$ whose mesh tends to zero. The second addend in the outer parentheses in (14.37) is a "regularizer" ensuring that the limit often exists even in this strong sense. If F is a smooth bounded function and $\mathbf{X}$ satisfies, for some $\alpha \in (1/3, 1/2]$ and $t \in [0, \infty)$,

$$\sup_{s,u \in [0,t]: s \neq u} \frac{|X_u - X_s|}{|u - s|^\alpha} < \infty, \qquad \sup_{s,u \in [0,t]: s \neq u} \frac{|\mathbb{X}_{s,u}|}{|u - s|^{2\alpha}} < \infty,$$

and (14.36) for $s \le u \le t$, Lyons's result [156, Theorem 4.4] says that the rough integral (14.37) exists (and enjoys various regularity properties). Perkowski and Prömel [301, section 4] show that a rough path can be associated in a natural way with almost all price paths (in the sense of game-theoretic probability) and give a simple representation of the rough integral in this context.

Timer Options

Timer options were suggested by Avi Bick [37] and trading in them started in 2007 [329]. The payoff of such options is $\mathcal{I}$-measurable, and hence they can be priced using Corollary 14.22 (or directly using Theorem 13.9) and standard measure-theoretic results. Examples are the *timer call* and *timer put*, which are like the usual call and put options but exercised when the *relative quadratic variation* $\Sigma_t^X := \int_0^t d[X]_s / X_s^2$ (to be discussed in greater detail in Chapter 16, e.g. (16.2)) reaches a prespecified value, called the *variance budget*. According to Corollary 14.22 and Theorem 13.9, their prices are given by the Black–Scholes formula with the time to maturity replaced by the variance budget. This result was obtained by Bick [37], who emphasized that his assumptions about the underlying security were minimal (continuity and positivity) but still assumed that the security's price followed a measure-theoretic semimartingale. He remarks that it is possible to reformulate his result in an entirely nonprobabilistic framework but does not elaborate.

Local Time

The theory of local time was originated by Lévy in [247]. Pathwise theories of local time are developed in a 1980 MSc thesis [433] and in recent articles [95, Appendix B], [300], and [96]. Perkowski and Prömel give two proofs of the existence of local time [300, Theorem 3.5]: a direct proof and a proof (on pp. 12–13) using the game-theoretic Dubins–Schwarz theorem, the latter forming the basis for Exercise 14.13.

Continuous-Time Game-Theoretic Probability for Discontinuous Processes

The pathwise Itô integral constructed in [301] was extended to càdlàg integrands in [254, Theorem 4.2]. Our definition of Itô integration (based on Proposition 14.11) is adapted to càdlàg integrands in [414]. Both [301] and [414] consider integrands that are more general than càdlàg, but they consider different generalizations.

Applications Outside Finance

This chapter's results are stated for idealized financial markets, but their potential applicability is much wider. In domains outside finance, we replace the appeal to market efficiency by another statement of Cournot's principle: we postulate that a nonnegative supermartingale chosen in advance will not multiply its value manyfold. This was called the "P-interpretation" in [393], where a continuous process W_t having drift B_t and quadratic variation A_t was modeled by the martingale space $(W_t - B_t, (W_t - B_t)^2 - A_t)$ and a counting process N_t having drift A_t was modeled by the martingale space $(N_t - A_t)$. As a special case, Brownian motion W_t can be modeled by the martingale space $(W_t, W_t^2 - t)$, as in [403]; notice that in this case the basic martingales W_t and $W_t^2 - t$ are heavily dependent (one is a function of the other), which makes the setting (14.1) of this chapter more appropriate than, e.g. the purely pathwise setting of [420].

A particularly interesting direction for further research that would be widely applicable beyond finance is stochastic differential equations and their weak and strong solutions. This topic was briefly explored, using nonstandard analysis, in [349, chapter 14]. For a recent progress, not using nonstandard analysis, see [24] and [159]. Bartle et al. [24, Theorem 2.8] prove a theorem of existence and uniqueness of strong solutions to stochastic differential equations $dX_t = \mu(t, X_t)dt + \sigma(t, X_t)dS_t$ in the martingale space $(S, S^2 - [S])$, $[S]$ being the quadratic variation of the basic martingale S, under certain regularity conditions. Galane et al. [159] prove a similar result without assuming that $S^2 - [S]$ is also a basic martingale.

15

Numeraires in Market Spaces

The theory of martingale spaces developed in the preceding chapter implies that a constant is a capital process. This can be acceptable when the betting picture is used as a metaphor in scientific applications, but it is unrealistic for actual applications to finance, as it implies that the trader can borrow money without interest.

The standard way to avoid this unrealistic assumption is to insist that the trader's strategies be self-financing; each time the trader rebalances his portfolio, he must preserve its total value. In Section 15.1, we use this idea to define the concept of *market space*, a self-financing analog of the concept of martingale space. Market spaces can be considered more general than martingale spaces, because trading strategies in a martingale space become self-financing if we add cash as one of its basic securities, and the addition makes no difference in what the trader can accomplish.

The continuous martingales in martingale spaces have an analog in market spaces; we call them *continuous gales*. If a market space has a positive continuous gale X, then, as we explain in Section 15.2, we can turn the market space into a martingale space by dividing all the prices at time t by X_t. The positive continuous gale we choose for this purpose is then our *numeraire*; it is the unit in which we are denominating the prices of all the securities. From our point of view, much of finance theory can be understood in terms of the hypothesis that a particular numeraire produces a martingale space in which the trader cannot expect any strategy he chooses to multiply the capital he risks by a large factor, even in the long run. This will be the theme of Chapters 16 and 17.

Game-Theoretic Foundations for Probability and Finance, First Edition. Glenn Shafer and Vladimir Vovk.
© 2019 John Wiley & Sons, Inc. Published 2019 by John Wiley & Sons, Inc.

The key result of this chapter, proven in Section 15.3, is a game-theoretic version of Girsanov's theorem. This theorem tells us that the class of continuous semimartingales does not change when the positive continuous gale used as numeraire is changed. Covariations do not change either, and the theorem uses these invariant covariations to describe the process with finite variation that needs to be added to each continuous martingale to make it into a continuous martingale under the new numeraire. Girsanov's theorem is central to the applications to finance that we study in Chapter 16.

15.1 MARKET SPACES

A *market space* is a quintuple of the same form as a martingale space:

$$(\Omega, \mathcal{F}, (\mathcal{F}_t)_{t\in[0,\infty)}, J, (S^j)_{j\in J}).$$

We impose the same conditions on each element of the quintuple as in the definition of a martingale space, and we add the conditions that $|J| > 1$ and that at least one of the S^j is positive. We call the S^j *basic gales*.

The more significant difference of this chapter's framework is that we now require a simple capital process to be self-financing. A *simple trading strategy* ϕ in the market space consists of a number $c \in \mathbb{R}$ (the *initial capital*), N stopping times $\tau_1 \leq \cdots \leq \tau_N$ for some $N \in \mathbb{N}$, bounded $\mathcal{F}_{\tau_n}$-measurable variables $h_{n,j}$, $n \in \{1, \ldots, N\}$ and $j \in J$, such that only finitely many of $h_{n,j}$ are different from 0, and is required to be *self-financing* in the sense that, for any $n \in \{1, \ldots, N\}$,

$$\sum_{j\in J} h_{n,j} S^j_{\tau_n} = \begin{cases} c & \text{if } n = 1, \\ \sum_{j\in J} h_{n-1,j} S^j_{\tau_n} & \text{otherwise.} \end{cases} \tag{15.1}$$

The requirement of being self-financing reflects the fact that in the new picture no borrowing or saving are allowed (they should be done implicitly via investing in the available securities). The capital available at time t is

$$\mathcal{K}^\phi_t := \begin{cases} c & \text{if } t \leq \tau_1, \\ \sum_{j\in J} h_{n,j} S^j_t & \text{if } \tau_n \leq t \leq \tau_{n+1}, \end{cases} \tag{15.2}$$

where n ranges over $\{1, \ldots, N\}$ and τ_{N+1} is understood to be ∞. The self-financing property (15.1) implies that the expression (15.2) is well defined when $t = \tau_k$ for some $k \in \{1, \ldots, N\}$.

A *simple capital process* in the market space is a process that can be represented as (15.2) for some simple trading strategy. A process is a *nonnegative supergale* if it belongs to the smallest lim inf-closed class of nonnegative processes containing all nonnegative simple capital processes in the market space. A property E of t and ω is *instantly enforceable* in the market space if there exists a nonnegative supergale X such that $X_0 = 1$ and (14.4) holds for all $t \in [0, \infty)$ and $\omega \in \Omega$. A continuous process is a *continuous gale* in the market space if it is an element of the smallest lim-closed class of continuous processes that contains all simple capital processes in the market space.

The next proposition shows that a market space becomes a martingale space when we take a positive basic gale as the numeraire. Once it is the numeraire, the basic gale's price process is the constant 1, and so it makes no difference whether we leave it in the space or not when we treat the space as a martingale space. In statement of the proposition, we remove it.

Proposition 15.1 *Suppose*

$$\mathfrak{S} := (\Omega, \mathcal{F}, (\mathcal{F}_t)_{t\in[0,\infty)}, J, (S^j)_{j\in J})$$

is a market space. Choose $v \in J$ such that S^v is positive. Define a martingale space $\mathfrak{S}'$ by

$$\mathfrak{S}' := (\Omega, \mathcal{F}, (\mathcal{F}_t)_{t\in[0,\infty)}, J \setminus \{v\}, (S^j/S^v)_{j\in J\setminus\{v\}}).$$

1. *A process X is a simple capital process in $\mathfrak{S}$ if and only if X/S^v is a simple capital process in $\mathfrak{S}'$.*
2. *A process X is a nonnegative supergale in $\mathfrak{S}$ if and only if X/S^v is a nonnegative supermartingale in $\mathfrak{S}'$.*
3. *A process X is a continuous martingale in $\mathfrak{S}$ if and only if X/S^v is a continuous gale in $\mathfrak{S}'$.*

Proof Let us see what the simple capital process (15.2) becomes if we use S^v as our numeraire. By the condition (15.1) of being self-financing, the number $h_{n,v}$ of units of S^v chosen at time τ_n, $n \in \mathbb{N}$, should satisfy

$$\mathcal{K}^\phi_{\tau_n} = h_{n,v} S^v_{\tau_n} + \sum_{j\in J\setminus\{v\}} h_{n,j} S^j_{\tau_n},$$

which gives

$$h_{n,v} = \frac{\mathcal{K}^\phi_{\tau_n} - \sum_{j\in J\setminus\{v\}} h_{n,j} S^j_{\tau_n}}{S^v_{\tau_n}}.$$

Therefore, the capital $\mathcal{K}^{\phi}_{\tau_n}$ at time τ_n, $n \in \{2, \dots, N\}$, can be expressed via $\mathcal{K}^{\phi}_{\tau_{n-1}}$ as

$$\mathcal{K}^{\phi}_{\tau_n} = h_{n-1,v} S^v_{\tau_n} + \sum_{j \in J \setminus \{v\}} h_{n-1,j} S^j_{\tau_n}$$

$$= \frac{S^v_{\tau_n}}{S^v_{\tau_{n-1}}} \left(\mathcal{K}^{\phi}_{\tau_{n-1}} - \sum_{j \in J \setminus \{v\}} h_{n-1,j} S^j_{\tau_{n-1}} \right) + \sum_{j \in J \setminus \{v\}} h_{n-1,j} S^j_{\tau_n}.$$

If we put a bar over prices and amounts of capital to indicate that they are expressed in units of S^v – i.e. they have been divided by S^v, this becomes

$$\overline{\mathcal{K}}^{\phi}_{\tau_n} = \overline{\mathcal{K}}^{\phi}_{\tau_{n-1}} - \sum_{j \in J \setminus \{v\}} h_{n-1,j} \overline{S}^j_{\tau_{n-1}} + \sum_{j \in J \setminus \{v\}} h_{n-1,j} \overline{S}^j_{\tau_n}$$

$$= \overline{\mathcal{K}}^{\phi}_{\tau_{n-1}} + \sum_{j \in J \setminus \{v\}} h_{n-1,j} \left(\overline{S}^j_{\tau_n} - \overline{S}^j_{\tau_{n-1}} \right).$$

Since this is also true for any $t \in [\tau_{n-1}, \tau_n] \cap \mathbb{R}$, $n \in \{2, \dots, N+1\}$, in place of τ_n, and since $\mathcal{K}^{\phi}_{\tau_1} = c$, we obtain (14.2), with the bar "¯" added on top of $\mathcal{K}$, S, and c, for all $t \geq 0$ (where $\overline{c}$ is the initial capital c expressed in units of S^v). This proves statement 1.

A simple argument based on transfinite induction proves statement 2; let us, e.g. check that, for any nonnegative supergale X in $\mathfrak{S}$, the process X/S^v will be a nonnegative supermartingale in $\mathfrak{S}'$: the inductive step is based on the identity

$$\liminf_{k \to \infty} \frac{X^k_t(\omega)}{S^v_t(\omega)} = \frac{\liminf_{k \to \infty} X^k_t(\omega)}{S^v_t(\omega)}. \tag{15.3}$$

In checking statement 3, the inductive step is based on the identity (15.3) with lim (in the sense of the uniform convergence on compact intervals) in place of lim inf.   □

We asserted in the introduction that a martingale space is equivalent to a special kind of market space – one in which cash is one of the basic continuous gales. This is a corollary of Proposition 15.1.

Corollary 15.2 *Suppose*

$$\mathfrak{S} := (\Omega, \mathcal{F}, (\mathcal{F}_t)_{t \in [0, \infty)}, J, (S^j)_{j \in J})$$

is a martingale space. For $v \notin J$, define a market space $\mathfrak{S}'$ by

$$\mathfrak{S}' := (\Omega, \mathcal{F}, (\mathcal{F}_t)_{t \in [0, \infty)}, J \cup \{v\}, (S^j)_{j \in J \cup \{v\}}),$$

where $S^v_t := 1$ for all $t \in [0, \infty)$ (and S^j, $j \in J$, are the same in $\mathfrak{S}$ and $\mathfrak{S}'$).

1. *A process X is a simple capital process in $\mathfrak{S}$ if and only if it is a simple capital process in $\mathfrak{S}'$.*

2. *A process X is a nonnegative supermartingale in $\mathfrak{S}$ if and only if it is a nonnegative supergale in $\mathfrak{S}'$.*

3. *A process X is a continuous martingale in $\mathfrak{S}$ if and only if it is a continuous gale in $\mathfrak{S}'$.*

Proof Since $\mathfrak{S}$ can be obtained from $\mathfrak{S}'$ by using the procedure of Proposition 15.1, $\mathfrak{S} = \mathfrak{S}''$, Corollary 15.2 is a special case of Proposition 15.1. $\qquad\qquad\square$

15.2 MARTINGALE THEORY IN MARKET SPACES

Let us fix a market space

$$\mathfrak{S} := (\Omega, \mathcal{F}, (\mathcal{F}_t), J, (S^j)) \tag{15.4}$$

for the rest of this chapter. In view of Proposition 14.10, we can use any positive continuous gale I as our numeraire. Let us fix such an $I_t \in (0, \infty)$ and use it as our numeraire for measuring capital at time t. It is now easy to generalize many results of Chapters 13 and 14, which will now correspond to the case $I := 1$ (intuitively, using cash as the numeraire).

Let $\mathfrak{S}/I$ be the martingale space

$$\mathfrak{S}/I := (\Omega, \mathcal{F}, (\mathcal{F}_t), J, (S^j/I)). \tag{15.5}$$

We call processes of the type $X_t(\omega)/I_t(\omega)$, where X is a nonnegative supergale (respectively, a continuous gale), *nonnegative I-supermartingales* (respectively, *continuous I-martingales*); they are completely analogous to the nonnegative supermartingales (respectively, continuous martingales) of Chapters 13 and 14 but use I rather than cash as the numeraire. The processes X of the form $Y + A$, where Y is a continuous I-martingale and A is a finite variation continuous process, are *continuous I-semimartingales*. Whereas the first two notions very much depend on the choice of I, the game-theoretic Girsanov theorem in Section 15.3 will show that the notion of a continuous I-semimartingale is invariant.

Proposition 14.11 can be applied to the martingale space $\mathfrak{S}/I$, and it remains true if X is a continuous I-martingale and the notion of instant enforcement is defined using nonnegative supergales instead of nonnegative supermartingales; indeed, the notion of instant enforcement is defined in terms of becoming infinitely rich infinitely quickly and so does not depend on the monetary unit. This remark is applicable to all results in Chapter 14 asserting that some property holds with instant enforcement.

We define the *I-upper expected value* of a nonnegative extended variable F by

$$\overline{\mathbb{E}}^I(F) := \inf\{X_0 | \forall \omega \in \Omega : \liminf_{t \to \infty} X_t(\omega)/I_t(\omega) \geq F(\omega)\}, \tag{15.6}$$

X ranging over the nonnegative supergales in $\mathfrak{S}$, and specialize it to I-*upper proba-bility* as $\overline{\mathbb{P}}^I(E) := \overline{\mathbb{E}}^I(\mathbf{1}_E)$. The definition (15.6) can be rewritten as

$$\overline{\mathbb{E}}^I(F) = \inf\{X_0 | \forall \omega \in \Omega : \liminf_{t \to \infty} X_t(\omega) \geq F(\omega)\},$$

X ranging over the nonnegative I-supermartingales.

The results of Section 14.8 carry over to the case of a general positive contin-uous martingale I as numeraire. In particular, we have the following version of Corollary 14.22.

Corollary 15.3   *Let $F : C[0, \infty) \to [0, \infty]$ be I-measurable, and X be a continuous I-martingale. Then*

$$\overline{\mathbb{E}}^I(F(X)) \leq \int F \, dW.$$

15.3   GIRSANOV'S THEOREM

Now we state a game-theoretic version of Girsanov's theorem in our market space (15.4). It shows that the notion of a continuous semimartingale does not depend on the numeraire and gives the explicit decomposition (unique w.i.e. by Proposition 14.23) into a continuous martingale and a finite variation continuous process for a continuous martingale in a new numeraire. Now we consider two numeraires B and I in (15.4); formally, they are positive continuous gales in that market space.

Theorem 15.4   *Let B and I be positive continuous gales, and let M be a continuous B-martingale. The process*

$$M_t - \int_0^t \frac{B}{I} \, d\left[\frac{I}{B}, M\right] \tag{15.7}$$

is a continuous I-martingale.

The integral in (15.7) is Lebesgue–Stieltjes. Being the difference of two increas-ing functions (cf. Lemma 14.15), it has finite total variation over compact intervals. This shows that we have the same families of continuous semimartingales for both numeraires: a continuous process M is a continuous B-semimartingale if and only if it is a continuous I-semimartingale.

Equation (15.7) might appear ambiguous in that it does not specify the meaning of covariation $[\cdot, \cdot]$. We have two natural martingale spaces in our current setting: $\mathfrak{S}/I$ (see (15.5)) and $\mathfrak{S}/B$ (defined by (15.5) with I replaced by B). In the proof we will understand $[\cdot, \cdot]$ in the sense of $\mathfrak{S}/B$, which is natural since both I/B and M are continuous B-martingales; we will write $[\cdot, \cdot]^B$ to avoid any ambiguity. However, the definition of covariation between two semimartingales as the ucie limit of (14.20) (for a sequence of partitions that is sufficiently fine) shows that the notion of covariation does not depend on the choice of numeraire: (14.20) does not involve the numeraire at all, and the notion of ucie convergence is invariant with respect to the choice of numeraire (see also Lemma 14.17).

If $B = 1$, the market space (15.4) becomes, essentially, a martingale space (cf. Proposition 15.2). The theorem says that, if M is a continuous martingale in this martingale space,

$$M_t - \int_0^t \frac{d[I, M]}{I} \tag{15.8}$$

will be a continuous I-martingale. Equation (15.8) is the most standard version of Girsanov's theorem (see, e.g. [314, Theorem III.39]).

Proof of Theorem 15.4   Our proof will be standard (see, e.g. [314, proof of Theorem III.39]). By the integration by parts formula (see Lemma 14.14) applied to the martingale space $\mathfrak{S}/B$,

$$\frac{I_t}{B_t} M_t = \int_0^t \frac{I}{B} \, dM + \int_0^t M \, d\frac{I}{B} + \left[\frac{I}{B}, M\right]_t^B \quad \text{w.i.e.}$$

Since $\int (I/B) \, dM$ and $\int M \, d(I/B)$ are continuous B-martingales,

$$\frac{I_t}{B_t} M_t - \left[\frac{I}{B}, M\right]_t^B$$

is also a continuous B-martingale, which implies that

$$I_t M_t - B_t \left[\frac{I}{B}, M\right]_t^B$$

is a continuous gale, which in turn implies that

$$M_t - \frac{B_t}{I_t} \left[\frac{I}{B}, M\right]_t^B$$

is a continuous I-martingale. The integration by parts formula (Lemma 14.14, which is also applicable to continuous semimartingales: see Exercise 15.2) applied to the martingale space $\mathfrak{S}/I$ allows us to transform the subtrahend (the product of continuous I-semimartingales, namely of an I-martingale and a finite variation continuous process) as

$$\frac{B_t}{I_t} \left[\frac{I}{B}, M\right]_t^B = \int_0^t \frac{B}{I} \, d\left[\frac{I}{B}, M\right]^B + \int_0^t \left[\frac{I}{B}, M\right]^B d\frac{B}{I} + \left[\frac{B}{I}, \left[\frac{I}{B}, M\right]^B\right]_t^I \quad \text{w.i.e.} \tag{15.9}$$

The second addend on the right-hand side of (15.9) is a continuous I-martingale since B/I is, and the third addend is a continuous I-martingale since it is zero w.i.e. (see the argument at the end of the proof of Lemma 14.17); therefore, (15.7) is also a continuous I-martingale.   □

The notion of Doléans logarithm allows us to simplify the statement of Theorem 15.4 as follows.

Corollary 15.5 *Let B and I be positive continuous gales, M be a continuous B-martingale, and L be the Doléans logarithm of the continuous B-martingale I/B. The process*

$$M_t - [L, M]_t$$

is a continuous I-martingale.

Proof Set $X := I/B$; we know that $L = \mathcal{L}(X)$ in $\mathfrak{S}/I$ (see the definition (14.30) and remember the invariance of quadratic variation). It suffices to prove

$$\int_0^t \frac{d[X, M]}{X} = [L, M]_t \quad \text{w.i.e.,}$$

and this inequality is a special case of Lemma 15.6.   □

Lemma 15.6 *Let X and M be continuous martingales and H be a continuous process. Then*

$$\int_0^t H\, d[X, M] = \left[\int H\, dX, M \right]_t \quad \text{w.i.e.} \tag{15.10}$$

The integral on the left-hand side of (15.10) is the usual Lebesgue–Stieltjes integral, but the integral on the right-hand side is the Itô integral.

Proof For sequences of partitions finely covering relevant continuous processes (including H, X, M, and $\int H\, dX$), we have

$$\int_0^t H\, d[X, M] = \lim_{k \to \infty} \sum_{n=1}^{\infty} H_{T^k_{n-1} \wedge t}([X, M]_{T^k_n \wedge t} - [X, M]_{T^k_{n-1} \wedge t}) \tag{15.11}$$

$$= \lim_{k \to \infty} \sum_{n=1}^{\infty} H_{T^k_{n-1} \wedge t}(X_{T^k_n \wedge t} - X_{T^k_{n-1} \wedge t})(M_{T^k_n \wedge t} - M_{T^k_{n-1} \wedge t}) \tag{15.12}$$

$$= \lim_{k \to \infty} \sum_{n=1}^{\infty} (Y_{T^k_n \wedge t} - Y_{T^k_{n-1} \wedge t})(M_{T^k_n \wedge t} - M_{T^k_{n-1} \wedge t}) = [Y, M]_t \quad \text{w.i.e.,} \tag{15.13}$$

where $Y := \int H \, dX$. In the rest of the proof we will justify two key equalities: between (15.11) and (15.12) and between (15.12) and (15.13) (the equality in (15.11) follows from the equivalence of the Lebesgue–Stieltjes and Riemann–Stieltjes definitions in this context).

The equality between (15.11) and (15.12) follows from the distribution functions $F_k : [0, t] \to (-\infty, \infty)$ putting weight

$$(X_{T_n^k \wedge t} - X_{T_{n-1}^k \wedge t})(M_{T_n^k \wedge t} - M_{T_{n-1}^k \wedge t})$$

on $H_{T_{n-1}^k \wedge t}$ converging weakly to the continuous distribution function $[X, M]$ on $[0, t]$ (w.i.e.); the integrals $\int_0^t H \, dF_k$ of the continuous function H will then converge to $\int_0^t H \, dF$. The convergence is uniform on compact intervals (Exercise 15.3).

The transition from (15.13) to (15.12) requires refining the partitions: the expression after $\lim_{k \to \infty}$ in (15.13) for a given large value of k is approximately equal to the expression after $\lim_{k \to \infty}$ in (15.12) for much larger values K of k. Without loss of generality we assume that X, M, and H are bounded (a nonnegative supermartingale witnessing the instant enforcement for unbounded X, M, and H can be obtained combining nonnegative supermartingales witnessing the instant enforcement for bounded X, M, and H with quickly increasing bounds) and that the partitions are nested. We know that the limits in (15.12) and (15.13) both exist w.i.e., and the problem is to show that they coincide.

In the rest of this proof we will need the following uniform version of Proposition 14.11, which uses X being bounded (without loss of generality we will assume that $|X| \le 1$, $|M| \le 1$, and $|H| \le 1$). There is a nonnegative supermartingale Z with $Z_0 = 1$ such that, for any $\epsilon > 0$ and any $B > 0$, there is $K_{\epsilon, B} \in \mathbb{N}$ such that, for any $K \ge K_{\epsilon, B}$,

$$\sup_{s \in [0, t]} \left| \sum_{N=1}^{\infty} H_{T_{N-1}^K \wedge s}(X_{T_N^K \wedge s} - X_{T_{N-1}^K \wedge s}) - \int_0^s H \, dX \right| \le \epsilon \quad \text{or} \quad Z_t \ge B.$$

Let S be a nonnegative supermartingale with $S_0 = 1$ witnessing the ucie convergence in the definitions of $\int H \, dX$, $[Y, M]$, and $[X, M]$; later in the proof we will add further requirements on S. Letting $k \to \infty$ and, for each k, setting $K > k$ to a sufficiently large number (see below for details), we will have uniformly over $s \in [0, t]$ when $S_t < \infty$: $[Y, M]_s$ becomes arbitrarily close to

$$\sum_{n=1}^{\infty}(Y_{T_n^k \wedge s} - Y_{T_{n-1}^k \wedge s})(M_{T_n^k \wedge s} - M_{T_{n-1}^k \wedge s}), \tag{15.14}$$

because S witnesses the ucie convergence in the definition of $[Y, M]$; in turn, (15.14) becomes arbitrarily close to

$$\sum_{N=1}^{\infty} H_{T_{N-1}^K \wedge s}(X_{T_N^K \wedge s} - X_{T_{N-1}^K \wedge s})(M_{T_{n(N)}^k \wedge s} - M_{T_{n(N)-1}^k \wedge s}) \tag{15.15}$$

for a sufficiently large $K > k$, where $n = n(N)$ is defined by the condition $T_{n-1}^k \leq T_{N-1}^K < T_N^K \leq T_n^k$ (assuming $T_{N-1}^K < T_N^K$), as will be established later in the proof; in turn, (15.15) becomes arbitrarily close to

$$\sum_{N=1}^{\infty} H_{T_{N-1}^K \wedge s}(X_{T_N^K \wedge s} - X_{T_{N-1}^K \wedge s})(M_{T_N^K \wedge s} - M_{T_{N-1}^K \wedge s}), \tag{15.16}$$

as will be established later in the proof. And we already know (cf. (15.11) and (15.12)) that (15.16) becomes arbitrarily close to $\int_0^s H \, d[X, M]$.

Next we show that (15.14) becomes arbitrarily close to (15.15) when $S_t < \infty$, for a suitable definition of S. Subtracting (15.15) from (15.14) gives

$$\sum_{n=1}^{\infty} v_n^k (M_{T_n^k \wedge s} - M_{T_{n-1}^k \wedge s}), \tag{15.17}$$

where

$$v_n^k := (Y_{T_n^k \wedge s} - Y_{T_{n-1}^k \wedge s}) - \sum_N H_{T_{N-1}^K \wedge s}(X_{T_N^K \wedge s} - X_{T_{N-1}^K \wedge s}),$$

where N ranges over the numbers satisfying $T_{n-1}^k \leq T_{N-1}^K < T_N^K \leq T_n^k$. For each k choose $K = K(k)$ such that either $|v_n^k| \leq 2^{-k+1}$ for all n (remember that the sequence of partitions finely covers $\int H \, dX$) or a fixed nonnegative supermartingale $S^{k,1}$ with $S_0^{k,1} = 1$ satisfies $S_t^{k,1} \geq 2^k$. Let us concentrate on $\omega \in \Omega$ at which $|v_n^k| \leq 2^{-k+1}$ for all n. As in the proof of Proposition 14.11, we can use Kolmogorov's martingale (3.1) to show that, from some k on,

$$\sum_{n=1}^{\infty} (M_{T_n^k \wedge s} - M_{T_{n-1}^k \wedge s})^2 \leq k^4 \tag{15.18}$$

for all $s \in [0, t]$ unless a fixed nonnegative supermartingale $S^{k,2}$ with $S_0^{k,2} = 1$ satisfies $S_t^{k,2} \geq k^4$. Now (15.18) implies

$$\sum_{n=1}^{\infty} (M_{T_n^k \wedge s} - M_{T_{n-1}^k \wedge s})^2 (v_n^k)^2 \leq k^4 2^{-2k+2} \tag{15.19}$$

for all $s \in [0, t]$. Again using Kolmogorov's martingale (3.1) (cf. Exercise 14.8), we can see that (15.19) implies that (15.17) does not exceed $k^8 2^{-2k+2}$ from some k on unless a fixed nonnegative supermartingale $S^{k,3}$ with $S_0^{k,3} = 1$ satisfies $S_t^{k,3} \geq k^4$. Combining $S^{k,1}$, $S^{k,2}$, and $S^{k,3}$ with weights k^{-2}, we obtain a nonnegative supermartingale S' with $S_0' < \infty$ such that $S_t' = \infty$ unless (15.17) tends to 0 uniformly in $s \in [0, t]$. We can see that it suffices to replace S, as defined earlier, by $(S + S')/2$.

It remains to show that (15.15) becomes arbitrarily close to (15.16) when $S_t < \infty$, for a suitable definition of S. Our argument will be similar to the one in the previous paragraph. Subtracting (15.16) from (15.15) gives

$$\sum_{N=1}^{\infty} (M_{T_N^K \wedge s} - M_{T_{N-1}^K \wedge s}) \xi_N^K + (X_{T_N^K \wedge s} - X_{T_{N-1}^K \wedge s}) \mu_N^K, \tag{15.20}$$

where

$$\xi_N^K := \sum_{N'=N^*+1}^{N-1} H_{T_{N'-1}^K \wedge s} (X_{T_{N'}^K \wedge s} - X_{T_{N'-1}^K \wedge s}),$$

$$\mu_N^K := H_{T_{N-1}^K \wedge s} \sum_{N'=N^*+1}^{N-1} (M_{T_{N'}^K \wedge s} - M_{T_{N'-1}^K \wedge s})$$

$$= H_{T_{N-1}^K \wedge s} (M_{T_{N-1}^K \wedge s} - M_{T_{N^*}^K \wedge s}),$$

and N^* is defined by $T_{N^*}^K = T_{n-1}^k \leq T_{N-1}^K < T_N^K \leq T_n^k$. In the rest of this proof we will concentrate on the first addend

$$\sum_{N=1}^{\infty} (M_{T_N^K \wedge s} - M_{T_{N-1}^K \wedge s}) \xi_N^K \tag{15.21}$$

of (15.20); the argument for second term is similar (but simpler).

For each k, choose $K = K(k)$ such that either $|\xi_N^K| \leq 2^{-k+1}$ for all N or a fixed nonnegative supermartingale $S^{k,4}$ with $S_0^{k,4} = 1$ satisfies $S_t^{k,4} \geq 2^k$. (In the overall picture, the value $K(k)$ should be large enough to ensure that all three sums, (15.17), (15.21), and the second addend in (15.20), are small in absolute value as indicated.) Let us concentrate on ω at which $|\xi_N^K| \leq 2^{-k+1}$ for all N. We can use Kolmogorov's martingale (3.1) to show that

$$\sum_{N=1}^{\infty} (M_{T_N^K \wedge s} - M_{T_{N-1}^K \wedge s})^2 \leq k^4 \tag{15.22}$$

(cf. (15.18)) holds for all $s \in [0, t]$ from some k on unless a fixed nonnegative supermartingale $S^{k,5}$ with $S_0^{k,5} = 1$ satisfies $S_t^{k,5} \geq k^4$. Now (15.22) implies

$$\sum_{N=1}^{\infty} (M_{T_N^K \wedge s} - M_{T_{N-1}^K \wedge s})^2 (\xi_N^K)^2 \leq k^4 2^{-2k+2} \tag{15.23}$$

(cf. (15.19)) for all $s \in [0, t]$. Again using Kolmogorov's martingale (3.1), we can see that (15.23) implies that (15.21) does not exceed $k^8 2^{-2k+2}$, from some k on, unless a fixed nonnegative supermartingale $S^{k,6}$ with $S_0^{k,6} = 1$ satisfies $S_t^{k,6} \geq k^4$. Combining $S^{k,4}$, $S^{k,5}$, and $S^{k,6}$ with weights k^{-2}, we obtain a nonnegative supermartingale S'' such that $S_t'' = \infty$ unless (15.21) tends to 0 uniformly in $s \in [0, t]$. $\qquad \square$

15.4 EXERCISES

Exercise 15.1 State (15.8) in terms of the Doléans logarithm of I.   □

Exercise 15.2 Prove that Lemma 14.14 is applicable to any continuous semimartingales X and Y.   □

Exercise 15.3 Suppose a sequence of distribution functions $F_k : [0, t] \to [0, \infty)$ (formally, increasing right-continuous functions considered to be the distribution functions of finite measures on $[0, t]$) converges uniformly to a continuous distribution function $F : [0, t] \to [0, \infty)$. Let $H : [0, t] \to \mathbb{R}$ be a continuous function. Prove that $\int_0^s H \, dF_k \to \int_0^s H \, dF$ as $k \to \infty$ uniformly in $s \in [0, t]$. *Hint*: For each $\epsilon > 0$, there exists a piecewise-constant (with a finite number of the intervals of constancy) right-continuous function $H^* : [0, t] \to \mathbb{R}$ at the uniform distance at most ϵ from H, $\sup |H - H^*| \le \epsilon$.   □

Exercise 15.4 (associativity of Itô integration). Let X be a continuous martingale and H and G be continuous processes. Show that

$$\int_0^t H \, d(G \cdot X) = \int_0^t HG \, dX \quad \text{w.i.e.}$$

Hint: Follow the proof Lemma 15.6.   □

Exercise 15.5 Let X and M be continuous martingales and H and G be continuous processes. Generalize (15.10) to

$$\int_0^t HG \, d[X, M] = \left[\int H \, dX, \int G \, dM \right]_t \quad \text{w.i.e.}$$

Hint: Combine Exercise 15.4 with Lemma 15.6.   □

Exercise 15.6 (research project) Develop the theory of this chapter relaxing the requirement of the numeraire being positive to the requirement that it is nonnegative.   □

15.5 CONTEXT

Gales and Supergales

The notion of gale used in this chapter further generalizes Jack Lutz's generalization of the notion of martingale [257, section 3]. According to Lutz (who gave this definition in the context of the uniform probability measure on binary sequences), a nonnegative process S_t is an s-supergale if $2^{(1-s)t} S_t$ is a supermartingale. In particular, a supermartingale is a 1-supergale. If the numeraire is the bank account e^{rt} with a constant interest rate r, supergales in our sense are $(1 + r / \ln 2)$-supergales in the sense of Lutz.

The word *gale* is the last element of both *martingale* and *farthingale*. In the course of writing [102], Dawid introduced *farthingale* for what we here call a game-theoretic martingale.

Numeraires

Proposition 15.1 is a version of the standard numeraire-invariance result in mathematical finance (see, e.g. [127, section 2.2] for a very intuitive exposition in discrete time). The idea of numeraire invariance is often traced to Robert Merton's 1973 article [279].

Girsanov's Theorem

The first version of Girsanov's theorem, for the Wiener process modified by adding a deterministic drift, was obtained by Robert Cameron and William Martin in 1944 [60]. Stochastic drifts (still for the Wiener process) were considered by Kolmogorov's student Igor Girsanov in 1960 [165]. Girsanov's result was generalized to continuous local martingales by Eugene Wong in 1973 [431, section III.E] and to a wider class of local martingales by Jim Van Schuppen and Wong in 1974 [383]. The modern statement of the theorem is often attributed to a 1976 article by Paul-André Meyer [280]. Philip Protter [314, p. 149] calls it the "Girsanov–Meyer theorem" and also emphasizes the role of Gisirō Maruyama's 1954 article [271]. Our results (Theorem 15.4 and Corollary 15.5) are game-theoretic versions of Wong's [431].

16

Equity Premium and CAPM

We now apply the results of Chapters 14 and 15 to four closely related topics in finance: the equity premium, the capital asset pricing model (CAPM), the Sharpe ratio, and the theoretical performance deficit. The first two topics are standard; the other two have been studied in the context of log-optimal investment, discussed in Exercises 12.8 and 12.9. The usual treatments of these topics rely on exogenous probabilities and assumptions about investors' risk preferences, whereas the theory of this chapter uses only the game-theoretic mathematics of the preceding chapters and relies for its application to actual markets only on the assumption that a given numeraire is efficient in our game-theoretic sense.

Throughout this chapter, we consider a fixed market space $\mathfrak{S}$ with three positive continuous gales B, S', and I'. The mathematical results we obtain do not, of course, depend on any particular interpretation of these objects, but in the application we are discussing they are interpreted as follows:

- B is a trader's bank account, perhaps a money market account, that pays interest and allows the trader to borrow at the same rate;
- S' is a stock, or more generally a portfolio or trading strategy;
- I' is a market index that we take to be an efficient numeraire – a positive continuous gale that we do not expect the trader to outperform substantially.

Game-Theoretic Foundations for Probability and Finance, First Edition. Glenn Shafer and Vladimir Vovk.
© 2019 John Wiley & Sons, Inc. Published 2019 by John Wiley & Sons, Inc.

Table 16.1 Percentage of active funds outperformed by relevant market indexes

	1 year	3 years	5 years	10 years
S&P Composite 1500 outperforms (%)	63	83	87	87
S&P 500 outperforms (%)	63	81	84	90
S&P MidCap 400 outperforms (%)	44	86	85	96
S&P SmallCap 600 outperforms (%)	48	89	91	96
S&P Europe 350 outperforms (%)	47	59	73	85
S&P Eurozone BMI outperforms (%)	74	77	88	88

Each row represents the percentage of active funds in a particular class that are outperformed by the market index representing that class. For example, over a 10-year period, 87% of all active domestic funds in the United States are outperformed by the S&P Composite 1500. Source: Data from [362, Report 1] and [195, Report 1].

As usual, we have two ways of expressing the prediction that the trader will not outperform I'. When we want a simple picture, we predict that the trader will not multiply the capital he risks infinitely with respect to I'. When we want a more practical picture, we suppose that a small $\delta > 0$ has been fixed and predict that the trader will not multiply the capital he risks relative to I' by a factor of $1/\delta$ or more. The more practical prediction has substantial empirical support when the market is large and the index is a broad-based index of stocks with large capitalization, such as the S&P 500. As Malkiel has put it, "the strongest evidence suggesting that markets are generally quite efficient is that professional investors do not beat the market" [259] (see also [258, 260] and Table 16.1).

We take B as our numeraire for measuring prices. This means that we work primarily in the martingale space $\mathfrak{S}/B$; most of the processes we consider are measured with respect to B. The capital process of our stock or portfolio is $S := S'/B$ instead of S', and the capital process of our efficient numeraire is $I := I'/B$ instead of I'. We assume without loss of generality that $I_0 = 1$.

We express predictions based on I being an efficient numeraire either in terms of almost sure events with respect to $\overline{\mathbb{P}}^I$ or in terms of upper and lower I-probabilities, as defined in Section 15.2: $\overline{\mathbb{P}}^I(E) \le \delta$ is a prediction that E will not happen, $\underline{\mathbb{P}}^I(E) \ge 1 - \delta$ is a prediction that E will happen, and the strength of the prediction is determined by how small $\delta > 0$ is. Both kinds of predictions are often deduced from various processes being continuous I-martingales.

Section 16.1 applies the game-theoretic Girsanov theorem (Theorem 15.4 or Corollary 15.5) to obtain three continuous I-martingales that provide the basis for our further discussion, in Sections 16.2, 16.3, and 16.4, respectively, of the equity premium, CAPM, and the theoretical performance deficit. Section 16.5 discusses implications of our game-theoretic CAPM for the Sharpe ratio, a standard performance measure for investment managers.

The theory developed in this chapter does not derive the existence of an efficient numeraire from other principles or assumptions. Instead, it postulates that a particular process is an efficient numeraire and derives implications from this postulate, implications that can be used to test the postulate. Section 16.7 includes references to theoretical arguments for the existence of an efficient numeraire.

16.1 THREE FUNDAMENTAL CONTINUOUS *I*-MARTINGALES

Table 16.2 lists four continuous *I*-martingales that we use in this chapter. In this section, we verify that they are indeed continuous *I*-martingales; we only need to consider the first three of them since the fourth is the difference between the first two.

We begin with some definitions applicable to positive continuous martingales in any martingale space. Given a positive continuous martingale X, we define a continuous martingale M^X, X's *cumulative relative growth*, by

$$M_t^X := \int_0^t \frac{dX_s}{X_s} = \Lambda_t^X - \ln X_0, \tag{16.1}$$

where Λ^X is the X's Doléans logarithm. The *relative covariation* of two positive continuous martingales X and Y is the Lebesgue–Stieltjes integral

$$\Sigma_t^{X,Y} := \int_0^t \frac{d[X,Y]_s}{X_s Y_s} = [\Lambda^X, \Lambda^Y]_t = [M^X, M^Y]_t \quad \text{w.i.e.} \tag{16.2}$$

(cf. Lemma 14.15 and Exercise 15.5). When X is a positive continuous martingale, we write Σ^X for $\Sigma^{X,X}$ and call it X's *relative quadratic variation*.

The following corollary of our game-theoretic Girsanov theorem will be the starting point of our discussion of CAPM in Section 16.3. It is also the starting point from which the other results in this section are derived.

Proposition 16.1 *The process* $M_t^S - \Sigma_t^{S,I}$ *is a continuous I-martingale.*

Proof Applying Corollary 15.5 to the continuous *B*-martingale M^S, we find that

$$M_t^S - [M^S, \Lambda^I]_t = M_t^S - \Sigma_t^{S,I}$$

is a continuous *I*-martingale. □

Table 16.2 Continuous *I*-martingales used in this chapter; more can be obtained using $M^S - \ln S - \frac{1}{2}\Sigma^S$ and $M^I - \ln I - \frac{1}{2}\Sigma^I$ being constant w.i.e. (see Lemma 16.3)

Martingale	Related to	Introduced in
$M^S - \Sigma^{S,I}$	Simplified CAPM	Proposition 16.1
$M^I - \Sigma^I$	Equity premium	Corollary 16.2
$\ln S - \ln I + \frac{1}{2}\Sigma^{S\triangle I}$	TPD	Proposition 16.4
$M^S - M^I + \Sigma^I - \Sigma^{S,I}$	Black-type CAPM	(16.26)

Since Proposition 16.1 is applicable to any pair (S, I) of positive continuous B-martingales, we can replace S by I obtaining the following corollary, which will be the starting point of our discussion of the equity premium in Section 16.2.

Corollary 16.2 *The process* $M_t^I - \Sigma_t^I$ *is a continuous I-martingale.*

Lemma 16.3 *The cumulative relative growth* M^X *of a positive continuous martingale X satisfies*

$$M_t^X = \ln \frac{X_t}{X_0} + \frac{1}{2}\Sigma_t^X \quad \text{w.i.e.} \tag{16.3}$$

Proof Equation (16.3) is essentially a special case of (14.30): remember the definition (16.1) and use the equality

$$\Sigma_t^X = [\ln X]_t \quad \text{w.i.e.} \tag{16.4}$$

(see Exercise 16.1). □

Finally, to obtain the positive continuous I-martingale that will provide the basis for our discussion of the theoretical performance deficit in Section 16.4, we set

$$\Sigma_t^{S \triangle I} := [M^S - M^I]_t := [\Lambda^S - \Lambda^I]_t. \tag{16.5}$$

Proposition 16.4 *The process* $\ln S_t - \ln I_t + \frac{1}{2}\Sigma_t^{S \triangle I}$ *is a continuous I-martingale.*

Proof Using the equality

$$\Sigma_t^{S \triangle I} = \Sigma_t^S + \Sigma_t^I - 2\Sigma_t^{S,I} \quad \text{w.i.e.} \tag{16.6}$$

(see Exercise 16.2), we can see that it suffices to prove that

$$\ln S_t + \frac{1}{2}\Sigma_t^S - \Sigma_t^{S,I} \tag{16.7}$$

and

$$\ln I_t - \frac{1}{2}\Sigma_t^I \tag{16.8}$$

are continuous I-martingales. The former, (16.7), follows immediately from the previous results in this section, namely Lemma 16.3 and Proposition 16.1, and the latter, (16.8), is Corollary 16.5 (proved independently of this proposition). □

16.2 EQUITY PREMIUM

Stock prices tend to grow faster than money on interest. This is the *equity premium.* The cleanest statement of the existence of an equity premium in this chapter's framework is Corollary 16.2, which says that M^I can be expected to grow at about the same rate as Σ^I. The following corollary tells what this means in terms of the capital I_t.

Corollary 16.5   *The process* $\ln I_t - \frac{1}{2}\Sigma_t^I$ *is a continuous I-martingale.*

Proof Combine Corollary 16.2 with Lemma 16.3 applied to the continuous *B*-martingale $X := I$. □

We now elaborate the approximate equality $\ln I_t \approx \frac{1}{2}\Sigma_t^I$ in two ways: as a central limit theorem in Corollary 16.6 and as a law of the iterated logarithm in Corollary 16.7.

Corollary 16.6   *If $\delta > 0$ and $\tau_C := \inf\{t|\Sigma_t^I > C\}$ for some constant $T > 0$,*

$$\underline{\mathbb{P}}^I\{|\ln I_{\tau_C} - C/2| < z_{\delta/2}\sqrt{C}\} \ge 1 - \delta, \tag{16.9}$$

where $z_{\delta/2}$ is the upper $\delta/2$-quantile of the standard Gaussian distribution and the inequality "$<$" in (16.9) is regarded as true when $\tau_C = \infty$.

Proof Combine Corollaries 16.5 and 15.3. □

Corollary 16.7   *Almost surely with respect to $\overline{\mathbb{P}}^I$,*

$$\Sigma_t^I \to \infty \implies \left(\limsup_{t\to\infty} \frac{\ln I_t - \frac{1}{2}\Sigma_t^I}{\sqrt{2\Sigma_t^I \ln \ln \Sigma_t^I}} = 1 \text{ and } \liminf_{t\to\infty} \frac{\ln I_t - \frac{1}{2}\Sigma_t^I}{\sqrt{2\Sigma_t^I \ln \ln \Sigma_t^I}} = -1 \right). \tag{16.10}$$

Proof Combine Corollaries 16.2 and 15.3 with the law of the iterated logarithm for measure-theoretic Brownian motion. □

Corollary 16.6 says that there is a nonnegative supergale that beats the index by a factor of nearly $1/\delta$ unless I_{τ_C} is close to $e^{C/2}$ in the sense

$$I_{\tau_C} \in (e^{C/2 - z_{\delta/2}\sqrt{C}}, e^{C/2 + z_{\delta/2}\sqrt{C}}).$$

In other words, the efficient index can be expected to outperform the bank account $e^{C/2}$-fold. This is a more precise way of saying that stocks are better than bonds for investors with a long-term investment horizon: the upward trend in I is likely to be much more important than the random fluctuations. Notice, however, that we measure the investment horizon by intrinsic rather than physical time.

If we are interested only in a lower bound on I or in an upper bound on I, we can use the following corollary instead of Corollary 16.6.

Corollary 16.8 *Let δ and τ_C be as before. Then*

$$\underline{\mathbb{P}}^I\{\ln I_{\tau_C} - C/2 > -z_\delta \sqrt{C}\} \geq 1 - \delta \quad \text{and} \quad \underline{\mathbb{P}}^I\{\ln I_{\tau_C} - C/2 < z_\delta \sqrt{C}\} \geq 1 - \delta.$$

A Crude Comparison with Data

The observed average return from stocks, as calculated from long-term market data, exceeds the average interest rate by more than predicted by standard economic theory. Rajnish Mehra and Edward Prescott dubbed this phenomenon the *equity premium puzzle* [275]. There is no consensus on how to explain the puzzle, or even on whether there is a puzzle, but our results can be interpreted as providing a partial solution.

Assuming that the bank account B is a money market account that pays the risk-free interest rate, data on average returns can be related to the cumulative relative growth M^I. Consider this restatement of Corollary 16.6 in terms of M^I:

Corollary 16.9 *If $\delta > 0$ and $\tau_C := \inf\{t|\Sigma_t^I > C\}$ for some constant $C > 0$,*

$$\underline{\mathbb{P}}^I\{|M_{\tau_C}^I - C| < z_{\delta/2}\sqrt{C}\} \geq 1 - \delta. \tag{16.11}$$

To relate this statement to data, we fix a time horizon T and neglect the fact that it has not been chosen using a stopping time τ_C. Under this loose interpretation, (16.11) says that

$$M_T^I - \Sigma_T^I \in \left(-z_{\delta/2}\sqrt{\Sigma_T^I}, z_{\delta/2}\sqrt{\Sigma_T^I}\right) \tag{16.12}$$

unless a given nonnegative supergale outperforms the efficient numeraire by a factor of $1/\delta$.

Because B is not a martingale, it does not follow from our results that it has a cumulative relative growth process M^B. Let us assume that B_t is slowly varying as a function of t, so that M^B does exist. (If we go further and assume that B grows according to a fixed interest rate r, then $B_t = B_0 e^{rt}$ for all $t \in [0, \infty)$, and $M_T^B = rT$.) Set

$$r := \frac{M_T^B}{T}, \quad \mu_I := \frac{M_T^I}{T} + r, \quad \sigma_I^2 := \frac{\Sigma_T^I}{T}, \tag{16.13}$$

where r is the mean interest rate over $[0, T]$, μ_I is the rate of growth of the index over $[0, T]$ in dollars, and σ_I^2 is its mean relative volatility over $[0, T]$. In this notation, we can rewrite (16.12) as

$$
\mu_I - r - \sigma_I^2 \in \left(-\frac{z_{\delta/2}\sigma_I}{\sqrt{T}}, \frac{z_{\delta/2}\sigma_I}{\sqrt{T}} \right). \tag{16.14}
$$

The difference $\mu_I - r$ is usually called the *equity premium*. According to (16.14), it should approximate σ_I^2 if I is efficient. The annual volatility of the S&P 500 is approximately 20% (see, e.g. [276, chapter 1, p. 3] or [274, p. 8]). So the theory of this chapter predicts an equity premium of about 4%. The standard theory, as applied by Mehra and Prescott, predicts an equity premium of at most 1% ([275, p. 146], [274, p. 11]).

The empirical study by Mehra and Prescott reported in [276, chapter 1, Table 2] estimates the equity premium over the period 1889–2005 as 6.36%. If the years 1802–1888 are also taken into account (as done by Jeremy Siegel [359], whose data were updated to 2004 by Mehra and Prescott [276, chapter 1, Table 2]), the equity premium (over the period 1802–2004) goes down to 5.36%. Our figure of 4% is below 6.36% and 5.36%, but the difference is much less significant than for Mehra and Prescott's predictions.

Equation (16.14) gives us the accuracy of our estimate σ_I^2 of the equity premium, and it holds unless a prespecified nonnegative gale beats the index by a factor of $1/\delta$. Plugging $\delta := 0.1$ (to obtain a reasonable accuracy), $\sigma := 0.2$, and $T := 2005 - 1888$, we evaluate $z_{\delta/2}\sigma/\sqrt{T}$ in (16.14) to 3.04% for the period 1889–2005, and changing T to $T := 2004 - 1801$, we evaluate it to 2.31% for the period 1802–2004. For both periods, the observed equity premium falls well within the prediction interval. It should also be remembered that, strictly speaking, Eq. (16.14) is not applicable in this situation: the stopping times used in the empirical studies that we report were different from waiting until the cumulative volatility reaches some threshold; more realistic prediction intervals would be slightly wider than 6.36 ± 3.04 and 5.36 ± 2.31 (although not as wide as those implied by the law of the iterated logarithm).

16.3 CAPITAL ASSET PRICING MODEL

We can generalize Corollaries 16.9 and 16.7 as follows.

Corollary 16.10   *If $\delta > 0$ and $\tau_C := \inf\{t | \Sigma_t^S > C\}$ for some constant $C > 0$,*

$$
\underline{\mathbb{P}}^I\{|M_{\tau_C}^S - \Sigma_{\tau_C}^{S,I}| < z_{\delta/2}\sqrt{C}\} \geq 1 - \delta. \tag{16.15}
$$

Proof Combine Proposition 16.1 and Corollary 15.3.   □

Corollary 16.11 *Almost surely with respect to* $\overline{\mathbb{P}}^I$,

$$\Sigma_t^S \to \infty \Longrightarrow \limsup_{t \to \infty} \frac{|M_t^S - \Sigma_t^{S,I}|}{\sqrt{2\Sigma_t^S \ln \ln \Sigma_t^S}} = 1. \qquad (16.16)$$

Proof Combine Proposition 16.1 and Corollary 15.3 with the law of the iterated logarithm for measure-theoretic Brownian motion. $\qquad \square$

Let us verify that Corollaries 16.6, 16.7, 16.10, and 16.11 imply a version of the standard CAPM. The standard CAPM says, in the one-period setting and framework of measure-theoretic probability, that

$$\mu_S - r = \frac{\sigma_{S,I}}{\sigma_I^2}(\mu_I - r), \qquad (16.17)$$

where μ_S is the expected return of an asset S, μ_I is the expected return of the market, σ_I^2 is the variance of the return of the market, $\sigma_{S,I}$ is the covariance between the returns of the asset S and the market, and r is the interest-free rate. For a derivation and discussion, see, e.g. [256, section 7.3].

Replacing the theoretical expected values (including those implicit in σ_I^2 and $\sigma_{S,I}$) by the empirical averages and using the notation (16.13) (including the underlying assumptions) and

$$\mu_S := \frac{M_T^S}{T} + r, \qquad \sigma_{S,I} := \frac{\Sigma_T^{S,I}}{T}, \qquad (16.18)$$

we obtain an approximate equality

$$M_T^S \approx \frac{\Sigma_T^{S,I}}{\Sigma_T^I} M_T^I. \qquad (16.19)$$

This approximate equality is also true in our game-theoretic framework under the further assumptions that $\Sigma_T^I \gg 1$ and $\Sigma_T^S \gg 1$. Indeed, our equity premium results, Corollaries 16.6 and 16.7, imply $M_t^I \approx \Sigma_t^I$, which makes (16.19) equivalent to $M_t^S \approx \Sigma_t^{S,I}$, which is asserted by Corollaries 16.10 and 16.11.

We will call Corollaries 16.10 and 16.11 a *simplified CAPM*, in the form of a central limit theorem and a law of the iterated logarithm, respectively. In the notation (16.13) and (16.18), we can rewrite them as

$$\mu_S - r \approx \sigma_{S,I}, \qquad (16.20)$$

which in combination with its special case

$$\mu_I - r \approx \sigma_I^2 \qquad (16.21)$$

for $S := I$ implies (16.17).

Corollaries 16.10 and 16.11 both have disadvantages. The former depends on the stopping time τ_C, which is even more awkward than the corresponding stopping time in Corollary 16.6, and the latter is asymptotic. We can deduce quantitatively weaker but conceptually more satisfying versions of Corollaries 16.10 and 16.11 using the following version of the Doléans supermartingale (3.3):

Corollary 16.12 *For each $\epsilon \in \mathbb{R}$, the process*

$$\exp\left(\epsilon(M_t^S - \Sigma_t^{S,I}) - \frac{\epsilon^2}{2}\Sigma_t^S \right) \qquad (16.22)$$

is a continuous I-martingale.

Proof Combine Propositions 16.1 and 14.20. □

The key advantage of the continuous I-martingale (16.22) is that it is nonnegative, and potentially it can detect M_t^S being too large (for $\epsilon > 0$) or too small (for $\epsilon < 0$) as compared with $\Sigma_t^{S,I}$. This observation can be used directly or packaged as the following corollary.

Corollary 16.13 *For any $\epsilon > 0$ and $\delta > 0$,*

$$\underline{\mathbb{P}}^I\left\{ \forall T \in [0, \infty) : \left| M_T^S - \Sigma_T^{S,I} \right| < \frac{\ln\frac{2}{\delta}}{\epsilon} + \frac{\epsilon}{2}\Sigma_T^S \right\} \geq 1 - \delta. \qquad (16.23)$$

Using the notation (16.13) and (16.18), and so leaving the dependence on the time horizon T implicit, we can rewrite (16.23), somewhat informally, as

$$\underline{\mathbb{P}}^I\left\{ \forall T \in [0, \infty) : |\mu_S - r - \sigma_{S,I}| < \frac{\ln\frac{2}{\delta}}{\epsilon T} + \frac{\epsilon}{2}\sigma_S^2 \right\} \geq 1 - \delta.$$

This can be interpreted as asserting that $\mu_S \approx r + \sigma_{S,I}$ when ϵ and δ are small and $T \gg \frac{1}{\epsilon}\ln\frac{2}{\delta}$.

Proof of Corollary 16.13 Fix $\epsilon > 0$ and $\delta > 0$. By Corollary 16.12, the definition of $\overline{\mathbb{P}}^I$, and Exercise 14.10, the nonnegative I-supermartingale (16.22) is always below $2/\delta$ with lower I-probability at least $1 - \delta/2$. In other words,

$$\forall t \in [0, \infty) : M_t^S - \Sigma_t^{S,I} < \frac{\ln \frac{2}{\delta}}{\epsilon} + \frac{\epsilon}{2} \Sigma_t^S$$

with lower I-probability at least $1 - \delta/2$. Applying the same argument with $-\epsilon$ in place of ϵ, we can see that

$$\forall t \in [0, \infty) : -M_t^S + \Sigma_t^{S,I} < \frac{\ln \frac{2}{\delta}}{\epsilon} + \frac{\epsilon}{2} \Sigma_t^S$$

with lower I-probability at least $1 - \delta/2$. Combining the last two inequalities shows that

$$\forall t \in [0, \infty) : \left| M_t^S - \Sigma_t^{S,I} \right| < \frac{\ln \frac{2}{\delta}}{\epsilon} + \frac{\epsilon}{2} \Sigma_t^S \tag{16.24}$$

with lower I-probability at least $1 - \delta$. This is (16.23). □

Black-Type CAPM

Subtracting (16.21) from (16.20), we obtain the approximate equality

$$\mu_S \approx \mu_I - \sigma_I^2 + \sigma_{S,I}, \tag{16.25}$$

which is a game-theoretic CAPM not involving r, in the spirit of Black [41] (as discussed in Section 16.7). It is obtained from the standard CAPM (16.17) by replacing the interest rate r by $\mu_I - \sigma_I^2$. This version of the CAPM can be formalized by

$$M_t^S - M_t^I + \Sigma_t^I - \Sigma_t^{S,I} \tag{16.26}$$

being a continuous I-martingale. The following two corollaries state implications of this process being a continuous I-martingale in the form of a central limit theorem and a law of the iterated logarithm. The first one is in the spirit of Corollary 16.10 but the stopping time is more natural: now we wait until I and S sufficiently diverge from each other (see (16.5) for the definition of $\Sigma^{S\triangle I}$).

Corollary 16.14 *If $\delta > 0$ and $\tau_C := \inf\{t | \Sigma_t^{S\triangle I} > C\}$ for some constant $C > 0$,*

$$\underline{\mathbb{P}}^I \left\{ \left| M_{\tau_C}^S - M_{\tau_C}^I + \Sigma_{\tau_C}^I - \Sigma_{\tau_C}^{S,I} \right| < z_{\delta/2} \sqrt{C} \right\} \geq 1 - \delta. \tag{16.27}$$

As usual, our convention is that the event in the curly braces in (16.27) happens when $\tau_C = \infty$ (the security S essentially coincides with the index I).

Corollary 16.15 *Almost surely with respect to* $\overline{\mathbb{P}}^I$,

$$\Sigma_t^{S\triangle I} \to \infty \Longrightarrow \limsup_{t \to \infty} \frac{\left| M_t^S - M_t^I + \Sigma_t^I - \Sigma_t^{S,I} \right|}{\sqrt{2\Sigma_t^{S\triangle I} \ln \ln \Sigma_t^{S\triangle I}}} = 1.$$

Specializing the Black-type CAPM of Corollary 16.15 to a constant S (such as the bank account $S = 1$), we obtain our equity premium results; in particular, the continuous I-martingale (16.26) essentially becomes the martingale of Corollary 16.2 and Corollary 16.14 becomes Corollary 16.9.

The following corollary of (16.26) being a continuous I-martingale is in the spirit of Corollary 16.13.

Corollary 16.16 *For any $\epsilon > 0$ and $\delta > 0$,*

$$\underline{\mathbb{P}}^I \left\{ \forall t \in [0, \infty) : \left| M_t^S - M_t^I + \Sigma_t^I - \Sigma_t^{S,I} \right| < \frac{\ln \frac{2}{\delta}}{\epsilon} + \frac{\epsilon}{2} \Sigma_t^{S\triangle I} \right\} \geq 1 - \delta.$$

Proof Analogously to Corollary 16.12, we find that, for each $\epsilon \in \mathbb{R}$, the process

$$\exp \left(\epsilon(M_t^S - M_t^I + \Sigma_t^I - \Sigma_t^{S,I}) - \frac{\epsilon^2}{2} \Sigma_t^{S\triangle I} \right)$$

is a continuous I-martingale. Proceed as in the proof of Corollary 16.13. $\square$

Setting $S := 1$ we obtain an equity premium result (identical to what we obtain by setting $S := I$ in Corollary 16.13).

16.4 THEORETICAL PERFORMANCE DEFICIT

The quadratic variation of the continuous I-martingale of Proposition 16.4 is $\Sigma_t^{S\triangle I}$, and so we have our usual two corollaries, CLT-type and LIL-type; they are very similar to Corollaries 16.14 and 16.15.

Corollary 16.17 *If $\delta > 0$ and $\tau_C := \inf\{t | \Sigma_t^{S\triangle I} > C\}$ for some constant $C > 0$,*

$$\underline{\mathbb{P}}^I \left\{ \left| \ln S_{\tau_C} - \ln I_{\tau_C} + \frac{1}{2} \Sigma_{\tau_C}^{S,I} \right| < z_{\delta/2} \sqrt{C} \right\} \geq 1 - \delta.$$

Corollary 16.18 *Almost surely with respect to* $\overline{\mathbb{P}}^I$,

$$\Sigma_t^{S\triangle I} \to \infty \implies \limsup_{t\to\infty} \frac{\left| \ln S_t - \ln I_t + \frac{1}{2}\Sigma_t^{S,I} \right|}{\sqrt{2\Sigma_t^{S\triangle I} \ln \ln \Sigma_t^{S\triangle I}}} = 1.$$

Therefore, for large T we have, with high probability,

$$\ln S_T \approx \ln I_T - \frac{1}{2}\Sigma_T^{S\triangle I}.$$

The subtrahend $\frac{1}{2}\Sigma_T^{S\triangle I}$ can be interpreted as measuring the lack of diversification in S as compared with the index I, and we call it the *theoretical performance deficit*.

16.5 SHARPE RATIO

Let us define the *Sharpe ratio* at time $t \in [0, \infty)$ of a continuous B-martingale X as

$$\text{Sh}_t^X := \frac{\mathrm{M}_t^X / t}{\sqrt{\Sigma_t^X / t}}. \tag{16.28}$$

The numerator is usually interpreted as a measure of reward brought by the portfolio or trading strategy represented by X, and the denominator as a measure of X's risk.

Let us first check, very informally, that, for a fixed t, the largest Sharpe ratio is approximately attained by I:

$$\text{Sh}_t^S \lesssim \text{Sh}_t^I. \tag{16.29}$$

The approximate equalities $\mathrm{M}_t^S \approx \Sigma_t^{S,I}$ and $\mathrm{M}_t^I \approx \Sigma_t^I$ (cf. (16.20) and (16.21)) allow us to deduce (16.29) from

$$\Sigma_t^{S,I} \le \sqrt{\Sigma_t^S \Sigma_t^I}, \tag{16.30}$$

which is a special case of the Cauchy–Schwarz inequality (Exercise 16.4).

The following proposition is a LIL-type formalization of (16.29).

Proposition 16.19 *Almost surely with respect to* $\overline{\mathbb{P}}^I$,

$$(\Sigma_t^S \wedge \Sigma_t^I) \to \infty \implies \limsup_{t\to\infty} \frac{\sqrt{t/2}}{\sqrt{\ln \ln \Sigma_t^S} + \sqrt{\ln \ln \Sigma_t^I}}(\text{Sh}_t^S - \text{Sh}_t^I) \le 1. \tag{16.31}$$

Proof By the laws of the iterated logarithm (16.10) (written in terms of M_t^I) and (16.16), we have, for each $\epsilon > 0$ and from some t on,

$$\frac{M_t^S}{\sqrt{\Sigma_t^S}} - \frac{M_t^I}{\sqrt{\Sigma_t^I}} \le \frac{M_t^S - \Sigma_t^{S,I}}{\sqrt{\Sigma_t^S}} - \frac{M_t^I - \Sigma_t^I}{\sqrt{\Sigma_t^I}}$$

$$\le (1+\epsilon)\frac{\sqrt{2\Sigma_t^S \ln \ln \Sigma_t^S}}{\sqrt{\Sigma_t^S}} + (1+\epsilon)\frac{\sqrt{2\Sigma_t^I \ln \ln \Sigma_t^I}}{\sqrt{\Sigma_t^I}}$$

$$= (1+\epsilon)\sqrt{2}\left(\sqrt{\ln \ln \Sigma_t^S} + \sqrt{\ln \ln \Sigma_t^I}\right)$$

given the antecedent of (16.31); the first inequality follows from (16.30). This gives the consequent of (16.31). □

According to Proposition 16.19,

$$Sh_t^S \le Sh_t^I + O\left(\sqrt{\frac{\ln \ln t}{t}}\right)$$

provided $\Sigma_t^S \to \infty$, $\Sigma_t^I \to \infty$, $\Sigma_t^S = O(t)$, and $\Sigma_t^I = O(t)$ as $t \to \infty$. (The last two conditions can be relaxed greatly; e.g. it suffices to require only that Σ_t^S and Σ_t^I grow only polynomially fast as $t \to \infty$.)

16.6 EXERCISES

Exercise 16.1 Prove Eq. (16.4). *Hint.* One way to do this is to apply Taylor's formula to the definitions. □

Exercise 16.2 Prove Eq. (16.6). □

Exercise 16.3 Show that Corollary 16.13 (in the form of (16.24)) implies the following version of Eq. (16.15) in Corollary 16.10:

$$\overline{\mathbb{P}}^I\left\{\left|M_{\tau_C}^S - \Sigma_{\tau_C}^{S,I}\right| \ge \sqrt{2\ln\frac{2}{\delta}}\sqrt{C}\right\} \le \delta.$$

(For the comparison between the functions z_δ and $\sqrt{2\ln\frac{1}{\delta}}$ of δ, see Figure 3.2.) *Hint.* Set $T := \tau_C$ and optimize ϵ. □

Exercise 16.4 Deduce (16.30) from the Cauchy–Schwarz inequality. □

16.7   CONTEXT

A Measure-Theoretic Argument for an Efficient Numeraire

Suppose, as usual in measure-theoretic finance, that a financial market is described as a set of adapted processes in a probability space with a filtration. Suppose further that the probability measure P is known to the trader. Then he can construct a trading strategy that optimizes his capital's growth rate. In discrete time, this idea goes back to John Kelly in 1956 [215] and Leo Breiman in 1960 [49, 50].

Under weak conditions, an optimal strategy can be obtained using the Kelly capital growth criterion, which maximizes the expected logarithmic return at each step; the resulting capital process is often referred to as growth-optimal (or log-optimal). In his proof of Theorem 3 of [50], Breiman made a key observation: the ratio of any capital process to the growth-optimal capital process is a supermartingale. Breiman's proof is based on a very natural idea: the expected logarithmic return of the growth-optimal strategy is at least as large as that of its perturbed version, the convex mixture of this strategy taken with weight $1 - \epsilon$ and a competing strategy taken with weight ϵ. Letting $\epsilon \to 0$ and simple analysis complete the proof. The argument has been developed in continuous time by Platen and Heath [310] and by Karatzas and Kardaras (see, e.g. [210] and their forthcoming book [209]). Karatzas and Kardaras's name for the growth-optimal portfolio is "numeraire portfolio"; they use the property discovered by Breiman as the definition, whereas Platen and Heath [310, Theorem 10.3.1] derive this property from a more traditional definition. A growth-optimal capital process is evidently an efficient numeraire.

The empirical argument for an efficient numeraire and the theoretical argument for a growth-optimal portfolio come together in Platen's Diversification Theorem (see [308] and [310, Theorem 10.6.5]). According to Platen and Heath, diversified portfolios can serve as proxies for the growth-optimal portfolio and are not expected to be outperformed. The move from a continuous price process S' to a continuous I-martingale S'/I, which they call "benchmarking," is the basis of their theory [310, section 10.3]. In a recent related work [140], Fernholz constructs an example of an "open market" in which the market portfolio is used as the numeraire.

Efficient Markets

The notion of market efficiency was introduced by Eugene Fama in 1965 (in [131]; see also his Nobel prize lecture [133]). According to Fama, a market is efficient if its prices reflect a market equilibrium that uses all available information. As Fama explained in 1970 [132, section IV], the hypothesis that a market is efficient cannot be tested by itself; we must also hypothesize a particular model of market equilibrium. This has come to be called the *joint hypothesis problem*. In Fama's picture, a model of market equilibrium involves assumptions about the risk preferences of investors. Fama did not use the earlier work of Kelly and Breiman, but if we take a bird's eye view, we can say that their picture simplifies Fama's picture by beginning with a probability measure for prices such as might result from a model of market equilibrium. From the same bird's eye view, our game-theoretic picture simplifies further, by beginning only with an efficient numeraire that plays the role of the growth-optimal portfolio determined by Breiman's probability distribution. A joint hypothesis problem remains; we can test whether a particular numeraire can be outperformed, not whether there exists some such numeraire.

The slogan "You cannot beat the market" is often taken to mean that a trader cannot do better than holding all stocks in proportion to their capitalization. But neither a growth-optimal portfolio in Breiman's theory nor an efficient numeraire in our theory is necessarily the capital-weighted portfolio. In fact, as we will see in Chapter 17, the hypothesis that the capital-weighted portfolio is an efficient numeraire can be problematic if there are finitely many stocks.

In 2001 [349], we called the hypothesis that a trader cannot multiply the capital he risks in a particular market by a large factor relative to a particular numeraire a *game-theoretic efficient market hypothesis*. In this book we have chosen instead to call the numeraire efficient; this seems simpler and less likely to lead readers to confuse our picture with Fama's.

The Equity Premium Puzzle

For a history of the discovery of this puzzle and the many suggestions for explaining it, see [274, 276]. For more recent empirical results, see, e.g. [293, Table 1]. Some researchers argue that the US results are affected by survivorship bias, because the US markets have been very successful over the past two centuries. Estimates for the equity premium over the period 1900–2014 given in [293, Table 3] are 6.1% for the United Kingdom, 7.5% for the United States, 5.7% for the world, 5.2% for the world excluding the United States, and 5.2% for Europe.

Our equity premium expression is well known for the growth-optimal portfolio in the context of measure-theoretic market models (see, e.g. [310, Theorem 11.1.3]).

Capital Asset Pricing Model

This model was first published by William Sharpe in 1964 [352, footnote 25] and discovered independently by John Lintner at about the same time [252]. Jan Mossin provided clarifications in 1966 [287]. In [153], Craig French traces the model's origins to unpublished work by Jack Treynor, also mentioned by Sharpe [352, footnote 7].

The standard derivation of CAPM, as in [352] and the textbook [256, section 7.3], relies on the mixing trick used earlier by Breiman [50] (see the beginning of this section) to prove that the growth-optimal portfolio is the numeraire portfolio (to use the modern terminology). The first derivations of a game-theoretic CAPM, expressed in the form (16.25), also used this trick [417, 421].

Our CAPM depends on a very different assumption (a given efficient numeraire) from the standard CAPM, which is a result about equilibrium under rather unrealistic assumptions about the beliefs and preferences of participants in the market. In discrete time, perhaps the first to obtain a CAPM of our kind was David Luenberger [256, section 7.3], who used the growth-optimal portfolio as the efficient numeraire. He refers to his CAPM as the log-optimal pricing formula and notices its striking similarity to the standard Sharpe–Lintner CAPM. A CAPM of this kind in continuous time was derived using the Black–Scholes model in [408, 407] and [410]. See also Platen [309] and Platen and Heath [310, Theorem 11.2.1], who call their continuous-time measure-theoretic result an ICAPM (intertemporal CAPM, following Merton [278]).

Fischer Black's version of the CAPM [41] is also known as the zero beta CAPM (see, e.g. [238, section 3]). Black was motivated by the empirical work by himself, Michael Jensen, and Myron Scholes [42], which showed that (16.17) fits empirical data much better when r is treated simply as another factor (sometimes referred to as the zero beta return) in addition to μ_I, dropping the standard CAPM's identification of r with the risk-free interest rate. Black shows that such a model arises in a natural way in the absence of a risk-free asset [41]. Our version is more specific: it says that the zero beta return is $\mu_I - \sigma_I^2$. But our basic gales include a bank account even when we discuss our Black-type CAPM.

Risk vs. Return

It is accepted wisdom in financial theory that there is a trade-off between risk and return. In our game-theoretic picture, the trade-off for the efficient numeraire is approximately an equality when we interpret the quadratic variation Σ^I as risk and the cumulative relative growth M^I as return: $\Sigma_t^I \approx M_t^I$. The law of the iterated logarithm is one way of making this precise:

$$\limsup_{t\to\infty} \frac{M_t^I - \Sigma_t^I/2}{\sqrt{2\Sigma_t^I \ln\ln\Sigma_t^I}} = 1 \quad \text{and} \quad \liminf_{t\to\infty} \frac{M_t^I - \Sigma_t^I/2}{\sqrt{2\Sigma_t^I \ln\ln\Sigma_t^I}} = -1$$

$\overline{\mathbb{P}}^I$-a.s. (combine Corollary 16.7 and Lemma 16.3).

In a different sense, risk is controlled in our framework, because predictions are tested by nonnegative capital processes. If a trader has become very rich relative to the efficient numeraire starting with a modest stake and without risking going into debt, we cannot explain his success by the risk he took; he is either skilled or lucky.

Sharpe Ratio

Investment managers often use this ratio as a risk-return characteristic of a trading strategy, the numerator of (16.28) representing return and the denominator representing risk. Sharpe introduced the ratio in 1966 [353, p. 123]. It is widely accepted that "in the world of the CAPM the market portfolio will have the highest possible Sharpe Ratio" [354, section 4.19]. In the context of the growth-optimal portfolio, this result is stated in [310, Theorem 11.1.3] and called the Portfolio Selection Theorem.

Returns vs. Logarithmic Returns

Lemma 16.3 can be interpreted as quantifying the difference between using returns and using logarithmic returns. We can define M^X as the ucie limit of the approximations

$$M_t^k := \sum_{n=1}^{\infty} \frac{X_{T_n^k \wedge t} - X_{T_{n-1}^k \wedge t}}{X_{T_{n-1}^k \wedge t}}, \quad k \in \mathbb{N}. \tag{16.32}$$

The ratios

$$\frac{X_{T_n^k \wedge t} - X_{T_{n-1}^k \wedge t}}{X_{T_{n-1}^k \wedge t}} \quad \text{and} \quad \ln \frac{X_{T_n^k \wedge t}}{X_{T_{n-1}^k \wedge t}}$$

are known as the *return* and *logarithmic return*, respectively, over the time period $[T_{n-1}^k \wedge t, T_n^k \wedge t]$. Replacing the returns in (16.32) by logarithmic returns, we obtain $\ln(X_t/X_0)$. If X is a security's price path, $\ln(X_t/X_0)$ is a direct measure of its performance, and Lemma 16.3 tells us it is overoptimistic to use M^X as a proxy, especially for a volatile X.

Arbitrage

The notion of arbitrage is often considered fundamental in mathematical finance. The notion remains important in the game-theoretic framework, but the word is not always helpful, because it has been interpreted in a number of ways, and often none of the interpretations match our definitions precisely.

We have used the notion of *coherence* in several ways. Our axioms for upper expectations and offers imply certain kinds of coherence (Sections 6.1 and 6.4), and we have characterized our protocols for finance as coherent when they give upper probability 1 to the entire sample space (Sections 13.3 and 14.9). Considered as conditions of no-arbitrage, these notions of coherence are particularly weak.

Freddy Delbaen and Walter Schachermayer's *no free lunch with vanishing risk* (*NFLVR*), perhaps the most popular notion of no arbitrage in the standard theory [103], is relatively strong. For simplicity, consider the case of one security whose price path S is a continuous semimartingale. A stochastic process H is called an *admissible strategy* if the stochastic integral $H \cdot S$ (analogous to the Itô integral of Section 14.3) exists and is bounded below. The end point T of the time interval $[0, T]$ may be finite or infinite; in the latter case the relevant limits are assumed to exist. Let C be the class of random variables f that can be superhedged, in the sense that $f \leq (H \cdot S)_T$ a.s. for an admissible strategy H. The most traditional notion of no arbitrage (NA) for S is that C does not contain a random variable that is nonnegative and positive with a positive probability. Delbaen and Schachermayer's NFLVR requires that the closure $\overline{C}$ of C in the uniform metric does not contain such a random variable. It is well known that NFLVR is equivalent to the conjunction of NA with the absence of arbitrage of the first kind (NA1), a concept that [213] attributes to [194]. An *arbitrage of the first kind* is a random variable $f \geq 0$ such that $f > 0$ with positive probability and for each $\epsilon > 0$ there exists an admissible strategy H such that the process $\epsilon + H \cdot S$ is nonnegative and $\epsilon + (H \cdot S)_T \geq f$ a.s. This resembles our definition of null events, as pointed out by Perkowski and Prömel [301, section 2].

It appears that some but not all of our game-theoretic results can be expressed in terms of arbitrage. One game-theoretic result that can be expressed in terms of arbitrage is the special case of Theorem 13.6 that says that, a.s., for all T, $\mathrm{vi}(\omega|_{[0,T]}) = 2$ or ω is constant over $[0, T]$. We can reformulate this by saying that adding to our model the requirement that $\mathrm{vi}(\omega|_{[0,T]}) \neq 2$ and ω is not constant over $[0, T]$ for some T leads to an opportunity similar to a violation of NA1: given any $\epsilon > 0$ there is a strategy that risks no more than ϵ and becomes as rich as you please. This is akin to the indirect interpretation of theorems in mathematical logic: a statement is a theorem if adding its negation to our axioms leads to a contradiction. This move does not seem to work, however, for results asserting the existence of strategies that make the trader merely very rich rather than infinitely rich. The approximate equality $\ln I_t \approx \frac{1}{2}\Sigma_t^I$ given in Corollary 16.7 might conceivably be expressed in terms of an NA1-type arbitrage, but we see no way to formulate even a crude form of Corollary 16.6 in terms of arbitrage.

Even in the standard theory, a formulation in terms of arbitrage sometimes restricts the applicability of a mathematical result unnecessarily. Black and Scholes's first article on option pricing [43] argued that if a European option is traded but not priced in accordance with their formula, the trader can create a risk-free portfolio bringing profit with probability 1. In fact, their argument (as perfected by further research) is applicable in greater generality: even if the option is not traded, we can perfectly replicate its payoff with a dynamic portfolio. Such limitations of no-arbitrage arguments become much more difficult to neglect in the game-theoretic framework.

Game-Theoretic Portfolio Theory

In an idealized market with a finite number of securities, the capital-weighted market portfolio can be outperformed by a long-only portfolio that gives greater weights to securities with smaller capitalization, provided that the securities' price paths are positive and trading never dies out, in some sense. This mathematical fact has been emphasized by E. Robert Fernholz, who has made it a centerpiece of his measure-theoretic *stochastic portfolio theory* [138, 139, 141]. In this chapter we show that Fernholz's picture can be reshaped into a relatively simple and transparent game-theoretic picture in the martingale space that uses the total market capitalization as numeraire. In particular, our way of formalizing trading never dying out is that the sum of the quadratic variations of the securities is unbounded.

In order to apply our game-theoretic portfolio theory to an actual market, we need to assume that transaction costs and complications arising from dividends can be neglected. We also need to assume that the martingale space we construct is valid in the trading-in-the-small sense; a trader cannot become very rich very suddenly. But application of the theory does not require the long-run efficiency assumption used in the preceding chapter: we need not assume that the total market capitalization is an efficient numeraire.

Whereas Fernholz has contended that his strategies can outperform the market, our exposition takes a more neutral stance. Under the assumption that quadratic variations are unbounded, we do have a way to outperform the market. But if the total market capitalization is in fact an efficient numeraire, so that it cannot be outperformed, then

Game-Theoretic Foundations for Probability and Finance, First Edition. Glenn Shafer and Vladimir Vovk.
© 2019 John Wiley & Sons, Inc. Published 2019 by John Wiley & Sons, Inc.

our assumptions will be violated: price paths will hit zero or at least trading in them will die out.

The assumption that there are only finitely many securities is essential to our theory and to Fernholz's. The portfolios that outperform the market invest more heavily in securities with smaller market capitalization than the market portfolio does. If there were infinitely many basic securities, then these portfolios could not be defined. The assumption that there are only a finite number of basic securities may seem commonsensical, but if we are modeling a large national market, such the market formed by NYSE and NASDAQ stocks in the United States, where some securities do disappear and diversity is maintained by new entries, a model with infinitely many securities may be more useful idealization for some purposes.

In Section 17.1, we construct a game-theoretic version of Fernholz's phenomenon using portfolios whose capital processes are game-theoretic Stroock–Varadhan martingales. In Section 17.2, we show that the best known of the capital processes studied in stochastic portfolio theory are also game-theoretic martingales; we call them *Fernholz martingales*. As in stochastic portfolio theory, these capital processes emerge from different measures of the diversity of capitalization of the securities. As we also show, they can be obtained by boosting Stroock–Varadhan martingales. Fernholz martingales eventually outperform Stroock–Varadhan martingales (and therefore the market as well) exponentially, but Stroock–Varadhan martingales may do better than Fernholz martingales over time horizons that are practical (see Figure 17.1).

Chapter 13's Dubins–Schwarz theorem points to another way to take advantage of positive price paths and unbounded quadratic variations, requiring so much time that it is clearly impractical yet enlightening because Dubins–Schwarz is so fundamental. The Brownian behavior implied by Dubins–Schwarz means that the upper

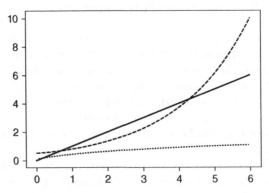

Figure 17.1 Lower bounds on the Stroock–Varadhan martingale generated by (17.3) (solid line), the Fernholz martingale generated by (17.8) (dashed line), and Section 17.3's Dubins–Schwarz martingale with $J = 2$ (dotted line) as functions of the total quadratic variation A. These lower bounds are given by (17.4), (17.12), and (17.20), respectively.

probability of the price paths remaining positive and having unbounded quadratic variations is zero, and by the definition of upper probability, this implies that the trader has a strategy that multiplies his capital without bound under these assumptions. In Section 17.3, we use the theory of Brownian motion to construct such a strategy.

In Section 17.4, we relate the possibility of profiting from the diversity of the performance of different securities to Jeffreys's law (see also Section 10.7).

Fernholz's work has attracted considerable notice, and his belief that his methods can succeed in practice has received some support. But stochastic portfolio theory and our game-theoretic alternative lead to clear-cut conclusions about the possibility of beating the market only when ignoring important features of real markets, including the existence of transaction costs, dividends, and securities leaving and entering the market. So there are many ways to explain why the trading strategies described in this chapter might fail to outperform the market in practice.

17.1 STROOCK–VARADHAN MARTINGALES

Consider a market space with a finite set of basic securities, which we number 1 to J, and their basic gales $S^1, \ldots, S^J$. Following Fernholz, we assume that all the S^j are positive, but we do not make Fernholz's assumptions implying that they have unbounded quadratic variation.

For convenience, we identify S_t^j with j's total market capitalization at time t. If the security j is a stock, we can suppose that there is only one share, and S_t^j is its price. The process $\sum_{j=1}^J S_t^j$, $t \in [0, \infty)$, the *total capitalization of the market*, is a positive gale. We form a martingale space by taking it as our numeraire. The basic martingales in this martingale space are $\mu^1, \ldots, \mu^J$, where

$$\mu_t^j := \frac{S_t^j}{\sum_{i=1}^J S_t^i} \quad \text{and thus} \quad \sum_{j=1}^J \mu^j = 1.$$

We call μ_t^j security j's *market weight* at time t. It is the price of j's single share, or the price of one *unit of j*, in terms of the numeraire.

For a positive constant A, we define a stopping time τ_A by

$$\tau_A := \inf \left\{ t \,\bigg|\, \sum_{j=1}^J [\mu^j]_t > A \right\}. \tag{17.1}$$

We say that the market *stays active* if $\sum_j [\mu^j]_t \to \infty$ as $t \to \infty$; equivalently, if $\tau_A < \infty$ for all A. As we will see, the central point of game-theoretic portfolio theory is that you can outperform a market that stays active.

We write μ_t for the vector $(\mu_t^1, \ldots, \mu_t^J)$, which is in the interior of the standard simplex (cf. (12.45)),

$$\Delta^J := \left\{ x \in (0, \infty)^J \,\middle|\, \sum_{j=1}^J x_j = 1 \right\}.$$

Let f be a C^2 function defined on an open neighborhood of Δ^J in $\mathbb{R}^J$. We write D_j for its jth partial derivative,

$$D_j f(x) = \frac{\partial f}{\partial x_j}(x), \quad x \in \mathrm{dom}\, f \subseteq \mathbb{R}^J,$$

and D_{ij} for its second partial derivative,

$$D_{ij} f(x) = \frac{\partial^2 f}{\partial x_i \partial x_j}(x).$$

The game-theoretic Itô formula (Theorem 14.19) implies that

$$f(\mu_t) - f(\mu_0) - \frac{1}{2} \sum_{i=1}^J \sum_{j=1}^J \int_0^t D_{ij} f(\mu) \, d[\mu^i, \mu^j]$$

$$= \sum_{j=1}^J \int_0^t D_j f(\mu) \, d\mu^j \quad \text{w.i.e.,} \tag{17.2}$$

and so the left-hand side of (17.2) is a continuous martingale, which we call the game-theoretic *Stroock–Varadhan martingale* generated by f. (Formally, Theorem 14.19 requires that f be defined on the whole of $\mathbb{R}^J$, but it is also applicable in this situation: see Exercise 17.1.) The game-theoretic Stroock–Varadhan martingales are analogous to the measure-theoretic martingales used by Daniel Stroock and S.R. Srinivasa Varadhan in their classical study of diffusion processes [212, (5.4.2)]. Kolmogorov's martingale, which we studied in Section 3.1, is a discrete-time analog. Setting

$$f(x) := -\frac{1}{2} \sum_{j=1}^J x_j^2, \tag{17.3}$$

we can rewrite the continuous martingale on the left-hand side of (17.2) as

$$Y_t := \frac{1}{2} \sum_{j=1}^J (\mu_0^j)^2 - \frac{1}{2} \sum_{j=1}^J (\mu_t^j)^2 + \frac{1}{2} \sum_{j=1}^J [\mu^j]_t \ge -\frac{1}{2} + \frac{1}{2} \sum_{j=1}^J [\mu^j]_t.$$

So $X := 2Y + 1$ is a nonnegative continuous martingale satisfying $X_0 = 1$ and

$$X_{\tau_A} \geq A, \tag{17.4}$$

with the convention that $X_\infty := \infty$.

The trading strategy in the martingale space that produces the martingale X satisfying (17.4) can be computed from the right-hand side of (17.2) using the definition (17.3): starting from initial capital 1, hold $-2\mu_t^j$ units of j at time t.

17.2 BOOSTING STROOCK–VARADHAN MARTINGALES

To boost the growth rate of Stroock–Varadhan martingales, we use the game-theoretic notion of Doléans exponential defined in Chapter 14. Recall from Section 14.6 that the Doléans exponential of a continuous martingale X is the continuous martingale $\mathcal{E}(X) = \exp(X - [X]/2)$. The important byproduct is that the corresponding trading strategies become long-only.

Let $\mathbf{S}$ be a positive C^2 function defined on an open neighborhood of Δ^J. The following is a game-theoretic version of a basic result of stochastic portfolio theory (see, e.g. [139, Theorem 3.1.5]).

Theorem 17.1 *The continuous process*

$$\frac{\mathbf{S}(\mu_t)}{\mathbf{S}(\mu_0)} \exp\left(-\frac{1}{2} \sum_{i=1}^{J} \sum_{j=1}^{J} \int_0^t \frac{D_{ij}\mathbf{S}(\mu)}{\mathbf{S}(\mu)}\, d[\mu^i, \mu^j] \right) \tag{17.5}$$

is a continuous martingale.

Proof When $f := \ln \mathbf{S}$, the Doléans exponential of the Stroock–Varadhan martingale on the left-hand side of (17.2) is

$$\frac{\mathbf{S}(\mu_t)}{\mathbf{S}(\mu_0)} \exp\left(-\frac{1}{2} \sum_{i,j} \int_0^t D_{ij}f(\mu)\, d[\mu^i, \mu^j] - \frac{1}{2} \sum_{i,j} \int_0^t D_i f(\mu) D_j f(\mu)\, d[\mu^i, \mu^j] \right).$$

By the identity

$$D_{ij}f = \frac{D_{ij}\mathbf{S}}{\mathbf{S}} - \frac{D_i\mathbf{S}}{\mathbf{S}}\frac{D_j\mathbf{S}}{\mathbf{S}} = \frac{D_{ij}\mathbf{S}}{\mathbf{S}} - D_i f D_j f,$$

this is equal to (17.5). $\qquad\square$

We call (17.5) the *Fernholz martingale generated by* $\mathbf{S}$.

Let us spell out the trading strategy whose capital process is the Fernholz martingale (17.5). You get the Stroock–Varadhan martingale X for $f := \ln \mathbf{S}$ by starting with 0 and always holding

$$D_j \ln \mathbf{S}(\mu) \tag{17.6}$$

units of j. So by (14.28), you obtain the Fernholz martingale $Y = \mathcal{E}(X)$ by starting with 1 and always holding a number of units of j equal to (17.6) times your current capital. Taking into account the fraction of your current capital in the numeraire, we find that the total fraction held in j is

$$\pi^j := \left(D_j \ln \mathbf{S}(\mu) + 1 - \sum_{k=1}^{J} \mu_k D_k \ln \mathbf{S}(\mu) \right) \mu^j. \tag{17.7}$$

The main part of the expression in the parentheses is (17.6); the rest is simply the constant c, depending on μ but not on j, that makes $(D_j \ln \mathbf{S}(\mu) + c)\mu^j$ sum to 1 over j. The factor μ^j turns the number of units of j into the capital invested in j.

Next we consider three standard special cases of the general formula (17.5) for Fernholz martingales. A positive C^2 function $\mathbf{S}$ defined on an open neighborhood of Δ^J is a *diversity measure* if it is symmetric and concave. The three special cases are produced by three different measures of diversity. Portfolios (17.7) corresponding to any measure of diversity, including the three special cases, are always long-only: $\pi^j \geq 0$ (see Exercise 17.2).

A Very Tame Portfolio

In [139, section 3.3], Fernholz describes what can be accomplished using the diversity measure

$$\mathbf{S}(x) := 1 - \frac{1}{2} \sum_{j=1}^{J} x_j^2 \tag{17.8}$$

(cf. (17.3)). Theorem 17.1 gives the following game-theoretic version of [139, Example 3.3.3].

Corollary 17.2   *The Fernholz martingale generated by (17.8) is*

$$\frac{\mathbf{S}(\mu_t)}{\mathbf{S}(\mu_0)} \exp\left(\frac{1}{2} \sum_{j=1}^{J} \int_0^t \frac{d[\mu^j]}{\mathbf{S}(\mu)} \right). \tag{17.9}$$

Proof   It suffices to plug $D_{ij}\mathbf{S}(x) = -\mathbf{1}_{i=j}$ into (17.5).   □

The following corollary simplifies the expression (17.9) so that we can see more clearly the effect of the accumulation of the variation in the market weights. It replaces the first factor by its lower bound and the denominator in the second factor by its upper bound, thus neglecting any effect from changes in the diversity of the market weights.

Corollary 17.3 *The Fernholz martingale Z generated by (17.8) satisfies*

$$Z_t \geq \frac{1}{2} \exp \left(\frac{1}{2} \sum_{j=1}^{J} [\mu^j]_t \right). \tag{17.10}$$

Proof It suffices to notice that $1/2 \leq \mathbf{S} \leq 1$. $\qquad\square$

By (17.7), we obtain the Fernholz martingale (17.9) by holding the fraction

$$\pi^j = \left(\frac{2 - \mu^j}{\mathbf{S}(\mu)} - 1 \right) \mu^j \tag{17.11}$$

of current capital in μ^j (Exercise 17.4). This portfolio is particularly tame (or admissible, in Fernholz's [139, section 3.3] terminology): it is long-only, it never loses more than 50% of its value relative to the market portfolio (by (17.10)), and it never invests more than three times what the market portfolio invests in any of the stocks (by (17.11) and $\mathbf{S} \in [1/2, 1]$).

From (17.10), we obtain

$$Z_{\tau_A} \geq \frac{1}{2} e^{A/2} \quad \text{w.i.e.,} \tag{17.12}$$

where A is any positive constant, τ_A is defined by (17.1), and $Z_\infty := \infty$. Figure 17.1 compares this with (17.4), the non-exponential lower bound that we obtained for a nonnegative continuous martingale based on a Stroock–Varadhan martingale. Quantitatively, (17.4) appears weaker: it does not feature the exponential growth rate in A. However, there is a range of A (roughly between 0.7 and 4.3), where the Stroock–Varadhan martingale X performs better.

Entropy-Weighted Portfolio

The archetypal diversity measure is the entropy function

$$\mathbf{S}(x) := - \sum_{j=1}^{J} x_j \ln x_j. \tag{17.13}$$

The components of the corresponding *entropy-weighted portfolio* can be obtained from (17.7):

$$\pi^j = -\frac{\mu^j \ln \mu^j}{S(\mu)}.$$

This portfolio, studied by Fernholz in [138, Theorem 4.1] and [139, Examples 3.1.2 and 3.4.3], is again long-only. Theorem 17.1 yields the following game-theoretic version of [139, Theorem 2.3.4].

Corollary 17.4   *The Fernholz martingale generated by (17.13) is*

$$\frac{S(\mu_t)}{S(\mu_0)} \exp\left(\frac{1}{2} \sum_{j=1}^{J} \int_0^t \frac{d[\mu^j]}{\mu^j S(\mu)} \right) \geq \frac{S(\mu_t)}{\ln J} \exp\left(\frac{1}{2 \ln J} \sum_{j=1}^{J} [\mu^j]_t \right). \tag{17.14}$$

Proof   For the equality in (17.14), it suffices to plug $D_{ij}S(x) = -\mathbf{1}_{i=j}/x_j$ into (17.5). For the inequality, use $S \leq \ln J$ and $\mu^j \in (0, 1)$.   □

It is standard in stochastic portfolio theory to assume both that the market does not become concentrated, or almost concentrated, in a single stock and that there is a minimal level of stock volatility; precise versions of these assumptions are called diversity and nondegeneracy, respectively. In place of nondegeneracy, we are using continued activity: $\sum_j [\mu^j]_\infty = \infty$. The inequality in Corollary 17.4 says that the entropy-weighted portfolio eventually outperforms the market exponentially if the market maintains both diversity and activity. If the market maintains its diversity, the first factor on the right-hand side of (17.14) will be bounded below, and if the market maintains a minimal level of activity, in the sense of $\sum_j [\mu^j]_t$ growing linearly in t, the second factor will grow exponentially fast. The tame portfolio based on the simple diversity measure (17.8), in contrast, relies on continued activity alone to guarantee exponential growth; we learned this from Corollary 17.3.

Diversity-Weighted Portfolios

Our third diversity measure, used by Fernholz for active portfolio management and going back at least to 1999 [137], is

$$D_p(x) := \left(\sum_{j=1}^{J} x_j^p \right)^{1/p}, \tag{17.15}$$

where the parameter p is in $(0, 1)$ (see also [139, Example 3.4.4] and [141, section 7]). The function D_p is sometimes called the *diversity measure with parameter p*. By

(17.7), the corresponding portfolio has positive components

$$\pi_t^j := \frac{(\mu_t^j)^p}{\sum_{k=1}^J (\mu_t^k)^p}.$$

This is the *diversity-weighted portfolio with parameter p.*

The following corollary is a game-theoretic version of [139, Example 3.4.4].

Corollary 17.5 *The Fernholz martingale generated by (17.15) is*

$$\frac{D_p(\mu_t)}{D_p(\mu_0)} \exp((1-p)\Gamma_t^\pi), \tag{17.16}$$

where

$$\Gamma^\pi = \frac{1}{2}\sum_{j=1}^J \int \pi^j \, d[\ln \mu^j] - \frac{1}{2}\left[\sum_{j=1}^J \int \pi^j \, d\ln \mu^j\right]. \tag{17.17}$$

The quantity Γ^π defined by (17.17) is the *excess growth term.*

Proof Evaluating $D_{ij}D_p$, we can rewrite (17.5) as

$$\frac{D_p(\mu_t)}{D_p(\mu_0)} \exp\left(\frac{1-p}{2}\sum_j \int_0^t (\mu^j)^{p-2}\left(\sum_k (\mu^k)^p\right)^{-1} d[\mu^j]\right.$$

$$\left. + \frac{p-1}{2}\sum_{i,j}\int_0^t (\mu^i\mu^j)^{p-1}\left(\sum_k (\mu^k)^p\right)^{-2} d[\mu^i,\mu^j]\right)$$

$$= \frac{D_p(\mu_t)}{D_p(\mu_0)} \exp\left(\frac{1-p}{2}\left(\sum_j\int_0^t \pi^j\frac{d[\mu^j]}{(\mu^j)^2} - \sum_{i,j}\int_0^t \pi^i\pi^j\frac{d[\mu^i,\mu^j]}{\mu^i\mu^j}\right)\right)$$

$$= \frac{D_p(\mu_t)}{D_p(\mu_0)} \exp\left(\frac{1-p}{2}\left(\sum_j\int_0^t \pi^j \, d[\ln \mu^j] - \sum_{i,j}\int_0^t \pi^i\pi^j \, d[\ln \mu^i, \ln \mu^j]\right)\right).$$

Combining Exercises 15.5 and 17.5 shows that the last expression is equal to (17.16). □

Corollary 17.5 immediately implies the following corollary

Corollary 17.6 *The Fernholz martingale Z generated by (17.15) satisfies*

$$Z_t \geq J^{-(1-p)/p}\exp((1-p)\Gamma_t^\pi). \tag{17.18}$$

Proof Since $D_p \in [1, J^{(1-p)/p}]$, we have

$$\frac{D_p(\mu_t)}{D_p(\mu_0)} \geq J^{-(1-p)/p};$$

plugging this into (17.16) gives (17.18). □

To see the intuition behind Corollaries 17.5 and 17.6, we will interpret twice the excess growth term (17.17),

$$2\Gamma_t^\pi = \sum_{j=1}^J \int_0^t \pi^j \, d[\ln \mu^j] - \left[\sum_{j=1}^J \int \pi^j \, d \ln \mu^j \right]_t$$

as a kind of variance. Let a sequence of partitions $T^1, T^2, \ldots$ finely cover all processes used in this chapter and set, for a given partition $T^k = (T_n^k)_{n=0}^\infty$,

$$\mu_n^j := \mu^j(T_n^k \wedge t), \qquad\qquad n = 0, 1, \ldots,$$

$$\Delta \ln \mu_n^j := \ln \mu_n^j - \ln \mu_{n-1}^j, \qquad\qquad n = 1, 2, \ldots,$$

$$\pi_n^j := \pi^j(T_n^k \wedge t), \qquad\qquad n = 0, 1, \ldots,$$

with the dependence on k suppressed in our notation. We can then regard

$$2\Gamma_t^{\pi,k} := \sum_{n=1}^\infty \sum_{j=1}^J \pi_{n-1}^j (\Delta \ln \mu_n^j)^2 - \sum_{n=1}^\infty \left(\sum_{j=1}^J \pi_{n-1}^j \Delta \ln \mu_n^j \right)^2 \qquad (17.19)$$

as the kth approximation to $2\Gamma_t^\pi$; it can be shown that $2\Gamma_t^{\pi,k} \to 2\Gamma_t^\pi$ ucie. Rewriting (17.19) as

$$2\Gamma_t^{\pi,k} = \sum_{n=1}^\infty \sum_{j=1}^J \pi_{n-1}^j \left(\Delta \ln \mu_n^j - \sum_{i=1}^J \pi_{n-1}^i \Delta \ln \mu_n^i \right)^2,$$

we can see that this expression is the cumulative variance of the logarithmic returns $\Delta \ln \mu_n^j$ over the time interval $[T_{n-1}^k \wedge t, T_n^k \wedge t]$ with respect to the "portfolio probability measure" $Q(\{j\}) := \pi_{n-1}^j$.

The remarks made above about the relation between Corollaries 17.2 and 17.3 are also applicable to Corollaries 17.5 and 17.6; the latter replaces the first factor in (17.16) by its lower bound. Corollary 17.5 decomposes the growth in the value of the diversity-weighted portfolio into two components, one related to the growth in the diversity $D_p(\mu)$ of the market weights and the other related to the accumulation of the diversity-weighted variance of the changes in market weights. Corollary 17.6

ignores the first component, which does not make it vacuous since D_p is bounded, always being between 1 (corresponding to a market concentrated in one stock) and $J^{(1-p)/p}$ (corresponding to a market with equal capitalizations of all J stocks).

17.3   OUTPERFORMING THE MARKET WITH DUBINS–SCHWARZ

As we mentioned in the introduction to this chapter, the game-theoretic Dubins–Schwarz theorem that we proved in Chapter 13 leads to another strategy for outperforming a persistently active market that has only a finite number of securities. Although this strategy has no hope of implementation because of the time it would require, its existence shows that the possibility of such outperformance is inherent in the Brownian character of security prices.

In Chapter 13, we used the properties of Brownian motion and that chapter's Dubins–Schwarz theorem to show that the upper probability is at most $1/\sqrt{c}$ that the market will choose a price path that begins at 1, stays positive, and achieves quadratic variation greater than c (Proposition 13.24). This suggests that there might exist a nonnegative supermartingale that starts at 1 and exceeds some function proportional to the square root of the price path's quadratic variation. The following proposition confirms this conjecture.

Proposition 17.7   *For any constant $A > 0$, there is a nonnegative supermartingale X such that $X_0 = 1$ and*

$$X_{\tau_A} \geq J^{-3/2} A^{1/2} \quad w.i.e., \tag{17.20}$$

where τ_A is the stopping time (17.1) and X_∞ is interpreted as ∞.

Proof   For each $j \in \{1, \dots, J\}$, we will construct a nonnegative supermartingale X^j satisfying $X_0^j = 1$ and

$$X_{\tau_j}^j \geq (A/J)^{1/2} \quad w.i.e., \tag{17.21}$$

where

$$\tau_j := \min\{t \mid [\mu^j]_t > A/J\}.$$

(In this case we can set X to the average of all J of X^j stopped at time τ_j; see Exercise 17.7.) According to [212, (2.6.2)], the probability that a Brownian motion starting from 1 (μ^j actually starts from $\mu_0^j < 1$) does not hit zero over the time period A/J is

$$1 - \sqrt{\frac{2}{\pi}} \int_{(J/A)^{1/2}}^{\infty} e^{-x^2/2} \, dx = \sqrt{\frac{2}{\pi}} \int_{0}^{(J/A)^{1/2}} e^{-x^2/2} \, dx \leq (J/A)^{1/2}.$$

In combination with the game-theoretic Dubins–Schwarz result (Theorem 13.9) applied to μ^j, this gives the existence of X satisfying (17.21).   □

The processes X in (17.4) and Z in (17.12) are nonnegative supermartingales in the sense of Chapter 14 (in fact, nonnegative continuous martingales). On the other hand, the process X in (17.20) is a nonnegative supermartingale in the sense of the more cautious definitions given in Chapter 13. This can be regarded as advantage of (17.20) over (17.4) and (17.12). But (17.20) is too weak to be interesting for practice. To guarantee doubling the strategy's money, it requires that the sum of the quadratic variations of the J securities reach $4J^3$. With volatilities usually observed for securities, say around 20% per annum, the sum of quadratic variations after T years will be less than $0.04\,JT$ (quite a bit less for a large J, since $\mu_j \ll 1$ for large j). So we would need more than $100J^2$ years to guarantee doubling: 400 years if there are two securities and 10 000 years if there are 10.

17.4 JEFFREYS'S LAW IN FINANCE

Consider two securities with price processes S^1 and S^2. The following corollary of Theorem 17.1 (in conjunction with Exercise 17.3) can be interpreted as a version of Jeffreys's law, which we discussed in Section 10.7.

Corollary 17.8 *For any two positive basic gales S^1 and S^2, the process*

$$\sqrt{S_t^1 S_t^2} \exp\left(\frac{1}{8}\left[\ln \frac{S^1}{S^2}\right]_t \right) \tag{17.22}$$

is a continuous gale.

Suppose we consider both S^1 and S^2 as efficient indexes. According to (17.22), if the indexes S^1 and S^2 disagree in the sense of the ratio S^1/S^2 being very volatile, we will be able to outperform greatly their geometric mean and, therefore, at least one of them. Therefore, efficient indexes must agree with each other.

Proof of Corollary 17.8 Choose one of the two securities, say S_2, as the numeraire. We get a martingale space with only one basic martingale, $\mu = S_1/S_2$, taking values in $(0, \infty)$. Setting $\mathbf{S}(\mu) := \mu^p$ for $p \in (0, 1)$ in (17.5) and using Exercise 17.3, we can see that

$$\frac{\mu_t^p}{\mu_0^p} \exp\left(-\frac{p(p-1)}{2} \int_0^t \mu^{-2}\, d[\mu] \right) = \left(\frac{\mu_t}{\mu_0} \right)^p \exp\left(\frac{p(1-p)}{2}[\ln \mu]_t \right)$$

is a continuous martingale in this martingale space. Setting $p := 1/2$ and multiplying by the numeraire S_2 gives us the continuous gale (17.22).   □

Corollary 17.8 sheds new light on the preceding chapter's CAPM: either a stock S will become very similar to the efficient index I or the index will outperform it greatly, in the sense that S_t/I_t will tend to 0. Quantitatively, for an efficient index I and constant $c \gg 1$, we expect

$$\sqrt{S_t I_t} \exp\left(\frac{1}{8}\left[\ln \frac{S}{I}\right]_t\right) \leq c I_t \tag{17.23}$$

(otherwise the positive gale on the left-hand side beats the index more than c-fold), and (17.23) can be rewritten as

$$\frac{S_t}{I_t} \leq c^2 \exp^2\left(-\frac{1}{8}[\ln \frac{S}{I}]_t\right).$$

17.5 EXERCISES

Exercise 17.1 Show that Theorem 14.19 can be applied to the function f to obtain (17.2) even when the open set $\mathrm{dom} f$ is different from $\mathbb{R}^J$. *Hint.* The Riemann–Stieltjes integral in (14.26) exists when the second partial derivatives are bounded, and this can be ensured by stopping when getting too close to the boundary of dom $\mathbf{S}$. □

Exercise 17.2 (Remark 11.1 in [141]) Suppose the positive C^2 function $\mathbf{S}$ is concave. Prove that the corresponding portfolio (17.7) is long-only. Show that the assumption that $\mathbf{S}$ is concave can be relaxed to the assumption that the Hessian $D_{ij}\mathbf{S}(x)$ has at most one positive eigenvalue for each $x \in$ dom $\mathbf{S}$ and, if a positive eigenvalue exists, the corresponding eigenvector is orthogonal to Δ^J. *Hint.* Show that the expression in the parentheses in (17.7) has the same sign as $\mathbf{S} + D_j\mathbf{S} - \sum_k \mu_k D_k \mathbf{S}$ and use the geometric interpretation of concavity (its graph lies below its tangent hyperplanes). □

The following exercise says that Theorem 17.1 continues to hold for numeraires different from $\sum_j S_j$.

Exercise 17.3 Show that Theorem 17.1 holds for any martingale space

$$(\Omega, \mathcal{F}, (\mathcal{F}_t), \{1, \ldots, J\}, (\mu^1, \ldots, \mu^J))$$

with positive $\mu^1, \ldots, \mu^J$. *Hint.* Check that the proof does not depend on the assumption that $\sum_j \mu^j = 1$ (in particular, Stroock–Varadhan martingales, given by the left-hand side of (17.2), remain continuous martingales when this assumption is omitted). □

Exercise 17.4 Show that (17.7) reduces to (17.11) for (17.8). *Hint.* Use the definition (17.8) of $\mathbf{S}$ more than once. □

Exercise 17.5 Show that the operation of covariation is additive in the sense of $[X + Y, Z] = [X, Z] + [Y, Z]$ w.i.e., for any continuous semimartingales X, Y, and Z. □

Exercise 17.6   Show that equivalent expressions for $2\Gamma^\pi$ (cf. (17.17)) are

$$2\Gamma^\pi_t = \int_0^t \sum_{j \in J} \pi^j_s \frac{d[\mu^j]_s}{(\mu^j_s)^2} - \left[\sum_{j \in J} \int \frac{\pi^j_s}{\mu^j} d\mu^j \right]_t$$

$$= \int_0^t \sum_j \pi^j_s \, d[\Lambda^j]_s - \left[\sum_j \int \pi^j_s \, d\Lambda^j_s \right]_t,$$

where $\Lambda^j := \mathcal{L}(\mu^j)$.   □

Exercise 17.7   Show that, for any nonnegative supermartingale X and stopping time τ, the process $X^\tau_t := X_{\tau \wedge t}$ is a nonnegative supermartingale. *Hint.* Use transfinite induction on the rank of X.   □

Exercise 17.8   (research project) Investigate theoretically the quantitative mechanism of leakage, as discussed in the next section. For example, state and prove a game-theoretic version of [139, Theorem 4.2.1]; remember that the game-theoretic notion of local time was discussed in Exercise 14.13 (and further in Section 14.11).   □

17.6   CONTEXT

Stochastic Portfolio Theory

The name *portfolio theory* goes back to Harry Markowitz's work in the early 1950s [265]. The key source on stochastic portfolio theory is Fernholz's 2002 book [139]. For the construction of long-only portfolios that outperform the capital-weighted market portfolio, see, e.g. [138, section 4], [139, chapters 2 and 3], and [141, section 7]. The articles [211] and [142] study the Stroock–Varadhan picture in depth and give numerous interesting examples, including (17.3).

Practical Difficulties

In this chapter, as in the previous chapters, we have assumed that our basic securities do not pay dividends. If they do, and securities with greater market weight pay greater dividends, then the portfolios constructed in this chapter will be disadvantaged. Fernholz has noted that this was indeed happening in the last decades of the twentieth century [139, Figure 7.4]. Fernholz has also pointed out the role of differential dividend rates in maintaining market diversity [138].

Another difficulty with investment strategies that emphasize small-cap stocks is these stocks' relative lack of liquidity. Aside from magnifying transaction costs, such strategies might be required to hold an impractically large portion of the capitalization of a very small stock. (This point has been made by Vassilios Papathanakos, quoted in [385, section 7].)

If we take our set of basic securities to be a fixed number of the largest stocks in a market (in the US market, this might be the 500 stocks in S&P 500), then the performance of all portfolios, and particularly those that emphasize small-cap stocks, will be affected by "leakage" [139, Example 4.3.5 and Figure 7.5]. Consider what happens when stock A is replaced by stock B in the S&P 500 because B's market share has overtaken A's. Since B has been performing better than A recently and is more expensive at the moment, replacing A by B can be interpreted as taking a loss, which will impact most portfolios that invest into the smaller of the 500 stocks more heavily. This effect will be exacerbated by the extensively documented fact that, because of the proliferation of tracker funds, newly added stocks are bought at a premium and deleted stocks are sold at a discount (see [69, Table 1] and Exercise 17.8).

Benchmark Approach to Finance

In Section 16.7 we mentioned the benchmark approach to finance [310], whose basic result, Platen's diversification theorem, can be interpreted as showing that diversified portfolios are good approximations to the growth-optimal portfolio. To a certain degree this contradicts the main finding of stochastic portfolio theory, whose basic results can be interpreted as showing that one can outperform the market. Our explanation is that those results of stochastic portfolio theory depend on the number of stocks in the market being finite and fixed, whereas the diversification theorem is asymptotic and lets the number of securities traded in the market go to infinity.

An Alternative Pathwise Approach

In contrast with the pathwise treatment of stochastic portfolio theory in this chapter, the interesting recent treatment in [330] begins by postulating a suitable "refining sequence of partitions" and the existence of a continuous covariation, in Föllmer's sense [146], between each pair of price paths with respect to this refining sequence of partitions. A possible extension to nonsmooth portfolio generating functions (as in [139, chapter 4]) would require further postulating the existence of local times, perhaps along the lines of [433].

Terminology and Notation

Though we use relatively standard mathematical terminology and notation in this book, our practices may depart in a few details from those of some readers.

Set Theory

We use the standard symbols $\mathbb{N} := \{1, 2, \ldots\}$ for the set of all natural numbers, $\mathbb{Z} := \{\ldots, -1, 0, 1, \ldots\}$ for the set of all integer numbers, $\mathbb{Q}$ for the set of all rational numbers, $\mathbb{R} := (-\infty, \infty)$ for the set of all real numbers, and $\mathbb{C}$ for the set of all complex numbers.

We write $|A|$ for the number of elements in the set A: if A is a finite set, $|A|$ in an integer; if A is an infinite set, $|A| := \infty$.

If A and B are two sets, $A \subseteq B$ stands for the inclusion of A in B allowing $A = B$, and $A \subset B$ means that $A \subseteq B$ and $A \neq B$. The set-theoretic difference of A and B is denoted $A \setminus B$ (it consists of the elements of A that are not in B). When working with a subset E of a set $\mathcal{Y}$, we write E^c for its complement: $E^c := \mathcal{Y} \setminus E$. We sometimes enclose a condition in curly braces to denote a subset; if f is a function on $\mathcal{Y}$, for example, then $\{f > 0\} = \{y \in \mathcal{Y} | f(y) > 0\}$.

We use $\implies$ for material implication: $(E \implies F) := E^c \cup F$. If E and F are thought of as events, then $E \implies F$ is the event that F holds if E holds.

Vector, topological, and metric spaces are always assumed to be nonempty.

Game-Theoretic Foundations for Probability and Finance, First Edition. Glenn Shafer and Vladimir Vovk.
© 2019 John Wiley & Sons, Inc. Published 2019 by John Wiley & Sons, Inc.

In Part IV we use the notion of a countable ordinal and transfinite induction (see, e.g. [128, section 1.3] and [105, section 0.8] for excellent summaries and [176] for details). In Section 13.5 we will need the notion of a well-order: a linear order is a *well-order* if every nonempty subset of its domain has a smallest element.

Sequences

We often consider finite and infinite sequences of elements of a set $\mathcal{Y}$, including the empty sequence $\square$. The set of all finite sequences is denoted by $\mathcal{Y}^*$, the set of all finite sequences of length n by $\mathcal{Y}^n$, and the set of all infinite sequences by $\mathcal{Y}^\infty$.

If ω is an infinite sequence, we write ω^n for its *prefix* of length n and say that ω *passes through* ω^n. If $\omega = y_1 y_2 \ldots$, then $\omega^n = y_1 \ldots y_n$ and $\omega^0 = \square$. We also sometimes write ω_n for y_n. If Ω is a set of infinite sequences and s is a prefix of at least one element of Ω, then we write Ω_s for the set of all elements of Ω of which s is a prefix.

We write $|s|$ for the length of a finite sequence s; thus $|y_1 \ldots y_n| = n$ and $|\square| = 0$. If s is a finite sequence, we write sy for the finite sequence obtained by concatenating s and $y \in \mathcal{Y}$; if $s = y_1 \ldots y_n$, then $sy := y_1 \ldots y_n y$. More generally, st is the concatenation of a finite sequence s and a finite or infinite sequence t.

If V is a function on $\mathcal{Y}^*$, $\omega \in \mathcal{Y}^\infty$, and $n \in \{0, 1, \ldots\}$, we often let $V_n(\omega)$ stand for $V(\omega^n)$. This makes V_n a function on $\mathcal{Y}^\infty$ that depends only on the first n elements of ω.

Numbers and Vectors

The precedence of the arithmetic operation is standard; the multiplication has higher precedence than division when the multiplication sign is omitted: a/bc is equivalent to $a/(bc)$.

We write $\overline{\mathbb{R}}$ for the extended real numbers: $\overline{\mathbb{R}} := [-\infty, \infty]$. We extend the operations of addition and multiplication in $\mathbb{R}$ to commutative and associative operations by supplementing the usual rules for addition and multiplication involving infinity with these two rules:

- $0 \times \infty = 0$ and $0 \times (-\infty) = 0$;
- $\infty - \infty = \infty$.

The expression $+(-\infty)$ is taken to mean $-\infty$ (so that $\infty + (-\infty) = \infty$), and the expression $-(-\infty)$ is taken to mean $+\infty$ (so that $(-\infty) - (-\infty) = \infty$). Notice that if $x_1 \in \mathbb{R}$, then $x_2 = x_1 + (x_2 - x_1)$, as we expect, but if $x_1 \in \{-\infty, \infty\}$, then $x_1 + (x_2 - x_1) = \infty$, regardless of the value of x_2.

The inner product between two vectors x and y in $\mathbb{R}^K$ is denoted

$$\langle x, y \rangle := \sum_{k=1}^{K} x_k y_k.$$

The (Euclidean) length of a vector $x \in \mathbb{R}^K$ is $|x| := \sqrt{\langle x, x \rangle}$.

We identify a vector x in $\mathbb{R}^K$ with a $K \times 1$ matrix, so that by default we regard vectors as column vectors; the corresponding row vector is x', where the prime stands for the transposition.

Analysis

Our notation for the natural logarithm is $\ln$; $\log$ stands for the logarithm whose base is given or irrelevant.

We write $f : A \to B$ or $f \in B^A$ to indicate that f is a mapping from the set A to the set B. The domain A of a mapping $f : A \to B$ is sometimes denoted $\operatorname{dom} f$. We reserve the name *function* for a mapping to the extended real numbers $\overline{\mathbb{R}}$ or some subset of $\overline{\mathbb{R}}$. A function f *dominates* a function g on the same domain if $f \geq g$. We reserve the name *functional* for a mapping that maps functions to $\overline{\mathbb{R}}$ or some subset of $\overline{\mathbb{R}}$ (therefore, functionals are a kind of functions). The restriction of a mapping $f : A \to B$ to a set $C \subseteq A$ is denoted $f|_C$; sometimes we use this notation even when C is not a subset of A, in which case $f|_C$ is understood to be $f|_{A \cap C}$.

We define functions and other mappings in various standard ways. We might, for example, define the function $f(x) = x^2$ by writing $f : x \in \mathbb{R} \mapsto x^2 \in \mathbb{R}$. Given a function of two or more variables, we sometimes use $\cdot$ to define a function of one variable. For example, when $f(x, y) = x^2 y$, we write $f(x, \cdot)$ for the function g given by $g(y) := x^2 y$.

For the indicator function of a subset E of a set $\mathcal{Y}$, we write $\mathbf{1}_E^{\mathcal{Y}}$:

$$\mathbf{1}_E^{\mathcal{Y}}(y) := \begin{cases} 1 & \text{if } y \in E, \\ 0 & \text{if } y \in \mathcal{Y} \setminus E. \end{cases}$$

When the set $\mathcal{Y}$ is clear from the context, we may omit it, writing simply $\mathbf{1}_E$. If A is a property, $\mathbf{1}_A$ stands for $\mathbf{1}_E$, where E is the set of all elements of $\mathcal{Y}$ that satisfy A. We sometimes write $\mathbf{1}$ for $\mathbf{1}_{\mathcal{Y}}^{\mathcal{Y}}$, the function on $\mathcal{Y}$ identically equal to 1.

Some clarifications may be in order with respect to our use of terminology in the context of the extended real numbers.

- When we say that a function f on $\mathcal{Y}$ is *bounded below*, we mean that $\inf_{y \in \mathcal{Y}} f(y) > -\infty$.

- When we say that a sequence of numbers in $\mathbb{R}$ or $\overline{\mathbb{R}}$ *converges in* $\mathbb{R}$, we mean it converges to a finite limit, and when we say that it *converges in* $\overline{\mathbb{R}}$, the limit is also allowed to be ∞ or $-\infty$. When we say that a sequence of functions defined on the same domain *converges*, in $\mathbb{R}$ or $\overline{\mathbb{R}}$, we mean that it converges to a function at each point of the domain, in the same sense.

We use the usual notation $C(\mathcal{Y})$ for the class of all continuous functions on $\mathcal{Y}$ equipped with the uniform metric

$$d(f, g) := \sup_{y \in \mathcal{Y}} |f(y) - g(y)|,$$

abbreviating $C([0, \infty))$ to $C[0, \infty)$. By the uniform convergence of functions (not necessarily continuous) we mean their convergence in uniform metric. The class of k times continuously differentiable functions on an open set in Euclidean space is C^k.

If f is a function on $\mathcal{Y}$, then when we say that f is *positive* or write $f > 0$, we mean that $f(y) > 0$ for all $y \in \mathcal{Y}$. When we say that f is *nonpositive* or write $f \leq 0$, we mean that $f(y) \leq 0$ for all $y \in \mathcal{Y}$. We use the terms *negative* and *nonnegative* similarly. More generally, when $c \in \overline{\mathbb{R}}$, $f > c$ means that $f(y) > c$ for all $y \in \mathcal{Y}$, and when f and g are two functions on $\mathcal{Y}$, $f > g$ means $f(y) > g(y)$ for all $y \in \mathcal{Y}$. We use the relations $<$, $\leq$, and $\geq$ analogously.

When we say that a sequence of numbers or functions is *(monotonically) increasing* or *(monotonically) decreasing*, we mean this in a wide sense: $f_1, f_2, \ldots$ is a monotonically increasing sequence of functions if $f_1 \leq f_2 \leq \cdots$.

We follow the usual convention that an empty sum is zero.

Topology

We usually use the terminology of and refer to [128]. For example, compact topological spaces are Hausdorff and neighborhoods are open by definition. As already mentioned, we only consider nonempty topological and metric spaces. The interior of a set A in a topological space is denoted $\text{Int}\, A$ and its closure is denoted $\overline{A}$. We usually drop "topological" in the expression "metrizable topological space."

Measure-Theoretic Probability and Statistics

We use standard definitions for measure-theoretic probability. A set of subsets of a nonempty set $\mathcal{Y}$ is a *σ-algebra* if it contains $\mathcal{Y}$ and is closed under complementation, countable union, and countable intersection. A *measurable space* is a pair $(\mathcal{Y}, \mathcal{F})$, where $\mathcal{Y}$ is a nonempty set and $\mathcal{F}$ is a σ-algebra on $\mathcal{Y}$. The elements of $\mathcal{F}$ are called *events*. A function f on $\mathcal{Y}$ is *measurable* if $\{y | f(y) \leq a\} \in \mathcal{F}$ for every real number a. We sometimes refer to $\mathcal{Y}$ as a measurable space, leaving the σ-algebra $\mathcal{F}$ implicit.

A *probability measure* P on a measurable space $\mathcal{Y}$ assigns a probability $P(E)$ to each element E of $\mathcal{F}$ (and is nonnegative and countably additive). A *probability space* is a triplet $(\mathcal{Y}, \mathcal{F}, P)$, where $(\mathcal{Y}, \mathcal{F})$ is a measurable space and P is a probability measure on $\mathcal{Y}$. A measurable real-valued (respectively extended real-valued) function on a probability space is called a *random variable* (respectively *extended random variable*). We write $P(f)$ for the expected value of a random variable or extended random variable f.[1] The expected value $P(f)$ exists when f is measurable and the integrals $\int f^+ \, dP$ and $\int f^- \, dP$ are not both infinite; in this case $P(f) := \int f^+ \, dP - \int f^- \, dP$. We follow the convention that any assertion about the value of $P(f)$ implicitly asserts its existence. If $y \in \mathcal{Y}$, we sometimes abbreviate $P(\{y\})$ to $P\{y\}$.

---

[1]Most authors on measure-theoretic probability use some form of the letter E for expected value with respect to a probability measure, but in this book we usually reserve E for game-theoretic expected value. The only exception is the first paragraph of Chapter 8, where we use $\mathbb{E}_n(X)$ to quote a theorem of Paul Lévy's.

A *filtration* in a measurable space $(\mathcal{Y}, \mathcal{F})$ is a sequence or family of increasingly larger σ-algebras contained in $\mathcal{F}$ and indexed by time, discrete or continuous. In the discrete-time case, it is a sequence $\mathcal{F}_0 \subseteq \mathcal{F}_1 \subseteq \cdots \subseteq \mathcal{F}$. A *filtered measurable space* is a measurable space with a filtration; a *filtered probability space* is a probability space with a filtration. In discrete time, a sequence $V_0, V_1, \ldots$ of functions on a filtered measurable space $(\mathcal{Y}, \mathcal{F}, (\mathcal{F}_n)_{n=0}^\infty)$ is a *process* (more fully, *adapted process*) if V_n is measurable with respect to $\mathcal{F}_n$ for $n = 0, 1, \ldots$; it is *predictable* if V_n is measurable with respect to $\mathcal{F}_{n-1}$ for $n = 1, 2, \ldots$. A function $\tau : \mathcal{Y} \to \{0, 1, \ldots\}$ is a *stopping time* if $\{y | \tau(y) = n\} \in \mathcal{F}_n$ for $n \in \{0, 1, \ldots\}$.

Suppose a random variable f in a probability space $(\mathcal{Y}, \mathcal{F}, P)$ has an expected value $P(f)$, and suppose $\mathcal{G}$ is a σ-algebra contained in $\mathcal{F}$. Then there exists at least one extended random variable g that is measurable with respect to $\mathcal{G}$ and satisfies $P(f\mathbf{1}_E) = P(g\mathbf{1}_E)$ for every E in $\mathcal{G}$. Any such extended random variable g is called a *version* of the *conditional expectation* of f with respect to $\mathcal{G}$. Any two versions of the same conditional expectation are equal except on a set of probability 0.

We write $P(f|\mathcal{G}) = g$ a.s. to indicate that g is a version of the conditional expectation of f with respect to $\mathcal{G}$. Conditional variance is handled similarly; we write $V_P(f|\mathcal{G})$ for $P((f - P(f|\mathcal{G}))^2|\mathcal{G})$.

A process $V_0, V_1, \ldots$ in a filtered probability space $(\mathcal{Y}, \mathcal{F}, (\mathcal{F}_n)_{n=0}^\infty, P)$ is a *measure-theoretic martingale* if $P(|V_n|) < \infty$ for all n and V_{n-1} is a version of the conditional expectation of V_n given $\mathcal{F}_{n-1}$:

$$P(V_n|\mathcal{F}_{n-1}) = V_{n-1} \quad \text{a.s.} \tag{T.1}$$

for $n \in \mathbb{N}$. If $P(V_n^-) < \infty$ for all n and (T.1) holds with $\leq$ in place of $=$, then $V_0, V_1, \ldots$ is a *measure-theoretic supermartingale*.

We write $\mathcal{P}(\mathcal{Y})$ for the set of all probability measures on $\mathcal{Y}$. If $\mathcal{Y}$ is a metrizable space, we equip $\mathcal{P}(\mathcal{Y})$ with the topology of weak convergence [39, Appendix III].

The *Gaussian distribution* on the real line $\mathbb{R}$ is denoted $\mathcal{N}_{\mu,\sigma^2}$ and parameterized by its mean μ and variance σ^2. We use z_δ to denote the upper δ-quantile of the standard Gaussian distribution: $\mathcal{N}_{0,1}([z_\delta, \infty)) = \delta$; we always assume $\delta \geq 0$ and set $z_\delta := -\infty$ for $\delta \geq 1$.

If P and Q are probability measures on the same measurable space $(\mathcal{Y}, \mathcal{F})$, $P \ll Q$ means that P is absolutely continuous with respect to Q ($Q(E) = 0$ implies $P(E) = 0$ for all $E \in \mathcal{F}$).

The standard Wiener probability measure on $C[0, \infty)$ is denoted $\mathcal{W}$. It is the probability measure for a Brownian motion B on the time interval $[0, \infty)$ that starts with the value 0 at time 0. In this book we only consider standard Brownian motion, in the sense that $B_t - B_0$ is distributed as $\mathcal{N}_{0,t}$ for all $t \in [0, \infty)$. By default, $B_0 = 0$, but sometimes we say that B is *started at* $c \in \mathbb{R}$, which means that we redefine $B := c + B$. The existence of $\mathcal{W}$ is not established by the results in this book, but see, e.g. [212, chapter 2] or [319, section I.1].

Game Theory

In the general theory of games, a strategy can be pure or mixed. A *pure strategy* for a player tells him how to move in each situation, whereas a *mixed strategy* specifies probability distributions for these moves. Mixed strategies can be advantageous when the players move simultaneously or not all earlier moves are known. But in perfect-information protocols, where players move in turn and see each other's moves, there is no advantage in the randomization introduced by mixed strategies. Since most of the game protocols we study in this book are perfect information protocols, we use *strategy* to mean a pure strategy, except when we qualify it explicitly as *mixed or randomized*.

List of Symbols

Some symbols (those amenable to alphabetic ordering) are also given in the index.

$A := B$: A equals B by definition

Set Theory

$\emptyset$: empty set

$\mathbb{N}$: the positive integers $\{1, 2, \ldots\}$

$\mathbb{Z}$: the integers

$\mathbb{Q}$: the rational numbers

$\mathbb{R}$: the real numbers $(-\infty, \infty)$

$\overline{\mathbb{R}}$: the extended real numbers $[-\infty, \infty]$

$\mathbb{C}$: the complex numbers

$|A|$: size of set A

$A \subseteq B$: A is a subset of B

$A \subset B$: $A \subseteq B$ and $A \neq B$

$A \cup B$: union of A and B

$A \cap B$: intersection of A and B

$A \setminus B$: set-theoretic difference

E^c: complement of E

$E \Rightarrow F$: material implication, $E^c \cup F$

Sequences

$\square$: empty sequence

$\mathcal{Y}^*$: all finite sequences of elements of $\mathcal{Y}$

$\mathcal{Y}^n$: all sequences of elements of $\mathcal{Y}$ of length n

$\mathcal{Y}^\infty$: all infinite sequences of elements of $\mathcal{Y}$

ω_n: the nth element of infinite sequence ω

ω^n: prefix of length n of infinite sequence ω

$|s|$: length of finite sequence s

st: concatenation of s with t

$\bar{y}_N$: average of sequence $y_1 \ldots y_N$ of numbers

y_N^*: $\max(y_0, \ldots, y_N)$

ΔA_n: increment, $\Delta A_n := A_n - A_{n-1}$

Game-Theoretic Foundations for Probability and Finance, First Edition. Glenn Shafer and Vladimir Vovk.
© 2019 John Wiley & Sons, Inc. Published 2019 by John Wiley & Sons, Inc.

425

Numbers, Vectors, and Matrices

$a \wedge b$: maximum $\max(a, b)$ of numbers a and b

$a \wedge b$: minimum $\min(a, b)$ of numbers a and b

a^+: nonnegative part $a \vee 0$ of a

a^-: nonpositive part $(-a) \vee 0$ of a

$\lfloor a \rfloor$: greatest integer less than or equal to a

$\lceil a \rceil$: least integer greater than or equal to a

$\langle x, y \rangle$: inner product of x and y

Δ^J: interior of standard simplex in $\mathbb{R}^J$

A': matrix A transposed

Analysis

B^A: all mappings from A to B

$\mathrm{dom} f$: domain of a mapping f

$\mathbf{1}_E$: indicator function of set E

$\ln$: natural logarithm

$\log$: logarithm (when the base is given or irrelevant)

D_j: partial derivative in the jth variable

D_{ij}: second partial derivative

$\mathrm{var}_p(f)$: p-variation of f, $p \in (0, \infty)$

$\mathrm{var}_\phi(f)$: ϕ-variation of f, ϕ being function

$\mathrm{vi}(f)$: variation index of f

$C(\mathcal{Y})$: class of continuous functions on $\mathcal{Y}$ with uniform metric

$\|f\|_\infty$: uniform norm $\sup_y |f(y)|$ of f

C^k: class of k times continuously differentiable functions on open set

k_z: representer of z

$\|\mathcal{H}\|_\infty$: supremum of representers' norms in $\mathcal{H}$

Topology

$\overline{A}$: closure of A

$\mathrm{Int}\, A$: interior of A

∂A: boundary of A

$\dim \mathcal{X}$: dimension of $\mathcal{X}$

Measure-Theoretic Probability and Statistics

$P(E)$: probability measure P's probability for event E

$P(X)$: probability measure P's expected value for random variable X

$\mathrm{V}_P(X)$: probability measure P's variance for X

$P(X|\mathcal{F})$: P's conditional expectation for X given σ-algebra $\mathcal{F}$

$\mathrm{V}_P(X|\mathcal{F})$: P's conditional variance for X given σ-algebra $\mathcal{F}$

$\mathcal{P}(\mathcal{Y})$: all probability measures on $\mathcal{Y}$

$P \ll Q$: P is absolutely continuous with respect to Q

$\rho^2(P, Q)$: Hellinger distance between P and Q

$\mathcal{N}_{\mu, \sigma^2}$: Gaussian distribution with mean μ and variance σ^2

$\mathcal{W}$: standard Wiener probability measure on $C[0, \infty)$

z_δ: upper δ-quantile of $\mathcal{N}_{0,1}$

$\mathcal{F}_\tau$: all events preceding τ

Game Theory

$G(Z)$: game of removal

Game-Theoretic Probability

$\mathbb{S}$: situation space (all situations)

Ω: sample space (all paths or abstract)

Ω_s: paths that pass through s

$D \cdot S$: martingale transform or Itô integral

$\mathcal{K}_n$: Skeptic's capital at the end of round n

$\mathcal{K}^\phi$: strategy ϕ's capital process

$\mathcal{K}^{G,c}$: simple capital process in canonical space

ϕ^{stake}: ϕ's stake

$\phi^{\text{bet}}, \phi^{\text{M}}, \phi^{\text{V}}$: ϕ's bets

T: class of supermartingales

L: class of supermartingales that converge in $\overline{\mathbb{R}}$

M: class of martingales

$\overline{\mathbb{E}}$: global upper expectation

$\overline{\mathbb{E}}_s$: global upper expectation in situation s

$\overline{\mathbb{E}}_n$: supermartingale of upper expected values

$\underline{\underline{\mathbb{E}}}$: global lower expectation

$\overline{\mathbf{E}}$ and $\underline{\mathbf{E}}$: (local) upper and lower expectation

$\overline{\mathbb{P}}$ and $\underline{\mathbb{P}}$: global upper and lower probability

$\overline{\mathbf{P}}$ and $\underline{\mathbf{P}}$: (local) upper and lower probability

$\mathbb{E}$ and $\mathbb{P}$: global expectation and probability

E, **P**, and **V**: local expectation, probability, and variance

$\overline{\mathbb{E}}^I$ and $\overline{\mathbb{P}}^I$: I-upper expectation and probability

G: offer

$\overline{\mathbb{E}}_{\mathbf{G}}$: offer's upper expectation

$\mathbf{G}_{\overline{\mathbb{E}}}$: upper expectation's offer

$\mathcal{I}$: all LT-superinvariant events

∂: cemetery state

dom X: effective domain of X

Σ_X: frontier of dom X

$[X, Y]$: covariation of X and Y

$[X]$: quadratic variation of X

$\mathcal{E}(X)$: Doléans exponential of X

$\mathcal{L}(X)$: Doléans logarithm of X

Λ^X: alternative notation for Doléans logarithm of X

M^X: cumulative relative growth of X

$\Sigma^{X,Y}$: relative covariation of X and Y

Σ^X: relative quadratic variation of X

$\Sigma^{S \triangle I}$: $[\mathsf{M}^S - \mathsf{M}^I]$

References

Some of the authors listed here used different forms of their name over the course of their career. We have generally used their later choices. In the case of Russian authors, we use a modern transliteration. The acronym "GTPN" stands for Working Paper N at the game-theoretic probability project; its web site is http://www .probabilityandfinance.com/.

1. Aalen, O.O., Andersen, P.K., Borgan, Ø., Gill, R.D., and Keiding, N. (2009). History of applications of martingales in survival analysis. *Electronic Journal for History of Probability and Statistics* 5 (1).

2. Abramovich, S. and Nikitin, Y.Y. (2017). On the probability of co-primality of two natural numbers chosen at random: from Euler identity to Haar measure on the ring of adeles. *Bernoulli News* 24 (1): 7–13.

3. Abramowitz, M. and Stegun I.A. (eds) (1964). *Handbook of Mathematical Functions: With Formulas, Graphs, and Mathematical Tables*. Washington, DC: US Government Printing Office.

4. Adams, W.J. (1974). *The Life and Times of the Central Limit Theorem*. New York: Kaedmon.

5. Adams, R.A. and Fournier, J.J.F. (2003). *Sobolev Spaces*, 2e. Amsterdam: Academic Press.

6. Agarwal, R.P., Meehan, M., and O'Regan, D. (2001). *Fixed Point Theory and Applications*. Cambridge: Cambridge University Press.

7. Aizerman, M.A., Braverman, E.M., and Rozonoer, L.I. (1970). *Метод потенциальных функций в теории обучения машин.* Moscow: Наука.

8. Aldrich, J. (2005). Fisher and regression. *Statistical Science* 20 (4): 401–407.

9. Ananova, A. and Cont, R. (2017). Pathwise integration with respect to paths of finite quadratic variation. *Journal de Mathématiques Pures et Appliquées* 107 (6): 773–757.

10. Andersen, E.S. and Jessen, B. (1948). On the introduction of measures in infinite product sets. *Kongelige Danske Videnskabernes Selskab, Matematisk-fysiske Meddelelser* 25 (4).

11. Arbuthnot, J. (1710). An argument for divine providence, taken from the constant regularity observ'd in the births of both sexes. *Philosophical Transactions of the Royal Society of London* 27: 186–190.

12. Aronszajn, N. (1943). La théorie générale des noyaux reproduisants et ses applications, première partie. *Proceedings of the Cambridge Philosophical Society* 39 (3): 133–153 (additional note: p. 205). The second part of this article is [13].

13. Aronszajn, N. (1950). Theory of reproducing kernels. *Transactions of the American Mathematical Society* 68 (3): 337–404.

14. Artzner, P., Delbaen, F., Eber, J.-M., and Heath, D. (1999). Coherent measures of risk. *Mathematical Finance* 9 (3): 203–228.

15. Augustin, T., Coolen, F.P.A., de Cooman, G., and Troffaes, M.C.M. (2014). *Introduction to Imprecise Probabilities.* Chichester: Wiley.

16. Azéma, J. and Yor, M. (1979). Une solution simple au problème de Skorokhod. In: *Séminaire de Probabilités XIII*, Lecture Notes in Mathematics, vol. 721 (eds. C. Dellacherie, P.-A. Meyer, and M. Weil), 90–115. Berlin: Springer.

17. Azuma, K. (1967). Weighted sums of certain dependent random variables. *Tohoku Mathematical Journal* 68 (3): 357–367.

18. Bachelier, L. (1900). Théorie de la spéculation. *Annales scientifiques de l'École Normale Supérieure*, 3ᵉ série 17: 21–86. This was Bachelier's doctoral dissertation. An English translation appears in [[81], pp. 17–78]. Another is in [20].

19. Bachelier, L. (1912). *Calcul des probabilités.* Paris: Gauthier-Villars.

20. Bachelier, L. (2006). *Louis Bachelier's Theory of Speculation: The Origins of Modern Finance.* Princeton University Press. Translation with commentary by Mark Davis and Alison Etheridge.

21. Banach, S. and Tarski, A. (1924). Sur la décomposition des ensembles de points en parties respectivement congruentes. *Fundamenta Mathematicae* 6: 244–277.

22. Barbut, M., Locker, B., and Mazliak, L. (2014). *Paul Lévy and Maurice Frechet: 50 Years of Correspondence in 107 Letters.* London: Springer.

23. Bártfai, P. and Révész, P. (1967). On a zero-one law. *Zeitschrift für Wahrscheinlichkeitstheorie und verwandte Gebiete* 7 (1): 43–47.

24. Bartl, D., Kupper, M., and Neufeld, A. (2018). Stochastic Integration and Differential Equations for Typical Paths. Technical Report arXiv:1805.09652 [math.PR], arXiv.org e-Print archive.

25. Bartl, D., Kupper, M., Prömel, D.J., and Tangpi, L. (2017). Duality for Pathwise Superhedging in Continuous Time. Technical Report arXiv:1705.02933 [q-fin.MF], arXiv.org e-Print archive.

26. Bellhouse, D.R. (2011). *Abraham De Moivre: Setting the Stage for Classical Probability and Its Applications.* Boca Raton, FL: CRC Press.

27. Bellhouse, D.R. (2017). *Leases for Lives: Life Contingent Contracts and the Emergence of Actuarial Science in Eighteenth-Century England.* Cambridge: Cambridge University Press.

28. Bennett, G. (1962). Probability inequalities for the sum of independent random variables. *Journal of the American Statistical Association* 57: 33–45.

29. Bercu, B. and Touati, A. (2008). Exponential inequalities for self-normalized martingales with applications. *Annals of Applied Probability* 18 (5): 1848–1869.

30. Berger, J.O. and Delampady, M. (1987). Testing precise hypotheses (with discussion). *Statistical Science* 2 (3): 317–352.

31. Berlinet, A. and Thomas-Agnan, C. (2004). *Reproducing Kernel Hilbert Spaces in Probability and Statistics.* Boston, MA: Kluwer.

32. Bernoulli, D. (1735). Recherches physiques et astronomiques sur le problème proposé pour la seconde fois par l'Académie Royale des Sciences de Paris: Quelle est la cause physique de l'inclinaison des plans des orbites des planetes par rapport au plan de l'équateur de la révolution du soleil autour de son axe; Et d'où vient que les inclinaisons de ces orbites sont différentes entre elles. *Recueil des pièces qui ont remporté les prix de l'Académie Royale des Sciences* 3: 95–122. The original Latin text occupies pages 125–144 of this volume; according to the author's preface, his French translation on pages 95–122 contains small additions and clarifications.

33. Bernoulli, J. (1713). *Ars Conjectandi.* Basel: Thurnisius.

34. Bernoulli, J. (2006). *The Art of Conjecturing, together with Letter to a Friend on Sets in Court Tennis.* Baltimore, MD: Johns Hopkins University Press. Translation, with introduction and notes, by Edith Sylla of [33].

35. Bernstein, S.N. (1924). Об одном видоизменении неравенства Чебышева (On one modification of Chebyshev's inequality, in Russian, with French summary). *Ученые записки Н. И. кафедр Украины, Отдел математический*, 1.

36. Bernstein, S.N. (1927). *Теория вероятностей* (Theory of Probability, in Russian). Moscow: Государственное издательство (State Publisher).

37. Bick, A. (1995). Quadratic-variation-based dynamic strategies. *Management Science* 41 (4): 722–732.

38. Bienvenu, L., Shafer, G., and Shen, A. (2009). On the history of martingales in the study of randomness. *Electronic Journal for History of Probability and Statistics* 5 (1).

39. Billingsley, P. (1968). *Convergence of Probability Measures.* New York: Wiley. The second edition was published in 1999, but most of our references are to Appendix III of the first edition, which essentially disappeared in the second.

40. Billingsley, P. (1995). *Probability and Measure*, 3e. New York: Wiley. Previous editions appeared in 1979 and 1986.

41. Black, F. (1972). Capital market equilibrium with restricted borrowing. *Journal of Business* 45 (3): 444–455.

42. Black, F., Jensen, M.C., and Scholes, M. (1972). The capital asset pricing model: some empirical tests. In: *Studies in the Theory of Capital Markets* (ed. M.C. Jensen). New York: Praeger.

43. Black, F. and Scholes, M. (1973). The pricing of options and corporate liabilities. *Journal of Political Economy* 81 (3): 637–654.

44. Borel, E. (1909). Les probabilités dénombrables et leurs applications arithmétiques. *Rendiconti del Circolo Matematico di Palermo* 27: 247–270. Reprinted in [45], Volume 2, pp. 1055–1079.

45. Borel, E. (1972). *Œuvres de Émile Borel*, vol. 4. Paris: Éditions du Centre National de la Recherche Scientifique.

46. Boucheron, S., Lugosi, G., and Massart, P. (2013). *Concentration Inequalities: A Nonasymptotic Theory of Independence*. Oxford: Oxford University Press.

47. Bourbaki, N. (2004). *Elements of Mathematics. Functions of a Real Variable: Elementary Theory*. Berlin: Springer. French original published in 1976.

48. Bourbaki, N. (2004). *Elements of Mathematics. Integration*, in 2 volumes. Berlin: Springer. French originals published in 1952–1969.

49. Breiman, L. (1960). Investment policies for expanding businesses optimal in a long-run sense. *Naval Research Logistics* 7 (4): 647–651.

50. Breiman, L. (1961). Optimal gambling systems for favorable games. In: *Proceedings of the 4th Berkeley Symposium on Mathematical Statistics and Probability*, Contributions to Probability Theory vol. 2 (ed. J. Neyman), 65–78. Berkeley, CA: University of California Press.

51. Breiman, L. (1968). *Probability*. Reading, MA: Addison-Wesley. 2e: Philadelphia, PA: SIAM, 1992.

52. Bricmont, J. (2016). *Making Sense of Quantum Mechanics*. Springer.

53. Brouwer, L.E.J. (1918). Begründung der Mengenlehre unabhängig vom logischen Satz vom ausgeschlossenen Dritte. Erster Teil. Allgemeine Mengelehre. *Koninklijke Nederlandse Akademie van Wetenschappen Verhandelingen* 5: 1–43.

54. Bru, M.-F. and Bru, B. (2018). *Les jeux de l'infini et du hasard*, vol. 2. Besançon, France: Presses universitaires de Franche-Comté.

55. Bru, B. and Eid, S. (2009). Jessen's theorem and Lévy's lemma: a correspondence. *Electronic Journal for History of Probability and Statistics* 5 (1).

56. Bruneau, M. (1975). Mouvement brownien et F-variation. *Journal de Mathématiques Pures et Appliquées* 54 (1): 11–25.

57. Bruneau, M. (1979). Sur la p-variation d'une surmartingale continue. *Séminaire de probabilités de Strasbourg* 13: 227–232.

58. Burdzy, K. (1990). On nonincrease of Brownian motion. *Annals of Probability* 18 (3): 978–980.

59. Burkholder, D.L. (1966). Martingale transforms. *Annals of Mathematical Statistics* 37 (6): 1494–1504.

60. Cameron, R.H. and Martin, W.T. (1944). Transformations of Wiener integrals under translations. *Annals of Mathematics* 45 (2): 386–396.

61. Cannon, J.R. (1984). *The One-Dimensional Heat Equation*. Reading, MA: Addison-Wesley.

62. Cantelli, F. (1916). Sulla legge dei grandi numeri. *Memorie, Accademia dei Lincei, V* 11: 329–349.

63. Carathéodory, C. (1914). Über das lineare Mass von Punktmengen — eine Verallgemeinerung des Längenbegriffs. *Nachrichten von der Gesellschaft der Wissenschaften zu Göttingen, Mathematisch-Physikalische Klasse* 1914: 404–426.

64. Cesa-Bianchi, N. and Lugosi, G. (2006). *Prediction, Learning, and Games*. Cambridge: Cambridge University Press.

65. Cesàro, E. (1881). Questions proposées 75. *Mathesis* 1: 184.

66. Cesàro, E. (1883). Solutions de questions proposées 75. *Mathesis* 3: 224–225.

67. Chebyshev, P.L. (1846). Démonstration élementaire d'une proposition générale de la théorie des probabilités. *Journal für die reine und angewandte Mathematik* 33: 259–267.

68. Chebyshev, P.L. (1867). Des valeurs moyennes (translation from the Russian). *Journal de Mathématiques Pures et Appliquées* 12: 177–184.

69. Chen, H., Noronha, G., and Singal, V. (2006). Index changes and losses to index fund investors. *Financial Analysts Journal* 62 (4): 31–47.

70. Chernov, A., Shen, A., Vereshchagin, N., and Vovk, V. (2008). On-line probability, complexity and randomness. In: *Proceedings of the 19th International Conference on Algorithmic Learning Theory*, Lecture Notes in Artificial Intelligence, vol. 5254, 138–153. Berlin: Springer.

71. Chernov, A. and Vovk, V. (2009). Prediction with expert evaluators' advice. In: *Proceedings of the 20th International Conference on Algorithmic Learning Theory*, Lecture Notes in Artificial Intelligence, vol. 5809, 8–22. Berlin: Springer. First posted as GTP30 February 2009. Full version: arXiv:0902.4127 [cs.LG].

72. Chernov, A. and Vovk, V. (2014). Prediction with advice of an unknown number of experts. In: *Proceedings of the 26th Conference on Uncertainty in Artificial Intelligence*, August 2014. arXiv:1408.2040 [cs.LG].

73. Cherny, A. (2006). Some particular problems of martingale theory. In: Kabanov et al. [205], 109–124.

74. Choquet, G. (1954). Theory of capacities. *Annales de l'Institut Fourier* 5: 131–295.

75. Choquet, G. (1959). Forme abstraite du théorème de capacitabilité. *Annales de l'Institut Fourier* 9: 83–89.

76. Choquet, G. (1969). *Integration and Topological Vector Spaces*, Lectures on Analysis, vol. 1. New York: Benjamin.

77. Chow, Y.-S. (1965). Local convergence of martingales and the law of large numbers. *Annals of Mathematical Statistics* 36 (2): 552–558.

78. Chow, Y.-S. and Teicher, H. (1997). *Probability Theory: Independence, Interchangeability, Martingales*, 3e. New York: Springer.

79. Condorcet (1805). *Elémens du calcul des probabilités, et son application aux jeux de hasard, à la loterie, et aux jugemens des hommes*. Paris: Royez.

80. Cooke, W.E. (1906). Forecasts and verifications in Western Australia. *Monthly Weather Review* 34 (1): 23–24.

81. Cootner, P.H. (ed.) (1964). *The Random Character of Stock Market Prices*. Cambridge, MA: MIT Press.

82. Cornfeld, I.P., Fomin, S.V., and Sinai, Y.G. (1982). *Ergodic Theory*. New York: Springer. Russian original: Эргодическая теория. Nauka, Moscow, 1980.

83. Cournot, A.A. (1843). *Exposition de la théorie des chances et des probabilités*. Paris: Hachette. Reprinted in 1984 as Volume I (ed. B. Bru) of [84].

84. Cournot, A.A. (1973–2010). *Œuvres complètes*. Paris and Besançon: Vrin and Presses Universitaires de Franche-Comté. The volumes are numbered I through XI, but VI and XI are double volumes.

85. Cover, T.M. (1991). Universal portfolios. *Mathematical Finance* 1 (1): 1–29.

86. Cox, D.R. (1972). Regression models and life tables (with discussion). *Journal of the Royal Statistical Society, Series B* 34 (2): 187–220.

87. Cox, D.R. (1975). Partial likelihood. *Biometrika* 62 (2): 269–276.

88. Cox, D.R. and Oakes, D. (1984). *Analysis of Survival Data*. London: Chapman and Hall.

89. de Cooman, G. (2008). A one-sided game-theoretic version of Hoeffding's inequality, Unpublished manuscript.

90. Cramér, H. (1946). *Mathematical Methods in Statistics*. Princeton, NJ: Princeton University Press.

91. Crépel, P. (2009). Jean Ville's recollections about martingales (1984–85). *Electronic Journal for History of Probability and Statistics* 5 (1).

92. Dambis, K.E. (1965). On the decomposition of continuous submartingales. *Theory of Probability and Its Applications* 10 (3): 401–410.

93. Daston, L. (1988). *Classical Probability in the Enlightenment*. Princeton, NJ: Princeton University Press.

94. Davis, M. (1964). Infinite games with perfect information. In: *Advances in Game Theory* (eds. M. Dresher, L.S. Shapley, and A.W. Tucker), 85–101. Princeton, NJ: Princeton University Press.

95. Davis, M., Obłój, J., and Raval, V. (2014). Arbitrage bounds for prices of weighted variance swaps. *Mathematical Finance* 24 (4): 821–854.

96. Davis, M., Obłój, J., and Siorpaes, P. (2018). Pathwise stochastic calculus with local times. *Annales de l'Institut Henri Poincaré (B) Probabilités et Statistiques* 54 (1): 1–21.

97. Dawid, A.P. (1984). Statistical theory: the prequential approach (with discussion). *Journal of the Royal Statistical Society, Series A* 147 (2): 278–292.

98. Dawid, A.P. (1985). Calibration-based empirical probability (with discussion). *Annals of Statistics* 13 (4): 1251–1285.

99. Dawid, A.P. (1985). Self-calibrating priors do not exist: comment. *Journal of the American Statistical Association* 80: 340–341.

100. Dawid, A.P., de Rooij, S., Grunwald, P. et al. (2011). Probability-Free Pricing of Adjusted American Lookbacks. Technical Report arXiv:1108.4113 [q-fin.PR], arXiv.org e-Print archive, August 2011. First posted as GTP37 August 2011.

101. Dawid, A.P., de Rooij, S., Shafer, G. et al. (2011). Insuring against loss of evidence in game-theoretic probability. *Statistics and Probability Letters* 81 (1): 157–162. First posted as GTP34 May 2010 and also available on arXiv.

102. Dawid, A.P. and Vovk, V. (1999). Prequential probability: principles and properties. *Bernoulli* 5 (1): 125–162.

103. Delbaen, F. and Schachermayer, W. (1994). The fundamental theorem of asset pricing. *Mathematische Annalen* 300 (1): 463–520.

104. Dellacherie, C. (1972). *Ensembles analytiques, capacités, mesures de Hausdorff*, Lecture Notes in Mathematics, vol. 295. Berlin: Springer.

105. Dellacherie, C. and Meyer, P.-A. (1978). *Probabilities and Potential*, Chapters I–IV. Amsterdam: North-Holland. French original: 1975; reprinted in 2008.

106. Dellacherie, C. and Meyer, P.-A. (1988). *Probabilities and Potential. C: Potential Theory for Discrete and Continuous Semigroups*, Chapters IX–XIII. Amsterdam: North-Holland. French original: 1983; reprinted in 2008.

107. Demidov, S.S. and Levshin, B.V. (eds) (2016). *The Case of Academician Nikolai Nikolaevich Luzin*. Providence, RI: American Mathematical Society. Russian original published in 1999.

108. De Moivre, A. (1733). Approximatio ad summam terminorum binomii $(a + b)^n$ in seriem expansi. Self-published pamphlet, 7 pages. See, e.g., [[4], p. 12] for a description.

109. De Moivre, A. (1756). *The Doctrine of Chances: Or, A Method of Calculating the Probabilities of Events in Play*, 3e. London: Millar. Previous editions appeared in 1718 and 1738.

110. DeSantis, A., Markowsky, G., and Wegman, M.N. (1988). Learning probabilistic prediction functions. In: *Proceedings of the 29th Annual IEEE Symposium on Foundations of Computer Science*, 110–119. Los Alamitos, CA: IEEE Computer Society.

111. Devroye, L., Györfi, L., and Lugosi, G. (1996). *A Probabilistic Theory of Pattern Recognition*. New York: Springer.

112. Dolinsky, Y. and Soner, H.M. (2014). Robust hedging and martingale optimal transport in continuous time. *Probability Theory and Related Fields* 160 (1–2): 391–427.

113. Dolinsky, Y. and Soner, H.M. (2015). Martingale optimal transport in the Skorokhod space. *Stochastic Processes and their Applications* 125 (10): 3893–3931. Corrigendum: 126 (1): 312–313, 2016.

114. Doob, J.L. (1938). Stochastic processes with an integral-valued parameter. *Transactions of the American Mathematical Society* 44 (1): 87–150.

115. Doob, J.L. (1940). Regularity properties of certain families of chance variables. *Transactions of the American Mathematical Society* 47 (3): 455–486.

116. Doob, J.L. (1953). *Stochastic Processes.* New York: Wiley.

117. Doob, J.L. (1961). Notes on martingale theory. In: *Proceedings of the 4th Berkeley Symposium on Mathematical Statistics and Probability*, Contributions to Probability Theory, vol. 2 (ed. J. Neyman), 95–102. Berkeley, CA: University of California Press.

118. Doob, J.L. (1984). *Classical Potential Theory and Its Probabilistic Counterpart.* New York: Springer.

119. Dubins, L.E. and Savage, L.J. (2014). *How to Gamble if You Must: Inequalities for Stochastic Processes.* New York: Dover. The first edition was published by McGraw-Hill in 1965.

120. Dubins, L.E. and Schwarz, G. (1965). On continuous martingales. *Proceedings of the National Academy of Sciences of the United States of America* 53 (3): 913–916.

121. Dudley, R.M. (2002). *Real Analysis and Probability*, revised edition. Cambridge: Cambridge University Press.

122. Dvoretzky, A., Erdős, P., and Kakutani, S. (1961). Nonincrease everywhere of the Brownian motion process. In: *Proceedings of the 4th Berkeley Symposium on Mathematical Statistics and Probability*, Contributions to Probability Theory, vol. 2, 103–116. Berkeley, CA: University of California Press.

123. Edwards, A.W.F. (1987). *Pascal's Arithmetical Triangle.* London: Charles Griffin.

124. El-Yaniv, R., Fiat, A., Karp, R.M., and Turpin, G. (2001). Optimal search and one-way trading online algorithms. *Algorithmica* 30 (1): 101–139.

125. Elliott, P.D.T.A. (1979). *Probabilistic Number Theory I: Mean-Value Theorems.* New York: Springer.

126. Elliott, P.D.T.A. (1980). *Probabilistic Number Theory II: Central Limit Theorems.* New York: Springer.

127. Elliott, R.J. and Kopp, P.E. (2005). *Mathematics of Financial Markets*, 2e. New York: Springer.

128. Engelking, R. (1989). *General Topology*, 2e. Berlin: Heldermann.

129. Erdős, P. (1942). On the law of the iterated logarithm. *Annals of Mathematics 2* 43 (3): 419–436.

130. Ethier, S.N. (2010). *The Doctrine of Chances: Probabilistic Aspects of Gambling.* Heidelberg: Springer.

131. Fama, E.F. (1965). Random walks in stock market prices. *Financial Analysts Journal* 21 (5): 55–59.

132. Fama, E.F. (1970). Efficient capital markets: a review of theory and empirical work. *Journal of Finance* 25 (2): 383–417.

133. Fama, E.F. (2014). Two pillars of asset pricing. *American Economic Review* 104 (6): 1467–1485.

134. Feller, W. (1943). The general form of the so-called law of the iterated logarithm. *Transactions of the American Mathematical Society* 54 (3): 373–402.

135. Feller, W. (1968). *An Introduction to Probability Theory and Its Applications*, vol. 1, 3e. New York: Wiley.

136. Feller, W. (1971). *An Introduction to Probability Theory and Its Applications*, vol. 2, 2e. New York: Wiley.

137. Fernholz, E.R. (1999). Diversity-weighted equity indexes. *ETF Journal of Indexes*. http://www.etf.com/publications/journalofindexes/joi-articles/1074.html.

138. Fernholz, E.R. (1999). On the diversity of equity markets. *Journal of Mathematical Economics* 31 (3): 393–417.

139. Fernholz, E.R. (2002). *Stochastic Portfolio Theory*. New York: Springer.

140. Fernholz, E.R. (2018). Numeraire Markets. Technical Report arXiv:1801.07309 [q-fin.MF], arXiv.org e-Print archive.

141. Fernholz, E.R. and Karatzas, I. (2009). Stochastic portfolio theory: an overview. In: *Handbook of Numerical Analysis*, vol 15, *Mathematical Modelling and Numerical Methods in Finance*, 89–167. Amsterdam: North-Holland.

142. Fernholz, E.R., Karatzas, I., and Ruf, J. (2018). Volatility and arbitrage. *Annals of Applied Probability* 28 (1): 378–417.

143. de Finetti, B. (1970). *Teoria Delle Probabilità*. Turin, Italy: Einaudi. An English translation was published as *Theory of Probability* by Wiley, London, in two volumes in 1974 and 1975.

144. Fisher, R.A. (1922). On the mathematical foundations of theoretical statistics. *Philosophical Transactions of the Royal Society of London, Series A* 222: 309–368.

145. Fischer, H. (2011). *A History of the Central Limit Theorem: From Classical to Modern Probabilty Theory*. New York: Springer.

146. Föllmer, H. (1981). Calcul d'Itô sans probabilités. In: *Séminaire de Probabilités XV*, Lecture Notes in Mathematics, vol. 850 (eds. J. Azema and M. Yor), 143–150. Berlin: Springer.

147. Föllmer, H. and Schied, A. (2002). Convex measures of risk and trading constraints. *Finance and Stochastics* 6 (4): 429–447.

148. Föllmer, H. and Schied, A. (2011). *Stochastic Finance: An Introduction in Discrete Time*. Berlin: De Gruyter.

149. Foster, D.P. and Vohra, R.V. (1998). Asymptotic calibration. *Biometrika* 85 (2): 379–390.

150. Fréchet, M. (1930). Sur la convergence "en probabilité". *Metron* 8 (4): 3–50.

151. Fréchet, M. (1939). The diverse definitions of probability. *Journal of Unified Science (Erkenntnis)* 8 (1): 7–23.

152. Freedman, D.A. (1975). On tail probabilities for martingales. *Annals of Probability* 3 (1): 100–118.

153. French, C.W. (2003). The Treynor capital asset pricing model. *Journal of Investment Management* 1 (2): 60–72.

154. Freudenthal, H. (1980). Huygens' foundations of probability. *Historia Mathematica* 7: 113–117.

155. Friedman, W.A. (2014). *Fortune Tellers: The Story of America's First Economic Fore-casters*. Princeton, NJ: Princeton University Press.

156. Friz, P.K. and Hairer, M. (2014). *A Course on Rough Paths: With an Introduction to Regularity Structures*. Cham: Springer.

157. Fujiwara, A. (2008). Randomness criteria in terms of α-divergences. *IEEE Transactions on Information Theory* 54 (3): 1252–1261.

158. Gács, P. (2005). Uniform test of algorithmic randomness over a general space. *Theoretical Computer Science* 341 (1–3): 91–137.

159. Galane, L.C., Łochowski, R.M., and Mhlanga, F.J. (2018). On SDEs with Lipschitz Coefficients, Driven by Continuous, Model-Free Price Paths. Technical Report arXiv:1807.05692v4 [q-fin.MF], arXiv.org e-Print archive.

160. Galane, L.C., Łochowski, R.M., and Mhlanga, F.J. (2018). On the quadratic variation of the model-free price paths with jumps. *Lithuanian Mathematical Journal* 58 (2): 141–156.

161. Gale, D. and Stewart, F.M. (1953). Infinite games with perfect information. In: *Contributions to the Theory of Games, II* (eds. H.W. Kuhn and A.W. Tucker), 245–266. Princeton, NJ: Princeton University Press.

162. Galmarino, A.R. (1963). A test for Markov times. *Revista de la Unión Matemática Argentina y de la Asociación Física Argentina* 21 (4): 173–178.

163. Gamow, G. and Stern, M. (1958). *Puzzle-Math*. London: MacMillan.

164. Gardner, M. (1959). Mathematical games: "brain-teasers" that involve formal logic. *Scientific American* 200 (2): 136–140.

165. Girsanov, I.V. (1960). On transforming a certain class of stochastic processes by absolutely continuous substitution of measures. *Theory of Probability and Its Applications* 5 (3): 285–301.

166. Gnedenko, B.V. (1997). *Theory of Probability*. St. Leonards: Gordon and Breach. English translation, with revisions by the author, of the sixth Russian edition of *Курс теории вероятностей*, published in 1988 by Nauka, Moscow.

167. Gnedenko, B.V. and Kolmogorov, A.N. (1954). *Limit Distributions for Sums of Independent Random Variables*. Cambridge, MA: Addison-Wesley. Original Russian edition: *Предельные распределения для сумм независимых случайных величин*, 1949.

168. Gneiting, T. and Katzfuss, M. (2014). Probabilistic forecasting. *Annual Review of Statistics and Its Applications* 1 (1): 125–151.

169. Gneiting, T. and Raftery, A.E. (2007). Strictly proper scoring rules, prediction, and estimation. *Journal of the American Statistical Association* 102: 359–378.

170. Gorroochurn, P. (2016). *Classic Topics on the History of Modern Mathematical Statistics from Laplace to More Recent Times*. New York: Wiley.

171. Grattan-Guinness, I. (1994). Heat diffusion. In: *Companion Encyclopedia of the History and Philosophy of the Mathematical Sciences*, vol. 2 (ed. I. Grattan-Guinness), 1165–1170. London: Routledge.

172. Gruenhage, G. (2006). The story of a topological game. *Rocky Mountain Journal of Mathematics* 36 (6): 1885–1914.

173. Gurevich, Y. and Vovk, V. (2017). p-values. Technical Report arXiv:1702.02590v2 [stat.ME], arXiv.org e-Print archive.

174. Hald, A. (1998). *A History of Mathematical Statistics from 1750 to 1930*. New York: Wiley.

175. Hallenbeck, C. (1920). Forecasting precipitation in percentages of probability. *Monthly Weather Review* 48 (11): 645–647.

176. Halmos, P.R. (1960). *Naive Set Theory*. New York: Van Nostrand Reinhold.

177. Hardin, C.S. and Taylor, A.D. (2008). A peculiar connection between the Axiom of Choice and predicting the future. *American Mathematical Monthly* 115 (2): 91–96.

178. Hardin, C.S. and Taylor, A.D. (2013). *The Mathematics of Coordinated Inference: A Study of Generalized Hat Problems*. Cham: Springer.

179. Hardy, G.H. and Littlewood, J.E. (1914). Some problems of Diophantine approximation. *Acta Mathematica* 37: 155–239.

180. Harper, P.S. (2008). *A Short History of Medical Genetics*. Oxford: Oxford University Press.

181. Hartman, P. and Wintner, A. (1941). On the law of the iterated logarithm. *American Journal of Mathematics* 63: 169–176.

182. Hastings, C. Jr. (1955). *Approximations for Digital Computers*. Princeton, NJ: Princeton University Press.

183. Hausdorff, F. (1914). *Grundzüge der Mengenlehre*. Leipzig: Von Veit.

184. Hawkins, T. (1974). *Lebesgue's Theory of Integration: Its Origins and Development*, 2e. New York: Chelsea.

185. Healey, R. (2017). Quantum-Bayesian and pragmatist views of quantum theory. In: *Stanford Encyclopedia of Philosophy* (ed. E.N. Zalta). Stanford, CA: Center for the Study of Language and Information, Stanford University, Spring 2017 edition.

186. Heams, T. (2014). Randomness in biology. *Mathematical Structures in Computer Science* 24 (3).

187. Hobson, D.G. (1998). Robust hedging of the lookback option. *Finance and Stochastics* 2 (4): 329–347.

188. Hoeffding, W. (1963). Probability inequalities for sums of bounded random variables. *Journal of the American Statistical Association* 58: 13–30.

189. Hoffmann-Jørgensen, J. (1987). The general marginal problem. In: *Functional Analysis II*, Lecture Notes in Mathematics, vol. 1242 (eds. S. Kurepa, H. Kraljević, and D. Butković), 77–367. Berlin: Springer.

190. Homer, S. and Sylla, R. (2005). *A History of Interest Rates*, 4e. Hoboken, NJ: Wiley.

191. Hull, J.C. (2018). *Options, Futures, and Other Derivatives*, 10e. New York: Pearson.

192. Hunt, G.A. (1960). Markoff chains and Martin boundaries. *Illinois Journal of Mathematics* 4 (3): 313–340.

193. Huygens, C. (1888–1950). *Œuvres complètes de Christiaan Huygens*, vol. XIV. La Haye: Nijhoff.

194. Ingersoll, J.E. Jr. (1987). *Theory of Financial Decision Making*. Totowa, NJ: Rowman & Littlefield.

195. Innes, A. and Cabrer, L. (2017). *SPIVA® Europe Scorecard*. S&P Dow Jones Indices.

196. Ionescu Tulcea, C.T. (1949). Mesures dans les espaces produits. *Atti della Accademia nazionale dei Lincei. Classe di scienze fisiche, matematiche e naturali. Rendiconti Series VIII* 7: 208–211.

197. Itô, K. (1944). Stochastic integral. *Proceedings of the Imperial Academy of Tokyo* 20 (8): 519–524. Reprinted in [[198], pp. 85–90].

198. Itô, K. (1987). *Selected Papers* (eds. D.W. Stroock and S.R. Srinivasa Varadhan). New York: Springer.

199. Jacod, J. and Shiryaev, A.N. (2003). *Limit Theorems for Stochastic Processes*, 2e. Berlin: Springer, First edition: 1987.

200. Jammer, M. (1974). *The Philosophy of Quantum Mechanics: The Interpretations of Quantum Mechanics in Historical Perspective*. New York: Wiley.

201. Jarrow, R. and Protter, P. (2004). A short history of stochastic integration and mathematical finance: the early years, 1880–1970. In: *A Festschrift for Herman Rubin*, (ed. A. Das Gupta), 75–91. Beachwood, OH: Institute of Mathematical Statistics.

202. Jech, T. (2003). *Set Theory: The Third Millennium Edition, revised and expanded*. Berlin: Springer.

203. Jessen, B. (1934). The theory of integration in a space of an infinite number of dimensions. *Acta Mathematica* 63: 249–323.

204. Kabanov, Y.M., Liptser, R.S., and Shiryaev, A.N. (1977). On the question of absolute continuity and singularity of probability measures. *Mathematics of the USSR—Sbornik* 33 (2): 203–221. The Russian original: К вопросу об абсолютной непрерывности и сингулярности вероятностных мер. *Математический сборник* 104: 227–247.

205. Kabanov, Y.M., Liptser, R.S., and Stoyanov, J. (eds.) (2006). *The Shiryaev Festschrift: From Stochastic Calculus to Mathematical Finance*. Berlin: Springer.

206. Kadane, J.B., Schervish, M.J., and Seidenfeld, T. (1999). *Rethinking the Foundations of Statistics*. New York: Cambridge University Press.

207. Kakade, S.M. and Foster, D.P. (2008). Deterministic calibration and Nash equilibrium. *Journal of Computer and System Sciences* 74 (1): 115–130.

208. Kalnishkan, Y. and Vyugin, M.V. (2008). The weak aggregating algorithm and weak mixability. *Journal of Computer and System Sciences* 74 (8): 1228–1244.

209. Karatzas, I. and Kardaras, C. Arbitrage Theory via Numéraires. Book in preparation.

210. Karatzas, I. and Kardaras, C. (2007). The numéraire portfolio in semimartingale financial models. *Finance and Stochastics* 11 (4): 447–493.

211. Karatzas, I. and Ruf, J. (2017). Trading strategies generated by Lyapunov functions. *Finance and Stochastics* 21 (3): 753–787.

212. Karatzas, I. and Shreve, S.E. (1991). *Brownian Motion and Stochastic Calculus*, 2e. New York: Springer.

213. Kardaras, C. (2012). Market viability via absence of arbitrage of the first kind. *Finance and Stochastics* 16 (4): 651–667.

214. Kechris, A.S. (1995). *Classical Descriptive Set Theory*. New York: Springer.

215. Kelly, J.L. Jr. (1956). A new interpretation of information rate. *Bell System Technical Journal* 35 (4): 917–926.

216. van Kesteren, E.-J. and Wagenmakers, E.-J. Exploring the diagnosticity of the p-value. https://www.shinyapps.org/apps/vs-mpr/ (accessed 6 June 2018).

217. Khinchin, A.Y. (1924). Über einen Satz der Wahrscheinlichkeitsrechnung. *Fundamenta Mathematicae* 6: 9–20.

218. Khinchin, A.Y. (1928). Sur la loi forte des grands nombres. *Comptes rendus des séances de l'Académie des Sciences* 186: 285–287.

219. Kitaev, A.Y., Shen, A., and Vyalyi, M.N. (2002). *Classical and Quantum Computation*. Providence, RI: American Mathematical Society. Russian original: *Классические и квантовые вычисления*, Издательство Московского центра непрерывного математического образования, Moscow, 1999.

220. Kolmogorov, A.N. (1927). Sur la loi des grands nombres. *Comptes rendus des séances de l'Académie des Sciences* 185: 917–919.

221. Kolmogorov, A.N. (1929). Sur la loi des grands nombres. *Atti della Reale Accademia Nazionale dei Lincei. Classe di scienze fisiche, matematiche e naturali. Rendiconti Serie VI* 9: 470–474.

222. Kolmogorov, A.N. (1929). Über das Gesetz der iterierten Logarithmus. *Mathematische Annalen* 101: 126–135. Appears in English in [[226], pp. 32–42] as "On the law of the iterated logarithm".

223. Kolmogorov, A.N. (1930). Sur la loi forte des grands nombres. *Comptes rendus des séances de l'Académie des Sciences* 191: 910–912. Appears in English in [[226], pp. 60–61] as "On the strong law of large numbers".

224. Kolmogorov, A.N. (1933). *Grundbegriffe der Wahrscheinlichkeitsrechnung*. Berlin: Springer. An English translation appeared under the title *Foundations of the Theory of Probability* (Chelsea, New York) in 1950, with a second edition in 1956. A Russian translation appeared under the title *Основные понятия теории вероятностей* (Nauka, Moscow) in 1936, with a second edition, slightly expanded by Kolmogorov with the assistance of Albert N. Shiryaev, in 1974. Third Russian edition: *Основные понятия теории вероятностей*. ФАЗИС, Moscow, 1998.

225. Kolmogorov, A.N. (1965). Three approaches to the quantitative definition of information. *Problems of Information Transmission* 1 (1): 1–7. Republished in [[227], pp. 184–193] as "Three approaches to the definition of the notion of amount of information".

226. Kolmogorov, A.N. (1992). *Selected Works of A. N. Kolmogorov: Probability Theory and Mathematical Statistics*, vol. II. Dordrecht, Netherlands: Kluwer. Translated from the Russian original, *Теория вероятностей и математическая статистика*, edited by A.N. Shiryaev and published in 1986 by Nauka, Moscow.

227. Kolmogorov, A.N. (1993). *Selected Works of A. N. Kolmogorov. Information Theory and the Theory of Algorithms*, vol. III. Dordrecht, Netherlands: Kluwer. Translated from the Russian original, *Теория информации и теория алгоритмов*, edited by A.N. Shiryaev and published in 1987 by Nauka, Moscow.

228. Koolen, W.M. and Vovk, V. (2014). Buy low, sell high. *Theoretical Computer Science* 558: 144–158.

229. Kumon, M. and Takemura, A. (2008). On a simple strategy weakly forcing the strong law of large numbers in the bounded forecasting game. *Annals of the Institute of Statistical Mathematics* 60 (4): 801–812.

230. Kumon, M., Takemura, A., and Takeuchi, K. (2007). Game-theoretic versions of strong law of large numbers for unbounded variables. *Stochastics* 79 (5): 449–468.

231. Kumon, M., Takemura, A., and Takeuchi, K. (2011). Sequential optimizing strategy in multi-dimensional bounded forecasting games. *Stochastic Processes and their Applications* 121 (1): 155–183.

232. Kunita, H. and Watanabe, S. (1967). On square-integrable martingales. *Nagoya Journal of Mathematics* 30: 209–245.

233. Lacroix, S.F. (1816). *Traité élémentaire du calcul des probabilités*. Courcier, Paris, 1816. Further editions in 1822, 1833, 1864, and later.

234. Lai, T.L., Robbins, H., and Wei, C.Z. (1978). Strong consistency of least squares estimates in multiple regression. *Proceedings of the National Academy of Sciences of the United States of America* 75 (7): 3034–3036.

235. Lai, T.L., Robbins, H., and Wei, C.Z. (1979). Strong consistency of least squares estimates in multiple regression II. *Journal of Multivariate Analysis* 9 (3): 343–361.

236. Lai, T.L. and Wei, C.Z. (1982). Least squares estimates in stochastic regression models with applications to identification and control of dynamic systems. *Annals of Statistics* 10 (1): 154–166.

237. Lamperti, J.W. (1996). *Probability*, 2e. New York: Wiley. The first edition appeared in 1966.

238. Lehmann, B.N. (2005). Fischer Black on valuation: the CAPM in general equilibrium. In: *The Legacy of Fischer Black* (ed. B.N. Lehmann). New York: Oxford University Press.

239. Lepingle, D. (1976). La variation d'ordre p des semi-martingales. *Zeitschrift für Wahrscheinlichkeitstheorie und verwandte Gebiete* 36 (4): 295–316.

240. Leung-Yan-Cheong, S.K. and Cover, T.M. (1978). Some equivalences between Shannon entropy and Kolmogorov complexity. *IEEE Transactions on Information Theory* 24 (3): 331–338.

241. Levin, L.A. (1976). Uniform tests of randomness. *Soviet Mathematics Doklady* 17 (2): 337–340. Russian original: Равномерные тесты случайности. *Доклады АН СССР* 227 (1): 33–35, 1976.

242. Levin, L.A. (1984). Randomness conservation inequalities; information and independence in mathematical theories. *Information and Control* 61 (1): 15–37.

243. Lévy, P. (1925). *Calcul des probabilités*. Paris: Gauthier-Villars.

244. Lévy, P. (1935). Propriétés asymptotiques des sommes de variables aléatoires enchaînées. *Bulletin des Sciences Mathématiques* 59: 84–96 and 109–128.

245. Lévy, P. (1937). *Théorie de l'addition des variables aléatoires.* Paris: Gauthier-Villars. Second Edition: 1954.

246. Lévy, P. (1940). Le mouvement brownien plan. *American Journal of Mathematics* 62 (1): 487–550.

247. Lévy, P. (1940). Sur certains processus stochastiques homogènes. *Compositio Mathematica* 7: 283–339.

248. Lewis, D.K. and Richardson, J.S. (1966). Scriven on human unpredictability. *Philosophical Studies: An International Journal for Philosophy in the Analytic Tradition* 17 (5): 69–74.

249. Lichtenstein, S. and Fischhoff, B. (1977). Do those who know more also know more about how much they know? *Organizational Behavior and Human Performance* 20 (2): 159–183.

250. Lindeberg, J.W. (1920). Über das Exponentialgesetz in der Wahrscheinlichkeitsrechnung. *Annales Academiae Scientiarum Fennicae, Series A* 16 (1).

251. Lindeberg, J.W. (1922). Eine neue Herleitung des Exponentialgesetzes in der Wahrscheinlichkeitsrechnung. *Mathematische Zeitschrift* 15 (1): 211–225.

252. Lintner, J. (1965). The valuation of risk assets and the selection of risky investments in stock portfolios and capital budgets. *Review of Economics and Statistics* 47 (1): 13–37.

253. Littlestone, N. and Warmuth, M.K. (1994). The weighted majority algorithm. *Information and Computation* 108 (2): 212–261.

254. Łochowski, R.M., Perkowski, N., and Prömel, D.J. (2018). A superhedging approach to stochastic integration. *Stochastic Processes and their Applications* 128 (12): 4078–4103.

255. Locker, B. (2009). Doob at Lyon. *Electronic Journal for History of Probability and Statistics* 5 (1).

256. Luenberger, D.G. (1998). *Investment Science.* New York: Oxford University Press.

257. Lutz, J.H. (2003). Dimension in complexity classes. *SIAM Journal on Computing* 32 (5): 1236–1259.

258. Malkiel, B.G. (1995). Returns from investing in equity mutual funds 1971 to 1991. *Journal of Finance* 50 (2): 549–572.

259. Malkiel, B.G. (2005). Reflections on the efficient market hypothesis: 30 years later. *Financial Review* 40 (1): 1–9.

260. Malkiel, B.G. (2016). *A Random Walk Down Wall Street,* 11th revised edition. New York: Norton.

261. Mansuy, R. (2009). The origins of the word "martingale". *Electronic Journal for History of Probability and Statistics* 5 (1).

262. Marcinkiewicz, J. and Zygmund, A. (1937). Remarque sur la loi du logarithme itéré. *Fundamenta Mathematicae* 29: 215–222.

263. Marcinkiewicz, J. and Zygmund, A. (1937). Sur les fonctions indépendantes. *Fundamenta Mathematicae* 29: 60–90.

264. Markov, A.A. (1900). *Исчисление вероятностей.* St. Petersburg: Типография Императорской Академии Наук. The second edition, which appeared in 1908, was translated into German as *Wahrscheinlichkeitsrechnung.* Leipzig: Teubner, 1912.

265. Markowitz, H. (1952). Portfolio selection. *Journal of Finance* 7 (1): 77–91.

266. Martin, D.A. (1975). Borel determinacy. *Annals of Mathematics* 102 (2): 363–371.

267. Martin, D.A. (1990). An extension of Borel determinacy. *Annals of Pure and Applied Logic* 49 (3): 279–293.

268. Martin, D.A. (1998). The determinacy of Blackwell games. *The Journal of Symbolic Logic* 63 (4): 1565–1581.

269. Martin-Löf, P. (1966). The definition of random sequences. *Information and Control* 9 (6): 602–619.

270. Martin-Löf, P. (1970). *Notes on Constructive Mathematics.* Stockholm: Almqvist & Wiksell.

271. Maruyama, G. (1954). On the transition probability functions of the Markov process. *Natural Science Report, Ochanomizu University* 5 (1): 10–20.

272. Mazliak, L. (2009). How Paul Lévy saw Jean Ville and martingales. *Electronic Journal for History of Probability and Statistics* 5 (1).

273. McDiarmid, C. (1989). On the method of bounded differences. In: *Surveys in Combinatorics: Invited Papers at the Twelfth British Combinatorial Conference,* 148–188. Cambridge: Cambridge University Press.

274. Mehra, R. (2006). The equity premium puzzle: a review. *Foundations and Trends in Finance* 2 (1): 1–81.

275. Mehra, R. and Prescott, E.C. (1985). The equity premium: a puzzle. *Journal of Monetary Economics* 15 (2): 145–161.

276. Mehra, R. and Prescott, E.C. (eds.) (2008). *Handbook of the Equity Risk Premium.* Amsterdam: Elsevier.

277. Mermin, N.D. (2017). Why QBism is not the Copenhagen interpretation and what John Bell might have thought of it. In: *Quantum [Un]Speakables II,* The Frontiers Collection (eds. R. Bertlmann and A. Zeilinger), 83–93. Cham: Springer.

278. Merton, R.C. (1973). An intertemporal capital asset pricing model. *Econometrica* 41 (5): 867–888.

279. Merton, R.C. (1973). Theory of rational option pricing. *Bell Journal of Economics and Management Science* 4 (1): 141–183.

280. Meyer, P.-A. (1976). Un cours sur les intégrales stochastiques. *Séminaire de probabilités de Strasbourg* 10: 245–400.

281. Meyer, P.-A. (2009). Stochastic processes from 1950 to the present. *Electronic Journal for History of Probability and Statistics* 5 (1). French original published in 2000.

282. von Mises, R. (1928). *Wahrscheinlichkeitsrechnung, Statistik und Wahrheit.* Vienna: Springer. The second edition appeared in 1936 and the third in 1951. A posthumous fourth edition, edited by his wife Hilda Geiringer, appeared in 1972. English editions, under the title *Probability, Statistics and Truth,* appeared in 1939 and 1957.

283. Miyabe, K. and Takemura, A. (2012). Convergence of random series and the rate of convergence of the strong law of large numbers in game-theoretic probability. *Stochastic Processes and their Applications* 122 (1): 1–30.

284. Miyabe, K. and Takemura, A. (2013). The law of the iterated logarithm in game-theoretic probability with quadratic and stronger hedges. *Stochastic Processes and their Applications* 123 (8): 3132–3152.

285. Miyabe, K. and Takemura, A. (2015). Derandomization in game-theoretic probability. *Stochastic Processes and their Applications* 125 (1): 39–59.

286. Mörters, P. and Peres, Y. (2010). *Brownian Motion.* Cambridge: Cambridge University Press.

287. Mossin, J. (1966). Equilibrium in a capital asset market. *Econometrica* 34 (4): 768–783.

288. Murphy, A.H. (1998). The early history of probability forecasts: some extensions and clarifications. *Weather and Forecasting* 13 (1): 5–15.

289. Mycielski, J. and Świerczkowski, S. (1964). On the Lebesgue measurability and the axiom of determinateness. *Fundamenta Mathematicae* 54: 67–71.

290. von Neumann, J. (1955). *Mathematical Foundations of Quantum Mechanics.* Princeton, NJ: Princeton University Press. First German edition: (1932). *Mathematische Grundlagen der Quantenmechanik.* Berlin: Springer.

291. Neumann, P.M., Stoy, G.A., and Thompson, E.C. (1994). *Groups and Geometry.* Oxford: Oxford University Press. Reprinted in 2002.

292. Nielsen, M.A. and Chuang, I.L. (2010). *Quantum Computation and Quantum Information,* 10th anniversary edition. Cambridge: Cambridge University Press.

293. Norges Bank Investment Management (2016). The equity risk premium. Discussion Note 1.

294. Novak, E., Ullrich, M., Woźniakowski, H., and Zhang, S. (2018). Reproducing kernels of Sobolev spaces on $\mathbb{R}^d$ and applications to embedding constants and tractability. *Analysis and Applications* 16 (5): 693–715.

295. Obłój, J. (2006). A complete characterization of local martingales which are functions of Brownian motion and its maximum. *Bernoulli* 12 (6): 955–969.

296. Obłój, J. and Yor, M. (2006). On local martingale and its supremum: harmonic functions and beyond. In: Kabanov et al. [205], 517–533.

297. Osborne, M.F.M. (1959). Brownian motion in the stock market. *Operations Research* 7 (2): 145–173.

298. Paley, R.E.A.C., Wiener, N., and Zygmund, A. (1933). Note on random functions. *Mathematische Zeitschrift* 37 (1): 647–668.

299. Paris, J.B. (1972). ZF ⊢ Σ_4^0 determinateness. *Journal of Symbolic Logic* 37 (4): 661–667.

300. Perkowski, N. and Prömel, D.J. (2015). Local times for typical price paths and pathwise Tanaka formulas. *Electronic Journal of Probability* 20 (46): 1–15.

301. Perkowski, N. and Prömel, D.J. (2016). Pathwise stochastic integrals for model free finance. *Bernoulli* 22 (4): 2486–2520.

302. Petrov, V.V. (1975). *Sums of Independent Random Variables*. Berlin: Springer.

303. Petrov, V.V. (1995). *Limit Theorems of Probability Theory: Sequences of Independent Random Variables*. Oxford: Oxford University Press.

304. Petrovsky, I.G. (1935). Zur ersten Randwertaufgabe der Wärmeleitungsgleichung. *Compositio Mathematica* 1 (3): 383–419. A Russian translation appears in Petrovsky's Collected Papers [305] along with commentary by Landis and Molchanov, pp. 325–333.

305. Petrovsky, I.G. (1987). Избранные труды. Дифференциальные уравнения. Теория вероятностей (Collected Works. Differential Equations. Probability Theory). Moscow: Nauka.

306. Pietruska, J.L. (2017). *Looking Forward: Prediction and Uncertainty in Modern America*. Chicago: University of Chicago Press.

307. Plackett, R.L. (1972). The discovery of the method of least squares. *Biometrika* 59 (2): 239–251.

308. Platen, E. (2005). Diversified portfolios with jumps in a benchmark framework. *Asia-Pacific Financial Markets* 11 (1): 1–22.

309. Platen, E. (2005). On the role of the growth optimal portfolio in finance. *Australian Economic Papers* 44 (4): 365–388.

310. Platen, E. and Heath, D. (2006). *A Benchmark Approach to Quantitative Finance*. Berlin: Springer.

311. von Plato, J. (1994). *Creating Modern Probability: Its Mathematics, Physics, and Philosophy in Historical Perspective*. Cambridge: Cambridge University Press.

312. Pólya, G. (1920). Über den zentralen Grenzwertsatz der Wahrscheinlichkeitsrechnung und das Momentenproblem. *Mathematische Zeitschrift* 8 (3–4): 171–181.

313. Porter, T. (2018). *Genetics in the Madhouse: The Unknown History of Human Heredity*. Princeton, NJ: Princeton University Press.

314. Protter, P.E. (2005). *Stochastic Integration and Differential Equations*. Berlin: Springer. Corrected third printing of the second edition. First edition: 1990. Second edition: 2003.

315. Pukelsheim, F. (1986). Predictable criteria for absolute continuity and singularity of two probability measures. *Statistics and Decisions* 4 (2–3): 227–236.

316. Putnam, H. (1963). "Degree of confirmation" and inductive logic. In: *The Philosophy of Rudolf Carnap*, Chapter 24 (ed. P.A. Schilpp), 761–783. La Salle, IL: Open Court.

317. Regazzini, E. (2013). The origins of de Finetti's critique of countable additivity. In: *Advances in Modern Statistical Theory and Applications: A Festschrift in honor of Morris L. Eaton* (eds. G. Jones and X. Shen), 63–82. Beachwood, OH: Institute of Mathematical Statistics.

318. Regnault, J. (1853). *Calcul des chances et philosophie de la bourse.* Paris: Mallet-Bachelier et Castel.

319. Revuz, D. and Yor, M. (1999). *Continuous Martingales and Brownian Motion*, 3e. Berlin: Springer.

320. Robinson, R.M. (1947). On the decomposition of spheres. *Fundamenta Mathematicae* 34: 246–260.

321. Rogers, L.C.G. and Williams, D. (1994). *Diffusions, Markov Processes, and Martingales:* vol. 1: Foundations, vol. 1, 2e. Chichester: Wiley. Reissued by Cambridge University Press in Cambridge Mathematical Library, 2000.

322. Ross, S. (2014). *First Course in Probability*, 9e. New York: Pearson.

323. Rudin, W. (1991). *Functional Analysis*, 2e. Boston, MA: McGraw-Hill.

324. Ruszczyński, A. and Shapiro, A. (2006). Optimization of convex risk functions. *Mathematics of Operations Research* 31 (3): 433–452.

325. Samuelson, P.A. (1965). Proof that properly anticipated prices fluctuate randomly. *Industrial Management Review* 6 (2): 41–50.

326. Sasai, T., Miyabe, K., and Takemura, A. (2014). A Game-Theoretic Proof of Erdös-Feller-Kolmogorov-Petrowsky Law of the Iterated Logarithm for Fair-Coin Tossing. Technical Report arXiv:1408.1790v2 [math.PR], arXiv.org e-Print archive.

327. Sasai, T., Miyabe, K., and Takemura, A. (2015). Erdös-Feller-Kolmogorov-Petrowsky law of the iterated logarithm for self-normalized martingales: A game-theoretic approach. Technical Report arXiv:1504.06398v1 [math.PR], arXiv.org e-Print archive. Journal version (to appear): *Annals of Probability*.

328. Sato, R., Miyabe, K., and Takemura, A. (2018). Relation between the rate of convergence of strong law of large numbers and the rate of concentration of Bayesian prior in game-theoretic probability. *Stochastic Processes and their Applications* 128 (5): 1466–1484.

329. Sawyer, N. (2007). SG CIB launches timer options. *Risk* 20 (7).

330. Schied, A., Speiser, L., and Voloshchenko, I. (2018). Model-free portfolio theory and its functional master formula. *SIAM Journal on Financial Mathematics* 9 (3): 1074–1101.

331. Schneider, I. (1980). Christiaan Huygens's contribution to the development of a calculus of probabilities. *Janus* LXVII: 269–279.

332. Schnorr, C.P. (1970). Klassifikation der Zufallsgesetze nach Komplexität und Ordnung. *Zeitschrift für Wahrscheinlichkeitstheorie und verwandte Gebiete* 16 (1): 1–21.

333. Schnorr, C.P. (1971). *Zufälligkeit und Wahrscheinlichkeit: Eine algorithmische Begründung der Wahrscheinlichkeitstheorie*, Lecture Notes in Mathematics, vol. 218. Berlin: Springer.

334. Schölkopf, B. and Smola, A.J. (2002). *Learning with Kernels*. Cambridge, MA: MIT Press.

335. Schwarz, G. (1968). Time-free continuous processes. *Proceedings of the National Academy of Sciences of the United States of America* 60 (4): 1183–1188.

336. Schwarz, G. (1972). On time-free functions. *Transactions of the American Mathematical Society* 167: 471–478.

337. Scriven, M. (1965). An essential unpredictability in human behavior. In: *Scientific Psychology: Principles and Approaches* (eds. B.B. Wolman and E. Nagel), 411–425. New York: Basic Books.

338. Sellke, T., Bayarri, M.J., and Berger, J. (2001). Calibration of p-values for testing precise null hypotheses. *American Statistician* 55 (1): 62–71.

339. Shafer, G. (1976). *A Mathematical Theory of Evidence*. Princeton, NJ: Princeton University Press.

340. Shafer, G. (1996). *The Art of Causal Conjecture*. Cambridge, MA: MIT Press.

341. Shafer, G. (1998). Mathematical foundations for probability and causality. In: *Mathematical Aspects of Artificial Intelligence*, Symposia in Applied Mathematics, vol. 55 (ed. F. Hoffman), 207–270. Providence, RI: American Mathematical Society.

342. Shafer, G. (2001). Nature's possibilities and expectations. In: *Probability Theory: Philosophy, Recent History and Relations to Science* (eds. V.F. Hendriks, S.A. Pedersen, and K.F. Jørgensen), 147–166. Dordrecht: Kluwer.

343. Shafer, G. (2007). From Cournot's principle to market efficiency. In: *Augustin Cournot: Modelling Economics* (ed. J.-P. Touffut), 55–95. Cheltenham: Edward Elgar. First posted as GTP15 November 2005.

344. Shafer, G. (2009). The education of Jean André Ville. *Electronic Journal for History of Probability and Statistics* 5 (1).

345. Shafer, G. (2015). When to call a variable random, GTP41, first posted June 2015.

346. Shafer, G. (2018). Marie-France Bru and Bernard Bru on dice games and contracts. *Statistical Science* 33 (2): 277–284.

347. Shafer, G., Gillett, P.R., and Scherl, R.B. (2000). The logic of events. *Annals of Mathematics and Artificial Intelligence* 28 (1–4): 315–389.

348. Shafer, G., Shen, A., Vereshchagin, N., and Vovk, V. (2011). Test martingales, Bayes factors, and p-values. *Statistical Science* 26 (1): 84–101. First posted as GTP33 December 2009 and also available on arXiv.

349. Shafer, G. and Vovk, V. (2001). *Probability and Finance: It's Only a Game!* New York: Wiley.

350. Shafer, G. and Vovk, V. (2006). The origins and legacy of Kolmogorov's *Grundbegriffe*. Technical Report arXiv:1802.06071 [math.HO], arXiv.org e-Print archive, February 2018. First posted as GTP4 February 2003. Abridged version published as The sources of Kolmogorov's *Grundbegriffe*, *Statistical Science* 21 (1): 70–98.

351. Shafer, G., Vovk, V., and Takemura, A. (2012). Lévy's zero-one law in game-theoretic probability. *Journal of Theoretical Probability* 25 (1): 1–24. First posted as GTP29 May 2009.

352. Sharpe, W.F. (1964). Capital asset prices: a theory of market equilibrium under conditions of risk. *Journal of Finance* 19 (3): 425–442.

353. Sharpe, W.F. (1966). Mutual fund performance. *Journal of Business* 39 (1): 119–138 (in Part 2: Supplement on Security Prices).

354. Sharpe, W.F. (2007). *Investors and Markets: Portfolio Choices, Asset Prices, and Investment Advice.* Princeton, NJ: Princeton University Press.

355. Shawe-Taylor, J. and Cristianini, N. (2004). *Kernel Methods for Pattern Analysis.* Cambridge: Cambridge University Press.

356. Sheynin, O. (1978). S. D. Poisson's work in probability. *Archive for History of Exact Sciences* 18 (3): 245–300.

357. Sheynin, O. (1979). C. F. Gauss and the theory of errors. *Archive for History of Exact Sciences* 20 (1): 21–72.

358. Shiryaev, A.N. (1996). *Probability*, 2e. New York: Springer. Translated from the Russian original, Вероятность, published by Nauka, Moscow, in 1989. Third Russian edition: Издательство Московского центра непрерывного математического образования, Moscow, 2004, in two volumes.

359. Siegel, J.J. (1992). The equity premium: stock and bond returns since 1802. *Financial Analysts Journal* 48 (1): 28–38.

360. Snell, J.L. (1952). Application of martingale system theorems. *Transactions of the American Mathematical Society* 73 (2): 293–312.

361. Snell, J.L. (1997). A conversation with Joe Doob. *Statistical Science* 12 (4): 301–311.

362. Soe, A.M. and Poirier, R. (2017). *SPIVA® U.S. Scorecard.* S&P Dow Jones Indices, Year-End.

363. Steinwart, I. and Christmann, A. (2008). *Support Vector Machines.* New York: Springer.

364. Sterkenburg, T.F. (2018). Universal prediction. PhD thesis. University of Groningen.

365. Steuding, J. (2002). Probabilistic number theory, Lecture notes available on the Internet.

366. Stigler, S.M. (1981). Gauss and the invention of least squares. *Annals of Statistics* 9 (3): 465–474.

367. Stigler, S.M. (1986). *The History of Statistics: The Measurement of Uncertainty before 1900.* Cambridge, MA: Harvard University Press.

368. Stigler, S.M. (2007). Chance is 350 years old. *Chance* 20 (4): 33–36.

369. Stone, C.J. (1977). Consistent nonparametric regression (with discussion). *Annals of Statistics* 5 (4): 595–645.

370. Stout, W.F. (1970). A martingale analogue of Kolmogorov's law of the iterated logarithm. *Zeitschrift für Wahrscheinlichkeitstheorie und verwandte Gebiete* 15 (4): 279–290.

371. Stratonovich, R.L. (1968). *Conditional Markov Processes and Their Application to the Theory of Optimal Control.* New York: Elsevier. Russian original: Условные марковские процессы и их применение к теории оптимального управления, Издательство Московского университета, Moscow, 1966.

372. Stricker, C. (1979). Sur la *p*-variation des surmartingales. *Séminaire de probabilités de Strasbourg* 13: 233–237.

373. Takazawa, S.-I. (2011). An exponential inequality and the convergence rate of the strong law of large numbers in the unbounded forecasting game. *Stochastics* 83 (2): 117–125.

374. Takazawa, S.-I. (2012). Exponential inequalities and the law of the iterated logarithm in the unbounded forecasting game. *Annals of the Institute of Statistical Mathematics* 64 (3): 615–632.

375. Takemura, A., Vovk, V., and Shafer, G. (2011). The generality of the zero-one laws. *Annals of the Institute of Statistical Mathematics* 63 (5): 873–885.

376. Takeuchi, K., Kumon, M., and Takemura, A. (2009). A new formulation of asset trading games in continuous time with essential forcing of variation exponent. *Bernoulli* 15 (4): 1243–1258.

377. Takeuchi, K., Kumon, M., and Takemura, A. (2010). Multistep Bayesian strategy in coin-tossing games and its application to asset trading games in continuous time. *Stochastic Analysis and Applications* 28 (5): 842–861.

378. Talagrand, M. (1995). Concentration of measure and isoperimetric inequalities in product spaces. *Publications Mathématiques de l'Institut des Hautes Etudes Scientifiques* 81 (1): 73–205.

379. Taylor, S.J. (1972). Exact asymptotic estimates of Brownian path variation. *Duke Mathematical Journal* 39 (2): 219–241.

380. Tenenbaum, G. (2015). *Introduction to Analytic and Probabilistic Number Theory*, 3e. Providence, RI: American Mathematical Society. Translation of the third edition of *Introduction à la théorie analytique et probabiliste des nombres*, 2008.

381. Troffaes, M.C.M. and de Cooman, G. (2014). *Lower Previsions*. Chichester: Wiley.

382. Trotter, H.F. (1958). A property of Brownian motion paths. *Illinois Journal of Mathematics* 2 (3): 425–433.

383. Van Schuppen, J.H. and Wong, E. (1974). Transformations of local martingales under a change of law. *Annals of Probability* 2 (5): 879–888.

384. Vapnik, V.N. (1998). *Statistical Learning Theory*. New York: Wiley.

385. Vervuurt, A. and Karatzas, I. (2015). Diversity-weighted portfolios with negative parameter. *Annals of Finance* 11 (3–4): 411–432.

386. Ville, J. (1936). Sur la notion de collectif. *Comptes rendus des séances de l'Académie des Sciences* 203: 26–27. 1936. Session of July 6, 1936.

387. Ville, J. (1939). *Étude critique de la notion de collectif*. Paris: Gauthier-Villars. This differs from Ville's dissertation, which was defended in March 1939, only in that a one-page introduction was replaced by a 17-page introductory chapter.

388. Vovk, V. (1987). The law of the iterated logarithm for random Kolmogorov, or chaotic, sequences. *Theory of Probability and Its Applications* 32 (3): 413–425.

389. Vovk, V. (1987). On a randomness criterion. *Soviet Mathematics Doklady* 35 (3): 656–660.

390. Vovk, V. (1988). Kolmogorov–Stout law of the iterated logarithm. *Mathematical Notes* 44: 502–507.

391. Vovk, V. (1990). Aggregating strategies. In: *Proceedings of the 3rd Annual Workshop on Computational Learning Theory* (eds. M. Fulk and J. Case), 371–383. San Mateo, CA: Morgan Kaufmann.

392. Vovk, V. (1992). Universal forecasting algorithms. *Information and Computation* 96 (2): 245–277.

393. Vovk, V. (1993). Forecasting point and continuous processes: prequential analysis. *Test* 2 (1–2): 189–217.

394. Vovk, V. (1993). A logic of probability, with applications to the foundations of statistics (with discussion). *Journal of the Royal Statistical Society, Series B* 55 (2): 317–351.

395. Vovk, V. (1998). A game of prediction with expert advice. *Journal of Computer and System Sciences* 56 (2): 153–173.

396. Vovk, V. (2005). Defensive prediction with expert advice. In: *Proceedings of the 16th International Conference on Algorithmic Learning Theory*, Lecture Notes in Artificial Intelligence, vol. 3734 (eds. S. Jain, H.U. Simon, and E. Tomita), 444–458. Berlin: Springer. Full version: Technical Report arXiv:cs/0506041v3 [cs.LG] "Competitive on-line learning with a convex loss function", arXiv.org e-Print archive, September 2005.

397. Vovk, V. (2007). Competing with wild prediction rules. *Machine Learning* 69 (2–3): 193–212. First posted as GTP16 December 2005 and also available on arXiv.

398. Vovk, V. (2007). Continuous and Randomized Defensive Forecasting: Unified View. Technical Report arXiv:0708.2353 [cs.LG], arXiv.org e-Print archive, August 2007. First posted as GTP21.

399. Vovk, V. (2007). Defensive Forecasting for Optimal Prediction with Expert Advice. Technical Report arXiv:0708.1503 [cs.LG], arXiv.org e-Print archive, August 2007. First posted as GTP20 August 2007.

400. Vovk, V. (2007). Hoeffding's Inequality in Game-Theoretic Probability. Technical Report arXiv:0708.2502 [math.PR], arXiv.org e-Print archive, August 2007.

401. Vovk, V. (2007). Non-asymptotic calibration and resolution. *Theoretical Computer Science* 387 (1): 77–89. First posted as GTP13 February 2005 and also available on arXiv.

402. Vovk, V. (2008). Continuous-time trading and the emergence of volatility. *Electronic Communications in Probability* 13 (32): 319–324. First posted as GTP25 December 2007 and also available on arXiv.

403. Vovk, V. (2008). Game-Theoretic Brownian Motion. Technical Report arXiv:0801.1309 [math.PR], arXiv.org e-Print archive, January 2008. First posted as GTP26 January 2008.

404. Vovk, V. (2015). Continuous-Time Trading and the Emergence of Probability. Technical Report arXiv:0904.4364 [math.PR], arXiv.org e-Print archive, May 2015. Earlier version published in (2012). *Finance and Stochastics* 16 (4): 561–609. First posted as GTP28 April 2009.

405. Vovk, V. (2009). Continuous-time trading and the emergence of randomness. *Stochastics* 81 (5): 455–466. First posted as GTP24 December 2007 and also available on arXiv.

406. Vovk, V. (2009). Merging of opinions in game-theoretic probability. *Annals of the Institute of Statistical Mathematics* 61 (4): 969–993. First posted as GTP19 August 2007 and also available on arXiv.

407. Vovk, V. (2011). The Capital Asset Pricing Model as a Corollary of the Black–Scholes model. Technical Report arXiv:1109.5144 [q-fin.PM], arXiv.org e-Print archive, September 2011. First posted as GTP39 September 2011.

408. Vovk, V. (2011). The Efficient Index Hypothesis and its Implications in the BSM Model. Technical Report arXiv:1109.2327 [q-fin.GN], arXiv.org e-Print archive, September 2011. First posted as GTP38 September 2011.

409. Vovk, V. (2011). Rough paths in idealized financial markets. *Lithuanian Mathematical Journal* 51 (2): 274–285. First posted as GTP35 May 2010 and also available on arXiv.

410. Vovk, V. (2011). A Simplified Capital Asset Pricing Model. Technical Report arXiv:1111.2846 [q-fin.PM], arXiv.org e-Print archive, November 2011.

411. Vovk, V. (2013). Kolmogorov's Strong Law of Large Numbers in Game-Theoretic Probability: Reality's Side. Technical Report. arXiv:1304.1074 [cs.GT], arXiv.org e-Print archive, March 2013. First posted as GTP40 March 2013.

412. Vovk, V. (2015). The fundamental nature of the log loss function. In: *Fields of Logic and Computation II: Essays Dedicated to Yuri Gurevich on the Occasion of His 75th Birthday*, Lecture Notes in Computer Science, vol. 9300 (eds. L.D. Beklemishev, A. Blass, N. Dershowitz et al.), 307–318. Cham: Springer.

413. Vovk, V. (2015). Itô calculus without probability in idealized financial markets. *Lithuanian Mathematical Journal* 55 (2): 270–290. First posted as GTP36 August 2011 and also available on arXiv.

414. Vovk, V. (2016). Purely pathwise probability-free Itô integral. *Matematychni Studii* 46 (1): 96–110. First posted as GTP42 December 2015 and also available on arXiv.

415. Vovk, V., Nouretdinov, I., Takemura, A., and Shafer, G. (2005). Defensive forecasting for linear protocols. In: *Proceedings of the 16th International Conference on Algorithmic Learning Theory*, Lecture Notes in Artificial Intelligence, vol. 3734, 459–473 (eds. S. Jain, H.U. Simon, and E. Tomita). Berlin: Springer. First posted as GTP10 February 2005 and also available on arXiv.

416. Vovk, V. and Shafer, G. (2005). Good randomized sequential probability forecasting is always possible. *Journal of the Royal Statistical Society. Series B* 67 (5): 747–763. First posted as GTP7 June 2003.

417. Vovk, V. and Shafer, G. (2008). The game-theoretic capital asset pricing model. *International Journal of Approximate Reasoning* 49 (1): 175–197. First posted as GTP1 March 2002.

418. Vovk, V. and Shafer, G. Game-theoretic probability. In: Augustin et al. [15], Chapter 6, 114–134.

419. Vovk, V. and Shafer, G. (2016). A Probability-Free and Continuous-Time Explanation of the Equity Premium and CAPM. Technical Report arXiv:1607.00830v2 [q-fin.MF], arXiv.org e-Print archive, July 2016. First posted as GTP44 June 2016.

420. Vovk, V. and Shafer, G. (2017). Towards a Probability-Free Theory of Continuous Martingales. Technical Report arXiv:1703.08715 [q-fin.MF], arXiv.org e-Print archive, March 2017. First posted as GTP45 July 2016.

421. Vovk, V. and Shafer, G. (2001). Game-Theoretic Capital Asset Pricing in Continuous Time. Technical Report arXiv:1802.01556 [q-fin.PR], arXiv.org e-Print archive, February 2018. First posted as GTP2 December 2001.

422. Vovk, V. and Shafer, G. (2003). A Game-Theoretic Explanation of the $\sqrt{dt}$ Effect. Technical Report arXiv:1802.01219 [q-fin.MF], arXiv.org e-Print archive, February 2018. First posted as GTP5 January 2003.

423. Vovk, V., Takemura, A., and Shafer, G. (2005). Defensive forecasting. In: *Proceedings of the 10th International Workshop on Artificial Intelligence and Statistics* (eds. R.G. Cowell and Z. Ghahramani), 365–372. Society for Artificial Intelligence and Statistics. Available electronically at http://www.gatsby.ucl.ac.uk/aistats/. First posted as GTP8 September 2004 and also available on arXiv.

424. V'yugin, V.V. (2018). *Математические основы машинного обучения и прогнозирования (Mathematical foundations of machine learning and forecasting, in Russian)*, Издательство Московского центра непрерывного математического образования, Moscow, second edition.

425. Wagon, S. (1985). *The Banach-Tarski Paradox*. Cambridge: Cambridge University Press.

426. Walley, P. (1991). *Statistical Reasoning with Imprecise Probabilities*. London: Chapman and Hall.

427. Whittle, P. (2000). *Probability via Expectation*, 4e. New York: Springer.

428. Williams, P.M. (1975). Coherence, strict coherence and zero probabilities. In: *Proceedings of the 5th International Congress of Logic, Methodology and Philosophy of Science*, vol. VI, 29–33.

429. Williams, P.M. (1976). Indeterminate probabilities. In: *Formal Methods in the Methodology of Empirical Sciences* (eds. M. Przełęcki, K. Szaniawski, and R. Wojcícki), 229–246. Wrocław, Poland: Ossolineum & Reidel.

430. Wolfe, P. (1955). The strict determinateness of certain infinite games. *Pacific Journal of Mathematics* 5: 841–847.

431. Wong, E. (1973). Recent progress in stochastic processes—a survey. *IEEE Transactions on Information Theory* 19 (3): 262–275.

432. Wong, W.H. (1986). Theory of partial likelihood. *Annals of Statistics* 14 (1): 86–123.

433. Würmli, M. (1980). Lokalzeiten für Martingale. Master's thesis. Universität Bonn. Supervised by Hans Föllmer.

434. Zermelo, E. (1913). Über eine Anwendung der Mengenlehre auf die Theorie des Schachspiels. In: *Proceedings of the Fifth International Congress of Mathematicians*, vol. II (eds. E.W. Hobson and A.E.H. Love), 501–504. Cambridge: Cambridge University Press.

435. Zygmund, A. (1960). Józef Marcinkiewicz (in Polish). *Wiadomości Matematyczne* 4: 11–41.

436. Zygmund, A. and Marcinkiewicz, J. (1964). *Józef Marcinkiewicz: Collected Papers* (ed. A. Zygmund), 1–33. Warsaw: Państwowe Wydawnictwo Naukowe. This is a revised English translation of [435].

Index

Game-Theoretic Foundations for Probability and Finance, First Edition. Glenn Shafer and Vladimir Vovk.
© 2019 John Wiley & Sons, Inc. Published 2019 by John Wiley & Sons, Inc.

Printed and bound by CPI Group (UK) Ltd, Croydon, CR0 4YY

16/04/2025